Urban Overheating—Progress on Mitigation Science and Engineering Applications

Urban Overheating—Progress on Mitigation Science and Engineering Applications

Special Issue Editors

Michele Zinzi
Mattheos Santamouris

MDPI • Basel • Beijing • Wuhan • Barcelona • Belgrade

MDPI

Special Issue Editors

Michele Zinzi
Energy and Sustainable Economic Development
Italy

Mattheos Santamouris
University of New South Wales
Australia

Editorial Office
MDPI
St. Alban-Anlage 66
4052 Basel, Switzerland

This is a reprint of articles from the Special Issue published online in the open access journal *Climate* (ISSN 2225-1154) from 2017 to 2019 (available at: https://www.mdpi.com/journal/climate/special_issues/real_urban)

For citation purposes, cite each article independently as indicated on the article page online and as indicated below:

LastName, A.A.; LastName, B.B.; LastName, C.C. Article Title. *Journal Name* **Year**, *Article Number, Page Range.*

ISBN 978-3-03897-636-3 (Pbk)
ISBN 978-3-03897-637-0 (PDF)

Contents

About the Special Issue Editors

Michele Zinzi, Phd in Energetics. Full time researcher at ENEA since 2000. Main field of interest and activities: expert in building physics, testing on building envelope and urban materials, development of methodologies for the energy performance of buildings and energy analyses, assessment and mitigation of urban overheating and heat island. Editorial board of Energy and Buildings, Climate and, formerly, Advances in Building Energy Research. Advisory and scientific committee member of many international conferences. Author of many papers published in scientific journals and in international conferences. ENEA scientific responsible in several national and European Projects. Former Executive Comittee member of IEA Solar Heating and Cooling Programme, and current National Executive Committee member in the Energy in Buildings and Communities Programme of IEA.

Mattheos Santamouris is a Scientia Professor of High Performance Architecture at UNSW, and past Professor in the University of Athens, Greece. Visiting Professor: Cyprus Institute, Metropolitan University London, Tokyo Polytechnic University, Bolzano University, Brunnel University and National University of Singapore. Past President of the National Center of Renewable and Energy Savings of Greece. Editor in Chief of the Energy and Buildings Journal, Past Editor in Chief of the Advances Building Energy Research, Associate Editor of the Solar Energy Journal and Member of the Editorial Board of 14 Journals. Editor of the Series of Book on Buildings, published by Earthscan Science Publishers. Editor and author of 14 international books published by Elsevier, Earthscan, Springer, etc. Author of 320 scientific articles published in journals. Reviewer of research projects in 29 countries including USA, UK, France, Germany, Canada, Sweden, etc.

climate

MDPI

Editorial

Introducing Urban Overheating—Progress on Mitigation Science and Engineering Applications

Michele Zinzi [1],* and Mattheos Santamouris [2]

[1] Agenzia Nazionale per le Nuove Tecnologie, L'energia e lo Sviluppo Economico Sostensibile (ENEA),
   Via Anguillarese 301, 00123 Rome, Italy
[2] The Anita Lawrence Chair in High Performance Architecture, School of Built Environment,
   University of New South Wales, Office room #RC2002, Sydney, NSW 2052, Australia;
   m.santamouris@unsw.edu.au
* Correspondence: michele.zinzi@enea.it

Received: 27 December 2018; Accepted: 16 January 2019; Published: 19 January 2019

1. Introduction

Buildings and construction is the most important economic sector in the world after agriculture. It represents a total turnover of 8.2 trillion dollars while forecasts for 2025 predict a total budget that is close to 15 trillion dollars [1]. In parallel, it offers employment to more than 110 million people around the world [2].

Despite the huge contribution of the construction sector in the global economy, it is actually facing very important challenges such as:

- Local and Global climate change that increases the energy consumption for cooling, increases the concentration of pollutants, deteriorates indoor and outdoor thermal comfort conditions and has a tremendous impact on heat-related mortality and morbidity [3], while it increases the ecological footprint of cities and urban settlements
- Overpopulation and economic growth in the developing world put a serious strain on the construction sector as some billions of new houses, building and infrastructures must be designed and built in the immediate next years
- The number of low-income individuals in both the developed and developing countries is increasing tremendously as the price of energy is rising, employment rates are decreasing and social equity seems not to be a high priority for modern societies. This puts billions of low-income people under threat and increases the health budget enormously.

The impact of local climate change on the energy consumption of buildings is now very well documented. Several studies have proven that because of urban overheating the cooling energy consumption of buildings may double [4,5], while the peak electricity demand may increase considerably and oblige utilities to build new power plants continuously to satisfy the additional demand [6]. In parallel, several recent studies have shown that heat-related mortality and morbidity has increased substantially when the ambient temperature increases during the summer period [7].

To face the problem of local and global climate change and to provide adequate housing to the vulnerable population, advanced mitigation and adaptation technologies have been proposed, developed and implemented all around the world successfully [8]. Mitigation technologies aim to fight the sources of overheating and counterbalance its impact, while adaptation technologies mainly aim to provide additional protection to the residents and the dwellers of the buildings.

Some of the most successful proposed mitigation technologies involve the use of reflective, cool, materials for buildings and urban structures, other advanced materials to decrease the urban temperature like thermochromic or radiative cooling structures, intensive use of greenery, use of water for evaporation, as well as other dissipation methods and technologies like the use of the ground [9].

In particular, the development of advanced materials for the outdoor built environment, like highly reflective light color materials, infrared reflective materials, thermochromic and fluorescent materials as well as photonic and plasmonic structures for radiative cooling contribute highly to decrease the peak ambient temperature up to 1.5–2.0 °C [10–13].

Advanced materials for mitigation as well as a combination of all the other mitigation technologies are implemented in hundreds of large scale urban rehabilitation projects. Monitoring and theoretical results show that it is possible to decrease the peak ambient temperature up to 3 °C. However, given the amplitude of the local overheating that exceeds 10 °C in many cases, there is a profound need to develop more efficient mitigation technologies able to provide a higher temperature reduction

2. Aim and Scope

The sim of the present special issue on "Urban Overheating—Progress on Mitigation Science and Engineering Applications" was to collect papers that are able to represent the newest information relating to the science, technology, application and policy perspective of urban environment overheating and its potential mitigation. Several hot topics were indicated for contribution:

- Studies and monitoring of urban overheating and mitigation technologies and strategies in real urban conditions;
- Modeling of the urban climate for an accurate assessment of mitigation solutions;
- Development of methodologies and tools to detect, predict and mitigate urban overheating;
- Implementation of policies and instruments to support the thermal rehabilitation of urban areas;
- Exemplary cases and demonstration projects.

These topics and also additional ones were deeply investigated in this paper's collection.

3. Presentation of the Published Papers

This Special Issue collects 18 relevant studies coming from 4 continents (South and North America, Asia, Europe, Oceania) and depicts this complex and multi-disciplinarily topic. The issues covered by the following contributions: four papers deal with urban overheating and urban heat island observation, with coupled methodological approaches; four papers deal with the outdoor microclimatic analysis and eventual mitigation strategies; three papers deal with the thermal interaction of buildings and the urban environment; three papers deal with the role and the impact of fabric materials in affecting the urban climates; two papers deal with urban greenery and mitigation; finally, one paper is about outdoor thermal comfort and one is about urban geometry. The incidence of each category, expressed in percentages, is reported in the following graph (Figure 1).

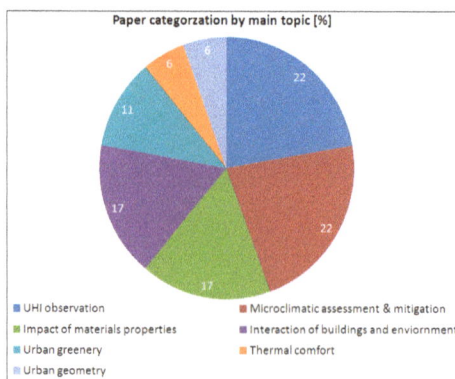

Figure 1. The paper categorization according to the main topic investigated.

3.1. Urban Overheating and Heat Island Observation and Analysis

A study aimed at identifying and quantifying an urban heat island index is presented in Reference [14]. The study is implemented at the census-tract level and intends to provide a reliable tool for health agencies to assess issues such as public overheating and air quality issues. By multi-scale modeling sorted by observation data, the index, based on several metrics, provides details of the urban heat island as a function of the urban seize area. The observation of the thermal environment, crucial to developing a mitigation plan, is carried out by high and medium spatial resolution satellite data from Reference [15]. The objective of this study is the implementation of a methodology that is able to identify "hot spots" and "cold spots" in a large metropolis such as Athens, and to quantify the thermal stress in the different zones of the city. Again the city of Athens is the field for a detailed analysis of urban climate in Reference [16], where the issue of understanding and manipulating big data coming from extensive environmental monitoring is carried out. In particular, a power spectral density analysis is performed over time scales spanning 10 min to several days, evidencing that air temperature data exhibit turbulent-like intermittent properties with multi-fractal statistics. Thermal datasets acquired for the city of Basel are used to validate a multiple linear regression model that is able to model the urban temperature distribution in a continuous way [17]. The model is based on the factors that mostly affect the urban heat island at night and the main outcome is that different datasets can be used to predict the heat island with comparable results.

3.2. The Role of the Materials in Urban Overheating Mitigation

It is well known that construction materials severely affect urban overheating because of their solar absorbing and heat storing properties. Increasing the albedo of cities is hence a major mitigation strategy. The impact of a relevant increase of albedo for roofs, facades and pavements is analyzed for Sacramento, Houston and Chicago in Reference [18], with the objective of estimating thermal mitigation, as well as air quality and human health. Starting from the measured datasets and combining chemistry and multi-layer canopy models, a decrease of ambient temperature up to 2.3 °C is calculated, as well as a decrease of PM2.5 and O_3 concentrations up to 2.7 $\mu g/m^3$ and 6.3 ppb. The role of materials, as well as of the canopy covers, is analyzed in research carried out in Los Angeles, which couples datasets acquired by shielded sensors mounted atop automobiles and WRF meteorological modeling [19]. It was found that the increase of albedo may lead to temperature reductions of up to 2.8 °C at the neighborhood scale. Higher albedo also means a higher luminous reflectance, which can provide an up to 75% reduction of electricity use for lighting, as proven in Reference [20]. In this paper, cool coatings for existing asphaltic surfaces are analyzed and, despite energy savings, it is evidenced that their optical behavior may affect the uniformity of visions for drivers and pedestrians, thus requiring ad-hoc installations.

3.3. The Role of the Green in Urban Overheating Mitigation

The role of greenery on urban mitigation is well explored and estimating the cooling potential is important when designing or rehabilitating green areas. Mobile measurements and computed fluid-dynamics analyses are carried out for a green area in the city of Kobe [21]. The numerical results, confirmed by monitoring, evidence that at a 30 m distance, the urban air temperature is not affected by the beneficial contribution of the green zone due to the sharp increase of the air temperature when entering the urban area. General information on the proper design of urban parks and areas are provided in Reference [22], where the cooling power is analyzed as a function of several variables relating to the physical parameters and characteristics of urban geometry. The analysis shows that the main parameter that can positively affect the urban climate in the surrounding area is the size of the park, with all the other parameters being less relevant.

3.4. The Interaction between Buildings and the Urban Environment

Buildings heavily affect urban overheating because of the characteristics of envelope materials and the ambient heat released because of active cooling and heating systems. The solar reflectance index, which takes into account the solar reflectance and thermal emittance of building materials, can be used to predict the thermal behavior of a surface subjected to solar irradiation. The correlation of the index with the heat released to the environment, as well as a sensitivity analysis of different input parameters, is analyzed in Reference [23]. While most studies focus on roofs, a lack of data exists for vertical walls and facades, which play a crucial role in the heat exchange between fabrics and ambient air under the canopy. A novel application of the empirical line method to calibrate a terrestrial low-cost multispectral sensor to recover spectral reflectance is developed in Reference [24]; promising results are achieved for this technique, able to characterize the building facades on site. The impact of buildings on energy uses is well recognized worldwide, an optimized procedure for cost optima renovation is carried out in Reference [25]. From among the results, the reduction of close to 30% of equivalent CO_2 emissions has to be mentioned, resulting in laying the groundwork for favorable conditions for urban climate mitigation.

3.5. Microclimatic Assessment and Mitigation

Microclimatic monitoring and calculation analyses help to understand the thermal response in specific areas that may be used to generalize the results at the city level, thus being important for policymakers. An accurate analysis of Athens is carried out in Reference [26], where urban temperatures were monitored in different zones of the city. The results showed the daily air temperature differences in the 5–8 °C range between the hottest and the coolest zones; the results also showed the dependence of local overheating as a function of climatic, geometrical and thermo-physical variables. Similar analyses are carried out for the city of Wien in Reference [27]. In this case, the numerical analysis followed-up the monitoring to verify the effectiveness of the selected mitigation strategies. Particular attention was given to tree planting, resulting in it being more efficient than green roof for urban mitigation and overheating prevention in buildings, thanks to the provided solar protection of windows. Additionally, in Rome, the monitoring and impact of mitigation technologies are carried out with a clear focus on architectural integration from the perspective of a holistic rehabilitation of a school campus [28]. Results take into account temperature mitigation, but also demonstrate the improvement of thermal comfort conditions using the Physiologically Equivalent Temperature as the driving indicator. The last paper on this topic, being the study focused on the rehabilitation of small public areas of the city of Beirut, couples numerical analyses of mitigation strategies with a strong policy analysis [29]. The results here show limited mitigation potential due to the size of the zones, but provide insights on the way municipalities may implement regenerative actions on the urban territory with a focus on sustainable issues.

3.6. Other Topics

The debated issue about the concordance between the outdoor thermal comfort indexes and the thermal sensation of users collected through questionnaires is the topic addressed in Reference [30]. The two methods are applied in Santa Maria, Brazil, and, thanks to a linear regression model, the results prove that there is a significant improvement in the agreement between instrumental assessment and subjective response to thermal comfort. The role of solar irradiation is crucial for the latter issue and it has a strong relationship to urban geometry. The last paper explores this aspect, analyzing and comparing two methods for assessing the sky view factor in densely built urban environments and identifying optimized calculation processes. The study is applied to calculate the sky view factor of each building block of Paris [31].

4. Conclusions

The papers included in this special issues show the complexity and multidisciplinary of the investigated topic, namely the urban overheating and its mitigation and adaptation. The presented body of work evidences the magnitude of the phenomenon, as well as potentials and limits of observation and detection instruments, and of the mitigation technologies and strategies. It also causes the role of the design and of policy instruments for the rehabilitation of thermally deteriorated urban areas to emerge.

Continuous efforts are needed in the future to properly address several open issues: methods and tools to accurately estimated urban overheating as a function of multiple variables affecting the phenomenon; innovative technologies and strategies to mitigate overheating in built urban environments (materials, green and blue technologies); accurate modeling of the urban environment with focus on fluid and thermodynamic aspects; and applications with a clear demonstration of the mitigation solutions in practice. Overheating and mitigation of the urban environment is, hence, a topic that will deserve attention and research activities in the next few years when its impact will be amplified by the consequence of climate change and global warming.

References

1. Santamouris, M. Innovating to zero the building sector in Europe: Minimising the energy consumption, eradication of the energy poverty and mitigating the local climate change. *Sol. Energy* **2016**, *128*, 61–94. [CrossRef]
2. WIEGO. Woman in Informal Economy: Globalizing and Organising: Construction Workers. 2015. Available online: http://wiego.org/informal-economy/occupational-groups/construction-workers (accessed on 12 January 2018).
3. Santamouris, M.; Kolokotsa, D. On the Impact of Urban Overheating and Extreme Climatic conditions on Housing Energy Comfort and Environmental Quality of Vulnerable Population in Europe. *Energy Build.* **2015**, *98*, 125–133. [CrossRef]
4. Santamouris, M. *Minimizing Energy Consumption, Energy Poverty and Global and Local Climate Change in the Built Environment Minimizing to Zero the Building Sector*; Elsevier: Amsterdam, The Netherlands, 2018.
5. Santamouris, M. On The Energy Impact of Urban Heat Island and Global Warming on Buildings. *Energy Build.* **2014**, *82*, 100–113. [CrossRef]
6. Santamouris, M.; Cartalis, C.; Synnefa, A.; Kolokotsa, D. On the Impact of Urban Heat Island and Global Warming on the Power Demand and Electricity Consumption of Buildings–A Review. *Energy Build.* **2015**, *98*, 119–124. [CrossRef]
7. Paravantis, J.; Santamouris, M.; Cartalis, C.; Efthymiou, C.; Kontoulis, N. Mortality Associated with High Ambient Temperatures, Heatwaves, and the Urban Heat Island in Athens, Greece. *Sustainability* **2017**, *9*, 606. [CrossRef]
8. Santamouris, M. Regulating the damaged thermostat of the Cities—Status, Impacts and Mitigation Strategies, Energy and Buildings. *Energy Build.* **2015**, *91*, 43–56. [CrossRef]
9. Akbari, H.; Cartalis, C.; Kolokotsa, D.; Muscio, A.; Pisello, A.L.; Rossi, F.; Santamouris, M.; Synnefa, A.; Wong, N.H.; Zinzi, M. Local Climate Change and Urban Heat Island Mitigation Techniques—The State of the Art. *J. Civ. Eng. Manag.* **2016**, *22*, 1–16. [CrossRef]
10. Santamouris, M.; Synnefa, A.; Karlessi, T. Using advanced cool materials in the urban built environment to mitigate heat islands and improve thermal comfort conditions. *Sol. Energy* **2011**, *85*, 3085–3102. [CrossRef]
11. Garshasbi, S.; Santamouris, M. Using Advanced Thermochromic Technologies in the Built Environment. Recent Development and Potential to Decrease the Energy Consumption and Fight Urban Overheating. *Sol. Energy Mater. Sol. Cells* **2019**, *191*, 21–32. [CrossRef]
12. Santamouris, M.; Feng, J. Recent Progress in Daytime Radiative Cooling: Is It the Air Conditioner of the Future? *Buildings* **2018**, *8*, 168. [CrossRef]
13. Berdahl, P.; Chen, S.S.; Destaillats, H.; Kirchstetter, T.W.; Levinson, R.M.; Zalich, M.A. Fluorescent cooling of objects exposed to sunlight—The ruby example. *Sol. Energy Mater. Sol. Cells* **2016**, *157*, 312–317. [CrossRef]

14. Taha, H. Characterization of Urban Heat and Exacerbation: Development of a Heat Island Index for California. *Climate* **2017**, *5*, 59. [CrossRef]

15. Mavrakou, T.; Polydoros, A.; Cartalis, C.; Santamouris, M. Recognition of Thermal Hot and Cold Spots in Urban Areas in Support of Mitigation Plans to Counteract Overheating: Application for Athens. *Climate* **2018**, *6*, 16. [CrossRef]

16. Karatasou, S.; Santamouris, M. Multifractal Analysis of High-Frequency Temperature Time Series in the Urban Environment. *Climate* **2018**, *6*, 50. [CrossRef]

17. Wicki, A.; Parlow, E.; Feigenwinter, C. Evaluation and Modeling of Urban Heat Island Intensity in Basel, Switzerland. *Climate* **2018**, *6*, 55. [CrossRef]

18. Jandaghian, Z.; Akbari, H. The Effect of Increasing Surface Albedo on Urban Climate and Air Quality: A Detailed Study for Sacramento, Houston, and Chicago. *Climate* **2018**, *6*, 19. [CrossRef]

19. Taha, H.; Levinson, R.; Mohegh, A.; Gilbert, H.; Ban-Weiss, G.; Chen, S. Air-Temperature Response to Neighborhood-Scale Variations in Albedo and Canopy Cover in the Real World: Fine-Resolution Meteorological Modeling and Mobile Temperature Observations in the Los Angeles Climate Archipelago. *Climate* **2018**, *6*, 53. [CrossRef]

20. Rossi, G.; Iacomussi, P.; Zinzi, M. Lighting Implications of Urban Mitigation Strategies through Cool Pavements: Energy Savings and Visual Comfort. *Climate* **2018**, *6*, 26. [CrossRef]

21. Takebayashi, H. Influence of Urban Green Area on Air Temperature of Surrounding Built-Up Area. *Climate* **2017**, *5*, 60. [CrossRef]

22. Bernard, J.; Rodler, A.; Morille, B.; Zhang, X. How to Design a Park and Its Surrounding Urban Morphology to Optimize the Spreading of Cool Air? *Climate* **2018**, *6*, 10. [CrossRef]

23. Muscio, A. The Solar Reflectance Index as a Tool to Forecast the Heat Released to the Urban Environment: Potentiality and Assessment Issues. *Climate* **2018**, *6*, 12. [CrossRef]

24. Fox, J.; Osmond, P.; Peters, A. The Effect of Building Facades on Outdoor Microclimate—Reflectance Recovery from Terrestrial Multispectral Images Using a Robust Empirical Line Method. *Climate* **2018**, *6*, 56. [CrossRef]

25. Ascione, F.; Bianco, N.; Mauro, G.; Napolitano, D.; Vanoli, G. A Multi-Criteria Approach to Achieve Constrained Cost-Optimal Energy Retrofits of Buildings by Mitigating Climate Change and Urban Overheating. *Climate* **2018**, *6*, 37. [CrossRef]

26. Georgakis, C.; Santamouris, M. Determination of the Surface and Canopy Urban Heat Island in Athens Central Zone Using Advanced Monitoring. *Climate* **2017**, *5*, 97. [CrossRef]

27. Vuckovic, M.; Maleki, A.; Mahdavi, A. Strategies for Development and Improvement of the Urban Fabric: A Vienna Case Study. *Climate* **2018**, *6*, 7. [CrossRef]

28. Laureti, F.; Martinelli, L.; Battisti, A. Assessment and Mitigation Strategies to Counteract Overheating in Urban Historical Areas in Rome. *Climate* **2018**, *6*, 18. [CrossRef]

29. Kaloustian, N.; Aouad, D.; Battista, G.; Zinzi, M. Leftover Spaces for the Mitigation of Urban Overheating in Municipal Beirut. *Climate* **2018**, *6*, 68. [CrossRef]

30. Gobo, J.; Galvani, E.; Wollmann, C. Subjective Human Perception of Open Urban Spaces in the Brazilian Subtropical Climate: A First Approach. *Climate* **2018**, *6*, 24. [CrossRef]

31. Bernard, J.; Bocher, E.; Petit, G.; Palominos, S. Sky View Factor Calculation in Urban Context: Computational Performance and Accuracy Analysis of Two Open and Free GIS Tools. *Climate* **2018**, *6*, 60. [CrossRef]

![climate logo]

climate

MDPI

Article

Characterization of Urban Heat and Exacerbation: Development of a Heat Island Index for California

Haider Taha

Altostratus Inc., 940 Toulouse Way, Martinez, CA 94553, USA; haider@altostratus.com; Tel.: +1-(925)-228-1573

Academic Editors: Michele Zinzi and Mattheos Santamouris
Received: 6 July 2017; Accepted: 2 August 2017; Published: 5 August 2017

Abstract: To further evaluate the factors influencing public heat and air-quality health, a characterization of how urban areas affect the thermal environment, particularly in terms of the air temperature, is necessary. To assist public health agencies in ranking urban areas in terms of heat stress and developing mitigation plans or allocating various resources, this study characterized urban heat in California and quantified an urban heat island index (UHII) at the census-tract level (~1 km^2). Multi-scale atmospheric modeling was carried out and a practical UHII definition was developed. The UHII was diagnosed with different metrics and its spatial patterns were characterized for small, large, urban-climate archipelago, inland, and coastal areas. It was found that within each region, wide ranges of urban heat and UHII exist. At the lower end of the scale (in smaller urban areas), the UHII reaches up to 20 degree-hours per day (DH/day; °C.hr/day), whereas at the higher end (in larger areas), it reaches up to 125 DH/day or greater. The average largest temperature difference (urban heat island) within each region ranges from 0.5–1.0 °C in smaller areas to up to 5 °C or more at the higher end, such as in urban-climate archipelagos. Furthermore, urban heat is exacerbated during warmer weather and that, in turn, can worsen the health impacts of heat events presently and in the future, for which it is expected that both the frequency and duration of heat waves will increase.

Keywords: heat health; meteorological modeling; urban climate; urban-climate archipelago; urban heat island; urban heat island index; Weather Research and Forecasting model (WRF)

---

1. Introduction

Urban heat, often quantified as urban heat island (UHI), can locally exacerbate the effects of regional climates on heat, emissions, and air quality [1–5]. The exacerbation has significant ramifications in terms of public health from both heat and air-quality pathways [6–8]. Because of urban heat, the cooling energy demand increases, emissions of anthropogenic and biogenic pollutants (e.g., ozone precursors) increase, the photochemical production of ozone accelerates, and air quality (O_3, $PM_{10/2.5}$, NO_x) deteriorates [9–13].

UHIs vary in intensity from one area to another because of differing causative factors. Some UHIs can be as high as 8 °C, e.g., in tropical regions [14,15], but this is rather atypical. More often, UHIs are in the order of 0.5 to 3 °C [8].

A range of urban-cooling measures, e.g., heat-island reduction strategies, have been proposed over the years as summarized, for example, in Taha, 2015 [8]. Some of the more common measures include increased urban albedo, by use of reflective roofs, and increased vegetation cover, e.g., ground cover or green roofs [16,17]. The potential benefits of implementing these measures on urban heat and air quality have been quantified in several studies [9,10,18–20].

To begin planning for the mitigation of urban heat in California, various efforts at city and regional levels have begun addressing and characterizing UHIs. Pursuant to the California Assembly Bill AB 296, which requires the quantification of the UHI, the California Environmental Protection Agency

(Cal/EPA) has initiated efforts to better understand urban heat and develop a UHI Index (UHII) for use in (1) assessing the heat health implications of urban land use; (2) identifying geographical areas where UHIs can exacerbate environmental health issues (heat and air quality); and (3) potentially updating information in the CalEnviroScreen tool [21]. CalEnviroScreen is an environmental health screening tool designed to help identify and assign a score to Californian communities that are burdened by multiple sources of pollution.

In this effort, the UHI and UHII were characterized via the meteorological modeling of California (Taha and Freed, 2015 [22]). It is to be noted that the UHII is not intended as a sole indicator of heat health. The UHII is developed to provide additional information for use in conjunction with CalEnviroScreen or other similar tools that include data on population and demographics so that a more accurate and complete picture can emerge on the role of urban heat in exacerbating public health issues.

2. Materials and Methods

The approach, models, resulting data, online interactive maps, and all related details for 40 urban areas in California are discussed in Taha and Freed, 2015 [22]. Here, in Section 2.1, Section 2.2, Section 2.3, Section 2.4, Section 2.5, Section 2.6, Section 2.7, Section 2.8, only some very brief highlights and pointers to the methodology are provided.

2.1. Study Area

This research and the development of the UHII were carried out for California, focusing on the southern, warmer two thirds of the state where heat and air quality aspects are relatively more significant. Because of urbanization trends, various agencies develop energy and environmental plans for addressing and mitigating the effects of warmer weather and heat events including, for example, UHI mitigation measures.

California is a U.S. west-coast state with a Mediterranean climate in parts of its southern two thirds. However, there are very large climatic variations among coastal areas, inland zones, deserts, higher elevations, and mountains. Even in relatively small regions, the climate can vary significantly with distance from the coast or with elevation changes. From a heat-health perspective, some of the concerns relate to the fact that California is among a few states with higher rates of long-term temperature increases in the summer. For example, the 2011–2014 average maximum temperature departure from the 20th century average was 1.1–2.2 °C [23]. There also is concern that heat events may become more common. During the California July 2006 heat wave, there were 16,166 excess emergency department visits and 1182 excess hospitalizations statewide [24]. During that heat wave, there was an increase of 9% in mortality for every 5.5 °C increase in the apparent temperature [25].

2.2. Definition of the UHI Index (UHII)

Several forms of the UHII with varying degrees of complexity were evaluated for use in this study. Equation (1) represents the simplest form of the UHII that was agreed upon in this effort, a form that also satisfies the AB 296 requirements in that the index captures both the *severity* (magnitude) and *extent* (duration) of the urban-nonurban temperature differential. In this equation, $T_{u(k),h}$ is the urban temperature at time step (hour) h, $T_{nu(k),h}$ is the nonurban temperature at time-step h, and H is the number of time-steps, in this case, the number of hours in the period June, July, and August of a given year. Here, k is a location index representing a pair of points, one urban and one reference, that is, $u(k)$ is the urban point of the pair k, and $nu(k)$ is the non-urban, reference point of the pair k. Note that there is no temperature threshold associated with this definition.

$$ UHII = \sum_{h=1}^{H(JJA)} \left[T_{u(k),h} - min\left(T_{u(k),h}, T_{nu(k),h} \right) \right] \tag{1} $$

Thus, the UHII is a cumulative metric, with units of degree-hours. In this paper, the UHII is also averaged per day, so that the units are degree-hours per day (DH/day). By contrast, an urban heat island (UHI) is an instantaneous temperature difference, or an average of such, and the units are degrees.

Four levels of UHI characterizations and modeling can be identified as follows:

Level 1: involves characterizing the UHI and computing the UHII regardless of heat transport from upwind sources, on-shore warming, or other non-local factors such as climate-archipelago effects that can contribute to the localized UHI. A Level-1 UHII characterizes the actual thermal environment, i.e., proportional to what a thermometer would indicate in the real world, regardless of the causative factors and heat sources that led to the creation of the UHI at any given location.

Level 2: includes Level 1 with the addition of weighting the UHII by (1) population density and (2) technical potential. Technical potential is an indicator of the availability of surface area for the deployment of measures such as cool roofs and pavements, urban forests, solar photovoltaics, and so on. It is also an indicator of the actual extent to which such measures can be implemented, i.e., the amount of albedo increase or change in surface properties. A Level-2 UHII can thus provide additional information in terms of health impacts and the potential for action to mitigate the local UHII (allocation of resources).

Level 3: includes Levels 1 + 2 but subtracts the amount of heat advected to each area (census tract) from the UHII by quantifying a length scale for transport. Heat is transported as a result of (1) on-shore flow warming in coastal areas; (2) the urban-climate archipelago effect; and (3) heat transport from adjacent upwind urban areas. Hence, the goal is to identify the upwind distance (e.g., the upwind census tracts) that affects the UHI at a certain location and that needs to be considered when implementing mitigation measures at that location. Thus, a Level-3 UHII can be used to "assign" urban-heat responsibility more accurately than Levels 1–2, that is, it allows each census tract or city to estimate the share of their UHI that they are actually responsible for (locally generated) and the potential for mitigating the corresponding local UHII.

Level 4: includes Levels 1 + 2 + 3 with the additional quantification of atmospheric impacts, both positive and negative, of mitigation measures. A Level-4 UHII can provide information as to how much of the local UHI a certain city or census tract can actually mitigate. The information that a Level-4 UHII can provide will, in turn, be important in tailoring the mitigation measures and scenarios for the site-specific attainment of the UHII, maximizing the positive effects and minimizing the negative ones at census-tract scales.

Thus, going from Level 1 towards Level 4, the UHII becomes less of a "characterization index" and gradually more of a "mitigation index". In this paper, only Level-1 modeling and characterizations are presented. A partial Level-2 example is also discussed.

As defined by Equation (1), the UHI and UHII are calculated based on air temperature, not skin-surface temperature. Among several reasons for using air temperature in developing the UHII is that an air-temperature UHII at a given location accounts for heat transport by (1) advection from adjacent upwind urban areas and sources of anthropogenic heat; (2) the transport of heat because of the urban climate archipelago effect; and (3) warming of the air with distance from the coastline (in coastal areas). These effects cannot be correctly captured by a skin-surface temperature UHII.

An air-temperature UHII may also be more relevant to public health assessments than skin-surface temperature, not just for heat, but also for air quality. Most often, the higher air temperatures and pollutant concentrations, e.g., ozone, are found in generally the same downwind locations where the UHII peaks, i.e., are displaced relative to the skin-surface temperature maxima. This is an important consideration because air quality is a compounding factor to heat in terms of public health. The displacement can be small in smaller urban areas but becomes significant in larger cities and climate-archipelago situations, as discussed in Section 3.

2.3. Models

This study was carried out with the WRF-ARW modeling system [26]. To better suit this application, several components of the model were modified or customized in this study. All model configurations, inputs, parameterizations, and modifications in this work are discussed in Taha and Freed, 2015 [22]. The model performance was evaluated for various approaches including the use of urbanized WRF modules such as UCM [27,28] and BEP / BEM [29], along with various boundary-layer schemes including Bougeault and Lacarrere [30] and NUDAPT/WUDAPT urban morphometric data [31]. Based on statewide data availability and model-performance evaluation results (see below), an approach by Taha [9,10,18–20] was adopted, as discussed briefly in the following sections.

2.4. Modeling Periods and Domains

The modeling periods presented in this paper encompass the months of June, July, and August (JJA) 2013 and JJA 2006, the latter of which was employed to capture the impacts of the California heat wave of 15 July through 1st August of that year [32,33]. Figure 1 depicts the nested-grids WRF configuration used in this study. The 3- and 1-km grids were positioned in relation to the CalEnviroScreen top 20%, 10%, and 5% score areas [21], so as to focus on the regions with an increased health risk and population vulnerability. Top scores in CalEnvioScreen indicate the worst conditions in terms of public health. Thus, for example, the top 5% score is worse than 10% and so on.

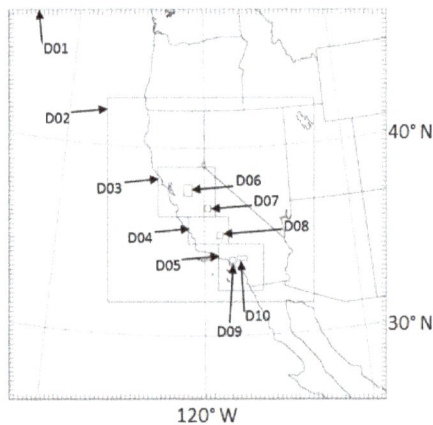

Figure 1. Modeling domains configuration: domain D01: 27-km resolution; D02: 9 km; D03–D05: 3 km; D06 through D10: 1 km. Domain D02 is centered on California and domains D03–D10 are positioned over various parts of the state.

2.5. Model Input

Four-dimensional NCEP-NCAR Reanalysis [34] for the years 2006 and 2013 was used for the model boundary conditions and data assimilation (FDDA). Additional NWS/NOAA datasets were obtained to complement the reanalysis, including the surface temperature, sea-surface temperature, soil moisture, and upper-air meteorology. Fine-resolution mesonet observations (e.g., NOAA/MADIS) from the areas of interest were also used in carrying out a quantitative model performance evaluation (MPE).

Land-use/land-cover (LULC): As is often the case, some California urban areas are extremely data-rich, whilst others are relatively more data-sparse. In order to ensure an even and comparable California-wide characterization, two LULC datasets were used: (1) 30-m United States Geological Survey (USGS) Level-II and Level-IV [35] classification representing years up to 2000; and (2) 30-m

National Land Cover Data (NLCD) representing year 2011 [36]. While these datasets may be less resolved than other local data when available, their even statewide coverage was one main criterion for selection. A crosswalk between the two datasets was carried out [22] to increase the number of useful LULC categories. NUDAPT/WUDAPT datasets [31] were not used directly in this effort because of their limited spatial coverage, but as templates, along with local climate zones [37] to map morphological characteristics onto LULC classes.

The LULC characterization of each domain provides a basis for the derivation of certain physical parameter inputs to the land-surface models. The approach in this study uses a bottom-up characterization of each model grid cell based on surface physical properties. That is, instead of scaling up the LULC classes and assigning lookup values of physical parameters to the dominant (by majority) LULC in each grid cell, as is done in the default input to WRF, this study scales up the physical properties and parameters (rather than LULC) so that details of physical characteristics are preserved. The simulation improvements gained from this approach can be seen in Figure 2 as an example (San Francisco Bay Area). The methodology is detailed in Taha and Freed, 2015 [22]—here, only some highlights are discussed.

In Figure 2A, an example LULC characterization of the San Francisco Bay Area, via NLCD 2011, is shown. The areas in red are various classes or urban LULC. Figure 2B–D present, respectively, the albedo, canopy shade factor, and roughness length derived based on the LULC characterizations from multiple datasets. In Figure 2E, surface-temperature differences between two cases are shown: (a) simulations pursued with study-specific WRF customizations and input modifications; and (b) WRF simulations with the standard input approach. The study-customized approach preserves the fine-scale urban details of relevance to UHI/UHII calculations. For example, as seen in Figure 2E, the major highways in the area (some of which are labeled on the figure), as well as detailed city boundaries, are clearly reflected in the simulated temperature pattern. Note that the temperature-difference scale in Figure 2E is intentionally selected to show two shades of green: the light green color roughly represents the areas classified as urban in standard WRF processing (temperature difference is between 0 and 1.5 °C). On the other hand, the dark-green color represents additional areas classified as urban in the modified approach (temperature difference of 2–4 °C, or more).

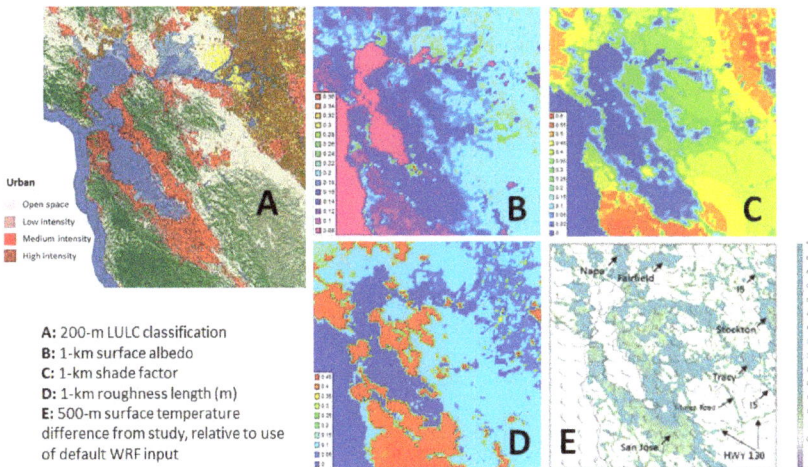

Figure 2. Example analysis for the San Francisco Bay Area (subdomain of D03). (**A**) NLCD 2011; (**E**) WRF-simulated surface temperature difference at 1400 PDT, 15 June 2013.

2.6. Model Modifications and Customizations

This study modified the process by which thermo-physical characterizations of the urban surface were carried out at various grid-nest levels for input into the land surface model in WRF (Noah LSM) and to the urban physics parameterizations discussed in Section 2.3. The goal of this modified approach was to preserve the fine-scale urban features and their effects that are important in modeling urban heat. The LULC and surface characterization is carried out independently at each grid level (for each nest). Unlike in standard WRF processing where LULC is scaled up and used in characterizing a grid cell, here the physical properties are scaled up instead, thus better preserving such properties at finer resolutions, including effective albedo, soil moisture, roughness length, canopy cover, shade factor, and urban morphometric and drag-related parameters, as discussed in Section 2.5. The basis for assigning thermo-physical properties at the surface-type level (e.g., roof, wall, street, tree level, etc.) in this project was the California urban-fabric analysis in several studies [9,18,19,38–40]. These aspects are discussed in Taha and Freed, 2015 [22] along with the model physics options, configurations, and parameterizations selected in this study, as well as methodologies for the computation of relevant gridded urban surface parameters.

The physical properties at urban points within each grid cell were then meshed with those of their non-urban surroundings, i.e., portions of cells that are non-urban, by weighing the respective physical properties by the fractions *urbfra* and *1-urbfra*, where *urbfra* is the cell's urban content. This is an improvement over the way in which urban and non-urban land uses within urban grid cells are characterized by default in urban WRF, where the non-urban part of an urban grid cell is typically reverted to some default category regardless of what existed in that grid cell originally. Furthermore, in implementing these modifications in the land-surface model, a distinction was made, in this study, between "above-canopy" and "below-canopy" vegetation cover [38,40]. In terms of model input parameters, the former is used in shade-factor calculations, whereas the latter is employed for adjusting the soil-moisture content.

Whereas the coarser grids were run with two-way feedback and analysis nudging FDDA, a nest-down approach was adopted for the finer grids of the configuration shown in Figure 1. In this study, nesting down to the finer domains was updated to directly ingest the output from the urban characterization steps discussed above.

2.7. UHI/UHII Reference Points

The interaction between UHI circulation and the background flow has been examined in several studies [9,41,42]. For an urban area, there is generally a certain wind-speed threshold that determines the relative dominance and impacts of background versus UHI-induced flows. This wind-speed threshold has been shown to depend on the city size, among other factors [43,44]. In this study, the dominance of background versus UHI circulation is determined hourly from the modeled fields by comparing the flow directions within and outside of each urban area.

Thus, unlike in simpler approaches where the UHI is computed relative to fixed reference point(s), this modeling study computes the UHI at each census tract relative to several time-varying (hourly) wind-dependent upwind reference points. That is, the UHI at each census tract is quantified dynamically based on the varying wind direction, such that the reference points are always upwind of the urban area. The dynamical calculations of the UHI/UHII and the criteria for selecting the reference points are discussed in Taha and Freed, 2015 [22].

Note that the combined effects of (1) time-varying reference points in calculating the UHI and (2) the inclusion of only time intervals when urban areas are warmer than non-urban reference points (see the "min" operator in Equation (1)) can result in spatial patterns of the UHI and UHII that are counter-intuitive, especially in climate archipelagos, as will be discussed in Section 3.1.

2.8. Model Performance Evaluation

As alluded to in Section 2.3, a rigorous model performance evaluation (MPE) was carried out in this study [22] and the performance was found to be satisfactory. The MPE was conducted using observational data from 151 NOAA meteorological monitors covering the fine-resolution California sub-domains (Figure 1) for periods JJA 2006 and 2013. The performance was compared against recommended benchmarks for temperature, wind speed, wind direction, and humidity, through metrics such as the forecast skill, mean bias, mean absolute error, root mean square error (including both systematic and unsystematic), index of agreement, and related variables [45]. For example, for D02, the mean wind speed bias is 0.08 m s^{-1}, root mean square error 1.73 m s^{-1}, and index of agreement 0.78. For wind direction, the bias is 0.92° and absolute error 35°. The temperature bias is 0.7 K, absolute error 2 K, and index of agreement 0.93. Finally, for humidity, the bias is -0.89 g kg^{-1}, absolute error 1.47 g kg^{-1}, and index of agreement 0.65. All of these values are well within or close to the recommended benchmarks for each variable. In addition to the statistical MPE of near-surface meteorology, a subjective evaluation of the upper-air model performance at 850, 700, and 500 hPa was undertaken to assess the model's ability to reproduce observed synoptic-scale features.

Furthermore, observations from ~1200 mesonet stations in Californian urban areas (NOAA/ MADIS) were also used for a more localized MPE in the study's domains. These observations were used in evaluating the model's ability to reproduce fine-scale observed temperature patterns and UHI. The example in Figure 3 (for the Los Angeles region) shows that the model can capture the observed UHI, temperature gradients and spatial patterns, and the archipelago effect. In this figure, observations and simulations for the month of June 2013 are shown.

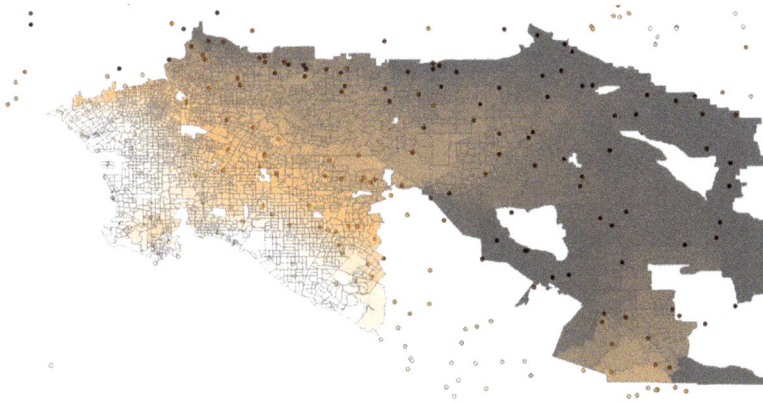

Figure 3. Los Angeles region model UHII (background) and mesonet observation (small circles) at 320 stations in this domain for June 2013. Both model UHII and observations are in DH/day (°C.h/day). Lightest color is 6 °C.hr/day; darkest color is 190 °C.hr/day. Compare the color changes of the small circles (mesonet observations) with those of the background UHII which shows a good agreement between the two.

3. Results and Discussion

3.1. UHII Calculations

The UHII, as defined in Equation (1), was computed for every census tract in 40 Californian urban areas and for each hour of JJA 2013 and 2006. Among several metrics examined in the study, the cumulative totals were converted to degree-hour per day (DH/day) averages, as in Table 1. The range of DH/day (column 3) reflects the variability of the UHI within each area. Column 2 identifies regions as either an "urban island" or collection of islands (UI) resulting in single-core or multi-core UHIs, or as an "urban-climate archipelago" (UA) (see discussion in Section 3.2). Column 4 represents the temperature difference between census tracts with higher UHIIs and those with lower UHIIs within each region. This represents an all-hours averaged UHI; however, in the datasets generated in this study [22], the temperature difference at each census tract is also provided individually, not just as the maximum region-wide value, as in column 4. Of note, in calculating the temperature difference in column 4 the non-urban reference points were not used. Rather, this difference is calculated between the census tracts with the highest UHII and urban areas with the lowest UHII.

Table 1. UHII range for a sample of urban areas in California. The full tabulation for all 40 urban areas is provided in Taha and Freed, 2015 [22].

1	2	3	4
Region	Type	DH/Day Range (°C.hr/day)	Largest ΔT (°C)
Davis	UI	5–35	1.3
Fairfield	UI	1–113	4.6
Napa	UI	1–69	2.8
Sacramento	UI	4–76	3.0
San Rafael	UI	1–43	1.7
Santa Rosa	UI	1–59	2.4
Fresno	UI	1–46	1.9
Merced	UI	6–45	1.7
Modesto	UI	1–45	1.8
Morgan Hill	UI	6–41	1.5
Livermore	UI	0–39	1.6
San Francisco	UA	0–122	5.0
San Jose	UA	2–49	2.0
Vallejo	UI	1–83	3.4
Walnut Creek	UI	1–96	3.9
Antioch	UI	1–46	1.9
Bakersfield	UI	0–34	1.4
East Bay	UA	0–121	5.0
Lancaster	UI	3–26	0.9
Mission Viejo	UI	0–109	4.6
Oceanside	UA	0–125	5.2
San Diego	UA	1–125	5.1
San Fernando	UA	1–42	1.7
San Luis Obispo	UI	9–58	2.0
Santa Clarita	UI	2–25	1.0
Santa Cruz	UI	2–42	1.6
Simi Valley	UI	0–59	2.4
Los Angeles West	UA	0–182	7.5

It is important to note that the UHII range (column 3) was computed independently for each area and not based on common UHI reference points across the regions. In other words, the absolute temperatures cannot be inter-compared across different areas in column 1. The UHII and temperature differences appear to be large in some instances (particularly UA regions) because they also include urban-climate archipelago effects (superimposed signals). This will be discussed further below.

In analyzing the spatial characteristics of the UHII in the 40 urban areas modeled in this study, it is possible to discern four types of spatial patterns:

i.   Small areas: where urban areas are relatively small and, thus, UHIs are small and localized.

ii.   Single cores: these are relatively larger urban areas where the UHI can develop fully and the downwind transport of heat can occur. There is one core, i.e., one main area where the UHI maximum can be defined and thus a single-core UHII can be identified.

iii.   Multiple cores: in this case, several UHI cores develop, although each core is still well defined.

iv.   Climate archipelagos: In this situation, urban land use covers a large geographical area often demarcated by coastlines or topography (in California). Thus, unlike single- or multi-core urban areas, where there is a clear beginning and end to urban land use, urban-climate archipelagos consist of sustained built-up cover, with minimal interruption. The only discontinuities in the urban expanse usually occur because of topography. Examples include San Francisco Bay Area and the Los Angeles basin.

Examples of these different UHI and UHII patterns (i–iv) for 40 urban areas, along with the maps and related datasets that were produced in this study can be found at https://calepa.ca.gov/wp-content/uploads/sites/34/2016/10/UrbanHeat-Report-Report.pdf and https://www.calepa.ca.gov/climate/urban-heat-island-index-for-california/urban-heat-island-interactive-maps/.

In the following discussion, an example of a single-core UHI (Fresno) and a climate archipelago (Los Angeles region) are presented (from 40 urban areas modeled in this study). These are shown in Figures 4 and 5, respectively, where Figure 4A is the Fresno UHII computed as degree hours per day (DH/day) and Figure 4B is the UHII weighted by population density. Similarly, Figure 5A shows the Los Angeles area UHII (DH/day), including the climate-archipelago effect, and Figure 5B is the UHII weighted by population density at each census tract (the western half is shown and will be discussed further in Figure 6). Population density is used in this discussion as an indirect proxy to urban density. Population weighting is done by multiplying each census tract's UHII by the ratio of its population density to that of the most populous tract in the area (highest density). Thus, the census tract with the highest population density has a weight of 1 for its UHII.

As discussed in Section 2.2, the UHI and UHII are based on 2-m air temperature, not skin-surface temperature (the latter typically results in patterns of hot and cold spots that are more familiar and similar to satellite thermal imagery). Further, the results presented here are for resolutions of 1–3 km. Therefore, the UHII calculations account for the effects of heat transport and mixing. Note that finer-scale modeling, e.g., at 500 m, shows more localized and more detailed temperature patterns, bu these are not relevant to the Level-1 and Level-2 UHII characterizations presented here (and defined in Section 2.2).

Figures 4A and 5A show a Level-1 UHII, whereas Figures 4B and 5B show a partial Level-2 product. It should be emphasized here that the type of information provided by Level-2 UHII, such as presented in Figures 4B and 5B, does not represent real-world temperature conditions like Level-1 UHII; they are only a useful tool to aid in the allocation of resources and in planning. As can be seen in these figures, the effects of the climate archipelago are evident. In the Fresno area, the population-weighted UHII spatial pattern is relatively similar to that of the UHII Level-1 (compare Figure 4A and 4B). That is because the region is relatively uniform geographically and, thus, the distribution of urban heat is relatively proportional to that of the population. However, in the Los Angeles region, the population-weighted UHII pattern differs from the Level-1 pattern (compare Figure 5A and 5B), because of the climate-archipelago effect and on-shore warming of the air. In this case, heat transport results in the highest temperatures in areas that are not necessarily the highest in population density.

In the Los Angeles region, the Level-2 population-weighted UHII (Figure 5B) may look more familiar and intuitive than Level-1 (Figure 5A) as the latter shows heat transport downwind throughout the climate archipelago, away from the coastline and towards the foothills and inland basin (east), and thus makes the localized variations in the UHI less noticeable and the pattern less intuitive. Among the observations that can be made with regard to Level-2 UHII in the Los Angeles region (i.e., Figure 5B relative to A) are: (1) Downtown Los Angeles (white oval) can be seen as a hotter area than its surroundings; (2) the hot areas in the north are displaced southward and away from the foothills (top part of the figure); and (3) the hotter areas in the eastern basin are more confined (not shown).

However, the UHII pattern in Figure 5A correctly reflects the actual conditions, that is, the model results are supported by observations. For example, the DH/day clustering (Figure 5C) based on the observed mesonet temperature from June 2013 clearly shows a spatial pattern and magnitudes that are similar to those from the model in Figure 5A (the clustering is done by grouping together weather stations (Figure 5D) that have similar DH/day values within given quantile intervals). Note that Figure 5A shows the UHII, i.e., differences relative to reference points, whereas Figure 5C shows the absolute DH/day at each station (not differences).

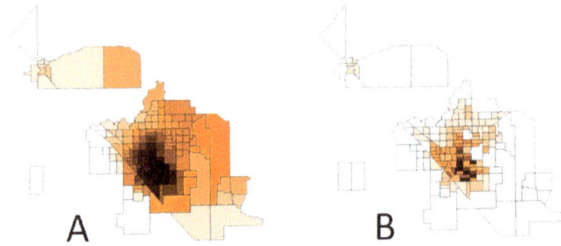

Figure 4. Fresno UHII as the difference from reference points. Lightest color is 2 DH/day; darkest color is 45 DH/day (°C.hr/day). (**A**): UHII; (**B**) UHII weighted by population density. Fresno is an example of a conventional, single-core UHI, as defined in item ii above. The geographical area shown in this figure is located in domain D07 in Figure 1.

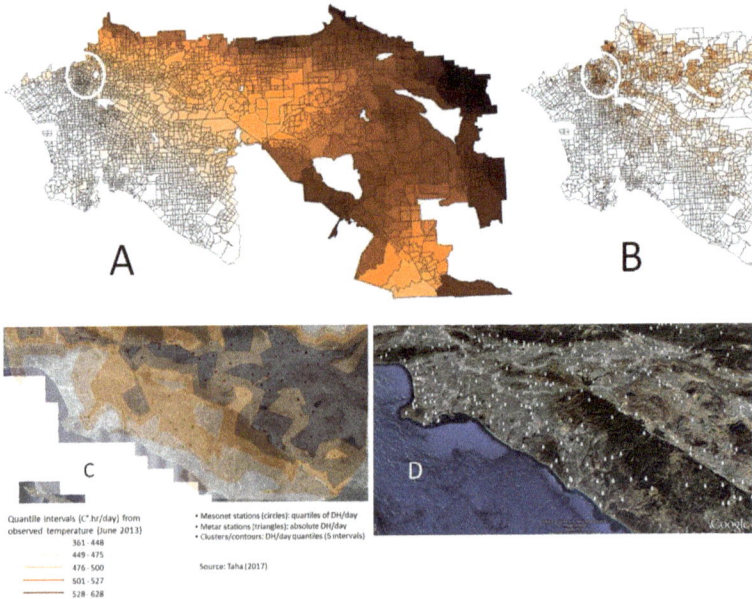

Figure 5. Los Angeles area UHII as the difference from reference points, showing the climate-archipelago effect. Lightest color is 6 DH/day; darkest color is 190 DH/day (°C.hr/day). (**A**) UHII; (**B**) UHII weighted by population density; (**C**) clustering of observations-based DH/day (absolute) at each monitor for June 2013; (**D**) locations of 329 monitors in the region (basis for 4C). The Los Angeles area is an example of the urban-climate archipelago as defined in item iv, above, that also includes fine-scale superimposed microclimate signals (see Figure 6). The geographical area shown in this figure is located in domain D05 in Figure 1, and further enhanced by D09 and D10.

In a climate archipelago, the range of UHII is usually large and such a region, when examined at a coarse scale, appears as if uniform in UHII values over large swaths (see Figure 6 for the west Los Angeles basin discussed earlier as an example). In reality, however, the model captures localized variations (fine-scale signals) in temperatures and locally-generated UHIs at fine scales, e.g., the census tract level, throughout such a region, in addition to the dominant on-shore warming and archipelago signals, e.g., as seen in Figure 6. These superimposed signals are discussed further in Section 3.2.

An additional analysis of the Los Angeles climate archipelago is provided here as an example. Census tracts numbered 1–5 in Figure 6 are in the Compton area, tracts numbered 6–8 are in Downtown Los Angeles, tracts 9–11 are in the San Pedro area, and tracts 12–15 are in the Anaheim-Orange area. These clusters of census tracts are also listed in Table 2 and the purpose of discussing them here is to show examples of micro-scale variations (locally-generated UHI and correlation with LULC) that are captured by the model but are embedded within the dominant signal of the climate archipelago and on-shore warming.

Figure 6. UHII in west Los Angeles Basin. Numbers indicate example census tracts listed in Table 2 and discussed in this section.

Table 2. Random samples (microclimate variations) in west Los Angeles Basin. (see Figure 6 for tract locations).

Compton			Downtown Los Angeles			San Pedro			Anaheim–Orange		
Tract	UHII	UHI	Tract	UHII	UHI	Tract	UHII	UHI	Tract	UHII	UHI
1	18.50	0.77	6	40.46	1.68	9	11.44	0.47	12	46.06	1.91
2	13.14	0.54	7	39.80	1.65	10	25.08	1.04	13	37.86	1.57
3	12.48	0.52	8	32.63	1.35	11	19.13	0.79	14	45.74	1.90
4	15.00	0.62							15	40.95	1.70
5	17.38	0.72									

UHII and UHI in the table are averages over the entire modeled periods. Units are °C.hr/day for the UHII and °C for UHI.

In the Compton area, the largest UHII is found at point 1 and the smallest at points 2 and 3. The corresponding average UHIs are 0.77, 0.54, and 0.52 °C, respectively. This represents a localized temperature difference (a locally-generated UHI) of 0.25 °C across a distance of less than 1 km. Tract 1 (with highest UHII) represents industrial-commercial land use, whereas tracts 2 and 3 (lowest UHII) represent residential land use with higher vegetation cover than in tract 1.

In Downtown Los Angeles, Financial-District census tract 6 (high-rise area of downtown) has the largest UHII among the three tracts considered. The locally-generated UHI in the high-rise area is 0.33 °C relative to tract 8 (South Park, a commercial area), which is less than 1 km to the south. The high-rise area is also similar to or slightly warmer than tract 7 (Arts District), despite the latter being (1) downwind of downtown and (2) in an industrial-commercial area.

In the San Pedro area, census tract 10 has a larger UHII than tracts to the west of it (tract 9), as well as to the east of it (tract 11). Both tracts 9 and 10 are residential; however, tract 9 has a lower density development than tract 10 and has significantly higher vegetation cover. Tract 10 is a high-density development with lower vegetation cover. As a result, tract 10 has a larger UHII than tract 9. Tract 11 is near golf courses and is cooler than tract 10. Just a few hundred meters south of tract 11, in a golf-course area, the UHII is 12.75 DH/day and is 0.5 °C cooler than tract 10.

It can also be observed that the Compton area is "counter-gradient" relative to San Pedro. In other words, whereas Compton is expected to be warmer than San Pedro, due to being further inland from the coast, it is actually cooler (thus counter gradient). It is also notable that relative to the west basin's UHI reference points, the downtown area (tract 6), while relatively more inland, has an average UHI of 1.68 °C, whereas tract 10 has an average UHI of 1.04 °C despite being closer to the coast. This highlights the influence of land-use and land-cover properties, as well as the ability of the model to detect these fine-scale effects that are embedded within the larger-scale signals (Figure 6).

In the Anaheim-Orange area, the largest UHIIs among the four tracts examined are 46.06 DH/day in tract 12 and 45.74 DH/day in tract 14. Relative to the lowest UHII of 37.86 at tract 13, this represents a localized average temperature difference of 0.34 °C across a distance of less than 1 km. Tract 12 is industrial-commercial with lower vegetation cover than the surrounding tracts and, thus, has the highest UHII. Tract 13 is residential, with significantly larger vegetation cover and the lowest UHII in this sample area. Tract 14 is also residential and is vegetated but is downwind of tract 12, hence the larger UHII, but still slightly smaller than in tract 12. Finally, tract 15 is mixed residential and commercial land with significant vegetation cover and lower UHII.

While a number of urban-climate archipelagos have been identified world-wide [46], including, in California, both the Los Angeles Basin and the San Francisco Bay Area, quantifying their atypical UHIs is not standard procedure and not much prior guidance is available in this respect. Thus, the work performed in this study could be considered as one attempt at characterizing archipelagos in terms of UHI/UHII.

3.2. Climate Archipelagos and Air Temperatures

An urban-climate archipelago acts as one large area-source of heat (no distinguishable core) and the highest temperatures are typically found downwind. In the case of the Los Angeles basin, the model results indicate an average air-temperature difference of 6–8 °C between inland areas (e.g., the towns of San Bernardino, Hemet, and Perris) and coastal areas, e.g., areas to the east of Los Angeles International (LAX), as seen in Figure 5A. Observational and analyzed data from PRISM [47] suggest an average temperature difference of 5–7 °C between these regions (Figure 7). The ~1 °C shift between modeled and PRISM UHI can be attributed to the finer resolution of the modeling, as well as the locations of the actual monitors and grids (in PRISM) versus the locations of the centroids (census tracts) used in the simulations. However, the example shows a good agreement between the modeled and analyzed temperature differentials in the archipelago.

The climate-archipelago and single/multi-core UHII situations can be further explained schematically with the aid of Figure 8. Using a perturbation analogy, the UHI (ΔT) at any point can be decomposed into a background component (leading up to that point) and a local component (at the point), such that:

$$UHI = \overline{UHI} + UHI' \tag{2}$$

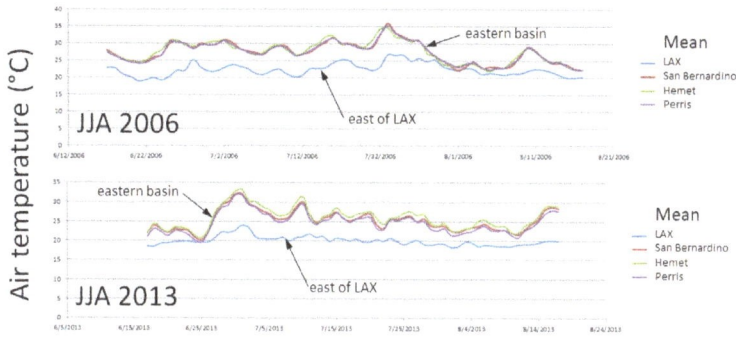

Figure 7. (**Top**) 2006; (**bottom**) 2013. Air temperatures at inland and coastal locations in the Los Angeles region based on PRISM data [47]. See text for discussion.

In Figure 8, line 1 in the top graph and line 3 in the bottom graph represent the background UHI component (e.g., resulting from urban-climate archipelago effects and the on-shore heating of air), whereas the multiple colored lines (in both top and bottom graphs) represent the locally-generated UHI, that is, the deviation of the local UHI (as totalized in line 4) from the background. Line 4 represents the sum of the localized UHI and the heat transported from the urban areas upwind of it. The top figure represents a relatively smaller urban area where no archipelago effect exists and, thus, line 1 has a slope of near zero. Relative to the upwind reference temperature (UHI ref point), a "zero plane" is defined (line 1 in top graph, line 3 in bottom graph) which can serve as a reference temperature that does not include the urban-archipelago or on-shore warming effects.

In the case of an urban island (UI), as in the top graph of Figure 8, such as in Fresno (see Figure 4A), the temperature (line 4) returns to about the value of the upwind reference point, i.e., the value at the zero plane (line 1) at some distance downwind of the trailing edge of the urban area (the trailing urban edge is marked with line 5). This tapering-off of the UHI is complete at the point marked with "X" (in the top graph), where conditions are relatively similar to those upwind (i.e., at the UHI ref point) if the urban area is relatively small. At the trailing edge of the urban area (line 5) there is still a UHI, as seen by the continuous red line and the dashed black line (line 4).

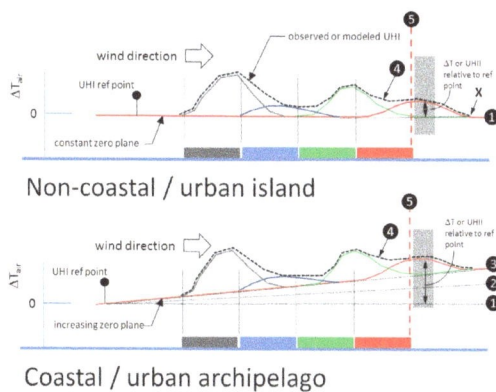

Figure 8. Single/multi core UHI versus climate archipelago. The colored blocks represent different urban land uses and land covers and the similarly-colored lines represent the temperature (localized UHI) corresponding to each block. The black dash line represents the sum of the local and downwind-displaced UHI at each point along the wind direction.

In coastal and/or urban-climate archipelago situations (bottom graph of Figure 8), such as Los Angeles Basin (see Figure 5A), the zero plane (line 3) increases in the on-shore direction and the urban area (archipelago) ends at topographical barriers (thus the archipelago ends at line 5). There is no downwind stretch past this trailing edge of the urban area (to the right of line 5) over which the temperature can readjust and return to upwind values. As a result, the total UHI at that point (at line 5 in bottom graph) consists of the following superimposed fields: (a) the onshore warming of air (line 2 minus line 1); (b) upwind urban warming of air by the urban archipelago (line 3 minus line 2) beyond that caused by the local urban land use; and (c) the locally-produced UHI (line 4 minus line 3). This explains why the highest UHII values in archipelagos and coastal areas are found further inland, near the downwind end of the air basin, close to the foothills (e.g., Figure 5A). At these barriers, which also are the trailing edges of the urban archipelagos, the temperature is generally higher than in other parts (the observational data also support these findings). Thus, line 4 and line 3 in the bottom graph of Figure 8 represent what is seen in Figure 6, i.e., variations in localized UHIs superimposed on a dominant west-to-east gradient in temperature.

As a result, areas with similar local UHIs (for example in the regions highlighted with grey vertical shade in Figure 8) will have a larger UHII in archipelago situations (bottom graph) than in urban islands (top graph) because of the superimposed on-shore warming and archipelago effects in the former. Or, conversely, an area with a certain UHII in the archipelago can have a smaller localized UHI than an area of a similar UHII in an urban island situation. Note that this discussion applies only to Level-1 UHII, not Levels 2–4.

The implications are that when computing the average temperature differences (ΔT) between the upwind and downwind parts of urban archipelagos/coastal areas (e.g., column 4 in Table 1), larger values are obtained than in urban islands (UI). For example, and as described earlier, the modeled ΔT across the LA basin is 6–8 °C. However, as shown in Figure 8, this ΔT, while correctly characterizing real-world conditions in this area (Figure 5A), is not solely a local UHI effect, but also includes other signals. On the other hand, for example, the average ΔT between the upwind and downwind parts in Fresno (1.9 °C) is mostly comprised of a local UHI effect, since the area is inland (no on-shore warming) and also relatively small, and thus no archipelago effect exists (flat line 1 in top graph of Figure 8). This suggests that in UI areas, the UHII is closer to being a mitigation indicator than in UA regions where it is solely a characterization of urban heat.

The cumulative urban-warming effect of an urban climate archipelago, especially in uninterrupted built-up areas like the Los Angeles basin, is a real part of the UHI and so it is correct to include it in calculating and mapping the UHII for heat-health assessments and evaluations of air quality, as was done in this study (Level-1 modeling), e.g., as shown in Figure 5A. The UHII, in this case, is proportional to what a thermometer would indicate in the field in these areas. However, the on-shore warming and archipelago effects should be subtracted in future efforts (e.g., Levels 3 and 4 modeling, as discussed in Section 2.2) if the goal is to develop localized (census-tract level) UHI mitigation guidelines or localized monitoring of the UHI at fine resolutions.

3.3. UHI Exacerbation during Hot Weather

An analysis was carried out in this study to evaluate the UHII pattern in each area during hotter weather. An arbitrary sample of regions is presented in Figure 9. In the left part of each figure, the average UHII (DH/day) for each region's reference points is shown for the JJA periods in 2013 and 2006 and used here as a cumulative indicator to and in lieu of the instantaneous absolute air temperature. The 2006 heat wave is marked with red arrows.

On the right side of each figure, two frequency distributions are shown: one for the 2006 heat wave (red line, corresponding to red arrow) and another for the period with the lowest average UHII in each region (blue line, corresponding to blue arrow). In the right part of the figures, the horizontal axis represents the UHII (in bins of 10 DH/day) and the vertical axis is the frequency, i.e., the percentage of census tracts within given UHII bins. Of note, each of the cooler and heat-wave periods are two weeks long and the coolest periods differ across the regions (blue arrows).

It can be seen in Figure 9 that the warmer weather shifts the UHII distribution towards larger values compared to the cooler periods. For example in Fresno, about 30% of census tracts in the UHII bin of 20 DH/day are shifted to the 30 DH/day bin. The 50 DH/day bin initially containing 5% of tracts in the cooler weather contains 18% of the tracts during the heat wave.

In Livermore (not shown in Figure 9), the 40 DH/day bin contained 14% of tracts during the cooler period, but 22% during the heat wave. This was shifted from the lower bin of 30 DH/day which contained 24% of tracts in cooler weather but decreased to 17% of tracts during the heat wave. In Sacramento (Figure 9), significant shifts occur such that the bin at 70 DH/day contained 0% of tracts during the cooler period but increased to 19% of tracts during the heat wave. The distribution of tracts in other bins also changed with some increasing and others decreasing.

In San Diego (Figure 9), the shift occurs through a range of bins, but noticeably, the number of tracts in the 20 DH/day bin is reduced and shifted to the higher bins. Those bins of 110–140 DH/day contained 0% census tracts in the cooler period that increased to 4% of the tracts (in each of these bins) during the heat wave. In San Francisco (not shown), the census tracts in bin 10 DH/day were reduced from 72% to 54% and shifted to higher bins, such that the bins 110–180 DH/day, containing 1% of tracts during the cooler periods, increased to 3–5% of the tracts during the heat wave. In Vallejo (not shown), census tracts in bins 40–60 DH/day were shifted to bins 70 and 80 DH/day, and the bins 90–120 DH/day that contained 0% of the tracts in the cooler period increased to 8–18% of the tracts during the heat wave.

In Antioch (not shown), bins were shifted upwards such that bin 50 DH/day, containing 0% of tracts during the cooler weather, contained 14% of tracts during the heat wave. In the East Bay (Figure 9), all bins smaller than 50 DH/day were shifted to bins higher than 50 DH/day. Furthermore, bins 120–170 DH/day containing 0% of tracts during cooler periods increased to contain 2–4% of the tracts in heat-wave conditions. In Fairfield (not shown), most tracts in the 10 DH/day bin were shifted to the 20 DH/day bin during the heat event. In addition, bins 90–140 DH/day with 0% tracts in cooler weather, increased so as to contain 2–6% of the tracts during heat wave conditions. In Manteca (Figure 9), tracts in bins 10 and 20 DH/day were shifted to higher bins, such that bins 30–60 DH/day contained larger numbers of census tracts under the heat wave conditions. Significant changes and shifts in the UHII are also seen in other regions throughout California and are discussed in Taha and Freed, 2015 [22].

In the Los Angeles urban-climate archipelago, the hotter weather also increases the number of census tracts affected by a high UHII, but the pattern is different from those in the regions discussed above. For example, in east Los Angeles basin (Figure 9, SoCABeast), the heat wave causes an increase in the number of census tracts in the mid-range bins of 150–210 DH/day, but a decrease in census tracts at both tail ends of the main distribution (the smaller distribution between 0 and 20 DH/day does not change). The number of census tracts in bins greater than 210 DH/day and in those bins smaller than 120 DH/day is reduced, such that the frequency distribution has a slightly smaller spread. That is, the archipelago now has a slightly more uniform temperature field (a smaller temperature differential across the basin). Revisiting Figure 7, one can see that the observational / analyzed data also suggest a smaller temperature differential around the heat-wave period (late July through early August 2006), thus lending further credibility to the modeled results, i.e., the smaller temperature differentials shown in Figure 9 (SoCABeast).

This reduction in the higher UHII values in the Los Angeles archipelago can be attributed to an effect akin to "reverse coastal cooling", a term coined by Bornstein and co-workers [48]. This occurs when inland areas, particularly deep basins with large catchment areas, warm up (e.g., during heat waves) and strengthen the sea breeze. The stronger venting, in turn, decreases the temperature differential across the archipelago or basin. A similar effect is also seen in Santa Clara Valley in the San Francisco Bay Area. Prior studies have identified and quantified a reverse coastal cooling effect albeit on longer time scales [48] in both the Los Angeles Basin and the San Francisco Bay Area. Similarly, a more recent study of Texas [49] found that warming caused by climate change has increased venting in the Houston region.

In this UHII study, however, it is not being suggested that the area becomes cooler (the absolute temperatures actually become predominantly higher during the heat wave), but that the temperature becomes slightly more uniform across the archipelago/basin, that is, a smaller temperature differential now exists across the region because of the relatively stronger venting. While this may only be a non-dominant, temporary mechanism, it can still affect the shape of the frequency distributions in these coastal areas.

The results discussed above suggest that the warmer weather can increase the UHI and shifts the UHII to larger values because of (1) reduced winds and mixing under high pressure systems, except in coastal basins to some extent; (2) a lower urban moisture content than in non-urban areas, which under such conditions allows urban areas to warm up more; and (3) reduced cloudiness and increased solar radiation at the surface [8,9,50].

Larger UHIs in warmer weather have also been reported in other regions [8]. For example, Li and Bou-Zeid, 2013 [51] used observational data from Baltimore, MD, to show that heat waves increase the UHI intensity. Their data demonstrate that the 2008 heat wave increased the nighttime UHI from 0.5 to 2.5°C and the daytime UHI from 0.25 to 1.5°C. Li et al., 2015 [52] also use observational data from China to show that heat waves increase the UHI. Thus if warmer weather increases the UHI, then mitigation measures such as cool cities will become more important in the future [53], when heat waves are expected to occur more frequently. On the other hand, some studies suggest that UHIs could decrease with higher background temperatures in the long term [54]. Thus, again, this highlights the region-specific nature of urban climates and heat-island responses to changes in background weather and synoptic conditions.

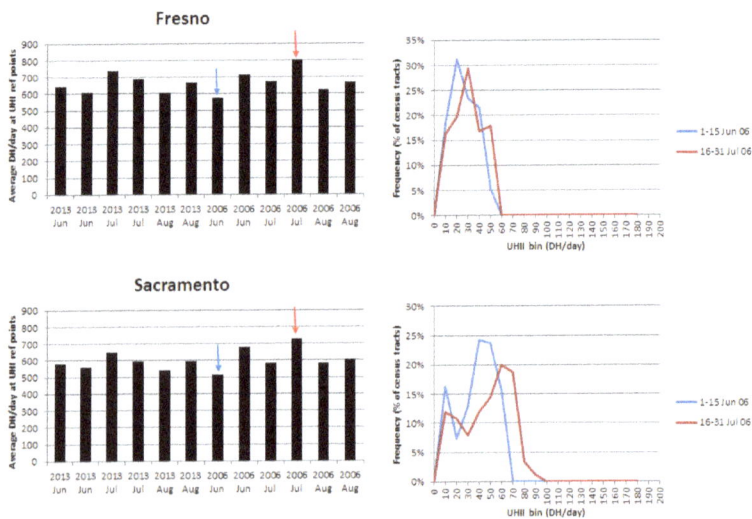

Figure 9. *Cont.*

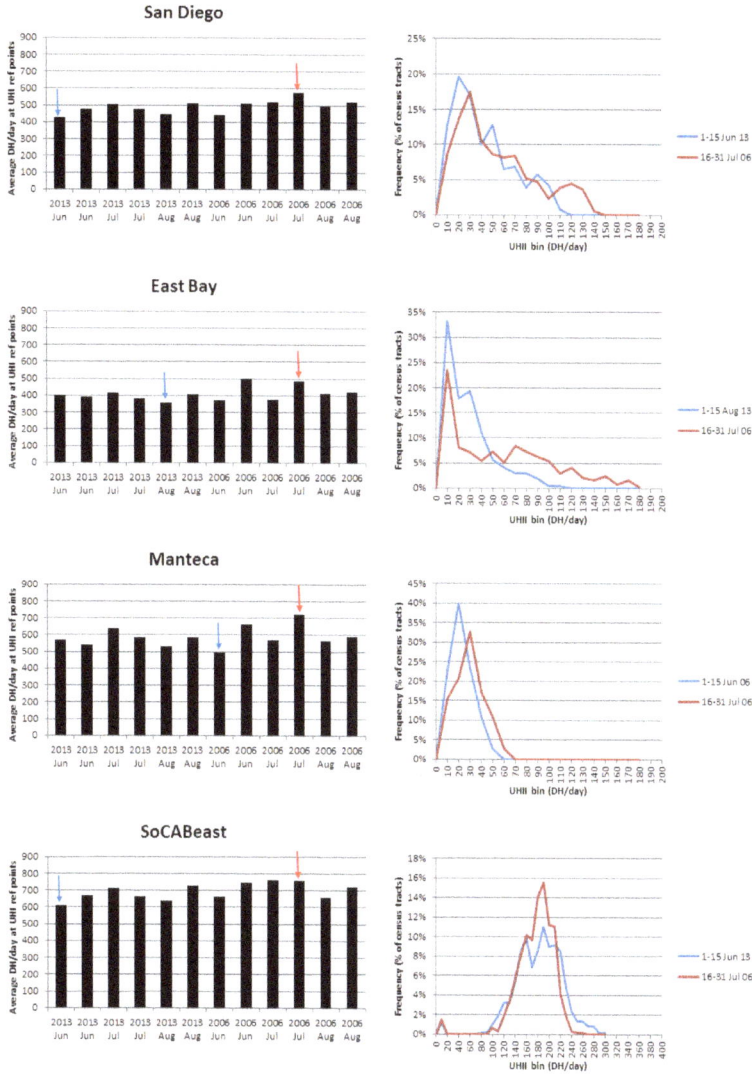

Figure 9. (**Left**) UHII (DH/day) at UHI reference points of each region (each month is divided into two halves: days 1–15 and 16–30 or 31); (**Right**) frequency distribution of census tracts in UHII bins for the selected periods.

While the meteorological model performance was thoroughly evaluated (as discussed in Section 2.8) and found to be satisfactory, there is no direct way to compare the model UHII to morbidity/mortality information at the fine resolutions presented in this paper. Such an effort may be undertaken in the future, but at this time, only some coarse-scale and qualitative assessments could be made. It is also important to recognize that the relationship between temperature, apparent temperature, and mortality is not a simple one: it is non-linear in nature and depends on a host of compounding factors, in addition to meteorology. As discussed earlier, the goal of developing the UHII is to provide an additional layer of information in decision-making tools such as CalEnviroScreen. In the following example, the binned UHII is compared to average daily deaths in four counties that were studied by Ostro et al., 2009 [25]. Both the model UHII and mortality data discussed here are for July 2006 which includes the heat wave period in the second half of the month.

Table 3 shows side by side the weighted UHII (wUHII) and mortality in those four counties. Note that the average daily deaths provided here are from all causes, not just heat, but since these occurred during the heat-wave month, it is plausible that a significant component is heat-related. The UHII is weighted by census tracts to make it more comparable to the total average deaths in each county (the number of tracts being a proxy to population). The weighted UHII (wUHII) is:

$$wUHII = \sum_{b=1}^{B} \{ tTRACT_b \times dhpd_b \} \tag{3}$$

where B is the total number of UHII bins (in a region) in increments of 10 DH/day, $tTRACT$ is the number of census tracts (in a region) that fall in UHII bin "b", and $dhpd$ is the number of DH/day in bin "b". It can be gleaned from Table 3 that the directionality of the $wUHII$ and the average daily deaths are similar. However, no attempt will be made here to develop any correlation on this basis alone.

Table 3. Weighted UHII (wUHII) and actual mortality in four counties during July 2006.

County/Urban Area	wUHII	Average Daily Deaths *
Fresno County/Fresno	4410	17.5
Kern County/Bakersfield	2180	13.0
Los Angeles County/East basin	105,000	162.5
Sacramento County/Sacramento	14,300	24.6

* Ostro et al., 2009 [25].

4. Conclusions and Future Research

Atmospheric modeling was performed in this study for the purpose of characterizing the UHI under present climate conditions and quantifying the UHII for California, including the superimposed effects of onshore warming and urban-climate archipelagos where they occur. The study identifies various patterns of UHI/UHII including single-core, multi-core, and urban-climate archipelago UHIs. The analysis shows that the UHII is shifted to larger values under the conditions of warmer weather. Thus, the potentially more frequent occurrences of heat waves in future climates could enhance UHIs and further exacerbate the health effects of hot weather.

From this perspective, it is important to continue research in (1) evaluating and quantifying the synergies between background weather, urban climates, and heat islands; (2) quantifying the potential exacerbation of UHIs by regional climate change; (3) evaluating the implications in terms of heat- and air-quality health; and (4) assessing the potential of mitigation measures (such as cool cities) in offsetting some or all of these negative effects, and evaluating the added importance of such measures in future years.

Acknowledgments: This paper is based on work that was supported by the California Environmental Protection Agency (Cal/EPA). The statements and conclusions in this paper are those of the author and do not necessarily reflect the views of the Cal/EPA. The mention of commercial products, their source, or their use in connection with material reported herein is not to be construed as actual or implied endorsement of such products. Spatial Informatics Group is acknowledged for GIS work in prior stages of this effort.

Conflicts of Interest: The authors declare no conflict of interest.

References

1. Goggins, W.B.; Chan, E.Y.Y.; Ng, E.; Ren, C.; Chen, L. Effect modification of the association between short-term meteorological factors and mortality by urban heat islands in Hong Kong. *PLoS ONE* **2012**, *7*, e38551. [CrossRef] [PubMed]

2. Adachi, S.A.; Kimura, F.; Kusaka, H.; Inoue, T.; Ueda, H. Comparison of the impact of global climate changes and urbanization on summertime future climate in the Tokyo metropolitan area. *J. Appl. Meteorol. Climatol.* **2012**, *51*, 1441–1454. [CrossRef]
3. Tan, J.; Zheng, Y.; Tang, X.; Guo, C.; Li, L.; Song, G.; Zhen, X.; Yuan, D.; Kalkstein, A.J.; Li, F. The urban heat island and its impact on heat waves and human health in Shanghai. *Int. J. Biometeorol.* **2010**, *54*, 75–84. [CrossRef] [PubMed]
4. Taha, H. *Potential Impacts of Climate Change on Tropospheric Ozone in California: A Preliminary Assessment of the Los Angeles Basin and the Sacramento Valley*; Lawrence Berkeley National Laboratory Report LBNL-46695; Lawrence Berkeley National Laboratory: Berkeley, CA, USA, 2001; Available online: http://escholarship. org/uc/item/5s41x609 (accessed on 3 March 2016).
5. Potchter, O.; Ben-Shalom, H.O. Urban warming and global warming: Combined effect on thermal discomfort in the desert city of Beer Sheva, Israel. *J. Arid Environ.* **2013**, *98*, 113–122. [CrossRef]
6. Vanos, J.K.; Cakmak, S.; Kalkstein, L.S.; Yagouti, A. Association of weather and air pollution interactions on daily mortality in 12 Canadian cities. *Air Qual. Atmos. Health* **2014**, *8*, 307–320. [CrossRef] [PubMed]
7. Kalkstein, L.; Sailor, D.; Shickman, K.; Sherdian, S.; Vanos, J. *Assessing the Health Impacts of Urban Heat Island Reduction Strategies in the District of Columbia*; Report DDOE ID#2013-10-OPS; Global Cool Cities Alliance: Washington, DC, USA, 2013.
8. Taha, H. Cool cities: Counteracting potential climate change and its health impacts. *Curr. Clim. Chang. Rep.* **2015**, *1*, 163–175. [CrossRef]
9. Taha, H. Meso-urban meteorological and photochemical modeling of heat island mitigation. *Atmos. Environ.* **2008**, *42*, 8795–8809. [CrossRef]
10. Taha, H. Meteorological, emissions, and air-quality modeling of heat-island mitigation: Recent findings for California, USA. *Int. J. Low Carbon Technol.* **2013**, *10*, 3–14. [CrossRef]
11. Taha, H. Meteorological, air-quality, and emission-equivalence impacts of urban heat island control in California. *Sustain. Cities Soc.* **2015**, *19*, 207–221. [CrossRef]
12. Papanastasiou, D.K.; Melas, D.; Kambezidis, H.D. Air quality and thermal comfort levels under extreme hot weather. *Atmos. Res.* **2015**, *251*, 4–13. [CrossRef]
13. Wang, X.; Chen, F.; Wu, Z.; Zhang, M.; Tewari, M.; Guenther, A.; Guenther, A.; Wiedinmyer, C. Impacts of weather conditions modified by urban expansion on surface ozone: Comparison between the Pearl River Delta and Yangtze River Delta regions. *Adv. Atmos. Sci.* **2009**, *26*, 962–972. [CrossRef]
14. Lombardo, M.A. *Ilhas de Calor nas Metropoles: O Example de Sao Paulo*; Editora Hucitec: Sao Paulo, Brazil, 1985; p. 244.
15. Rajagopalan, P.; Lim, K.C.; Jamei, E. Urban heat island and wind flow characteristics of a tropical city. *Sol. Energy* **2014**, *107*, 159–170. [CrossRef]
16. Santamouris, M. Cooling the cities—A review of reflective and green roof mitigation technologies to fight heat island and improve comfort in urban environments. *Sol. Energy* **2014**, *103*, 682–703. [CrossRef]
17. Skoulika, F.; Santamouris, M.; Kolokotsa, D.; Boemia, N. On the thermal characteristics and the mitigation potential of a medium size urban park in Athens, Greece. *Landsc. Urban Plan.* **2013**, *123*, 73–86. [CrossRef]
18. Taha, H. Episodic performance and sensitivity of the urbanized MM5 (uMM5) to perturbations in surface properties in Houston TX. *Bound.-Layer Meteorol.* **2008**, *127*, 193–218. [CrossRef]
19. Taha, H. Urban surface modification as a potential ozone air-quality improvement strategy in California: A mesoscale modeling study. *Bound.-Layer Meteorol.* **2008**, *127*, 219–239. [CrossRef]
20. Taha, H. *Multi-Episodic and Seasonal Meteorological, Air-Quality, and Emission-Equivalence Impacts of Heat-Island Control and Evaluation of the Potential Atmospheric Effects of Urban Solar Photovoltaic Arrays*; PIER Environmental Research Program; California Energy Commission: Sacramento, CA, USA, 2013. Available online: http://www.energy.ca.gov/2013publications/CEC-500-2013-061/CEC-500-2013-061.pdf (accessed on 3 March 2016).
21. OEHHA 2014. *California Communities Environmental Health Screening Tool*; version 2.0 (CalEnviroScreen 2.0) Guidance and Screening Tool; Office of Environmental Health Hazard Assessment Report; Office of Environmental Health Hazard Assessment: Sacramento, CA, USA, 2014; p. 136. Available online: https://oehha.ca.gov/media/CES20FinalReportUpdateOct2014.pdf (accessed on 3 March 2017).

22. Taha, H.; Freed, T. *Creating and Mapping an Urban Heat Island Index for California*; Report prepared by Altostratus Inc., Contract 13-001; California Environmental Protection Agency (Cal/EPA): Sacramento, CA, USA, 2015; Available online: https://calepa.ca.gov/wp-content/uploads/sites/34/2016/10/UrbanHeat-Report-Report.pdf and https://www.calepa.ca.gov/climate/urban-heat-island-index-for-california/urban-heat-island-interactive-maps/; (accessed on 3 March 2017).

23. National Oceanic and Atmospheric Administration (NOAA). State Annual and Seasonal Time Series. 2017. Available online: www.ncdc.noaa.gov/temp-and-precip/state-temps/ (accessed on 1 January 2016).

24. Knowlton, K.; Rotkin-Ellma, M.; King, G.; Margolis, H.G.; Smith, D.; Solomon, G.; Trent, R.; English, P. The 2006 California heat wave: Impacts on hospitalizations and emergency departments visits. *Environ. Health Perspect.* **2009**, *117*, 61. [CrossRef] [PubMed]

25. Ostro, B.D.; Roth, L.A.; Green, R.S.; Basu, R. Estimating the mortality effect of the July 2006 California heat wave. *Environ. Res.* **2009**, *109*, 614–619. [CrossRef]

26. Powers, G.; Huang, X.Y.; Klemp, B.; Skamarock, C.; Dudhia, J.; Gill, O.; Duda, G.; Barker, D.; Wang, W. *A Description of the Advanced Research WRF*; NCAR Technical Note NCAR/TN-475+STR; National Center for Atmospheric Research: Boulder, CO, USA, 2008.

27. Kusaka, H.; Kondo, H.; Kikegawa, Y.; Kimura, F. A simple single-layer urban canopy model for atmospheric models: Comparison with multi-layer and slab models. *Bound.-Layer Meteorol.* **2001**, *101*, 329–358. [CrossRef]

28. Chen, F.; Kusaka, H.; Bornstein, R.; Ching, J.; Grimmond, C.S.B.; Grossman-Clarke, S.; Loridan, T.; Manning, K.; Martilli, A.; Miao, S.; et al. The integrated WRF/urban modeling system: Development, evaluation, and applications to urban environmental problems. *Int. J. Climatol.* **2010**, *31*, 273–288. [CrossRef]

29. Salamanca, F.; Martilli, A.; Tewari, M.; Chen, F. A study of the urban boundary layer using different urban parameterizations and high-resolution urban canopy parameters with WRF. *J. Appl. Meteorol. Climatol.* **2011**, *50*, 1107–1128. [CrossRef]

30. Bougeault, P.; Lacarrere, P. Parameterization of orography-induced turbulence in a mesobeta-scale model. *Mon. Weather Rev.* **1989**, *117*, 1872–1890. [CrossRef]

31. Ching, J.; Brown, M.; Burian, S.; Chen, F.; Cionco, R.; Hanna, A.; Hultgren, T.; McPherson, T.; Sailor, D.; Taha, H.; et al. National urban database and access portal tool, NUDAPT. *Bull. Am. Meteorol. Soc.* **2009**, *90*, 1157. [CrossRef]

32. Trent, R.B. *Review of July 2006 Heat Wave Related Fatalities in California*; California Department of Health Services: Sacramento, CA, USA, 2007. Available online: http://www.cdph.ca.gov/HealthInfo/injviosaf/Documents/HeatPlanAssessment-EPIC.pdf (accessed on 3 March 2016).

33. Gershunov, A.; Cayan, D.R.; Iacobellis, S.F. The great 2006 heat wave over California and Nevada: Signal of an increasing trend. *J. Clim.* **2009**, *22*, 6181–6203. [CrossRef]

34. Kistler, R.; Kalnay, E.; Collins, W.; Saha, S.; White, G.; Woollen, J.; Kalnay, E.; Chelliah, M.; Ebisuzaki, W.; Kanamitsu, M.; et al. The NCEP-NCAR 50-year reanalysis: Monthly means CDROM and documentation. *Bull. Am. Meteorol. Soc.* **2001**, *82*, 247–267. [CrossRef]

35. Anderson, J.R.; Hardy, E.E.; Roach, J.T.; Witmer, R.E. *A Land Use and Land Cover Classification System for Use with Remote Sensor Data*; USGS Professional Paper 964; U.S. Government Printing Office: Washington, DC, USA, 2001.

36. Multi-Resolution Land-Characteristics Consortium (MRLC). National Land Cover Databases. Available online: http://www.mrlc.gov/nlcd2006.php (accessed on 1 January 2016).

37. Stewart, I.D.; Oke, T.R. Local climate zones for urban temperature studies. *Bull. Am. Meteorol. Soc.* **2012**, *93*, 1879–1900. [CrossRef]

38. Akbari, H.; Rose, S.; Taha, H. *Characterizing the Fabric of the Urban Environment: A Case Study of Sacramento, California*; Lawrence Berkeley National Laboratory Report LBNL-44688; Lawrence Berkeley National Laboratory: Berkeley, CA, USA, 1999.

39. Rose, S.; Akbari, H.; Taha, H. *Characterizing the Fabric of the Urban Environment: A Case Study of Greater Houston, Texas*; Lawrence Berkeley National Laboratory Report LBNL-51448; Lawrence Berkeley National Laboratory: Berkeley, CA, USA, 2003.

40. Taha, H. *Urban Surface Modification as a Potential Ozone Air-Quality Improvement Strategy in California—Phase 2: Fine-Resolution Meteorological and Photochemical Modeling of Urban Heat Islands*; PIER Environmental Research; California Energy Commission: Sacramento, CA, USA, 2007. Available online: http://www.energy.ca.gov/2009publications/CEC-500-2009-071/CEC-500-2009-071.PDF (accessed on 4 April 2015).

41. Boucouvala, D.; Bornstein, D. Analysis of transport patterns during an SCOS97 NARSTO episode. *Atmos. Environ.* **2003**, *37*, 73–94. [CrossRef]

42. Kim, Y.-H.; Baik, J.-J. Spatial and temporal structure of the urban heat island in Seoul. *J. Appl. Meteorol.* **2005**, *44*, 591–605. [CrossRef]

43. Oke, T.R.; Hannell, F.G. Urban Climates. *WMO Tech. Note* **1970**, *108*, 113–126.

44. Oke, T.R. City size and the urban heat island. *Atmos. Environ.* **1973**, *7*, 769–779. [CrossRef]

45. Tesche, T.W.; McNally, D.E.; Emery, C.A.; Tai, E. *Evaluation of the MM5 Model over the Midwestern U.S. for Three 8-Hour Oxidant Episodes, Prepared for the Kansas City Ozone Technical Workgroup*; Alpine Geophysics LLC and Environ Corp.: San Rafael, CA, USA, 2001; p. 23.

46. Shepherd, J.M.; Bounoua, L.; Mitra, C. *Urban Climate Archipelagos: A New Framework for Urban Impacts on Climate*; Earthzine: New York, NY, USA, 2013; Available online: http://earthzine.org/2013/11/29/urban-climate-archipelagos-a-new-framework-for-urban-impacts-on-climate/ (accessed on 3 March 2016).

47. Daly, C.; Halbleib, M.; Smith, J.I.; Gibson, W.P.; Doggett, M.K.; Taylor, G.H.; Vurtis, J.; Pasteris, P.P. Physiographically sensitive mapping of climatological temperature and precipitation across the conterminous United States. *Int. J. Climatol.* **2008**, *28*, 2031–2064. [CrossRef]

48. Lebassi-Habtezion, B.; Gonzalez, J.; Bornstein, R.D. Modeled large-scale warming impacts on summer California coastal-cooling trends. *J. Geophys. Res.* **2011**, *116*, D20. [CrossRef]

49. Liu, L.; Talbot, R.; Lan, X. Influence of climate change and meteorological factors on Houston's air pollution: Ozone case study. *Atmosphere* **2015**, *6*, 623. [CrossRef]

50. Taha, H.; Wilkinson, J.; Bornstein, R. *Urban Forest for Clean Air Demonstration in the Sacramento Federal Non-Attainment Area: Atmospheric Modeling in Support of a Voluntary Control Strategy*; Sacramento Metropolitan Air Quality Management District (SMAQMD): Sacramento, CA, USA, 2011.

51. Li, D.; Bou-Zeid, E. Synergistic interactions between urban heat islands and heat waves: The impact in cities is larger than the sum of its parts. *J. Appl. Meteorol. Climatol.* **2013**, *52*, 2051–2064. [CrossRef]

52. Li, D.; Sun, T.; Liu, M.; Yang, L.; Wang, L.; Gao, Z. Contrasting responses of urban and rural surface energy budgets to heat waves explain synergies between urban heat islands and heat waves. *Environ. Res. Lett.* **2015**, *10*, 054009. [CrossRef]

53. Georgescu, M.; Morefield, P.E.; Bierwage, B.G.; Weaver, C.P. Urban adaptation can roll back warming of emerging metropolitan regions. *Proc. Natl. Acad. Sci. USA* **2014**, *111*, 2909–2914. [CrossRef] [PubMed]

54. Lemonsu, A.; Kounkou-Arnaud, R.; Desplat, J.; Salagnac, J.-L.; Masson, V. Evolution of the Parisian urban climate under a global changing climate. *Clim. Chang.* **2013**, *116*, 679–692. [CrossRef]

climate

MDPI

Article

Influence of Urban Green Area on Air Temperature of Surrounding Built-Up Area

Hideki Takebayashi

Department of Architecture, Kobe University, Kobe 657-8501, Japan; thideki@kobe-u.ac.jp; Tel.: +81-78-803-6062

Academic Editors: Michele Zinzi and Mattheos Santamouris
Received: 17 July 2017; Accepted: 2 August 2017; Published: 7 August 2017

Abstract: In this investigation, a numerical model expressing advection and diffusion effects is used to examine air temperature rise in urban areas that are on the leeward side of green areas. The model results are then verified by comparison with measurement results. When the measurement point is at a distance of 30 m or more from a green area, the air temperature of the urban area is not affected by the green area. An isotropic diffusion model and a model incorporating buoyancy were applied for the vertical diffusion term. Results of air temperature rise with distance from the green area were compared for both calculated and measured values. The rise in air temperature due to the development of the urban boundary layer in the area near a green space is expressed using the sensible heat flux from the ground surface, the distance from the green area and the wind velocity. We considered an approximation of air temperature rise in order to express the following situation: when entering the urban area, air temperature rises sharply, and when reaching a certain distance from a green area, it becomes almost constant.

Keywords: green area; built-up area; air temperature; measurement; calculation

1. Introduction

Urban greenery is one of the main measures for mitigating the thermal environment in urban spaces. Givoni [1] has organized the functions and impacts of urban planted areas through a review of research papers and has presented climatic guidelines for hot-dry regions, hot-humid regions and cold regions. A summary of climatic guidelines for park design is as follows: it is to provide ample shade and to protect from dust for hot-dry regions; it is to provide shade, to minimize wind blockage, to improve the ventilation and to minimize floods for hot-humid regions; it is to provide wind protection without blocking the winter sun for cold regions. He summarized that the influence of city parks and open spaces on the urban climate is limited to the conditions prevailing within these areas themselves, and extends only a short distance into the surrounding, densely built, urban area. On the other hand, Honjo and Takakura [2] explained that the range of the effects of urban green areas extends to about 100 to 300 m into the surrounding urban area. They also explained that 300 m along the main wind direction is the ideal length for an urban green area, based on two-dimensional analysis results.

In recent years, interest in this field of study has increased. How to quantify the range of the air temperature reduction effect of an urban green area on the surrounding urban area is a question that has been frequently asked by administrative officials responsible for organizing urban green spaces. Moriyama et al. [3] have conducted numerical simulations to examine increases and decreases in air temperature in urban areas adjacent to green areas. They used the following conditions: an inflow upper wind velocity of 2 to 6 m/s at 50 m above the ground, a ground surface temperature difference of 1 to 5 °C between green and urban areas, and a roughness parameter of 0.1 to 1.0 m for green areas and 0.5 to 1.0 m for urban areas. The evaluation height was 3.25 m above the ground. They concluded that the influence of the green space extends to a distance of about 150 m from the urban-green boundary. The above-mentioned Honjo et al. [2] have carried out numerical simulations under the condition that an inflow upper wind velocity is 4 m/s at 200 m above the ground, a ground surface temperature difference is 4 °C between green and urban areas, and a roughness parameter is 0.2 m for both the green area and the urban area. The evaluation height in this case was 2 m above the ground. They concluded that even a green area with 100 m size affects the area within a distance of about 300 m from the urban-green boundary.

There are a few studies focusing on air temperature reduction in urban areas around a green area [4]. Ca et al. [5] have carried out field measurements to determine the cooling influence of a park on the surrounding area in the Tama New Town, a city in the west of Tokyo. With the size of 0.6 km^2, a park can reduce the air temperature by up to 1.5 °C at noon time in a leeward commercial area at distance of 1 km. Yu and Hien [6] have carried out temperature and humidity measurements in two big city green areas (36 ha and 12 ha) in Singapore. A three-dimensional non-hydrostatic model (Envi-met) was applied for the simulation of Surface-Plant-Air interactions inside urban environments. Horizontal air temperature profiles in both the green area and surrounding area are calculated by the Envi-met model.

Yagi and Takebayashi [7] have performed measurements at four urban areas in Kobe City. The spatial variation of the vertical air temperature gradient between 4.0 m and 1.5 m is large in urban areas, since air temperature reduction effect in urban areas is different depending on the circumstances around the measurement point. Since sea breezes dominate in summer days in many cities in Japan, air temperature reduction due to advection effects is expected in regions leeward of urban green areas. In this study, the characteristics of air temperature in the urban area on the leeward side of green areas are considered using a numerical model incorporating advection and diffusion, and verified by comparison with measurement. The objective of this study is to clarify the characteristics of air temperature rise in an urban area on the leeward side of a green area, as a contribution to the practical planning of urban greening.

2. Measurements

2.1. Study Site

Mobile measurements were carried out in Higashi-yuen Park (about 2.7 ha, green coverage rate, which is the ratio of the canopy area to the park area: about 45%) and a neighboring business area at 13:00 and 17:00 on 2 August 2012, in Ishiyagawa Park (about 4 ha, green coverage ratio: about 42%) and a residential area at 13:00 and 17:00 on 4 August 2012, and in Okurayama Park (about 7.9 ha, green coverage ratio: about 70%) and a residential area at 13:00 and 17:00 on 8 August 2012. These parks are all located in Kobe city, Japan. Mobile measurement points and aerial photographs are shown in Figure 1. The grid lines are spaced 50 m apart. The green color indicates green coverage.

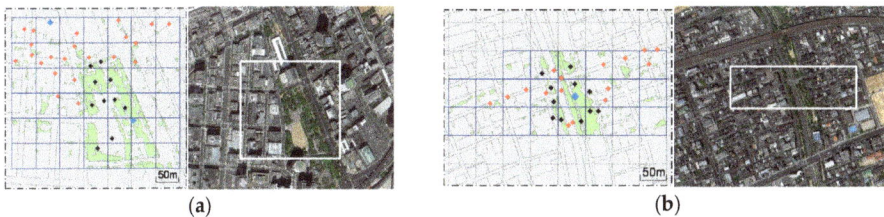

(a)   (b)

Figure 1. *Cont.*

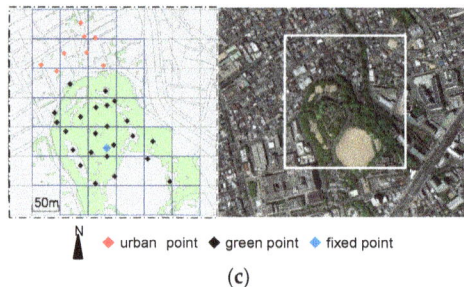

Figure 1. Mobile measurement points and aerial photograph, all located in Kobe city, Japan. (**a**) Higashi-yuen Park and business area; (**b**) Ishiyagawa Park and residential area; (**c**) Okurayama Park and residential area.

Higashi-yuen Park and the business area are located in the center of Kobe City. There are public buildings such as Kobe City Hall and general offices etc. in the business area. Middle-high-rise buildings are dominant. Ishiyagawa Park and the neighboring residential area are located in an urban area at the southern foot of Rokko Mountain on the east side of Kobe City. There are mainly detached houses and small scale collective houses in the residential area. Low-rise buildings are dominant. Okurayama Park and its residential area are located in the urban area at the southern foot of Rokko Mountain on the west side of Kobe City. There are detached houses, hospitals, etc. in the residential area. Low-rise building and middle-rise building are mixed.

2.2. Outline of Measurements

The elements measured are air temperature, wind direction, wind velocity at a height of 1.5 m, and surface temperature. The measuring device and method are shown in Table 1. Wind velocity was sampled every second at each mobile measurement point which is indicated as an urban point and green point in Figure 1 and the averaged value for 30 s was recorded. Wind direction was recorded based on the direction with the highest frequency in the 30 s. Measurement results for air temperature at the fixed measurement points are shown in Figure 2. It was continuously measured only at fixed points. Thermistor sensors were installed in a natural ventilation-type solar radiation shielding device and were set on the roof of the Kobe City Hall No. 3 building (47 m above the ground, flat concrete roof with the usual waterproof sheet finish) and the trunks of trees in the Higashi-yuen Park, Ishiyagawa Park, and Okurayama Park (3 m above the ground). Kobe City Hall and Higashi-yuen Park are close to each other. The distances from Ishiyagawa Park and Okurayama Park to Kobe City Hall are about 5.7 km and 2.1 km, respectively.

Although it took a maximum of 1.5 h for the mobile measurements at each site to be made, no sudden changes in weather were confirmed as compared with the results of the fixed-point measurements, so no correction was made to the results of the mobile measurements. For the analysis in the next section, I used the difference between air temperature from the mobile measurements in the urban area and the air temperature of the fixed-point measurements in the park at that time.

Table 1. Measuring device and method.

	Device	Method	Accuracy of Device
Air temperature	Thermistor with solar radiation shield	Averaged for 5 min by sampling every 5 s	±0.5 K
Wind direction	Windsock	Highest frequency in 30 s	by visual inspection
Wind velocity	Hot-wire anemometer	Averaged for 30 s by sampling every second	±2% of indicated value
Surface temperature	Infrared thermometer	Measured on ground and wall surface, a representative material surface at each measurement point was measured several times to obtain stable data	±1.0 K

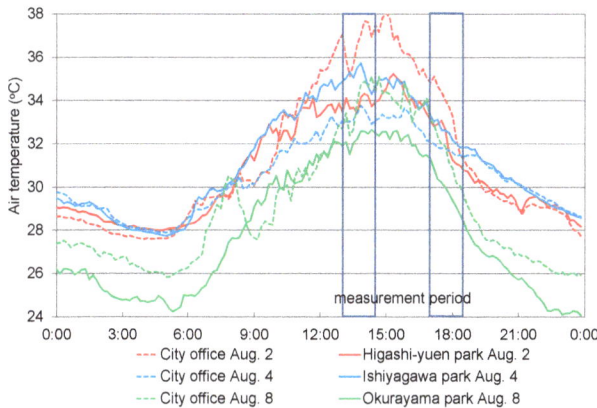

Figure 2. Measurement results of air temperature at the fixed measurement points.

The boundary between the green area and the urban area is set to 0, and the following analysis is carried out, focusing on the relationship between the horizontal distance from the boundary and the air temperature in the urban areas. Since the main wind direction was southwest in the case of measurements around Higashi-yuen Park, the measurement results in the northeastern urban area were used for analysis. The distance to the park was calculated by drawing a straight line in the southwest direction from each mobile measurement point. Similarly, since the main wind direction was east in the case of measurements around Ishiyagawa Park, the measurement results in the west urban area were used for analysis. In the case of measurements around Okurayama Park, the main wind direction was south-southeast, so the measurement results in the northern urban area were used for analysis. Figure 3 shows the distance from the green area to each mobile measurement point in the urban area and the air temperature rise. This is the difference to the air temperature measured in the windward side green area. Air temperature rise is large in a weak wind case. Strong wind and weak wind were classified by the upper wind velocity of 5.5 m/s measured at the Kobe meteorological observatory. A measurement point where the distance from the green area is about 30 m or more was considered representative of the urban area's air temperature, without being affected by the green area. It is considered that air temperature in the urban area in a weak wind case is fluctuating due to the influence of local ventilation and solar radiation shielding. The wind velocity was measured by mobile measurement at a height of 1.5 m above the ground. In the urban area, this was 1.0 to 1.3 m/s at 13:00 and 0.7 to 1.1 m/s at 17:00 in a strong wind case and 0.5 to 1.0 m/s at 13:00 and 0.4 to 1.1 m/s at 17:00 in a weak wind case. The wind velocity in the urban area fluctuated because of the influence of the surrounding buildings.

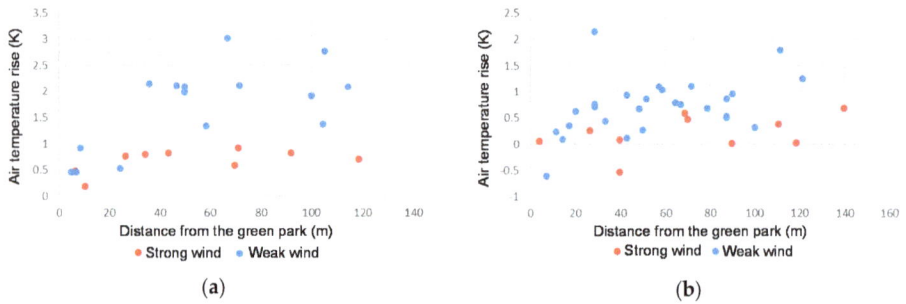

Figure 3. Distance from the green area to each mobile measurement point in urban area and air temperature rise (**a**) at 13:00; (**b**) at 17:00.

3. Results

3.1. Outline of Calculations

Calculations were carried out by Computational Fluid Dynamics (CFD). For the turbulence closure model, a standard k-ε model was used. This is the most common model used in CFD to simulate mean flow characteristics for turbulent flow. The outline of the calculation model is shown below.

$$\frac{\partial \overline{u_i}}{\partial x_i} = 0 \tag{1}$$

$$\frac{\partial \overline{u_i}}{\partial t} + \frac{\partial \overline{u_i u_j}}{\partial x_j} = -\frac{\partial \overline{\pi}}{\partial x_i} + \frac{\partial}{\partial x_j}\left\{ (\nu_t + \nu)\left(\frac{\partial \overline{u_i}}{\partial x_j} + \frac{\partial \overline{u_j}}{\partial x_i}\right) \right\} + \frac{g_i}{\Theta_0}(\Theta - \Theta_0) - 2\,\ni_{i,j,k}\,\Omega_j \overline{u_k} \tag{2}$$

$$\frac{\partial k}{\partial t} + \frac{\partial k\overline{u_i}}{\partial x_j} = \frac{\partial}{\partial x_j}\left\{ \left(\frac{\nu_t}{\sigma_1} + \nu\right)\left(\frac{\partial k}{\partial x_j}\right) \right\} + \nu_t\left(\frac{\partial \overline{u_i}}{\partial x_j} + \frac{\partial \overline{u_j}}{\partial x_i}\right)\frac{\partial \overline{u_i}}{\partial x_j} - \varepsilon - g\beta\frac{\nu_t}{Prt}\frac{\partial \Theta}{\partial x_k} \tag{3}$$

$$\frac{\partial \varepsilon}{\partial t} + \frac{\partial \varepsilon \overline{u_i}}{\partial x_j} = \frac{\partial}{\partial x_j}\left\{ \left(\frac{\nu_t}{\sigma_2} + \nu\right)\left(\frac{\partial \varepsilon}{\partial x_j}\right) \right\} + C_1\frac{\varepsilon}{k}\nu_t\left(\frac{\partial \overline{u_i}}{\partial x_j} + \frac{\partial \overline{u_j}}{\partial x_i}\right)\frac{\partial \overline{u_i}}{\partial x_j} - C_2\frac{k\varepsilon}{\nu_t} \tag{4}$$

$$\frac{\partial \overline{\theta}}{\partial t} + \frac{\partial}{\partial x_j}(\overline{\theta u_i}) = \frac{\partial}{\partial x_j}\left\{ \left(\frac{\nu_t}{Prt} + \nu\right)\left(\frac{\partial \overline{\theta}}{\partial x_j}\right) \right\} + \frac{\overline{Q}}{C_p\rho} \tag{5}$$

For the application of the turbulence model to urban space, Ashie and Ca [8] have proposed a model that expresses the eddy viscosity coefficient ν_t as a function of the flux Richardson number R_f (Equations (8)–(11)). They do this by aggregating the buoyancy effect into the vertical eddy viscosity model coefficient C_μ and the turbulent Prandtl number Prt. In this study, we used both Equation (7) and the conventional Equation (6). Equations (8)–(11) are used in calculating the vertical eddy viscosity coefficient ν_t on the right side of Equation (7). Calculation conditions and the outline of calculation conditions are shown in Table 2 and Figure 4.

(Conventional isotropic diffusion model)

$$\nu_t = C_D\frac{k^2}{\varepsilon}\ (\text{horizontal and vertical diffusion}) \tag{6}$$

(Model incorporating buoyancy effect)

$$\nu_t = C_D\frac{k^2}{\varepsilon}\ (\text{horizontal diffusion}), \nu_t = C_\mu\frac{k^2}{\varepsilon}\ (\text{vertical diffusion}) \tag{7}$$

C_D: eddy viscosity constant (0.09), C_μ: eddy viscosity coefficient

$$C_\mu = \frac{0.8\varnothing\gamma - 0.5\gamma(\varnothing_\theta - \gamma)}{0.8\gamma + \varnothing(\varnothing_\theta - \gamma)} \frac{0.53 - 0.94R_f}{1 - R_f} \tag{8}$$

$$P_{rt} = P_{rt0}\frac{1.59 - R_f(1.5\varnothing_\theta + 2.82)}{1.59 + R_f(3\varnothing - 5.22)} \tag{9}$$

$$\gamma = \varnothing_\theta\frac{1.59 - 5.22R_f}{1.59 - 2.82R_f}, \quad \varnothing = 0.2, \ \varnothing_\theta = 1/3.2 \tag{10}$$

$$R_g = \frac{\beta g \frac{\partial \Theta}{\partial z}}{\left[\left(\frac{\partial U}{\partial z}\right)^2 + \left(\frac{\partial V}{\partial z}\right)^2\right]} \quad \begin{array}{l} R_g \le 0.195 \ R_f = 0.6588\left[R_g + 0.1776 - \left(R_g{}^2 - 0.3221R_g + 0.03156\right)^{1/2}\right] \\ R_g \ge 0.195 \ R_f = 0.191 \end{array} \tag{11}$$

The composition of the model was set according to Moriyama et al. [3], and the calculation conditions were set based on weather conditions at the time of measurement. The mesh size in the horizontal direction was set to 50 m in correspondence with the selection policies of the measurement points. The calculation condition as shown in Figure 4 expresses the phenomenon flowing out from the green area to the urban area in three dimensions. The vertical air temperature profile in the green area was uniformly given for the inflow condition. The upper wind velocity at 50 m high was relatively large, as it was measured under conditions where a sea breeze was dominant.

Table 2. Calculation conditions.

	13:00	17:00
Inflow air temperature with uniform vertical profile	33 °C	31 °C
Inflow wind velocity at 50 m high with logarithmic vertical profile	Large: 5.6 m/s Small: 4.1 m/s	Large: 4.7 m/s Small: 4.2 m/s
Sensible heat from ground surface	314 W/m²	196 W/m²
Roughness parameter	0.5 m	
Horizontal mesh size	50 m	
Vertical mesh size	3 m	

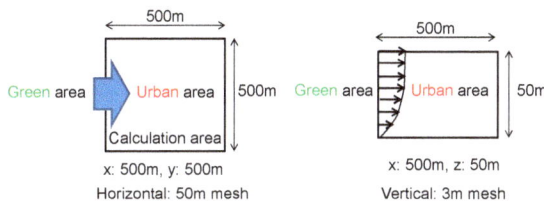

Figure 4. Outline of calculation conditions.

3.2. Results

Calculation and measurement results for air temperature rise with distance from the green area are shown in Figures 5 and 6. The results in both isotropic and non-isotropic diffusion models are shown. In the isotropic diffusion model, the horizontal and vertical eddy viscosity coefficients v_t are given by Equation (6). In the non-isotropic diffusion model, the eddy viscosity coefficient v_t in the vertical direction is given by the formula of Equation (7) when considering the buoyancy effect. Distance from the green area and the heat flux component of the calculation result, at 13:00 in the mesh near the ground surface, is shown in Figure 7. In the incorporated buoyancy model, the sensible heat flux supplied from the ground surface is transported in the vertical direction due to the vertical diffusion effect, so air temperature in the mesh near the ground surface does not rise. The part of the urban area more than 50 m from the green area is dominated by the diffusion effect in the vertical direction over the advection effect.

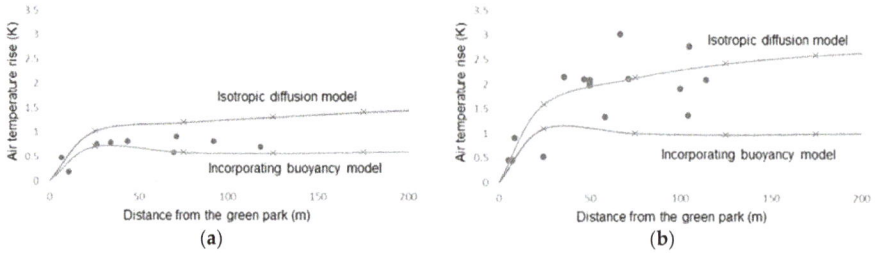

Figure 5. Calculation results and measurement results of air temperature rise at 13:00 according to the distance from the green area. (**a**) strong wind case; (**b**) weak wind case.

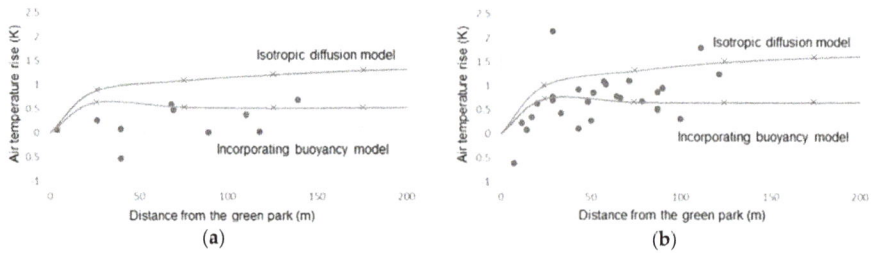

Figure 6. Calculation results and measurement results of air temperature rise at 17:00 according to the distance from the green area. (**a**) strong wind case; (**b**) weak wind case.

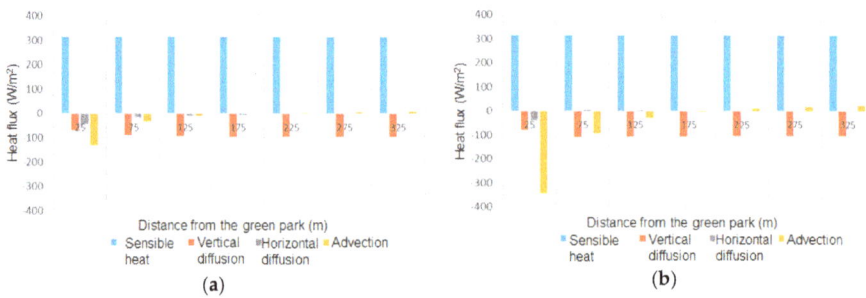

Figure 7. Distance from the green area and heat flux component of the calculation result at 13:00. (**a**) isotropic diffusion model; (**b**) incorporating buoyancy model.

When the inflow wind velocity is large, the calculation result in the incorporated buoyancy model tends to coincide with the measurement result of the air temperature. When the inflow wind velocity is small, the calculation result in the isotropic diffusion model, in which the diffusion effect in the vertical direction is not prominent, is close to the measurement result of the air temperature.

Therefore, the calculation result by the previous study, using the isotropic diffusion model, may be matched with the measurement result when the inflow wind velocity is small. This is shown in the right-hand panels of Figures 5 and 6. Even if the distance from the green area is 150 m or more, air temperature rises more and its effect may thus extend to over 200 m. Since the inflow wind velocity is small, the vertical diffusion effect is also small, and the air temperature rise in the urban area is thus larger than when the inflow wind velocity is large. At this time air temperature in the urban area varies considerably because of the influence of local ventilation, solar radiation shielding, etc. This can be seen from the measurement results in the right-hand panel of Figure 5.

4. Discussion

In addition to the measurement results in Kobe City, the calculation results in Figures 8 and 9 were also compared to the measurement results in the urban area around Koishikawa park in Tokyo by Kato et al. [9], and several parks in Osaka city by Moriyama et al. [10]. In the results measured in Tokyo and Osaka, air temperature does not rise as it enters the part of the urban area more than 50 m from the green area. On the other hand, Honjo and Takakura [2] explained that the range of the effects of urban green areas extends to about 100 to 300 m into the surrounding urban area. Since they used the isotropic diffusion model, it is recognized that it was a finding only in the case of weak wind.

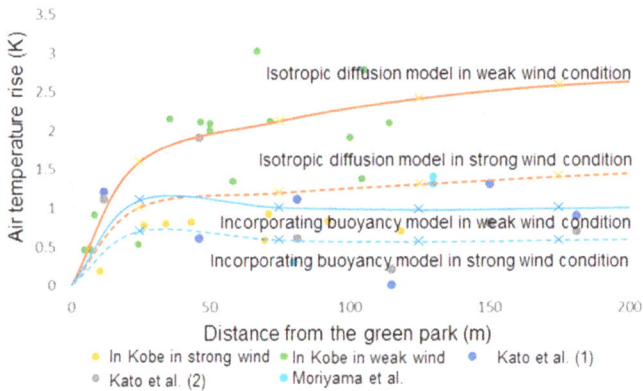

Figure 8. Distance from the green area and the air temperature rise in several urban areas in the daytime.

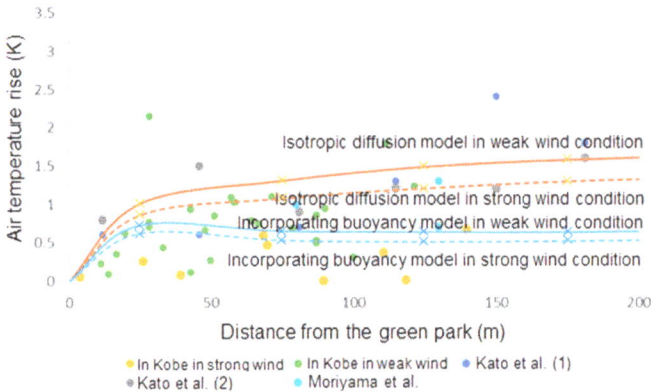

Figure 9. Distance from the green area and air temperature rise in several urban areas in the evening.

In order to discuss this in more detail, a recalculation was carried out, improving the spatial resolution in the urban area near the green area. An outline of the modified calculation conditions is shown in Figure 10. The horizontal mesh size was changed to 5 m from 50 m as in the above calculation. The other calculation conditions were not changed.

Calculation results of air temperature rise according to the distance from the green area are shown in Figure 11. Sensible heat flux from the ground surface in the urban area was assumed to be

236.7 W/m^2 for daytime and 28.3 W/m^2 for evening. A value of 132.5 W/m^2 was also assumed for their intermediate value. When entering the urban area air temperature rises sharply. The smaller the wind velocity, the larger the distance influenced by the green area, and the larger the air temperature rise. As the distance from the green area increases, air temperature becomes constant. When entering the part of the urban area more than 50 m from the green area, the air temperature near the ground surface is dominated by the diffusion effect in the vertical direction rather than the advection effect from the green area.

In general, air temperature rise ΔT (K) due to the development of the urban boundary layer is expressed by Equation (12).

$$\Delta T = \sqrt{\frac{2(1+k)HL\alpha}{C_p \rho U}} \tag{12}$$

where k is the ratio of entrainment (0 to 1), H is the sensible heat flux from the ground surface (W/m^2), L is the distance from the boundary (m), α is air temperature gradient (K/m), C_p is the specific heat of air (=1000 J/(kgK)), ρ is air density (=1.2 kg/m^3), and U is wind velocity (m/s). Assuming $\alpha = 0.006$ (K/m), it becomes Equation (13).

$$\Delta T = (0.0032 \sim 0.0045)\sqrt{H/U}\sqrt{L} \tag{13}$$

Air temperature rise ΔT, by Equation (12), when $k = 0$ is shown in Figure 12 together with the calculation results. Equation (12) is calculated using the boundary layer thickness $h = \Delta T/\alpha$. Actually, when the development of the boundary layer is not sufficient and h is small, α should be set to be large. Then, air temperature rise ΔT approximated by Equation (14) is shown in Figure 13. The coefficient a at this time is shown in Table 3. It is larger than the 0.0032 used in Equation (12).

$$\Delta T = a\sqrt{H/U}\sqrt{L} \tag{14}$$

As described above, the calculated air temperature near the ground surface rises sharply as it enters the urban area. This is because of the sensible heat flux from the ground surface, and when entering the area beyond about 50 m, it becomes almost constant. On the other hand, the approximate value of the air temperature due to the development of the boundary layer monotonically rises with the distance from the green area. Therefore, we considered an approximation based on the following equation where air temperature rise becomes constant as the distance goes above a certain value. Air temperature rise ΔT by Equation (15) is shown in Figure 14 together with the calculation results. When entering the urban area, air temperature rises sharply, and when entering the area beyond a certain distance it becomes almost constant.

$$\Delta T = \begin{cases} a\sqrt{H/U}\sqrt{L} & L < b \\ a\sqrt{H/U} & L > b \end{cases} \tag{15}$$
$$\approx a'\sqrt{H/U}\sqrt{L/(L+b)}$$

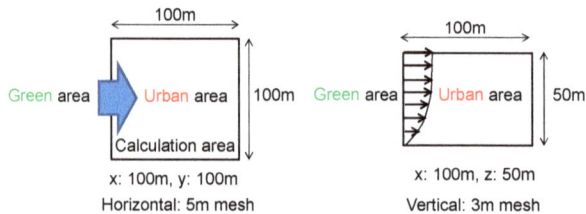

Figure 10. Outline of modified calculation conditions.

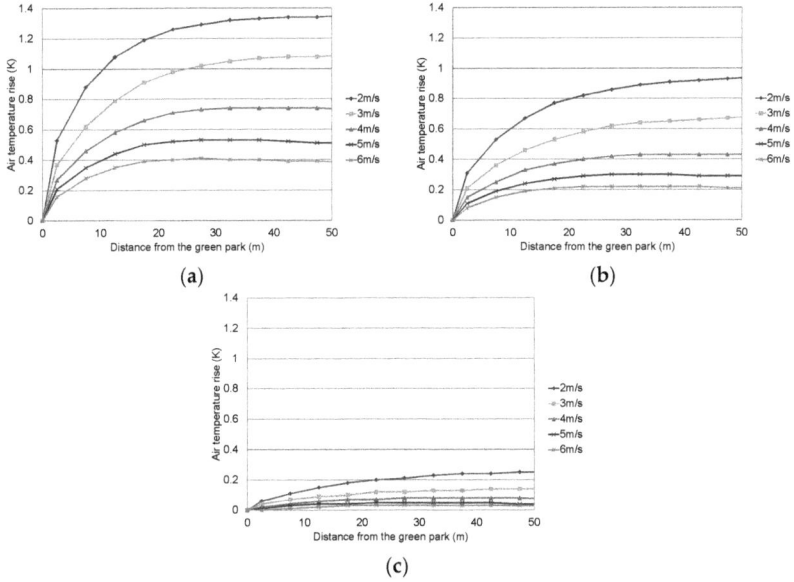

Figure 11. Calculation results of air temperature rise according to the distance from the green area. In the cases where the sensible heat flux is (**a**) 236.7 W/m^2; (**b**) 132.5 W/m^2; (**c**) 28.3 W/m^2.

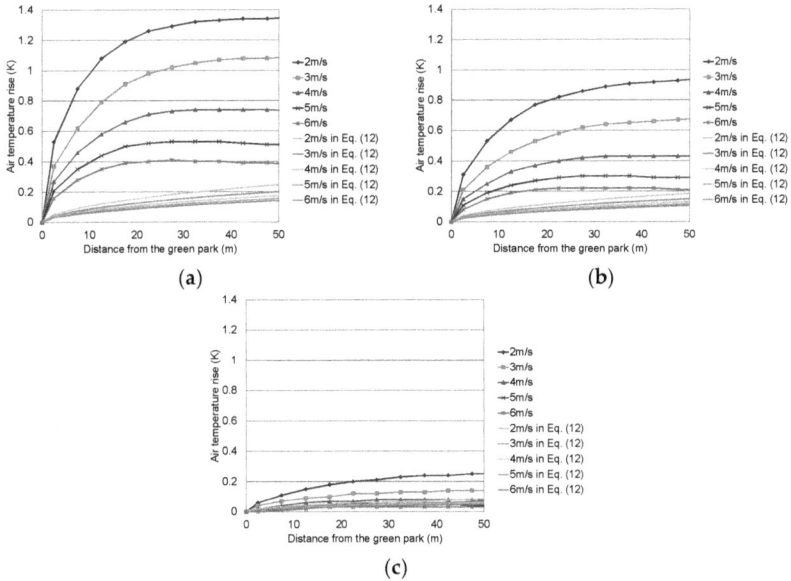

Figure 12. Air temperature rise by Equation (12). In the cases where the sensible heat flux is (**a**) 236.7 W/m^2; (**b**) 132.5 W/m^2; (**c**) 28.3 W/m^2.

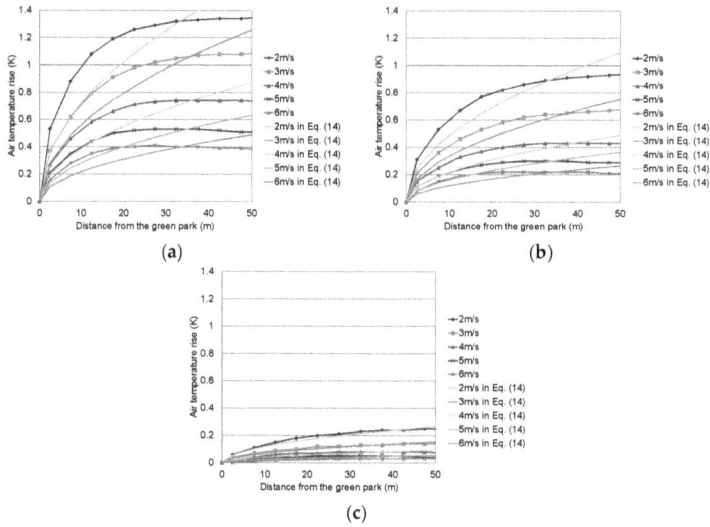

Figure 13. Air temperature rise by Equation (14). In the cases where the sensible heat flux is (a) 236.7 W/m^2; (b) 132.5 W/m^2; (c) 28.3 W/m^2.

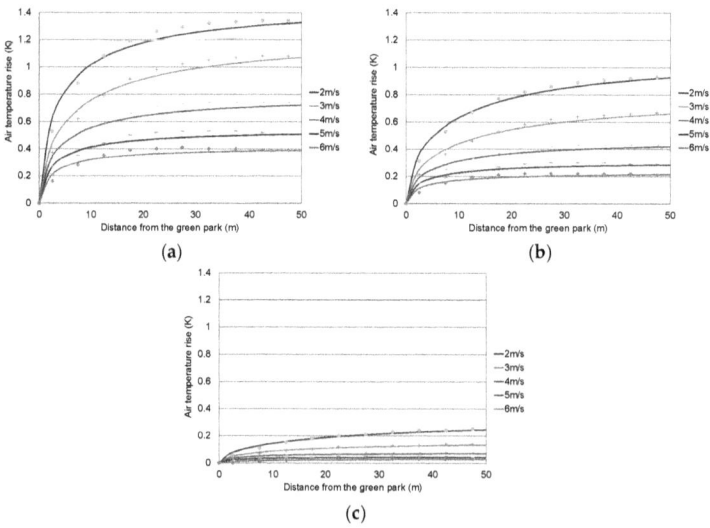

Figure 14. Air temperature rise by Equation (15). In the cases where the sensible heat flux is (a) 236.7 W/m^2; (b) 132.5 W/m^2; (c) 28.3 W/m^2.

Table 3. Coefficient *a* when it is approximated by Equation (14).

	2 m/s	3 m/s	4 m/s	5 m/s	6 m/s
236.7 W/m^2	0.021	0.020	0.016	0.013	0.011
132.5 W/m^2	0.019	0.016	0.012	0.010	0.008
28.3 W/m^2	0.010	0.007	0.005	0.003	0.002

5. Conclusions

In order to clarify the characteristics of air temperature rise in an urban area on the leeward side of a green area, mobile measurements and calculations expressing advection and diffusion effects are made. These calculations were then verified by comparison with the measurement results. The relationship between the distance from the green area to each mobile measurement point in the urban area and the air temperature rise is analyzed using the measurement results in Kobe city. At a measurement point where the distance from the green area is 30 m or more, the air temperature of the urban area becomes unaffected by the green area.

Calculation results and measurement results for air temperature rise with distance from the green area are compared when an isotropic diffusion model and an incorporated buoyancy model are applied for the vertical diffusion term. From the comparison with the measurement results in Kobe City, as well as in Tokyo and Osaka, it is considered that air temperature does not rise as it enters the part of the urban area beyond more than 50 m from the edge of the green area. The air temperature rise in the urban area near the green area, due to the development of the urban boundary layer, is expressed using the sensible heat flux from the ground surface, the distance from the green area and the wind velocity. We considered an approximation of air temperature rise in order to express the following situation: when entering the urban area, air temperature rises sharply, and when passing beyond a certain distance, it becomes almost constant.

Acknowledgments: I thank the urban planning bureau of Kobe city office for their cooperation with respect to our measurements. I used the modified source code by Moriyama et al. for the calculations.

Conflicts of Interest: The author declares no conflict of interest.

References

1. Givoni, B. Impact of planted areas on urban environmental quality: A review. *Atmos. Environ.* **1991**, *25B*, 289–299. [CrossRef]
2. Honjo, T.; Takakura, T. Simulation of thermal effects of urban green areas on their surrounding areas. *Energy Build.* **1990–1991**, *15*, 443–446. [CrossRef]
3. Moriyama, M.; Takebayashi, H.; Fukumoto, K. Effects of Green Areas on Urban Air Temperature by Numerical Solution. *Mem. Grad. Sch. Sci. Technol. Kobe Univ.* **1997**, *15*, 101–115.
4. Bowler, D.E.; Buyung-Ali, L.; Knight, T.M.; Pullin, A.S. Urban greening to cool towns and cities: A systematic review of the empirical evidence. *Landsc. Urban Plan.* **2010**, *97*, 147–155. [CrossRef]
5. Ca, V.T.; Asaeda, T.; Abu, E.M. Reductions in air conditioning energy caused by a nearby park. *Energy Build.* **1998**, *29*, 83–92. [CrossRef]
6. Yu, C.; Hien, W.N. Thermal benefits of city parks. *Energy Build.* **2006**, *38*, 105–120. [CrossRef]
7. Yagi, R.; Takebayashi, H. Study on the near ground surface temperature influenced by surface temperature and wind velocity. *Eighth Natl. Conf. Heat Isl. Inst. Int.* **2013**, *1*, 48–49.
8. Ashie, Y.; Ca, V.T. Developing a three-dimensional urban canopy model by space-averaging method: Development of the urban climate simulation system for urban and architectural planning Part 2. *J. Environ. Eng. (Trans. AIJ)* **2004**, *69*, 45–51. [CrossRef]
9. Kato, T.; Yamada, T.; Hino, M. Spatial structure of air temperature and humidityin urban park forest and its surrounding. *J. Inst. Sci. Eng. Chuo Univ.* **2016**, *12*, 63–71.
10. Moriyama, M.; Kono, H.; Yoshida, A.; Miyazaki, H.; Takebayashi, H. Data analysis on "cool spot" effect of green canopy in urban areas. *J. Archit. Plan. (Trans. AIJ)* **2001**, *66*, 49–56. [CrossRef]

climate

MDPI

Article

Determination of the Surface and Canopy Urban Heat Island in Athens Central Zone Using Advanced Monitoring

Chrissa Georgakis [1,*] and Mattheos Santamouris [2]

[1] Group of Building Environmental Physics, University of Athens, Building Physics 5, University Campus, 15784 Athens, Greece
[2] The Anita Lawrence Chair in High Performance Architecture, School of Built Environment, University of New South Wales, Sydney, NSW 2052, Australia; m.santamouris@unsw.edu.au
*   Correspondence: cgeorgakis@phys.uoa.gr; Tel.: +30-210-7276870

Received: 7 November 2017; Accepted: 14 December 2017; Published: 20 December 2017

Abstract: The present study aims to present all the findings of micro-climate measurements that were performed by the University of Athens in the center of Athens, during the summer period. The extended experimental campaign aimed to collect thermal and air flow measurements, in different measuring points along a main street in the city center, in order to estimate the surface and canopy heat island intensity. In this work, the methodology of collecting the data, the experimental procedure, the equipment used, and lastly, the results are being presented. Comparison with the meteorological conditions that are recorded in the National Observatory of Athens, for the same period, lead to important conclusions about the local microclimate in the center of Athens and specifically the magnitude of the heat island effect. Particularly, in the denser area of the city after midday, air temperature increases reaching values up to 5 degrees higher than the one recorded in the suburban area. On the contrary, early in the morning the air temperature of the "green area" of the city was found to be lower up to 2 degrees than the corresponding in the suburban area.

Keywords: urbanization; air and surface temperature measurements; outdoor thermal comfort; urban heat island; surface cool island effect

1. Introduction

1.1. The UHI and Its Intensity Impact

The urban heat island (UHI) is the most studied phenomenon relative to increased urbanization. Air temperatures in the densely built area are higher when compared to the ones in the surrounding sub-urban areas due to positive urban thermal balance. The maximum differences between urban air temperatures and the background rural or suburban temperatures define the UHI magnitude [1]. There are several parameters influencing the intensity of the urban heat island. These are the urban characteristics (size and population of the city), the local meteorological features, the topography, the type of urban materials, and the presence (or lack) of green areas [2]. Several studies around Europe and the United States (US) have been performed estimating the increased air temperature in the urban fabric [2–4].

The effect of UHI is well known in Athens, the capital city of Greece, a city with a population explosion during the last decades. Primary studies conducted in order to estimate the heat island effect in Athens, used data from routine standard fixed meteorological stations. Time series of the air temperature data recorded in routine standard fixed urban and rural meteorological stations around the Athens basin, for the period 1961–1982, were used in a study about UHI. The minimum mean monthly air temperature's differences between the urban and rural stations indicated that, the UHI intensity was up to 3 °C during that period [5].

Meso-scale and synoptic data covering the period 1990–2001 were analysed in order to estimate UHI [6]. The urban heat island was estimated using the minimum temperature differences between rural and urban areas of the city of Athens, in the morning. The UHI was detected in the 2/3 of the examined period, being strong enough for the 1/3 of these days, yearly. During this 1/3 of the days annually, the strength of the UHI was more than 3 °C [6].

Intelligent 'data-driven' methods have been used to assess the magnitude of the urban heat island phenomenon. Hourly ambient air temperature values, which were recorded at twenty-three stations in the Athens region, have been 'data-driven' analysed. It was found that the heat island intensity follows both periodic and non-periodic fluctuations, depending on the weather conditions, as well as on topographic and topoclimatic complexities and synoptic flow patterns. It was estimated that the mean seasonal values of the UHI for the fifteen most central urban stations around Athens, were close to 5.4 °C for summer, 3.2 °C for autumn, 2.1 °C for winter, and 3.1 °C for spring [7].

An extended experimental study about the urban heat island magnitude in Athens basin, conducted in thirty urban and suburban monitoring stations. The results were based on maximum air temperature differences between measurements of the most central urban stations to the rural ones. It was estimated that the heat island intensity for the city of Athens was up to 10 °C during daytime and 5 °C during night [8]. Measurements have been used to assess the impact of the urban climate on the energy consumption of buildings. The impact of the urban heat island to the microclimatic conditions is of great interest and increases the need for cooling. It was estimated that monthly cooling load in the dense built urban area is around 120% higher, while the heating load is around 38% lower when compared to the reference suburban areas [8].

Urban heat island phenomenon may occur during nighttime period, as a function of the local thermal balance, resulting from the delayed cooling of the city, in comparison to the temperatures in the surrounding rural areas. Measurements of air temperature have been carried out, within three deep urban canyons of different aspect ratios, during nighttime [9]. Under clear and calm climatic conditions, the variation of the median heat island intensity was found to be between 2.2 °C to 2.7 °C.

In the microclimatic analysis that was carried out in a medium size sub-urban city area, Acharnes, 10 km north of the center of Athens, heat island intensity was found to be strongly connected to the thermal properties of the materials used as coatings in the urban fabric [10]. Air temperature varied from 22.3 °C up to 32.3 °C in the streets of the studied areas, significantly high values when considering the measuring period (April-May). The thermal properties of the materials that were used as coatings in the studied areas were of low albedo and high thermal capacity. The daily average surface temperatures of the materials that were used in the urban fabric, varied between 37.4 °C for the pavements and 41.8 °C for the asphalt.

1.2. Advanced Mitigation Techniques

Nowadays, in order to counterbalance the urban heat island effects, several efficiency techniques have been proposed and assessed in a considerable amount of studies. Advanced mitigation techniques could possibly lead to a proportionate reduction in degrees of the peak ambient summer temperature. Urban regeneration leads to the sustainable development of the urban fabric through the rehabilitation of the existing urban fabric and the preservation of green spaces. Use of cool materials in paths, roads, and building rooftops in order to highly reflect solar radiation, together with a high emissivity factor can significantly contribute to the reduction of surface temperatures up to several degrees. Cool communities' strategies reroof and repave in lighter colors and cool materials in order to reduce air temperature in cities and reduce the increased heat island effect [11–13].

In the Maroussi area, a densely built neighborhood in Athens, detailed simulation techniques have permitted the evaluation of the rehabilitation of the area. Cool materials have been proposed as coatings for pavements and streets; the amount of green spaces was increased. Solar control devices and earth to air heat exchangers were implemented in the simulation. Computational analysis indicated a decrease of the peak ambient temperature in the built area up to 3.4 °C under peak summer conditions. The application of all the above techniques may decrease the surface temperature and improve the thermal comfort conditions [14].

Recently, one of the largest urban mitigation projects in Greece provided information about the rehabilitation of an urban area with increased ambient temperature during the last decay [15]. In a major traffic axis of Western Athens, covering a total zone of 37,000 m^2 design and experimental evaluation of a large scale implementation of cool asphaltic and concrete pavements, took place. An extended monitoring was performed in the area during the summer period in its entirety. The thermal impact of the application was evaluated by means of Computational Fluid Dynamics (CFD) simulations. Simulations indicated that the use of cool non-aged asphalt can reduce the ambient temperature by up to 1.5 °C, and the maximum surface temperature by close to 11.5 °C, while the thermal comfort conditions can be strongly positively affected [15].

1.3. The Study Objective

The basic objective of the present study was to collect proper data for the ambient temperature and the wind speed along heavy traffic streets in the center of Athens. An additional objective of this study was to collect proper information about the thermal and optical characteristics of the materials that are used in pavements and roads in the area.

The aim of this study was the estimation of the heat island effect, based on differences of hourly air temperature values, recorded in the center of Athens and in a sub-urban station. It was estimated that during midday, the air temperature in the center of the city was up to 5 °C higher than the corresponding one in the sub-urban station. On the contrary, early in the morning in the center of Athens near a large green area, ambient air temperature was close to 2 °C lower than the corresponding one in the reference sub-urban station. That was due to the 'negative' surface urban heat island.

Additionally, in this study a specific questionnaire depicted the unpleasantness of the city's dwellers about relevant thermal comfort conditions. Results support the application of advanced cool materials for pavements and streets, green spaces, and solar control devices in order to counterbalance the heat island phenomenon and improve the thermal comfort conditions in the city center. Studies like this provide comprehensive microclimatic data, which is of great importance for the rehabilitation of a city and the design of a sustainable urbanization to urban planners.

2. Description of the Measuring Plan and the Monitoring Campaign

The city of Athens is located on a basin surrounded by Penteli Mountain, (1107 m) in the North East, Parnitha Mountain (1426 m) in the North, Hymettus Mountain (1026 m) in the East and Egaleo Mountain (458 m) in the West. In the South of Athens' basin is the Aegean Sea. The Greater Athens basin (Figure 1) extends beyond its administrative municipal limits of Athens, with a population of 3,090,508 (over an area of 412 km^2). According to Eurostat in 2011, the functional urban area (FUA) of Athens was the 9th most populous FUA in the European Union (the 6th most populous capital city of European Union), with a population of 3,828,000. The center of Athens had a population of 664,046 (in 2011) within its administrative limits, and a land area of 38.96 km^2. Based on these data, the center of Athens is a highly dense city and measurements about UHI are of great importance.

2.1. The Monitoring Plan

A complete monitoring plan had been set and applied in order to depict the micro-climate conditions in the center of Athens during summer time. Measurements were performed on a continuous basis along Amalias Avenue, Panepistimiou Street and Patission Avenue in central Athens. Amalias is a major avenue with four traffic lanes, linking Panepistimiou Street with the Greek Parliament and Syntagma Square. Panepistimiou Street situates in the historical center of Athens. The total length of the street is about 1.2 km. It consists of six lanes, five of which are for traffic and one eastbound lane for transit buses only. Most of the street runs almost diagonally from southeast to northwest, with the long axis in a NE-SW direction. The geometrical characteristic of the axis is not continuous. Several are the historical buildings along the street, such as the University of Athens, the Academy of Athens, the National Library, the Numismatic Museum, and the Catholic Cathedral of Athens. Many buildings as high as ten to fifteen stories line this street. Both sides of Panepistimiou Street are covered with red concrete tile pavements, while the street is covered with conventional black asphalt.

Patission Avenue is one of the major streets in Athens, connecting Omonoia Square, which is located in the center of Athens, with residential areas. It consists of four heavy traffic lanes, with the long axis in an N-S direction. The geometrical characteristic of the axis H/W is close to 2.5. The street is covered with conventional black asphalt and the pavements are of grey tiles. Vegetation is limited across the street.

The daily experimental campaign lasted approximately nine hours (9:00LT–18:00LT). Air temperature and wind speed and direction were recorded at ten different points along the measuring route, at hourly basis. Each route lasted 60 min to 90 min, depending on the traffic conditions. The measuring campaign started on the 22 May and was completed on the 17 June.

2.2. Description of the Site

The monitoring procedure started from Amalias Avenue. The first three measuring points were selected to be on Amalias Avenue, the entrance for the measuring route facing the North. The 1st measuring point was in front of an archaeological site. The 2nd measuring point was close to the entrance of the National Park of Athens, a widely open space with tall trees. The 3rd measuring point was in front of the Greek Parliament, an open area that covered in marble and some greenery. These measuring points were selected to be on Amalias Avenue, since they were placed on open spaces and strongly differed from the ones in the densely built Panepistimiou Street. The 4th measuring point was on Panepistimiou Street close to the intersection with Voukourestiou Street. At the specific measuring point Panepistimiou Street can be considered as an urban canyon with aspect ratio equal to two (Height/Width = 2). The following measuring point, along Panepistimiou Street, was the green area in front of the Central Historical Building of the University of Athens. The 6th measuring point—along Panepistimiou Street—was in front of a densely built block with 30 m height buildings, while across there was a square. No aspect ratio was estimated on the specific measuring point.

Omonoia Square is at the end of the measuring route along Panepistimiou Street. Omonoia Square is an intensive traffic area lacking greenery in the center of Athens. There was selected to be the 7th measuring point. Pavement at most spots in the square was either red or grey tiles. The measuring route continued along Patission Avenue. The thermal contribution of this area was considered to be of great importance to the thermal phenomena examined in this study. The 8th measuring point was in front of a building block (20 m height), while across was the Historical Building of the National and Technical University of Athens, which placed in an extended green area. The 9th and the 10th measuring points were along Patission Avenue. At these spots, Patission Avenue was considered as an urban canyon, with aspect ratio equal to 2.5 (H = 25 and W = 12), while building blocks were from both sides. Detailed description of all the measuring points is presented in Table 1. The survey area is presented in the two-dimensional (2D) Map of Athens basin (Figure 1). All of the measuring points are depicted in the map of the center of Athens (Figure 2).

Table 1. The monitoring route.

Measuring Point	Description	Type of Vegetation	Pavements Coating	H/W	Orientation
M.P.1	Amalias Avenue (shelted)	Archeological site	Red tiles	-	40°
M.P.2	Amalias Avenue (shelted)	National Garden of Athens	Grey tiles	-	56°
M.P.3	Amalias Avenue (open)	No vegetation	White Marble	-	21°
M.P.4	Panepistimiou Street	Few tall trees	Red tiles	H/W = 2	297°
M.P.5	Panepistimiou Street	Garden in the one site	Red tiles	-	315°
M.P.6	Panepistimiou Street	Fee tall trees	Red tiles	-	307°
M.P.7	Omonoia Square	No vegetation	Grey and Red tiles	-	354°
M.P.8	Patission Avenue	Garden in the one site	Grey tiles	-	196°
M.P.9	Patission Avenue	Limited vegetation	Grey tiles	H/W = 2.5	196°
M.P.10	Patission Avenue	Limited vegetation	Grey tiles	H/W = 2.5	196°

Figure 1. Survey area in GOOGLE EARTH.

Figure 2. The measuring points in the center of Athens.

2.3. The Experimental Protocol

The Mobile Meteorological Station of the University of Athens was used for the air temperature and wind speed and direction measurements. The Mobile Meteorological Station consists of: (a) a vehicle and (b) a telescopic mast PT8 Combined Collar Mast Assembly with erection height equal to 3.5 m and maximum head load 15 kgr. All meteorological equipment was placed on the mast at 3.5 m height from ground level. Measurements carried out every 30 s. The equipment used was the following:

2.3.1. Air Temperature Miniature Thermometer (T351-PX 1/3 DIN Thermometer). The thermometers' accuracy is equal to ±0.5 °C under normal meteorological conditions and +2 °C under low wind speed value conditions.

2.3.2. Wind Speed Anemometers (A100K Pulse Output Anemometer). The anemometers' accuracy is ±0.05 m/s with threshold value of 0.15 m/s. The operation frequency is 10 Hz per knot and the sampling rate of the instrument is equal to 10 pulses per 1.69 feet.

2.3.3. Wind Direction Anemometer (W200 Porton Windwane, ±300° Range). The accuracy of the instrument is equal to ±4° for wind speed greater than 3 m/s. The instruments' threshold value is equal to 0.2 m/s with sampling analysis 15°.

Inside the van, data acquisition modules consisted of intelligent sensor-to-computer interface modules, which provided all of the measurements in a computer. A computer application was developed on LabVIEW engineering software, which is appropriate for the recording and the presentation of the measurements (www.ni.com).

2.3.4. On hourly basis, asphalt and pavement surface temperature, carried out for sunlit and shading conditions along the measuring path. For this purpose, an infrared thermometer equipped with a laser beam was used. The accuracy of the infrared thermometer was ±0.2 °C. For more accurate measurement of the surface temperature, an infrared thermal camera was used. By means of specific software (http://www.flir.eu/home), pictures of the infrared camera in visible and in infrared lead to the estimation of the accurate surface temperature in each of the measuring point according to their albedo values.

2.4. Undisturbed Meteorological Parametrs

The National Observatory of Athens (NOA) was preferred as reference meteorological station for this study, since it is located quite close to the measuring route. NOA is located at the top of the hill of the Nymphs a sub-urban open area close to the center of the city. The meteorological data was derived from the Environmental Research and Sustainable Development Institute measurements of the National Observatory of Athens.

3. Discussion on the Micro-Climatic Measurements

The climate in the city of Athens is typical 'Mediterranean', which is characterized by mild winters and dry hot summers. During the measuring procedure, an anticyclone circulation prevailed over the greater Athens basin. Anticyclone conditions favour the development of the heat island phenomenon [7,12,14].

During the experimental period of this study, the wind speed measured in the reference meteorological station was up to 10 m/s (Figure 3a). The reference meteorological station (NOA) is located at the top of the hill of the Nymphs 107 m from ground level. That explains the magnitude of the undisturbed wind speed values. Wind flow measured in the reference meteorological station was mainly from SW direction (Figure 3b). Simultaneously, wind speed and direction were measured in the center of the city, along the measuring path. The anemometers were placed on the telescopic mast at 3.5 m height from ground level. The magnitude of the wind speed values that were measured in the center of the city, during the experimental campaign, were extremely low. The 99% of the relative frequencies for the wind speed values, measured along the measuring path, were below the value of 2 m/s (Figure 4). Due to the variance of the extremely low wind speed values, no predominated wind direction, in the center of the city, was estimated. This comes into to agreement with wind speed measurements in the center of Athens in a previous study [16].

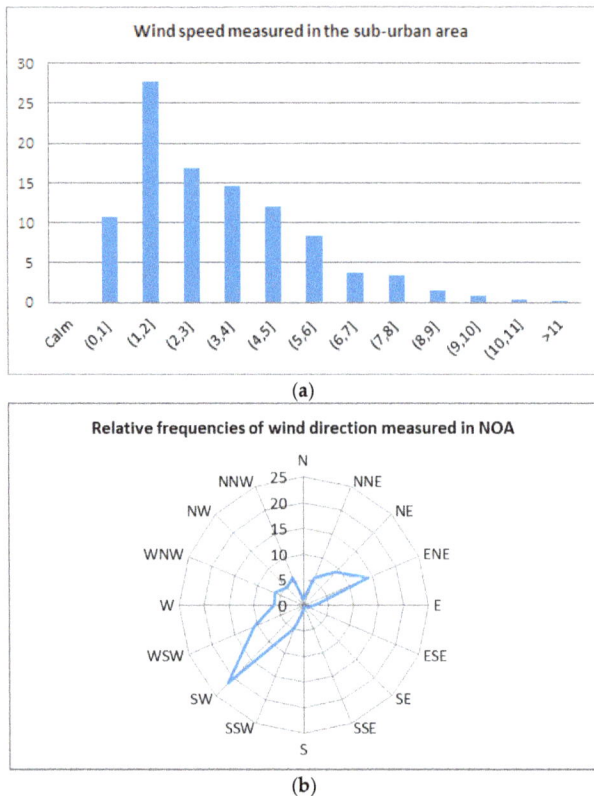

(a)

(b)

Figure 3. Relative Frequencies of the wind speed and wind direction valued recorded in National Observatory of Athens (NOA), during the experimental period.

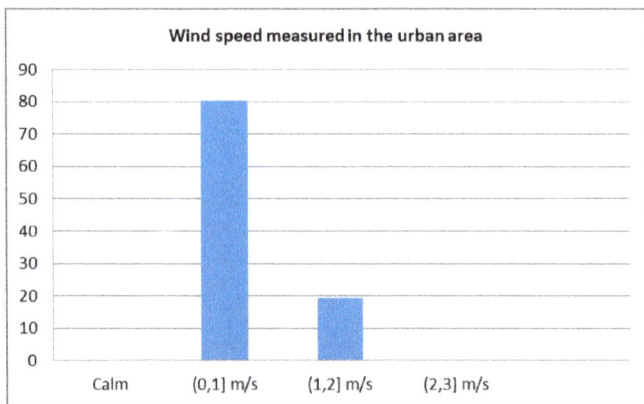

Figure 4. Relative Frequencies of the wind speed valued recorded in the center of Athens, during the experimental period.

The experimental period lasted from 22 May up to 17 June. Early in the morning, the air temperature measured in the center of the city, close to open areas such as the National Garden of Athens (M.P.2) or in front of the Greek Parliament (M.P.3), was lower than the one that was measured in the reference meteorological station. The minimum air temperature difference between values measured in the city center and the ones that were measured in the suburban area, where the reference meteorological station was located, depicts the 'cool' surface island intensity.

In the afternoon, air temperature that was measured in the center of the city, close to Omonoia square (M.P.7) or in Patission Avenue (M.P.8) was higher than the one measured in the suburban area. The maximum air temperature difference between values measured along the measuring path, in the city center, and the one measured in the sub-urban area depicts the UHI intensity. All the measurements in UHI and the surface urban 'cool' island intensity are presented in Table 2.

Table 2. Maximum air temperature differences between urban and sub-urban measurements.

Experimental Period	Air Temperature Differences between Urban and Sub-Urban Area	
	Surface Urban 'Cool' Island Intensity	Urban Heat Island Intensity
22 May: 10:00LT–16:00LT	−0.70 °C (M.P.2) at 10:00LT	2 °C (M.P.7) at 13:00LT
24 May: 9:00LT–18:00LT	−0.89 °C (M.P.1) at 10:00LT	1 °C (M.P.7) at 15:00LT
28 May: 9:00LT–16:00LT	−1.85 °C (M.P.1) at 09:00LT	1.52 °C (M.P.7) at 15:00LT
31 May: 10:00LT–17:00LT	−0.48 °C (M.P.4) at 10:00LT	2.12 °C (M.P.8) at 14:00LT
3 June: 9:00LT–18:00LT	−1.50 °C (M.P.2) at 09:00LT	1.23 °C (M.P.7) at 16:00LT
10 June: 9:00LT–16:00LT	−0.54 °C (M.P.3) at 10:00LT	2.12 °C (M.P.8) at 14:00LT
12 June: 9:00LT–14:00LT	−1.72 °C (M.P.3) at 10:00LT	4.5 °C (M.P.4) at 14:00LT
17 June: 9:00LT–19:00LT	−2.14 °C (M.P.3) at 10:00LT	3.81 °C (M.P.7) at 15:00LT

During the experimental period, the surface urban 'cool' intensity was always calculable. Some of the days it was less than one Celsius degree, but some others it exceed 2 °C. The UHI intensity was measured always more than one Celsius degree, and reached values close to 5 °C.

Representative, the UHI, and surface urban 'cool' intensity values, during 17 June are presented in Figures 5 and 6. Air temperature differences were depicted by the means of Matlab. A full grid system with two monotonically increasing grid vectors depicted all of the measuring points. The (x, y) vectors referred to the real longitude and latitude of the measuring points. The limits of the temperature scale in the colorbar, in the right part of the figure, corresponded to the minimum and the maximum air temperature differences that were calculated for the total experimental period. The specific plotting permitted the observation and the thorough understanding of the temporal evolution of the air temperature in the experimental region during the experimental campaign.

Regarding the heat island intensity, an important parameter is related to the thermal properties of the materials that the area is composed of. For this reason, the surface temperature of paving and asphalt was measured on hourly basis during the experimental campaign. As representative for the whole experimental period, the 3 and the 17 June were selected. These were the days that the lowest and the highest air temperatures recorded, both in the center of Athens and the reference meteorological station. On the 'cool' day of the experimental period (3 June), the differences of the mean surface temperature between the sunlit and the shaded asphalt were close to 16 °C. For the presentation of the surface temperature differences, the whisker chart was used, which was based on the boxplot analysis of the measurements (Figure 7). It is of great importance that the magnitude of the differences between the sunlit and the shaded asphalt on 17 June the 'warmest' day of the experimental period was exactly of the same magnitude (Figure 8). It seems that the degraded asphalt contributes a specific amount of energy to the air above due to the very low albedo, regardless the day. The albedo of the degraded asphalt, used in the center of Athens, was measured close to 0.04 [17].

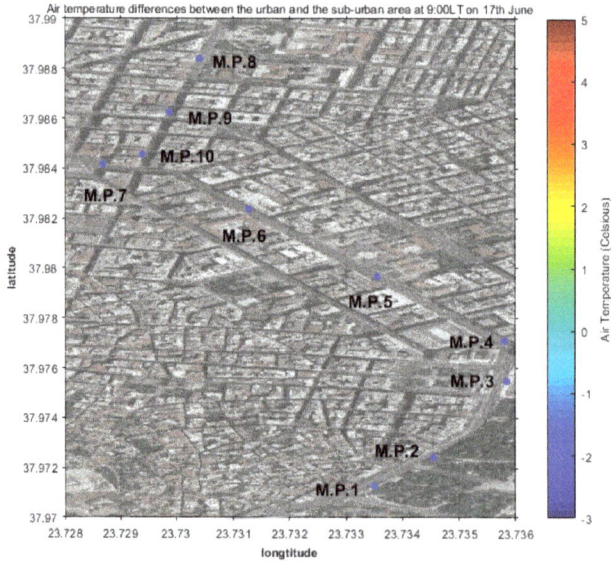

Figure 5. Air temperature differences between the ones recorded in the center of Athens at 9:00LT and the ones in the suburban station, during the 17 June.

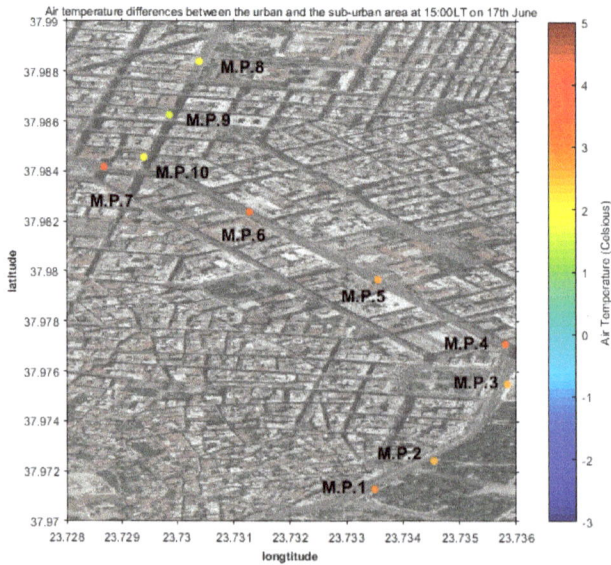

Figure 6. Air temperature differences between the ones recorded in the center of Athens at 15:00LT and the ones in the suburban station, during the 17 June.

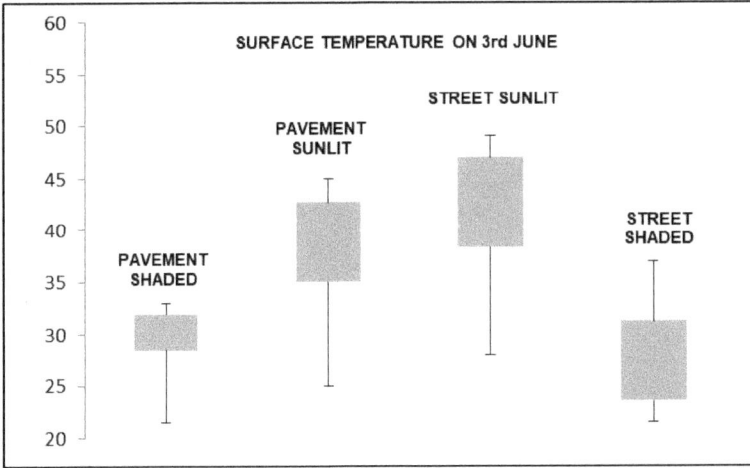

Figure 7. Surface temperature measured on sunlit and shaded pavement/asphalt/on the 3 June.

The maximum surface temperature differences between sunlit and shaded pavement were 8 °C during the 'cool' day and 12 °C during the 'warmest' day of the experimental period (Figures 7 and 8). The albedo of the dark paving materials was measured close to 0.22, which is five times higher than the one measured for asphalt [17]. Due to that, the surface differences for pavement are lower than the one measured for asphalt. Another important parameter is that the hourly surface temperature for pavement is strongly connected to the measuring point where measurement was held.

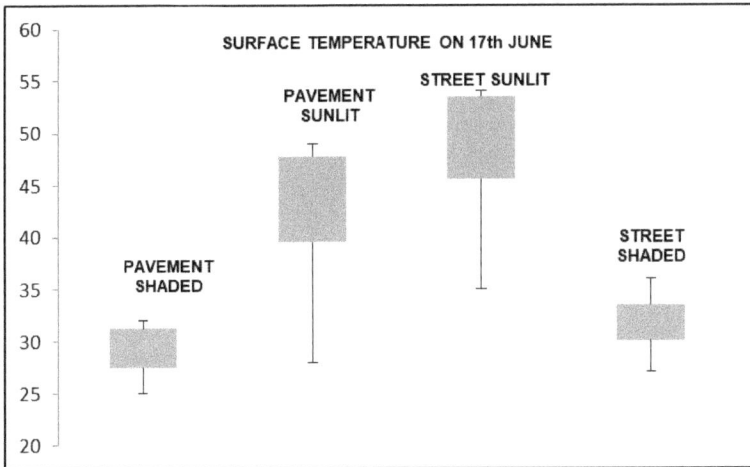

Figure 8. Surface temperature measured on sunlit and shaded pavement/asphalt/on the 17 June.

4. Pedestrians Estimation on Thermal Comfort Conditions in the Measuring Area

The existence of the urban heat island, because of the high urban temperatures, has a great impact to the quality of living in the city. People living in the densely built center of the city usually feel unpleasant because of the high summer air temperature situation. In the worst cases, it was referred that urban warming is increasing the number of hospital admissions [18].

Part of the experimental procedure was the distribution of a questionnaire that was related to the thermal comfort conditions, along the experimental path, in specific measuring points. Questionnaires were distributed to pedestrians in the National Garden of Athens (M.P.2), in Syntagma Square (M.P.3), Voukourestiou Street (M.P.5), and in Omonoia Square (M.P.7). In these questionnaires, pedestrians expressed their opinion on the experienced thermal conditions, the weather conditions and the street set-up.

The dissatisfaction was obvious on how pedestrians felt in Omonoia Square (M.P.7), which has no shading and the vegetation is limited when compared to the rest of the measuring points. The city dwellers around Omonoia Square had a completely negative opinion about the environmental quality. Apparent was the unpleasantness regarding temperature and solar radiation levels in the measuring point. At the same time, the questioned were dissatisfied with the quality of the materials that were used in Omonoia Square. Air pollution, noise pollution, and anthropogenic heat increased the unpleasantness levels of the people asked. Based on the answers that were given, the thermal discomfort feeling increased gradually from National Park area (M.P.1) up to Omonoia square in the center of Athens (M.P.7).

5. Main Conclusions and Suggestions

The aim of this study was to analyze and understand the thermal behavior of the city center of Athens. Based on the analysis of the micro-climatic measurements, the following conclusions were derived:

➢ The air temperature in the monitored zone presented important differences when compared to the undisturbed air temperature recorded in reference station in the National Observatory of Athens.

➢ Based on measurements of this study, during the morning hours, air temperature measured in the densely built area close to green areas was lower up to 2 °C than the ones that were measured at the reference meteorological station. The magnitude of the surface 'cool' island reached up to 2 °C, proving the great influence that greenery has on air temperature distribution within the city. This comes into agreement with a relative previous study [13,19,20].

➢ In the afternoon, the warmest parts of the monitoring area in the city center presented up to 5 °C higher temperatures than the ones measured in the reference meteorological station. The UHI intensity measured in this study was up to 5 °C and come with agreement with measurements of previous studies [6–8].

➢ The daily air temperature variance along the monitoring route, between the warmest and the coolest zones of the city center, varied from 5 °C up to 8 °C. The temperature variation strongly depended on the current meteorological conditions. This comes into agreement with the results of a previous study, where maximum air temperature differences between the most central urban stations to the rural ones were up to 10 °C during daytime [8].

➢ The highest ambient temperature was recorded in the most central measuring points (M.P.7 and M.P.8). This was mainly due to the increased thermal balance of the city's center. The anthropogenic heat released in the city's center, the lack of solar control, the lack of vegetation, and the type of materials used for the coating of pavements, contributed to enhancing the heat island effect.

➢ Huge was the contribution of the unsuitable pavement coatings and the degraded asphalt into the air temperature distribution in the center of Athens. Differences between sunlit and shaded parts of the asphalt was stable, regardless the day, and were measured close to 16 °C. Differences between sunlit and shaded pavements varied from 8 °C up to 12 °C. The existence of the materials presenting very high absorptivity, together with the very high thermal capacitance, increases the storage capacity of the monitoring area in the center of Athens and keeps air temperature in high levels for longer periods.

➢ As it was expected, wind speed and direction values within the city strongly differed from measurements in the suburban meteorological station. The lower wind speed values measured in the center of the city are a fact that increases the intensity of the heat island [10,16].

6. General Comments

Microclimatic conditions in the urban fabric are mainly influenced by the geometry of the built area, the greenery, and the optical and thermal properties of the materials that are used as coatings of the buildings. The local micro-climate was affected, as expected, by the morphology of the urban pattern (orientation of the measuring points; aspect ratio of the street). The most important factors for the existence of the heat island were vegetation/shading factors and the thermal impact of the surrounding surfaces. Higher air temperature values were recorded during midday, due to the urban heat island, possibly caused from the heavy traffic, the increased anthropogenic heat in the center of the city, and the lack of green spaces.

The positive effect of vegetation on the UHI is obvious, at least during the morning hours. That proves the importance of green spaces, since in the morning the evapotranspiration cooling mechanisms of a big green urban area (like the National Garden of Athens) reduce air temperature up to 2 °C when compared to the surrounding parts of the city. After midday, air temperature in the center of a city increases dramatically, reaching values up to 5 °C higher than air temperature in the suburban area. That means that the air temperature daily distribution increases abruptly. The contribution of the unsuitable pavement coatings and the asphalt into the air temperature distribution in the surrounding area is of great importance. Surface temperature differences between sunlit versus shaded asphalt measured close to 16 °C. This study can assist architects of outdoor spaces, building physicists, and engineers to improve the urban thermal behavior by means of cool materials. Additionally, the increase of urban greenery as a solar control technique is important on the regulation of the ambient temperature in a highly dense area, like the city of Athens.

Acknowledgments: Thanks are due to Institute for Environmental Research and Sustainable Development, of the National Observatory of Athens, for the permission to use meteorological data for the aims of this study.

Author Contributions: Chrissa Georgakis and Mattheos Santamouris contributed to the design and implementation of the experimental procedure, to the analysis of the measurements and to the writing of the manuscript.

Conflicts of Interest: The authors declare no conflict of interest.

References

1. Oke, T.R. *Boundary Layer Climates*, 2nd ed.; Routledge: London, UK; New York, NY, USA, 1987.
2. Akbari, H.; Cartalis, C.; Kolokotsa, D.; Muscio, A.; Pisello, A.; Rossi, F.; Santamouris, M.; Synnefa, A.; Wong, N.; Zinzi, M. Local Climate Change and Urban Heat Island Mitigation Techniques—The State of the Art. *J. Civ. Eng. Manag.* **2015**, *22*, 1–16. [CrossRef]
3. Santamouris, M. Heat island research in Europe-state of the art. *Adv. Build. Energy Res.* **2007**, *1*, 123–150. [CrossRef]
4. Santamouris, M.; Cartalis, C.; Synnefa, A. Local urban warming, possible impacts and a resilience plan to climate change for the historical center of Athens, Greece. *Sustain. Cities Soc.* **2015**, *19*, 281–291. [CrossRef]
5. Catsoulis, B.D.; Theoharatos, G.A. Indications of urban heat island in Athens, Greece. *J. Clim. Appl. Meteorol.* **1985**, *24*, 1296–1302. [CrossRef]
6. Kassomenos, P.A.; Katsoulis, B.D. Mesoscale and macroscale aspects of the morning urban heat island around Athens, Greece. *Meteorol. Atmos. Phys.* **2006**, *94*, 209–218. [CrossRef]
7. Mihalakakou, G.; Santamouris, M.; Papanikolaou, N.; Cartalis, C.; Tsangrassoulis, A. Simulation of the urban heat island phenomenon in Mediterranean cli-mates. *J. Pure Appl. Geophys.* **2004**, *161*, 429–451. [CrossRef]
8. Santamouris, M.; Papanikolaou, N.; Livada, I.; Koronakis, I.; Georgakis, C.; Argiriou, A.; Assimakopoulos, D.N. On the impact of urban climate on the energy consumption of buildings. *Sol. Energy* **2001**, *70*, 201–216. [CrossRef]

9. Giannopolulou, K.; Santamouris, M.; Livada, I.; Georgakis, C.; Caouris, Y. The Impact of Canyon Geometry on Intra Urban and Urban: Suburban Night Temperature Differences Under Warm Weather Conditions. *J. Pure Appl. Geophys.* **2010**, *167*, 1433–1449. [CrossRef]
10. Gaitani, N.; Santamouris, M.; Cartalis, C.; Pappas, I.; Xyrafic, F.; Mastrapostoli, E.; Karahaliou, P.; Efthymiou, C. Microclimatic analysis as a prerequisite for sustainable urbanisation: Application for an urban regeneration project for a medium size city in the greater urban agglomeration of Athens, Greece. *Sustain. Cities Soc.* **2014**, *13*, 230–236. [CrossRef]
11. Santamouris, M.; Synnefa, A.; Karlessi, T. Using advanced cool materials in the urban built environment to mitigate heat islands and improve thermal comfort conditions. *Sol. Energy* **2011**, *85*, 3085–3102. [CrossRef]
12. Santamouris, M.; Gaitani, N.; Spanou, A.; Saliari, M.; Giannopoulou, K.; Vassilakopoulou, K.; Kardomateas, T. Using cool paving materials to improve microclimate of urban areas—Design realization and results of the flisvos project. *Build. Environ.* **2012**, *53*, 128–136. [CrossRef]
13. Santamouris, M. Cooling the cities—A review of reflective and green roof mitigation technologies to fight heat island and improve comfort in urban environments. *Sol. Energy* **2014**, *103*, 682–703. [CrossRef]
14. Santamouris, M.; Xirafi, F.; Gaitani, N.; Spanou, A.; Saliari, M.; Vassilakopoulou, K. Improving the Microclimate in a Dense Urban Area Using Experimental and Theoretical Techniques—The Case of Marousi, Athens. *Int. J. Vent.* **2012**, *11*, 1–16. [CrossRef]
15. Kyriakodis, G.E.; Santamouris, M. Using reflective pavements to mitigate urban heat island in warm climates—Results from a large scale urban mitigation project. *Urban Clim.* **2017**, in press. [CrossRef]
16. Georgakis, C.; Santamouris, M. Experimental investigation of air flow and temperature stratification in deep urban canyons for natural ventilation purposes. *Energy Build.* **2006**, *38*, 367–376. [CrossRef]
17. Lontorfos, V.; Efthymiou, C.; Santamouris, M. On the Time Varying Mitigation Performance of Reflective Geoengineering Technologies in Cities. *Renew. Energy* **2017**, in press. [CrossRef]
18. Santamouris, M.; Kolokotsa, D. On the impact of urban overheating and extreme climatic conditions on housing energy comfort and environmental quality of vulnerable population in Europe. *Energy Build.* **2015**, *98*, 125–133. [CrossRef]
19. Kolokotroni, M.; Giridharan, R. Urban heat island intensity in London: An investigation of the impact of physical characteristics on changes in outdoor air temperature during summer. *Sol. Energy* **2008**, *82*, 986–998. [CrossRef]
20. Stathopoulou, M.; Synnefa, A.; Cartalis, C.; Santamouris, M.; Karlessi, T.; Akbari, H. A surface heat island study of Athens using high solution satellite imagery and measurements of the optical and thermal properties of commonly used building and paving materials. *Int. J. Sustain. Energy* **2009**, *28*, 59–76. [CrossRef]

climate

MDPI

Article

Strategies for Development and Improvement of the Urban Fabric: A Vienna Case Study

Milena Vuckovic [1,*]**, Aida Maleki** [2] **and Ardeshir Mahdavi** [1]

[1] Department of Building Physics and Building Ecology, TU Wien, Karlsplatz 13, 1040 Vienna, Austria;
    bpi@tuwien.ac.at
[2] Department of Architecture and Urban Planning, Tabriz Islamic Art University, 51368 Tabriz,
    East Azarbaijan, Iran; a.maleki@tabriziau.ac.ir
* Correspondence: milena.vuckovic@tuwien.ac.at

Received: 20 December 2017; Accepted: 25 January 2018; Published: 27 January 2018

Abstract: Numerous studies have shown that densely developed and populated urban areas experience significant anthropogenic heat flux and elevated concentrations of air pollutants and CO_2, with consequences for human health, thermal comfort, and well-being. This may also affect the atmospheric composition and circulation patterns within the urban boundary layer, with consequences for local, regional, and global climate. One of the resulting local implications is the increase in urban air temperature. In this context, the present contribution explores urban fabric development and mitigation strategies for two locations in the city of Vienna, Austria. Toward this end, the potential of specific planning and mitigation strategies regarding urban overheating was assessed using a state-of-the-art CFD-based (computational fluid dynamics) numeric simulation environment. The results display different levels of effectiveness for selected design and mitigation measures under a wide range of boundary conditions.

Keywords: urban overheating; urban microclimate; mitigation strategies; urban development

---

1. Introduction

1.1. Background

Current projections foresee a vast expansion of global urban population by 2050, with an increase from 60% of the global population living in urban settings in 2030 to 66% by 2050 [1–3]. This development is accompanied by the dynamic growth of cities and its implications in terms of environmental degradation, given the fact that cities and their inhabitants are major contributors to waste heat and CO_2 emissions [4,5]. Numerous studies have shown that densely developed and populated urban areas show significant anthropogenic heat flux [6–10]. This is in part due to the unbalanced integration of urban infrastructure into the urban fabric, namely the transportation network, and multiple systems for heating, cooling, ventilation, and air-conditioning of buildings. Dong et al. (2017) noted that relatively high values of anthropogenic heat flux may be found in large cities of eastern Asia, south and southeastern Asia, Eastern Europe, and the US. A majority of these cities are known to accommodate a dense population that consumes large amounts of energy and requires extensive transportation networks to accommodate cities' emerging needs. The highest value of anthropogenic heat flux (of 493 Wm^{-2}) was recorded within the Hong Kong metropolitan area for the year 2013, for an individual cell of a global model with a spatial resolution of 30 arc-seconds (1 km) and a temporal resolution of 1 h. Further environmental problems associated with higher anthropogenic heat emissions are the elevated concentrations of air pollutants and atmospheric CO_2, which have immediate consequences for human health, thermal comfort, and well-being [11,12]. Such large concentrations of air pollutants and CO_2 substantially influence air quality, and are found to affect the atmospheric composition and circulation patterns within the urban boundary layer, with potential repercussions for local, regional, and global climate [13–16]. One of the resulting local implications is the increase of outdoor urban air and surface temperature and consequential microclimatic development [8,13,17,18]. Chen et al. (2014) analyzed the distribution and magnitude of the global anthropogenic heat flux and concluded that the anthropogenic heat flux density is large enough to affect local climate change. They further observed the surface temperature increase (of 1–2 K) in the mid and high latitudes of Eurasia and North America, due to the elevated anthropogenic heat release. In turn, elevated outdoor air temperatures have repercussions for energy use for cooling, due to the extensive use of air-conditioning systems, which leads to even more waste heat in the built environment [19]. The extent of these impacts may be exacerbated by dense urban morphologies, higher thermal storage in the built environment, distinct surface cover, and poorly ventilated urban fabric [20,21]. For these reasons, the ongoing rapid urbanization and resulting environmental implications call for an unprecedented commitment to change the way cities evolve.

The spectrum of possibilities for the development and improvement of the urban fabric is rather wide. One approach relates to new urban developments on as yet empty building lots. The other relates to a variety of well-conceived and well-coordinated actions that are aimed at the transformation of existing urban domains. This set of actions is commonly referred to as the mitigation strategies, and they are believed to positively influence the negative phenomena associated with the urban overheating [22]. These efforts usually focus on the reduction of the fraction of energy that is stored within the urban fabric, promotion of the cooling potential of building materials in the physical environment, and enhancement of airflow through the city [23–26]. As these kinds of mitigation measures require substantial resources and major investments, the provision of timely and detailed information regarding the assessment of energy and environmental implications of these measures is, as such, of great importance. In recent years, the research in this field has substantially expanded. An increasing number of these efforts is concerned with the impact assessment analysis of the modification of the building stock, including cool building envelopes, green roofs, and green facades [27–31]. Pisello et al. (2015) [27], for example, observed that the local decrease of external surface temperature of roofs and facades, after the higher albedos materials were applied, was 19.8 K and 9.9 K, respectively, when compared to conventional materials. The combined effect of the two solutions led to the reduction of the indoor operative temperature of 3.1 K. Heusinger and Weber (2015) [28] reported, for a case study in Germany, that the summer surface temperature of a green roof may be 17.4 K lower than of that of a bitumen roof, with a significant reduction of ambient air temperature 0.5 m above roof level. Other efforts are directed toward assessing the environmental implications of specific interventions in the urban landscape, such as the increase in urban vegetation, namely parks and trees, the application of paving materials with higher albedo values, the application of pervious paving materials, and introducing the bodies of water [32–34]. Georgakis et al. (2014) [32] investigated the potential of high reflective coatings used for pavements and walls toward reducing the heat content in urban canyons. They noted a decrease of 8 K of surface temperature, and 1 K of the ambient air temperature inside the urban canyon. Wang and Akbari (2016) [34] discussed the effect of trees on the outdoor thermal environment within the urban canyon in Montreal and documented an air temperature reduction of 4 K at 20 m height from the ground level.

1.2. Overview

Given this background, the present contribution investigates the potential and the implications of specific planning and mitigation strategies regarding urban overheating for two study domains in the city of Vienna, Austria. For this purpose, the following steps were taken:

- First, high-resolution data streams across distinct urban and non-urban locations were obtained, structured, and analyzed. This facilitated the investigation of the microclimatic diversity across these locations. Additionally, this allowed for the identification of essential features of the built environment that are hypothesized to influence the extent of stored heat in the physical mass of the city.

- Subsequently, we investigated the potential of specific mitigation strategies to remedy the negative phenomena associated with urban overheating. For this purpose, three mitigation strategies were considered for the targeted high-density urban area: (i) planting trees, (ii) greening of the roofs, (iii) combination of both measures. These measures were selected based on their potential as a viable mitigation strategy that can be conveniently integrated into the existing urban fabric, their potential for generating both short- and long-term mitigation effects in urban areas, and their compatibility with local climatic conditions. To facilitate the environmental impact assessment of these measures, comprehensive simulations were carried out using the state-of-the-art CFD-based numeric simulation environment ENVI-met [35].

- Additionally, we investigated the microclimatic consequences of a proposed urban development in an existing abandoned industrial site using the same numeric simulation environment.

2. Methodology

2.1. Microclimate Development in Vienna

As stated at the outset, higher anthropogenic heat emissions, resulting from rapidly increasing urban population and emerging energy needs, may influence the atmospheric composition and circulation patterns within the urban boundary layer. In turn, this may have potential consequences for local climate, specifically for elevated ambient air temperature. To illustrate this development, we conducted a comprehensive microclimatic investigation of a number of distinct low-density and high-density urban and non-urban (outside the metropolitan area) segments within the city of Vienna (see Figure 1 and Table 1). In our previous research efforts, we developed advanced Python-based spatial algorithms implemented in a GIS environment to derive a set of morphological and physical parameters for these locations, for a spatial dimension of 400 m, as described in [36–41] and seen in Table 1.

In order to investigate the microclimatic behavior of selected urban and non-urban locations, we obtained hourly-based meteorological information pertaining to air temperature, wind speed, solar radiation, and precipitation from five weather stations centrally positioned within these areas. These stations are operated by the Central Institution for Meteorology and Geodynamics (Zentralanstalt für Meteorologie und Geodynamik, ZAMG) [42]. These stations provide continuous data at frequent intervals, where reliability of meteorological observations is assured via a thorough quality control using the QualiMET system and according to the WMO guidelines [43]. Additionally, the correction of the hourly-based data is performed using GEKIS (Geografisches Klimainformationssystem—Geographic Climate Information System). With approximately 250 semi-automatic weather stations (Teil Automatisches Wetter Erfassungs System—TAWES) spatially distributed throughout Austria, the coverage of the ZAMG network is rather extensive, with denser networks in populated areas [42].

The acquired meteorological information was further processed into four hourly-based reference days, representing typical weather conditions of each season for the year 2012. Thereby, the hourly data on air temperature and wind speed was averaged over a continuous three-week period identified for each meteorological season. These representative periods were characterized by a stable air mass free of excessive changes in pressure and airflow velocity (less than 3 m·s^{-1}), with little or no precipitation, thus allowing us to investigate micro-level changes in urban climate. It should be noted that the data selection process was performed for the area of the highest urban density (IS). Subsequently, the same time frame was applied for other areas. In order to compare wind speed readings taken from different heights, as seen in Table 1, we used the Hellmann exponential law to estimate the wind speed at street level (1.1 m above ground) [44]:

$$\frac{v}{v_n} = \left(\frac{H}{H_n}\right)\alpha \tag{1}$$

where v is the wind speed at height $H = 1.1$ m, v_n is the wind speed at height H_n (height of the observations), and α is the friction coefficient (Hellmann exponent). The friction coefficient used for this study is as follows: 0.25 for non-urban and low-density suburban areas (SEI and MB, respectively), 0.30 for mid-density suburban areas (HW and DF), and 0.40 for the high-density urban area (IS).

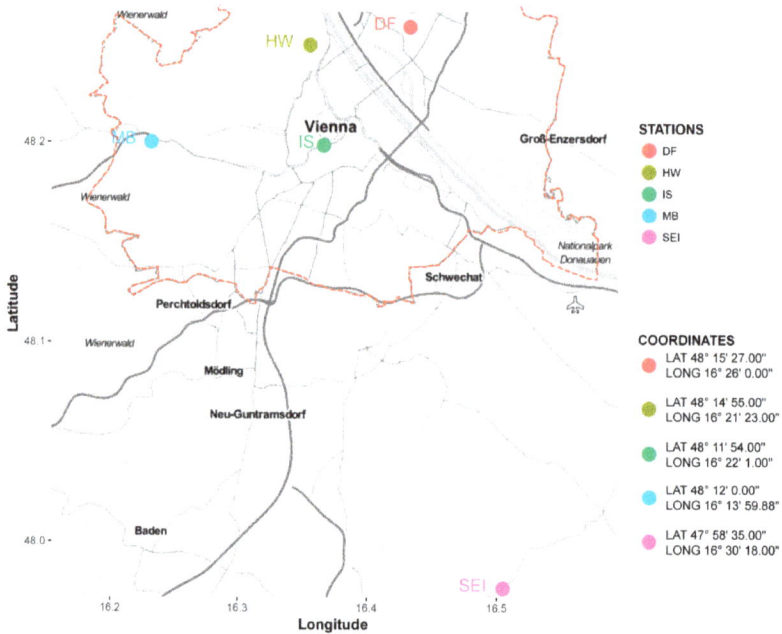

Figure 1. Selected locations and the respective weather stations within (DF, HW, IS, MB). And outside (SEI) Vienna (red dashed line marks the boundary of the metropolitan area).

Table 1. Information on selected weather stations.

Name	LCZ [1]	Temperature Sensor Height (m)	Wind Sensor Height (m)	Built Area Fraction [2]	Average Building Height [m]	Q_f [kWh/m²a¹]
IS	LCZ 2	9.3	52	0.41	23.35	350
HW	LCZ 6	1.9	35	0.18	8.00	177
DF	LCZ 6	2	13	0.20	6.15	89
MB	LCZ 9	2.1	9.5	0.04	5.23	70
SD	LCZ 8$_D$	2.1	15	0.08	5.29	67

[1] LCZ stands for Local Climate Zone, a classification system devised by Stewart and Oke (2012) [45]. [2] Q_f denotes the mean annual heat flux density from fuel combustion and human activity. More detailed information regarding the concerned parameters can be found in [37].

2.2. Strategies for Development and Improvement of the Urban Fabric

Current transformations of the physical environment, such as the increased building density, abundance of sealed surfaces, and reduced vegetation fraction lead to, among other things, significant heat storage in the urban fabric (i.e., building surfaces, pavements, and roads). As such, the study of strategies for improvement of the urban fabric holds great potential, with important sustainability implications. However, these strategies may prove effective only if a comprehensive body of scientific knowledge and expertise supports their realization.

As a contribution to the ongoing research efforts in this direction, we focus on a comprehensive environmental impact assessment of a number of mitigation and urban development strategies using a numerical model, ENVI-met version 4.0 [35]. ENVI-met is a 3-dimensional non-hydrostatic model specifically tailored for the simulation of surface-plant-air interactions within urban environments. The capabilities of this tool to facilitate the assessment of complex and non-linear interactions between the surrounding urban fabric and local climatic context have been broadly documented by the scientific community [46–48]. However, as with other areas of applied numerical modelling, certain issues related to the reliability of the model must be addressed. Thereby, in our previous research efforts we have documented the model calibration potential toward improved performance, as described in [49].

Three mitigation strategies were considered for the targeted area IS: (i) planting trees, (ii) greening of the roofs, (iii) combination of both measures (Figure 2). This area was selected as it is the most developed part of the city with high traffic intensity, hence it has a tendency toward higher urban overheating. For the trees, we considered deciduous trees of an average height of 13 m, C3-type, with an average albedo value of 0.2, average crown width of 9 m, and a LAD (Leaf Area Density) ranging from 0.5 to 2. More specifically, C3-type plants are referred to as the temperate or cool-season plants that are most efficient at photosynthesis in cool, wet climates [50]. Some examples include evergreen trees, deciduous trees and weed-like plants. For the green roofs we considered semi-intensive systems of an average vegetation height of 0.18 m, C3-type, and with an albedo value of 0.2. Four sets of simulation runs were conducted for a base case and for each scenario using the previously derived hourly-based seasonal reference days as boundary conditions. The input model assumptions are presented in Table 2. The possibility of user-defined diurnal variations of atmospheric boundary conditions (hourly forcing) offered in ENVI-met version 4.0 was used.

Once the high-resolution modelling output was generated, the sensitivity of ambient outdoor air temperature (sampled from the height of 1.8 m above the ground) to various mitigation strategies was investigated. For this purpose, the concept of Cumulative Temperature Decrease (CTD) [38] was used as the thermal performance indicator. CTD (Kh) denotes the sum of the hourly differences between the air temperature of a base case ($\theta_{B,i}$) and respective mitigation scenarios ($\theta_{M,i}$) over a specific period of time (e.g., over a day), whereby only positive differences ($\theta_{B,i} > \theta_{M,i}$) were considered:

$$\text{CTD} = \sum_{i=1}^{24} (\theta_{B,i} - \theta_{M,i}) \quad \text{for } \theta_{B,i} >_{M,i} \tag{2}$$

In order to investigate the temporal scale of the potential thermal benefits of each mitigation strategy, CTD values were further represented for nighttime (the period between the sunset and sunrise) and daytime (the period between sunrise and the sunset). The variation in daylight hours was adopted from the annual sun path diagram for the city of Vienna and rounded to the whole hour, as seen in Table 3 [51].

Additionally, we investigated the implications of a new large-scale design and renovation proposal for a local microclimate. The study area, called Nordbahnhof, represents an abandoned industrial site, currently a brownfield, located on the periphery of the urban center of the city of Vienna. The area is targeted for new urban redevelopment, specifically the construction of a new residential complex with a building height in the range of 10 to 70 m (Figure 3). The meteorological information for this area was provided by the Municipal Department of Environmental Protection in Vienna, MA22 [52]. The MA22 network comprises 17 monitoring weather stations distributed throughout the metropolitan area of the city of Vienna. These stations are calibrated following the ÖNORM standards (national standards published by the Austrian Standards Institute). Monitoring data storage and quality check procedures are based on ON/EN/ISO standards, which are recognized by WMO. The same input model assumptions as presented in Table 2, were applied for this case, except of the initial temperature at the upper soil layer, which was recalculated.

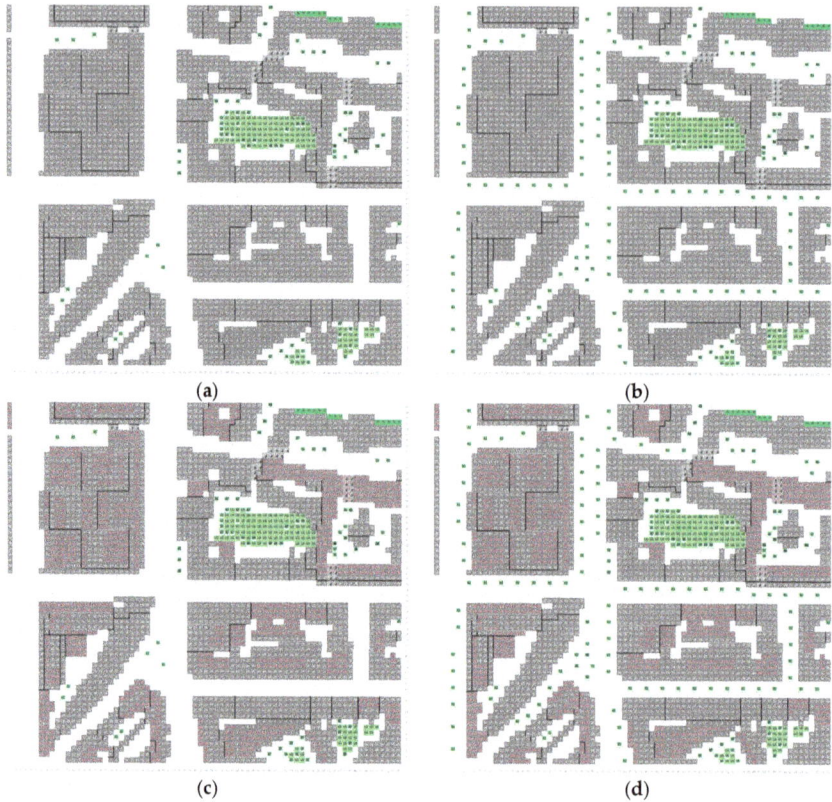

Figure 2. Schematic illustration of the envisioned mitigation strategies implemented in ENVI-met: (**a**) base case, (**b**) trees, (**c**) green roofs (in red), (**d**) combined.

Table 2. Input model assumptions for location IS.

Parameter	Unit	Winter	Spring	Summer	Autumn
Total simulation time	h	48	48	48	48
Grid size	m	4	4	4	4
Adjustment factor for solar radiation	-	0.82	0.82	0.82	0.82
Specific humidity at 2500 m	g Water/kg air	3	7	8	6
Initial temperature at the upper soil layer (0–20 cm)	K	275	290	293	281
Turbulence scheme for 1D reference model/3D main model	-	Prognostic (TKE closure)			
Roughness length z0 at reference point	m	0.1	0.1	0.1	0.1

For the parameters not listed in Table 2, default system settings were used.

Table 3. Daylight hours for each season.

Time	Winter	Spring	Summer	Autumn
Reference day	22 February	4 May	5 August	5 November
Daylight	07:00–17:00	05:00–20:00	06:00–20:00	07:00–16:00

Figure 3. Nordbahnhof area: (**a**) base case, (**b**) new urban development.

3. Results and Discussion

Figure 4 illustrates the ambient air temperature and wind speed for the winter and summer reference days for the selected locations. These results demonstrate significant variation in the distribution of observed meteorological parameters across time (season) and space (location). Generally, the central urban area (IS) had the smallest diurnal temperature range, with the highest daily maximum, median, and minimum values, for both winter and summer periods. The highest minimum value demonstrates the reduced nighttime cooling potential of densely developed urban areas. Moreover, looking at the inter-quartile range (IQR) for location IS for the summer period, the median is shifted toward the lower quartile with a wider spread of individual observations in the upper quartile compared to the clustering of observations in the lower quartile. This implies a somewhat larger tendency toward extreme values for higher temperatures.

The wind speed data shows a substantial decrease in airflow in location IS for both seasons. Although this may be considered positive in cold seasons (e.g., lower building heating demands, higher outdoor thermal comfort), it may have a negative impact in summer due to urban overheating. These results underline the significant potential of mitigation measures specifically in the IS location.

Figure 5 illustrates, for area IS, the mean hourly air temperature difference between the base case and three mitigation scenarios in the course of seasonal reference days. Figure 6 illustrates the computed daytime and nighttime CTD values of the envisioned mitigation strategies over four seasons. The results point to the different levels of impact of selected mitigation measures under a wide range of boundary weather conditions. In general, the consideration of trees prove to be quite beneficial for the improvement of thermal conditions in the urban canyon, especially during the summer period. The degree of improvement over day- and nighttime periods appears to be, however, dependent on the season. Higher daytime air temperature reductions (Δq) and CTD values were observed during warmer months, while the opposite was true for colder months. This occurrence may be explained, on one hand, by the overall higher solar gain and resulting increased effect of radiation shielding by trees during the summer period, thus pointing to the valuable role of tree shade. On the other hand, deciduous trees provide significantly less shade in the colder months due to the loss of leaves, allowing solar access to horizontal and vertical surfaces (mainly pavements, roads, and building walls), thus maximizing the absorption of heat from solar radiation. This in turn may have important positive consequences for building heating demands. Higher Δq and CTD values were also observed during the nighttime, for both colder and warmer months. As the overall solar gain was reduced during the day due to the shading effect of the trees, there was substantially less heat absorbed in the urban fabric, thus making the temperature difference from the base case larger.

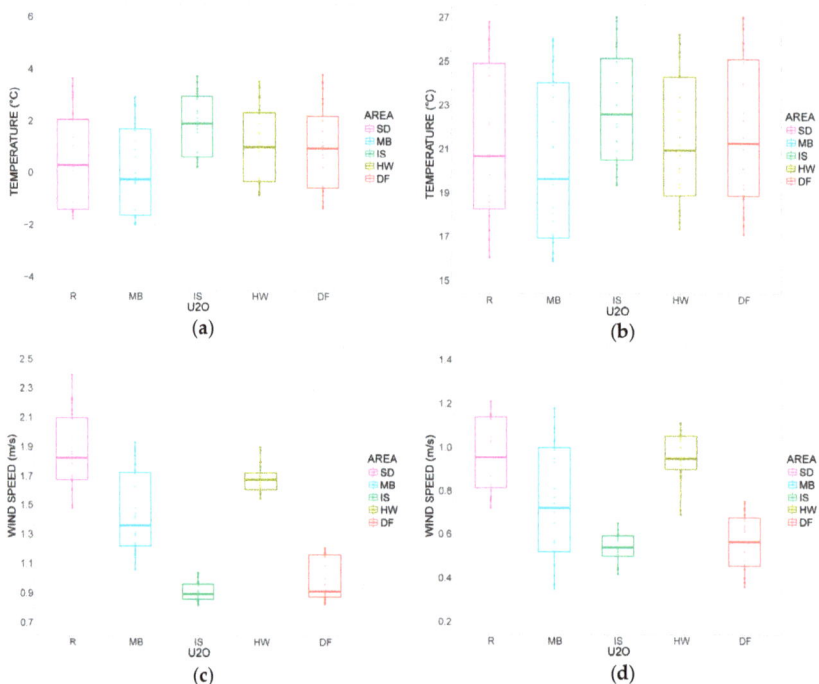

Figure 4. Boxplots of ambient air temperature and wind speed data for a reference day in summer and winter periods, with individual hourly data points: (**a**) winter temperature, (**b**) summer temperature, (**c**) winter wind speed, (**d**) summer wind speed.

The implementation of green roofs appears to have little or no effect on the thermal conditions within the urban canyon. This may be attributed to the relatively high elevation of vegetative elements, thus limiting the effect on data sampled from the street level. Green roofs are generally more important for their effect on the boundary layer conditions due to the overall lower storage heat flux than in conventional roof constructions. This, in turn, leads to less energy available for release back into the atmosphere and results in cooler air masses above the roof surface. This circumstance is specifically relevant for reductions of near-surface and surface roof temperatures. Additionally, green roofs provide better roof insulation, with a significant impact on a building's energy use.

As it could be expected, the results suggest that, in the case of the study area in Vienna, a combination of the two measures had the largest positive effect on urban overheating. It can thus be concluded that the concurrent deployment of multiple intervention measures whose mitigation mechanisms vary in temporal scale and magnitude may amplify the discrete effects of a single measure.

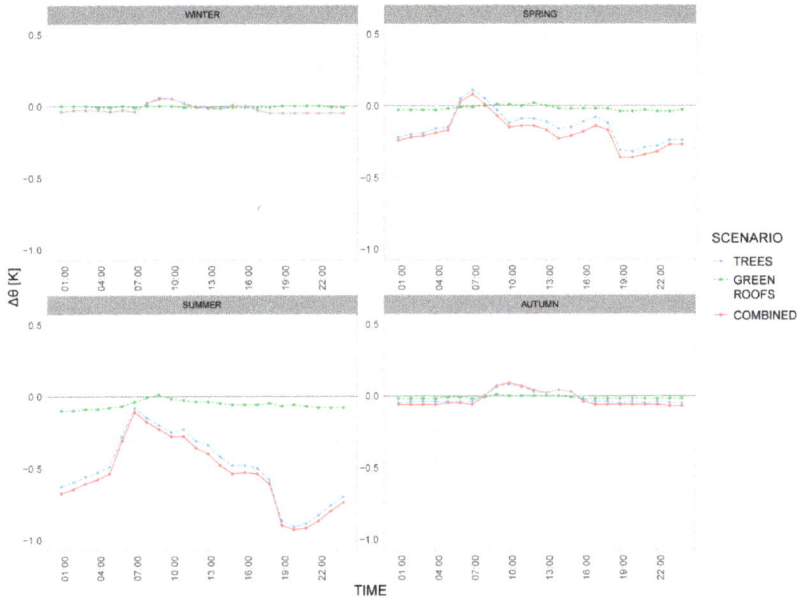

Figure 5. Mean hourly air temperature difference between the base case and three mitigation scenarios in the course of seasonal reference days, area IS.

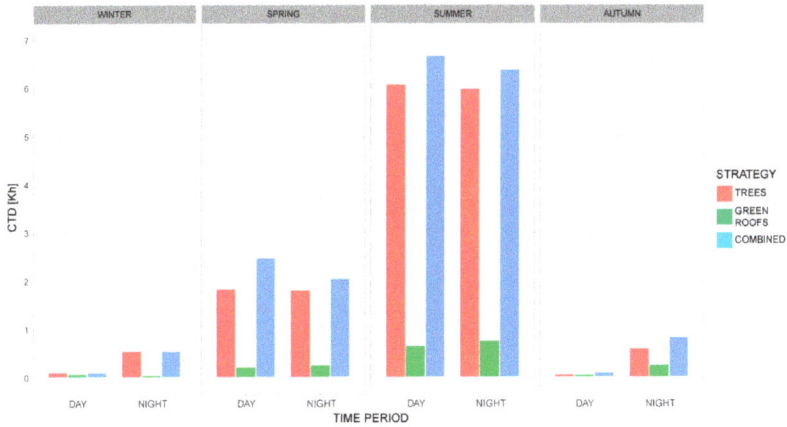

Figure 6. Computed daytime and nighttime Cumulative Temperature Decrease (CTD) values for envisioned mitigation strategies for the targeted area IS.

Lastly, we investigated the resulting climatic effect of distinct transformations of the urban landscape. The computed daytime and nighttime seasonal CTD values for the Nordbahnhof area are presented in Table 4. Figure 7 illustrates the mean seasonal hourly air temperature of the base case (brownfield) and the development scenario for the Nordbahnhof area. The results point to a varying effect on the thermal environment across both the day-night cycle and the season. Namely, a diurnal cooling effect may be observed during warmer months (up to 1.6 K). This is in part due to the shadowing effect caused by new tall buildings, limiting the incidence of direct solar radiation on the neighboring buildings and the ground surface. Consequently, less heat was stored in the physical mass of built structures and eventually re-emitted back into the environment, leading to cooler local surroundings. However, as this might be seen as an opportunity to increase the shade in summer, during the winter it might affect the insolation potential of the area.

Table 4. Computed daytime and nighttime CTD values for Nordbahnhof area.

CTD [Kh]	Winter	Spring	Summer	Autumn
Day	0	6.5	12.7	0
Night	0	0	0.33	0

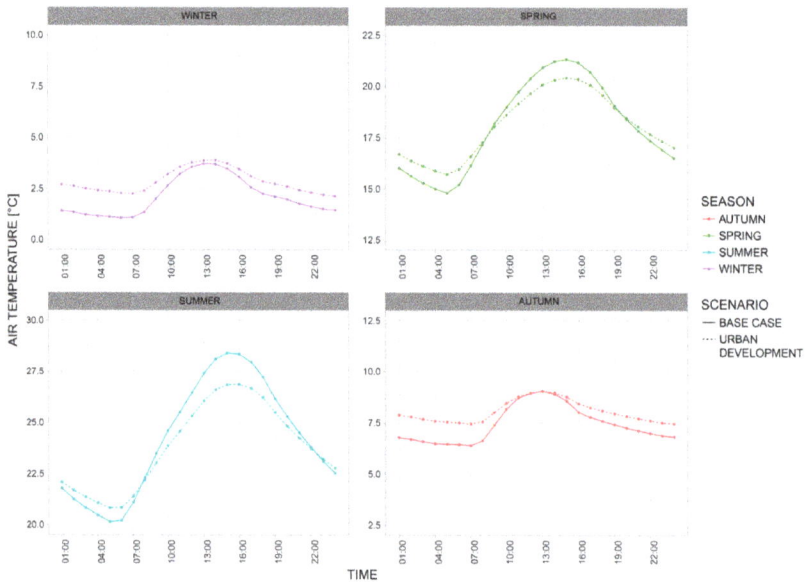

Figure 7. Mean hourly air temperature of the base case and development scenario for each season, Nordbahnhof area.

Additionally, a nocturnal heating effect was observed for each season, ranging from 0.7 to 1.3 K. This is to be expected due to the higher fraction of built surfaces with high thermal admittance and, therefore, the higher thermal mass of the area compared to the base case. This resulted in a greater fraction of heat being stored within the urban fabric. In turn, the heat loss from within the urban canyons was hampered due to the lower sky view factor (SVF). The effect appears to be more pronounced during the colder months. This might be due to the obstructed cold winter winds by buildings and vegetation, causing a greater departure from the base case conditions.

4. Conclusions

The identification of appropriate urban intervention strategies to mitigate the negative effects of urban overheating is a necessary step towards more sustainable urban environments. In this light, the first objective of our research effort was to contribute to the understanding of the very drivers behind the warming of the cities. It was concluded that areas of higher urban density and higher built surface fraction tend to store more heat during the day. Due to their dense arrangement of built structures, these areas also display a substantial decrease of airflow and are thus prone to more extreme and unfavorable thermal conditions.

One important aspect of our further inquiry was to investigate the potential of a number of mitigation strategies to remedy the negative effects of this process. Thereby, urban trees have been identified as a promising strategy for the improvement of thermal conditions in the urban canyon, leading to a maximum of 6 K summertime air temperature decrease due to their shading properties. Green roofs, on the other hand, had little (around 0.3 K decrease) or no effect on thermal conditions within the urban canyon. The combination of both intervention alternatives proved, in case of Vienna, to be the most effective strategy. Therefore, our results suggest that the application of multiple mitigation strategies may prove more effective if deployed concurrently, rather than in isolation.

Planting urban trees can be argued to be a relatively low-cost, easy-to-implement, and climatically efficient measure against urban overheating. The application of a vegetation layer to building envelopes, more specifically to roof surfaces, may be more financially intensive and, in practice, more difficult to implement due to the issues of private property ownership. Thus, the implementation of urban trees might be the strategy that local authorities would more readily adopt.

Lastly, we investigated the climatic response of a new urban development in an existing abandoned industrial site. The simulation results suggested a post-development diurnal cooling effect of up to 1.6 K (for the spring and summer periods) and an all-year nocturnal heating effect of up to 1.3 K. This is due to the obstructed solar access and airflow by new tall buildings and vegetation. Daytime temperature reduction can be attributed to the solar shielding effect, whereas reduced nighttime cooling potential results in higher temperatures.

There are many potential applications of the presented research and related insights. In general, the outcome of our research effort is expected to advance the local urban development concepts and techniques towards more effective planning practices. Specifically, it was shown that the consideration of climatic knowledge, and understanding the essential urban energy balance concepts can support decision making at the urban level.

Acknowledgments: The research work presented in this paper was supported in part within the framework of the EU-Project, "Development and application of mitigation and adaptation strategies and measures for counteracting the global Urban Heat Island phenomenon" (Central Europe Program, No. 3CE292P3).

Author Contributions: Milena Vuckovic prepared the input weather data for simulations, conducted data analysis and interpreted the results, and contributed substantially to the writing of the manuscript. Aida Maleki performed the simulations. Ardeshir Mahdavi provided general supervision and significant provision of ideas during the project run, and revised the manuscript critically prior the final submission.

Conflicts of Interest: The authors declare no conflict of interest.

References

1. World Health Organization (WHO). *Global Report on Urban Health: Equitable, Healthier Cities for Sustainable Development*; World Health Organization: Geneva, Switzerland, 2016; Available online: http://www.who.int/iris/handle/10665/204715 (accessed on 10 December 2017).

2. United Nations, Department of Economic and Social Affairs, Population Division. *World Urbanization Prospects: The 2014 Revision, Highlights (ST/ESA/SER.A/352)*; United Nations: New York, NY, USA, 2014.

3. United Nations, Department of Economic and Social Affairs, Population Division. *World Population Prospects: The 2015 Revision, Key Findings and Advance Tables*; Working Paper No. ESA/P/WP.241; United Nations: New York, NY, USA, 2015.

4. Wilson, B.; Chakraborty, A. The Environmental Impacts of Sprawl. *Sustainability* **2013**, *5*, 3302–3327. [CrossRef]

5. The European Parliament and the Council of the European Union (EUROPA). Directive 2010/31/EU on the Energy Performance of Buildings. Official Journal of the European Union L 153/13-33. Available online: http://www.buildup.eu/en/practices/publications/directive-201031eu-energy-performance-buildings-recast-19-may-2010 (accessed on 10 December 2017).

6. Harlan, S.L.; Ruddell, D.M. Climate change and health in cities: Impacts of heat and air pollution and potential co-benefits from mitigation and adaptation. *Curr. Opin. Environ. Sustain.* **2011**, *3*, 126–134. [CrossRef]

7. Dong, Y.; Varquez, A.C.G.; Kanda, M. Global anthropogenic heat flux database with high spatial resolution. *Atmos. Environ.* **2017**, *150*, 276–294. [CrossRef]

8.   Murray, J.; Heggie, D. From urban to national heat island: The effect of anthropogenic heat output on climate change in high population industrial countries. *Earth's Future* **2016**, *4*, 298–304. [CrossRef]

9.   Feng, J.-M.; Wang, Y.-L.; Ma, Z.-G.; Liu, Y.-H. Simulating the Regional Impacts of Urbanization and Anthropogenic Heat Release on Climate across China. *J. Clim.* **2012**, *25*, 7187–7203. [CrossRef]

10. Sailor, D.J.; Georgescu, M.; Milne, J.M.; Hart, M.A. Development of a national anthropogenic heating database with an extrapolation for international cities. *Atmos. Environ.* **2015**, *118*, 7–18. [CrossRef]

11. Ryu, Y.-H.; Baik, J.-J.; Lee, S.-H. Effects of anthropogenic heat on ozone air quality in a megacity. *Atmos. Environ.* **2013**, *80*, 20–30. [CrossRef]

12. Asikainen, A.; Pärjälä, E.; Jantunen, M.; Tuomisto, J.T.; Sabel, C.E. Effects of Local Greenhouse Gas Abatement Strategies on Air Pollutant Emissions and on Health in Kuopio, Finland. *Climate* **2017**, *5*, 43. [CrossRef]

13. Zhang, G.J.; Cai, M.; Hu, A. Energy consumption and the unexplained winter warming over northern Asia and North America. *Nat. Clim. Chang.* **2013**, *3*, 466–470. [CrossRef]

14. Stott, P.A.; Stone, D.A.; Allen, M.R. Human contribution to the European heatwave of 2003. *Nature* **2004**, *432*, 610–614. [CrossRef] [PubMed]

15. Krpo, A.; Salamanca, F.; Martilli, A.; Clappier, A. On the Impact of Anthropogenic Heat Fluxes on the Urban Boundary Layer: A Two-Dimensional Numerical Study. *Bound.-Layer Meteorol.* **2010**, *136*, 105–127. [CrossRef]

16. Chen, Y.; Jiang, W.M.; Zhang, N.; He, X F.; Zhou, R.W. Numerical simulation of the anthropogenic heat effect on urban boundary layer structure. *Theor. Appl. Climatol.* **2008**, *97*, 123–134. [CrossRef]

17. Erell, E.; Pearlmutter, D.; Williamson, T.J. *Urban Microclimate: Designing the Spaces between Buildings*, 1st ed.; Earthscan: London, UK, 2011.

18. Chen, B.; Dong, L.; Shi, C.; Li, L.-J.; Chen, L.-F. Anthropogenic Heat Release: Estimation of Global Distribution and Possible Climate Effect. *J. Meteorol. Soc. Jpn.* **2014**, *92*, 157–165. [CrossRef]

19. Crutzen, P. New directions: the growing urban heat and pollution "island" effect- impact on chemistry and climate. *Atmos. Environ.* **2004**, *38*, 3539–3540. [CrossRef]

20. Grimmond, C.S.B. Urbanization and global environmental change: local effects of urban warming. *Cities Glob. Environ. Chang.* **2007**, *173*, 83–88. [CrossRef]

21. Zhao, C.; Fu, G.; Liu, X.; Fu, F. Urban planning indicators, morphology and climate indicators: A case study for a north-south transect of Beijing, China. *Build. Environ.* **2011**, *46*, 1174–1183. [CrossRef]

22. Gago, E.J.; Roldan, J.; Pacheco-Torres, R.; Ordóñez, J. The city and urban heat islands: A review of strategies to mitigate adverse effects. *Renew. Sustain. Energy Rev.* **2013**, *25*, 749–758. [CrossRef]

23. Aleksandrowicz, O.; Vuckovic, M.; Kiesel, K.; Mahdavi, A. Current trends in urban heat island mitigation research: Observations based on a comprehensive research repository. *Urban Clim.* **2017**, *21*, 1–26. [CrossRef]

24. Santamouris, M. Heat Island Research in Europe: The State of the Art. *Adv. Build. Energy Res.* **2007**, *1*, 123–150. [CrossRef]

25. Santamouris, M. Cooling the cities—A review of reflective and green roof mitigation technologies to fight heat island and improve comfort in urban environments. *Solar Energy* **2014**, *103*, 682–703. [CrossRef]

26. Hebbert, M.; Webb, B. Towards a Liveable Urban Climate: Lessons from Stuttgart. In *Liveable Cities: Urbanising World*; Routledge: New York, NY, USA, 2012; pp. 132–150.

27. Pisello, A.L.; Castaldo, V.L.; Piselli, C.; Pignatta, G.; Cotana, F. Combined Thermal Effect of Cool Roof and Cool Façade on a Prototype Building. *Energy Procedia* **2015**, *78*, 1556–1561. [CrossRef]

28. Heusinger, J.; Weber, S. Comparative microclimate and dewfall measurements at an urban green roof versus bitumen roof. *Build. Environ.* **2015**, *92*, 713–723. [CrossRef]

29. Cameron, R.W.F.; Taylor, J.E.; Emmett, M.R. What's 'cool' in the world of green façades? How plant choice influences the cooling properties of green walls. *Build. Environ.* **2014**, *73*, 198–207. [CrossRef]

30. Berardi, U.; GhaffarianHoseini, A.H.; GhaffarianHoseini, A. State-of-the-art analysis of the environmental benefits of green roofs. *Appl. Energy* **2014**, *115*, 411–428. [CrossRef]

31. Akbari, H.; Kolokotsa, D. Three decades of urban heat islands and mitigation technologies research. *Energy Build.* **2016**, *133*, 834–842. [CrossRef]

32. Georgakis, Ch.; Zoras, S.; Santamouris, M. Studying the effect of "cool" coatings in street urban canyons and its potential as a heat island mitigation technique. *Sustain. Cities Soc.* **2014**, *13*, 20–31. [CrossRef]

33. Gunawardena, K.R.; Wells, M.J.; Kershaw, T. Utilising green and bluespace to mitigate urban heat island intensity. *Sci. Total Environ.* **2017**, *584*, 1040–1055. [CrossRef] [PubMed]

34. Wang, Y.; Akbari, H. The effects of street tree planting on Urban Heat Island mitigation in Montreal. *Sustain. Cities Soc.* **2016**, *27*, 122–128. [CrossRef]

35. Huttner, S.; Bruse, M. Numerical modelling of the urban climate—A preview on ENVI-met 4.0. In Proceedings of the 7th International Conference on Urban Climate ICUC-7, Yokohama, Japan, 29 June–3 July 2009.

36. Vuckovic, M.; Kiesel, K.; Mahdavi, A. The extent and implications of the microclimatic conditions in the urban environment: A Vienna case study. *Sustainability* **2017**, *9*, 177. [CrossRef]

37. Vuckovic, M.; Kiesel, K.; Mahdavi, A. Toward advanced representations of the urban microclimate inbuilding performance simulation. *Sustain. Cities Soc.* **2016**, *27*, 356–366. [CrossRef]

38. Mahdavi, A.; Kiesel, K.; Vuckovic, M. Methodologies for UHI analysis. In *Counteracting Urban Heat Island Effects in a Global Climate Change Scenario*; Musco, F., Ed.; Springer: Berlin, Germany, 2016; pp. 71–91. [CrossRef]

39. QGIS 2.10. Available online: www.qgis.org/ (accessed on 1 September 2017).

40. Glawischnig, S.; Kiesel, K.; Mahdavi, A. Feasibility analysis of open-government data for the automated calculation of the micro-climatic attributes of Urban Units of Observation in the city of Vienna. In Proceedings of the 2nd ICAUD International Conference in Architecture and Urban Design, Epoka University, Tirana, Albania, 8–10 May 2014.

41. Glawischnig, S.; Hammerberg, K.; Vuckovic, M.; Kiesel, K.; Mahdavi, A. A case study of geometry-based automated calculation of microclimatic attributes. In Proceedings of the 10th European Conference on Product and Process Modelling, Vienna, Austria, 17–19 September 2014.

42. ZAMG, Zentralanstalt für Meteorologie und Geodynamik. Available online: http://www.zamg.ac.at (accessed on 1 September 2017).

43. Svensson, P.; Björnsson, H.; Samuli, A.; Andresen, L.; Bergholt, L.; Tveito, O.E.; Agersten, S.; Pettersson, O.; Vejen, F. *Quality Control of Meteorological Observations, Description of potential HQC Systems*; met.no Report; Norwegian Meteorological Institute: Oslo, Norwegian, 2004.

44. Bañuelos-Ruedas, F.; Angeles-Camacho, C.; Rios-Marcuello, S. Methodologies Used in the Extrapolation of Wind Speed Data at Different Heights and Its Impact in the Wind Energy Resource Assessment in a Region. In *Wind Farm—Technical Regulations, Potential Estimation and Siting Assessment*; Suvire, G.O., Ed.; InTech: Rijeka, Croatia, 2011; p. 246.

45. Stewart, I.D.; Oke, T.R. Local Climate Zones for Urban Temperature Studies. *Bull. Am. Meteorol. Soc.* **2012**, *93*, 1879–1900. [CrossRef]

46. Salata, F.; Golasi, I.; De Lieto Vollaro, R.; De Lieto Vollaro, A. Urban microclimate and outdoor thermal comfort. A proper procedure to fit ENVI-met simulation outputs to experimental data. *Sustain. Cities Soc.* **2016**, *26*, 318–343. [CrossRef]

47. Huttner, S.; Bruse, M.; Dostal, P. Using ENVI-met to simulate the impact of global warming on the microclimate in central European cities. In *Berichte des Meteorologischen Instituts der Albert-Ludwigs-Universität Freiburg Nr. 18 (2008), Proceedings of the 5th Japanese-German Meeting on Urban Climatology, Albert-Ludwigs-University of Freiburg, Freiburg, Germany, 6–8 October 2008*; Mayer, H., Matzarakis, A., Eds.; Self-Publishing Company of the Meteorological Institute, Albert-Ludwigs-University of Freiburg: Freiburg, Germany, 2008; pp. 307–312.

48. Gusson, C.S.; Duarte, D.H.S. Effects of Built Density and Urban Morphology on Urban Microclimate—Calibration of the Model ENVI-met V4 for the Subtropical Sao Paulo, Brazil. *Procedia Eng.* **2016**, *169*, 2–10. [CrossRef]

49. Maleki, A.; Kiesel, K.; Vuckovic, M.; Mahdavi, A. Empirical and computational issues of microclimate simulation. *Inf. Commun. Technol. Lect. Notes Comput. Sci.* **2014**, *8407*, 78–85. [CrossRef]

50. Still, C.J.; Berry, J.A.; Collatz, G.J.; DeFries, R.S. Global distribution of C3 and C4 vegetation: Carbon cycle Implications. *Glob. Biogeochem. Cycles* **2003**, *17*, 1006. [CrossRef]

51. Yearly Sun Graph. Available online: www.timeanddate.com/sun/austria/vienna (accessed on 1 December 2017).

52. MA22, 2015. Magistrat der Stadt Wien—Die Wiener Umweltschutzabteilung (Municipal Department 22—Environmental Protection in Vienna). Available online: www.umweltschutz.wien.at (accessed on 1 Spetember 2017).

Article

How to Design a Park and Its Surrounding Urban Morphology to Optimize the Spreading of Cool Air?

Jérémy Bernard [1,†], **Auline Rodler** [2,3,†], **Benjamin Morille** [2,4,*,†] **and Xueyao Zhang** [2]

1   CNRS, Lab-STICC Laboratory UMR 6285, 56000 Vannes, France; jeremy.bernard@univ-ubs.fr
2   CNRS, Institut de Recherche en Sciences et Techniques de la Ville, FR 2488, École Centrale de Nantes, 44000 Nantes, France; auline.rodler@gmail.com (A.R.); xueyao.zhang@gmail.com (X.Z.)
3   Cerema Ouest, 44000 Nantes, France
4   UMR CNRS AAU CRENAU, Ecole Nationale Supérieure d'Architecture de Nantes, 44000 Nantes, France
*   Correspondence: benjamin.morille@soleneos.fr; Tel.: +33-6-32541716
†   These authors have all contributed equally to this work.

Received: 30 December 2017; Accepted: 30 January 2018; Published: 6 February 2018

Abstract: Green areas induce smaller increases in the air temperature than built-up areas. They can offer a solution to mitigating the urban heat island impacts during heat waves, since the cool air generated by a park is diffused into its immediate surroundings through forced or natural convection. The purpose of this study is to characterize the effect of several variables (park size, morphology of surrounding urban area, and wind speed) on the spreading of cool air. A parametric study is performed to run computational fluid dynamics simulations. The air temperature entering the computational domain was set at 35 °C, and the 2-m high surface included within the 34 °C isotherm was defined as an indicator of cool air spreading. The effects of park shape and orientation were negligible in comparison with size effects. The number of buildings was better correlated with the cooled surface area than the typical urban parameters identified in the literature (i.e., building density, aspect ratio, or mean building height). Since the number of buildings is obviously related to the number of streets, this result suggests that the greater the number of streets around a park, the wider the area that cool air spreads.

Keywords: park cool island; urban cooling; urban morphology; micro-climate simulations

---

1. Introduction

According to the Intergovernmental Panel on Climate Change (IPCC) reports, heat wave intensity and frequency should increase in the coming decades [1], which may lead to situations of outdoor and indoor discomfort as well as major health impacts. During the summer of 2003, European countries recorded 70,000 excess deaths, attributing them to an unusually hot summer [2].

Urban areas are especially vulnerable when facing such problems, for two reasons. First, heat waves are exacerbated in cities when compared to their surroundings due to the urban heat island (UHI) phenomenon [3]. Second, urban areas concentrate most of the human population (i.e., 70% in European Union countries [4]).

In order to cool cities, several solutions have been investigated: evaporative techniques (fountains, water ponds, street watering), green techniques (grass, trees, green facades or green roofs, etc.), and material techniques (reflective, water retentive, etc.). All of these techniques have shown cooling potential, but further research is still needed to improve their performance [5]. This article focuses on the use of parks as cooling solutions.

Several studies have shown that parks may create cool air during the day [6], but also at night [7]. Moreover, this cool air may be transported to the neighborhoods surrounding parks [8,9]. The ability of a park to cool the air and spread this cooler air to its vicinity depends mainly on three factors: park characteristics, wind speed conditions, and urban morphology of the peripheral areas. The characteristics of a park (tree species and density, presence of water ponds, soil types, etc.) are key parameters in explaining its cooling potential [9]. The cool air produced within the park is transported into the surrounding areas by diffusion, advection, and convection. For the sake of simplification, the term "spreading" will be used in the following sections. The spreading of cool air varies in intensity depending on both wind conditions and the urban layout of the park surroundings. This article focuses solely on the aspect of cool air spreading (i.e., on the effect of wind speed and urban layout on the level of spreading).

In 1991, Jauregui [7] established that park cooling may be measured up to one park's width away from its boundaries, which is consistent with several other studies [9–12]. However, the methodology used to assess this distance was not clearly stated, which makes the results difficult to replicate. To explain the relationship between air temperature drop and distance from the park, Shashua-Bar and Hoffman [12], Doick et al. [13] proposed an asymptotic nonlinear model, which was then used to define the Park Cool Island Distance (PCD) as the "distance [from the park] where 10% of the UHI is still present" [13]. Despite its interest, this method has been applied based on air temperature observations for a single park, hence the model is only valid for this specific park and cannot be used to draw general conclusions. We have previously seen that the production of cool air depends on park characteristics, whereas its spreading depends on both wind conditions and the layout of the urban surroundings.

The spreading of cool air is affected by wind speed value [14]. Under light winds, the temperature difference observed between the park and the streets is responsible for the advection phenomenon [15,16]. In the presence of high wind speeds, this phenomenon becomes negligible. Skoulika et al. [17] showed that for a wind speed above 5.5 m/s, the PCD decreases linearly as wind speed value increases, whereas Doick et al. [13], Upmanis et al. [15], Oke et al. [18] considered the PCD to be negligible whenever wind speed exceeds 2.3 m/s, 5 m/s, and 6 m/s, respectively. According to Doick et al. [13], this wind speed threshold is likely a function of street geometry, but they did not study this point in further detail.

Chandler [14] demonstrated that some urban structural parameters characterizing the surroundings of a park play a key role in the spreading of cool air. Chang and Li [8] analyzed the air temperature gradient around 60 parks relative to the urban canyon dimensions. Their results agreed with the previous literature regarding the average cooling distance from the park (i.e., approx. one park width). However, they were unable to establish any relationship between cool air spreading and the aspect ratio or building density. This shortcoming might be explained by the use of experimental data, which makes it difficult to differentiate the influence of street dimensions on both radiation trapping and the reduction of cool air flow originating from the park. Moreover, the performance of the shelter protecting the air temperature sensor from the sun was very sensitive to solar radiation conditions [19,20]. A temperature measurement conducted with a shelter located on a street exposed to direct solar radiation may be overestimated, thus complicating comparisons with a measurement performed on a shaded street.

The objective of this article is to better understand the phenomenon of cool air drainage from a park to its surroundings, particularly the effects of wind conditions and street dimensions on the spreading of air. Simulation rather than observation is used in order to better control the key variables, such as wind speed, urban layout, and both park and urban thermal characteristics. Simulation is also appropriate to facilitate the obtainment of a temperature field, thus avoiding having to deploy a large number of sensors, which could lead to numerous observation issues [21].

2. Methodology

2.1. Definition of Urban Form Parameters

A numerical approach is used herein to establish relationships between cool air spreading and both wind speed and urban form parameters. The urban form parameters are selected from a literature review, namely: building height, building density, and aspect ratio [8,13,22]. For the purposes of this analysis, a square park has been designed, and the street width (W), building footprint area (S_B), and reference area (S_{ref}) are all shown in Figure 1.

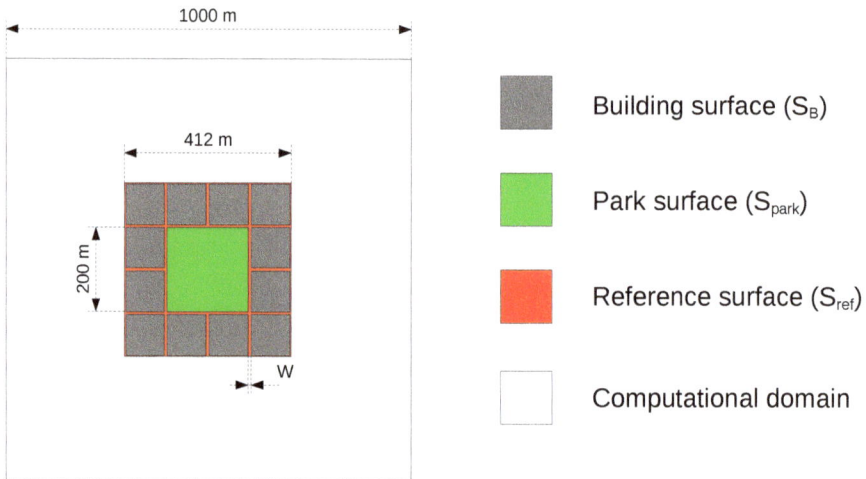

Figure 1. Study area and parameter definition—note that the reference surface consists of the entire square surface area.

The aspect ratio (H/W) and building density (D_B) are calculated by means of Equations (1) and (2):

$$H/W = \frac{H}{W},\qquad(1)$$

where H is the building height

$$D_B = \frac{S_B}{S_{ref} - S_{park}}.\qquad(2)$$

Since this investigation concerns the effect of the urban form on air spreading, the scenario used as a reference is represented by a park on its own (the reference area contains a park but no building, and consists of the entire square surface area).

2.2. Model and Implementation

The park is assumed to produce cool air that will then be spread over the urban area. Depending on the morphology of the urban area, radiation trapping may vary in intensity and thus interfere with the cool air originating from the park. To focus solely on air spreading and simplify the simulation analysis, we assume herein that the urban surfaces do not exchange any energy with the air, whereas the park surface cools the air at a constant energy rate of 300 W/m². This initial assumption is a major one: during a summer, Rodriguez [23] measured in Nantes (city in western France) both the latent and sensible heat fluxes according to an eddy-covariance method. At noon, they observed an average sensible heat flux value of 250 W/m² for a highly urban neighborhood (i.e., vegetation density less than 10% in a 200-m buffer circle around the station) versus just 100 W/m^2 for a more highly vegetated area (76% vegetation). The second assumption is much closer to reality, at least for cities located in the Cfb (temperate without dry season warm summer) or Dfb (cold without dry season warm summer) climate zones [24]. The cooling flux of 300 W/m^2 corresponds to the maximum latent heat flux measured around noon in the urban forested park of Chicago [25], and is consistent with observations recorded in the City of Nantes in a fairly green urban neighborhood (measurement performed using the eddy-covariance method with a vegetation density in the 200-m buffer circle around the station of 76%) [23]. The results of our study may not be replicable for any park type (differences in the proportion of grass, tree, concrete, etc.) or climate zone (differences in the amount of solar radiation, air humidity, air temperature, etc.), since the cooling intensity may be affected [5]. Several other latent heat flux values (100 and 200 W/m^2) have been tested, but the natural convection effect was harder to observe. We thus decided to apply the method for the most sensible case (300 W/m^2). The air temperature was also set very high (35 °C) to reflect the peak temperature that Europe has had to face during its most recent heat waves.

To assess the spreading phenomenon of the cooled air generated by a park, the computational fluid dynamics (CFD) tool Code_Saturne was used in the environment of SOLENE-microclimat (Morille et al. [26], Musy et al. [27])—a numerical tool dedicated to urban climate modeling. Thanks to Code_Saturne, the airflow was computed by resolving Navier–Stokes equations (momentum, mass continuity, energy, species transport) using a $k - \epsilon$ turbulence model. The tool configuration was similar to that described in Malys [28], except that the buoyancy forces were modeled in order to obtain realistic airflow when natural convection is predominant (i.e., under low wind speed and high temperature differences). These forces were modeled using the Boussinesq approximation, which considers air density to be a function of air temperature.

$$\rho = \rho_r \cdot (1 - \beta \cdot (T - T_r)) \tag{3}$$

where

T_r is the reference air temperature (=300 K)

T the air temperature

ρ_r the air density at temperature T_r (=1.18 kg·m^{-3})

ρ the air density

β the coefficient of thermal expansion (=$\frac{1}{T_r}$)

The computational domain represents a 100-ha square area discretized using 270,000 tetrahedral meshes (Figure 2). The mesh size varied within the domain. The minimum size of 1.5 m for the tetrahedron side was set in the region of interest (the park and its close surroundings under 2 m high). The mesh size then increased linearly up to the limits of the domain (10 m at surface level, and 50 m at the top of the domain).

A sensitivity analysis was performed in order to evaluate the effect of mesh size on air temperature and wind speed values calculated at a height of 2 m. Several meshing scenarios with different mesh sizes were tested. For each case, air temperature and wind speed were calculated according to the following procedure:

1. An interpolation is performed to calculate the air temperature of the vertex from the air temperature in the tetrahedron
2. A 2-m-high horizontal plane on the ground is generated and intersects the tetrahedron vertex
3. The air temperature at these points is the result of a linear interpolation between the segment ends.

Results of this analysis revealed that between the chosen scenario (270,000 tetrahedrons) and the most accurate one (507,000 tetrahedrons), the temperature error never exceeded 0.5 °C inside the park and 0.02 °C outside the park.

For each time step, the calculation stopped under a convergence criterion that needed to be set. This threshold was chosen based on several simulations, where both park size and wind speed varied, but not wind direction. For each simulation, 2000 iterations were performed, and the temperature convergence parameter always stabilized above 10^{-4}. This threshold was thus chosen as the convergence criterion.

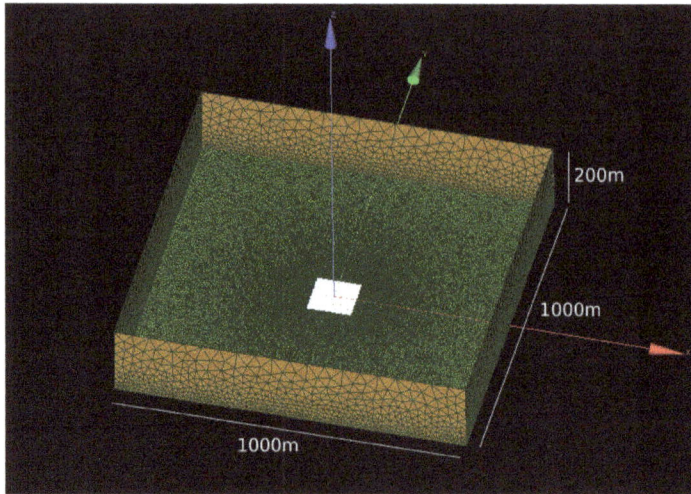

Figure 2. Meshed computational domain.

2.3. Indicators of Air Spreading Efficiency

To evaluate how spreading efficiency varies according to wind speed and urban form, performance indicators prove to be necessary. Several studies employed the concept of PCD, which is defined as the distance where the cooling induced by the park is still noticeable. In this study, the cooling effect of a park was considered to end where the 2-m-high temperature drops below 1 °C from the initial temperature. The PCD was then defined as the mean distance between the park boundary and the 34 °C isoline at a height of 2 m above ground. The distance between each point of the isoline and the park was calculated differently depending on the specific location (Figure 3). Concerning zones 1, 2, 3, and 4, it was defined as the distance between the point and the closest corner of the park. As for zones 5, 6, 7, and 8, it was defined as the minimum distance between the point and a side of the park.

However, the PCD is a 1-dimensional indicator: in the case of a long but narrow isotherm shape, the average distance of the 34 °C isotherm can be high, whereas the surface benefiting from the cooled air may be small. The surface of the area included in the 34 °C isoline is then also calculated in order to overcome the shortcomings of the PCD. However, these results show that both indicators exhibit similar behavior for all of the following analyses. For the sake of simplification, only the results obtained with the cooled surface area will be presented and analyzed.

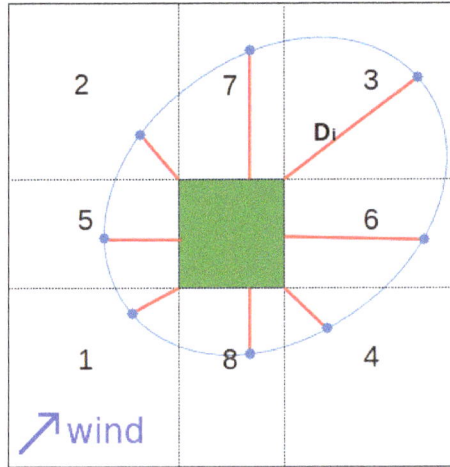

Figure 3. Zoning (dashed lines and numbers) and example of calculation (red lines) used to calculate the distance between any point on the 34 °C isoline and the park boundary.

2.4. Parametric Study

The cooling surface is impacted by several kinds of parameters. Three of them will be investigated herein: park size (with the shape remaining square), the urban form of its surrounding areas, and wind conditions.

The park width was set at 50, 100, and 200 m. One simulation was run for each park width without any adjacent buildings.

Several urban form (UF) scenarios were adopted to investigate the effect of urban parameters on the spreading of cool air. Each scenario was established to define a building height, building size, and street width in order to obtain a building density and aspect ratio values (Table 1). Values were chosen so as to highlight the separate effect of H/W, D_B, and H on the cool air spreading, but also to optimize the number of simulations to be carried out. Seven urban forms are proposed (Table 2). First, the building density influence was investigated by changing either the building size (UF 1 to 3—method 1) or the street width (UF 2 to 4—method 2). Three density levels were obtained and defined as low, medium, and high. Next, the aspect ratio influence was investigated by adopting either a low density (UF 1 and 4) or high density (UF 3 and 5). Lastly, the height was varied while the building density remained constant at a low value (UF 1 and 6) or high value (UF 3 and 7). For each scenario, the resulting number of buildings is also given in the Table.

Table 1. Parameters set for each of the seven scenarios investigated.

UF	Height (m)	Building Width (m)	Street Width (m)	D_B (%)	H/W	Number of Buildings
1	10	8	8	25 (low)	1.25	672
2	10	18	8	48 (medium)	1.25	260
3	10	96	8	85 (high)	1.25	12
4	10	18	18	25 (low)	0.56	160
5	10	18	5	95 (high)	2.00	360
6	17	8	8	25 (low)	2.12	672
7	4	96	8	85 (high)	0.50	12

Table 2. Geometry of each urban form (UF).

UF 1/UF 6	UF 2	UF 3/UF 7

UF 4	UF 5

The influence of wind speed was investigated for each of these scenarios: values were selected every 0.5 m/s, from 0.5 m/s to 5 m/s. The scenario combinations derived from urban form and wind speed led to a total of 70 simulations. For all simulations, wind crossed the park along its diagonal (Figure 3). The effect of wind direction and park shape have been investigated, but only the conclusions will be briefly presented in the following section.

3. Results

3.1. Influence of Park Size

For each park size (50, 125, and 200 m wide), the variation in the cooled surface area (i.e., where temperature is lower than 34 °C) was evaluated for various wind speeds. It is obvious that a larger park will cool a larger area. The cooled surface was then divided by the park area in order to obtain a cooling efficiency (Ce) indicator. This value may be interpreted as an equivalent surface cooled by each square meter of park.

The simulations show that Ce increased with park size or as wind speed decreased (Figure 4).

The first observation implies that one square meter of a large park is more efficient than one square meter of a small park. The second observation may be attributed to the assumption made for the simulation: the cooling rate of the air located in the park is constant regardless of its air temperature or wind speed. One result of this assumption is that the longer the air stays in the park, the cooler it leaves the park. Thus, when wind speed was very low (0.5 m/s), the air temperature difference between the park and its surroundings was high (8 °C). In this case, the air temperature of 34 °C was easily reached, and natural convection dominated: the cool air flow was emitted in all directions, including the opposite direction of the incoming wind, and the 34 °C isotherm was quite large (Figure 5a). For higher wind speeds (2 m/s), the air temperature difference between the park and its surroundings was lower (2 °C), in which case the air temperature barely reached 34 °C and spreading was mainly driven by the wind. As a result, the flow was channeled in a single direction and the 34 °C isotherm was small (Figure 5b). Overall, upon analysis of the temperature distribution around a park, we can conclude that the lower the wind speed, the further the spreading of cooled air in the direction opposite the incoming air flow.

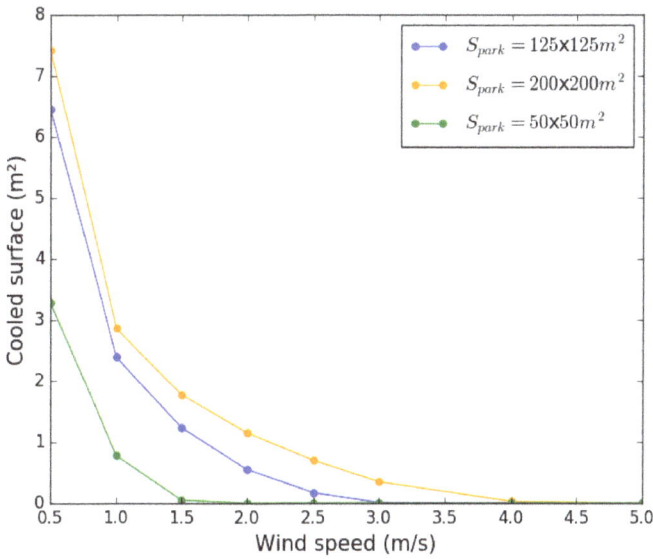

Figure 4. Cooled surface evolution with wind speed.

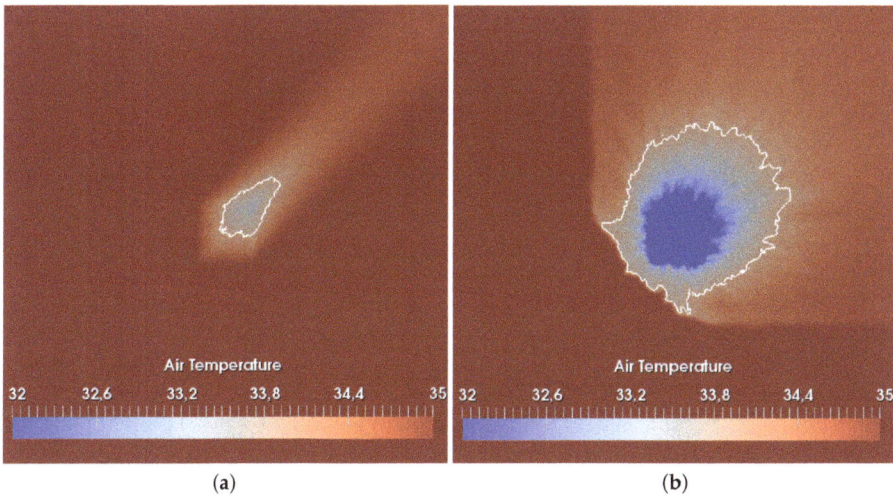

Figure 5. Two-meter-high air temperature distribution inside and around a park with a width of 125 m: (**a**) Exposed to a wind speed of 2 m/s; (**b**) Exposed to a wind speed of 0.5 m/s—the white line is the 34 °C isoline.

The shape and orientation of the park (with respect to wind speed) have been investigated. For a given surface area, they both exerted a very limited influence compared to park size (which is why the results are not shown in this study).

Overall, two main observations can be drawn:

- The larger the park, the greater the cooling intensity;
- The higher the wind speed, the smaller the cooled surface area.

In conclusion, the longer the air stays in the park, the cooler it is and the larger the size of the 34 °C isotherm.

3.2. Influence of the Neighboring District

Adding buildings around the park has a direct impact on the cooled surface shape. To investigate the influence of the neighboring district on cool air spreading, the larger park was used (200 m wide). The surface generated by the 34 °C isotherm had nearly the same size for both the reference urban form (without any buildings) and UF 1 (with buildings). However, the former was located further downstream than the latter (Figure 6). The buildings can be seen as contributing to slowing the wind speed. In agreement with the previous analysis, this finding leads to increasing the natural convection influence, diffusing the air further toward the incoming airflow direction.

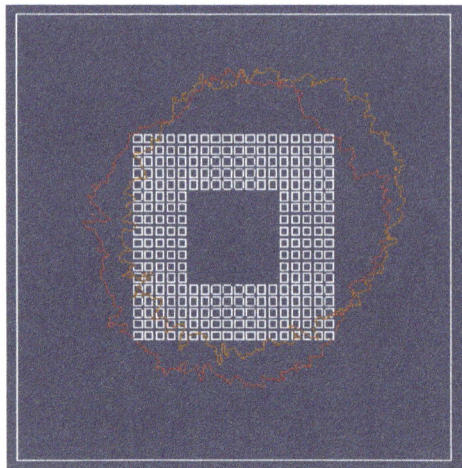

Figure 6. The 34 °C isotherm at a 0.5 m/s wind speed for the reference urban form (no buildings—orange line) and for UF 1 (red line)—the small white squares represent building footprints.

The influence of the urban form of the district surrounding the park is now investigated. Three parameters were considered: building density, aspect ratio, and building height. As described in the Methodology section, the building density was increased using two distinct methods (either the building size was modified—method 1—or the street width was modified—method 2). These two methods are considered separately and results are shown on separate plots. As previously observed, the cooled surface decreased as wind speed increased under all scenarios (Figure 7). For a given wind speed, when building density was increased by increasing the building size, the cooled surface area decreased (Figure 7a). In contrast, when building density was increased by increasing street width, the cooled surface area expanded (Figure 7b).

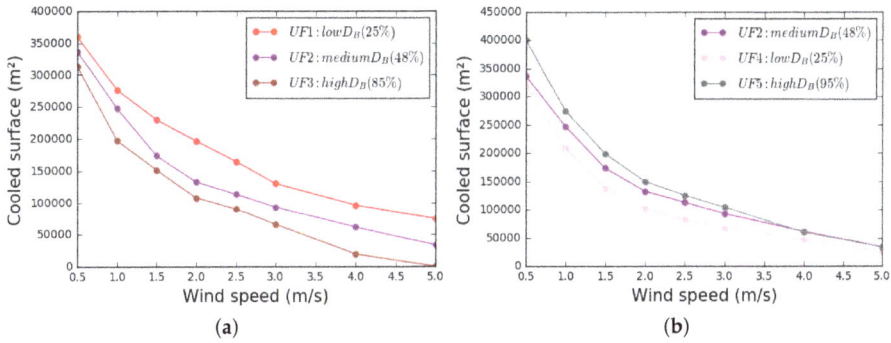

Figure 7. Influence of wind speed on the cooled surface for various building density values: (**a**) When building density was increased by modifying the building size; (**b**) When building density was increased by modifying the street width.

Building density cannot therefore be used as the lone parameter explaining the influence of the built environment on the cooled surface of a park.

The effect of the aspect ratio on the cooled surface was tested by decreasing the street width. In this manner, the aspect ratio was increased, thus keeping building density and building height constant. For a given wind speed, the size of the cooled surface increased as the aspect ratio increased (also true for a given building density, regardless of its value (Figure 8).

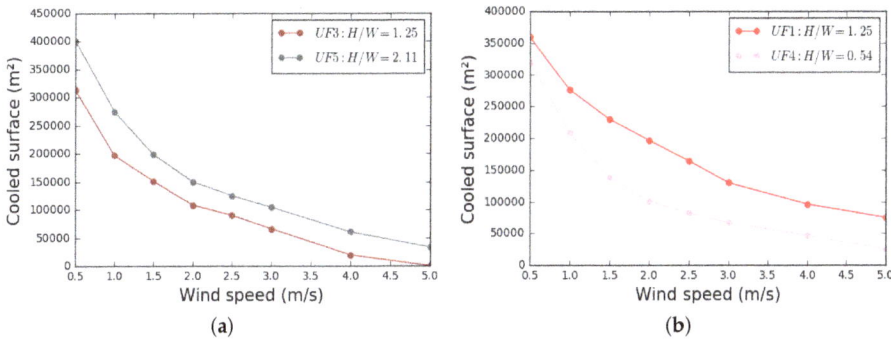

Figure 8. Influence of wind speed on the cooled surface for different aspect ratio values. (**a**) For high building density; (**b**) For low building density.

Thirdly, the building height was increased for a given building density with an increasing aspect ratio. For high building density (Figure 9a) at a given wind speed, the cooled surface area was similar when the building height increased from 10 m to 17 m. For the low building density case (Figure 9b), it increased as the building height rose from 4 m to 10 m. According to an initial analysis, we could assume that building height—like building density—is not a key parameter in explaining the cooled surface, while the aspect ratio was a key parameter. It is also possible that the effect of building height only appears at a certain threshold value. This assumption will be further analyzed in the following section.

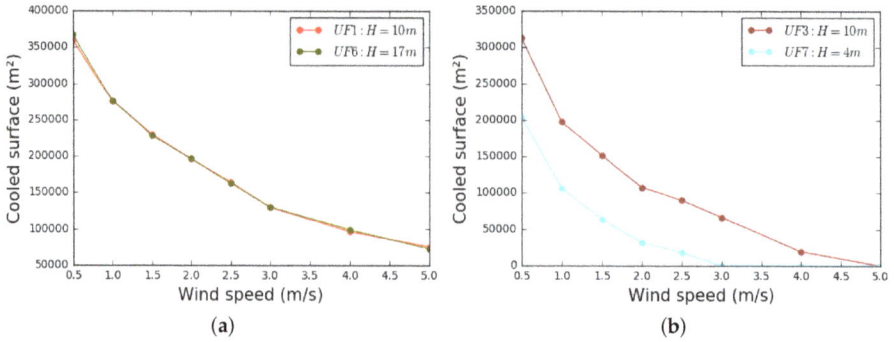

Figure 9. Influence of wind speed on the cooled surface for various building heights. (**a**) For a low building density district; (**b**) For a high building density district.

The three urban form parameters (building density, aspect ratio, and height) investigated herein do not impact the air spreading pattern in a way that could have been expected:

- For a given building height and aspect ratio, the building density increase enhances or reduces the spreading of cooled air, depending on how the density is modified. Density does not seem to be a key parameter affecting air spreading.
- For a given building height and building density, the aspect ratio appears to enhance the cooling process, whereas it is not often identified in the literature as a key parameter.
- Higher buildings may increase air spreading under a certain threshold value, which has not yet been identified (note that this threshold may be dependent on wind speed conditions).

4. Discussion

The results presented above lead to a more detailed investigation of the simulation results and then a discussion of these results. This discussion section will focus on two points:

- An understanding of the influence of building height on air spreading based on the configuration
- The identification of other parameters that could be relevant for the air spreading characterization.

The air temperature field for UF 3 and UF 7 (buildings 10 m and 4 m high, respectively, in a high-density district due to large buildings—few streets) is presented in Figure 10. In the low building height case (UF 7), the cool air diffused over the buildings (Figure 10a). In the tall building case (UF 3), the buildings seemed to be sufficiently high to prevent air from spreading above them (Figure 10b). As a result, the cool air was channeled into the streets, reaching a further distance from the park and thus covering a wider area.

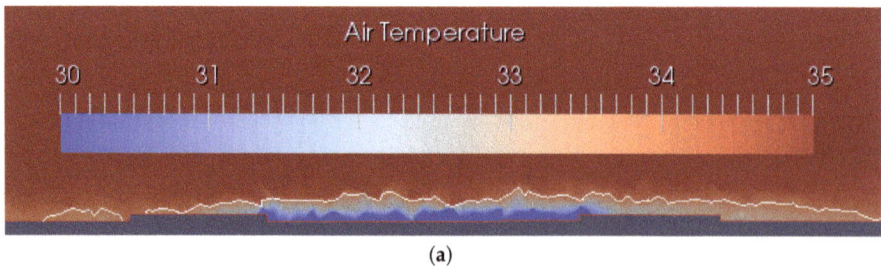

Figure 10. *Cont.*

(b)

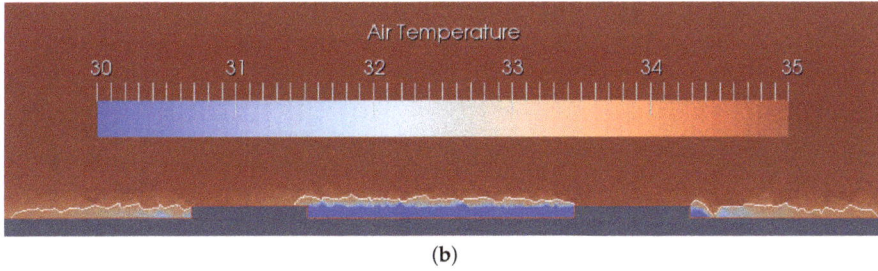

Figure 10. Vertical slice of the air temperature field for a high building density district. (**a**) UF 7: buildings are 4 m high; (**b**) UF 3: buildings are 10 m high—the white line is the 34 °C isotherm.

The air temperature field for UF 1 and UF 6 (buildings 10 m and 17 m high, respectively, in a low-density district—many streets) is presented Figure 11. For the UF 1, the cool air did not pass over the buildings; it remained channeled in the streets. It is obvious that in this case building height increases did not affect the spreading of cool air.

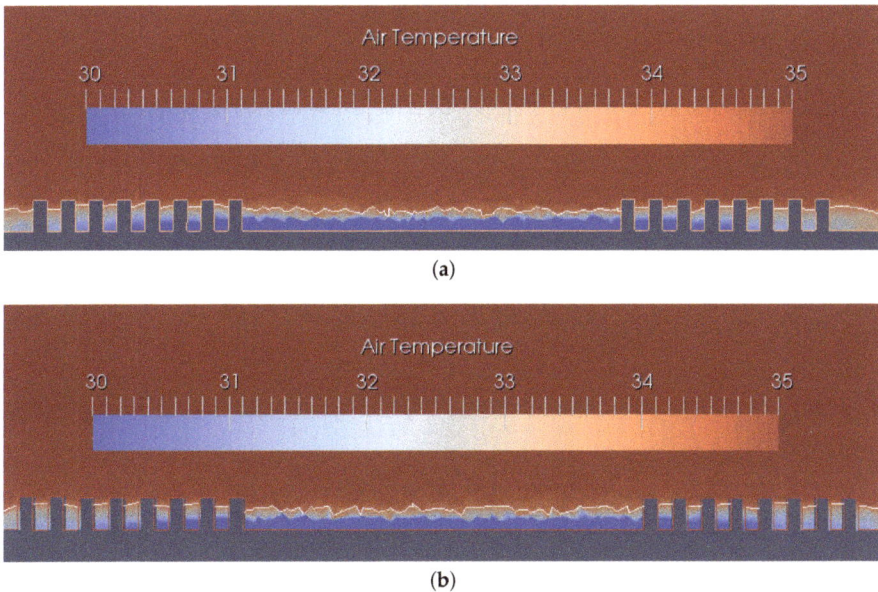

(a)

(b)

Figure 11. Vertical slice of the air temperature field for a low building density district. (**a**) UF 1: buildings are 10 m high; (**b**) UF 6: buildings are 17 m high—the white line is the 34 °C isotherm.

Above a certain building height threshold, the cool air can no longer spread over the buildings. Further increasing the building height would then have no effect on the cooled surface since the air is only spreading under the urban canopy.

The issue of building height relevance calls into question the relevance of two other indicators, namely: urban fragmentation (e.g., number of buildings), and street width. We previously showed that the aspect ratio is positively correlated with the cooled surface area. The aspect ratio is defined as the ratio of building height to street width. Since past results have demonstrated that building height only affects the cool air spreading in very specific configurations, we might wonder whether street width is an influential parameter. Urban fragmentation was not initially identified as an important parameter as regards the cool air spreading. However, the number of buildings varies along the chosen urban form, and thus its cooling potential warrants investigation. The cooled surface area is plotted versus both the number of buildings (Figure 12a) and the street width (Figure 12b). The number of buildings appears to be much more closely correlated with the cooled surface area than street width. A large variability in the cooled surface area existed for the UF with 8-m width streets, thus making this parameter insignificant (Figure 12b). In Figure 12a, the cooled surface area clearly increased as the number of buildings or streets increased. Moreover, the distance to the regression line remained low (except when the number of buildings was very small).

Ultimately, urban fragmentation (here the number of buildings) seemed to be a more relevant variable than building density, aspect ratio, or building height for characterizing the cooled air spreading from a park through a district.

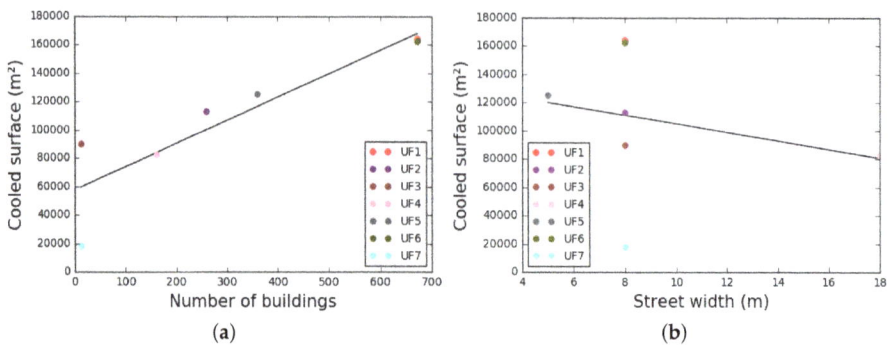

Figure 12. Cooled surface area evolution versus: (**a**) the number of buildings; (**b**) street width—wind speed set at 2.5 m/s.

5. Conclusions

The parameters impacting the spreading of cool air produced by a park have been investigated herein. Simulations were performed to observe the cooling distance (surface included inside the 1 °C temperature decrease isoline) induced by the park and by the form of its surrounding urban district. Some of the studied parameters are intrinsic to the park (shape and size), while others describe the urban morphology of the surrounding neighborhoods (building density, building height, and aspect ratio). The simulations were run for various wind speeds and under a constant cooling flux of 300 W/m^2 generated by the green area.

The analysis of our results led to the following conclusions:

- A larger park will cool more efficiently: it will generate a greater surface area of cooled air per square meter than a smaller park.
- The building density variation can enhance or reduce the spreading of cooled air. This parameter is therefore irrelevant as regards cool air spreading.
- For a given building height and building density, the aspect ratio seems to enhance the cooling process, even though it has not often been identified in the literature as a key parameter.
- Building height does not exert any influence on air spreading above a certain threshold value.
- The number of buildings seems to be a very relevant parameter for characterizing the cooled air spreading from a park through a district. This parameter and all other fragmentation indicators (the number of buildings is an indicator of fragmentation intensity, but the direction and shape of fragmentation may also be of interest) should be further analyzed by the research community.

In this paper, a constant cooling flux was considered in order to focus solely on the spreading of the cool air being produced. Some bias was thus introduced when using this method for analysis (if air remains for an infinite time in the park, its temperature should be infinitely negative). In reality, the flux should have been modified by the temperature difference between the ground and the air and by the wind speed. This hypothesis could be revised in future work by employing a coupled heat transfer simulation (thermo radiative and CFD) in order to generalize the methodology proposed herein and consider a cooling flux that varies in both space and time.

This paper has highlighted new findings, but it is based on strong hypotheses. Further investigation is thus required (empirical data should be used) to verify that these assumptions do not affect the veracity of the main findings. However, only a few studies regarding the topic of cool park air spreading have so far been performed using a numerical approach. The academic approach proposed in this paper has led to both a better understanding of the physical phenomena involved and an identification of the key parameters affecting cool air spreading. For these reasons, it is complementary to the experimental studies available in the literature.

Author Contributions: This work was been conducted within the scope of Xueyao Zhang's internship. The work program was managed by Jérémy Bernard, Benjamin Morille and Auline Rodler. Xueyao Zhang studied the bibliography and carried out all of the simulation runs. Jérémy Bernard, Benjamin Morille and Auline Rodler wrote the paper and contributed to this effort in equal measure.

Conflicts of Interest: The authors declare no conflict of interest.

References

1. Revel, D.; Füssel, H.M.; Jol, A. *Climate Change, Impacts and Vulnerability in Europe 2012*; Office of Official Publication of the European Union: Luxembourg, 2012.
2. Robine, J.M.; Cheung, S.L.K.; Le Roy, S.; Van Oyen, H.; Griffiths, C.; Michel, J.P.; Herrmann, F.R. Death toll exceeded 70,000 in Europe during the summer of 2003. *C. R. Biol.* **2008**, *331*, 171–178.
3. Oke, T.R. *Boundary Layer Climates*; Routledge: London, UK, 2002.
4. Eurostat. *Eurostat Regional Yearbook 2014: Focus on European Cities*; Office of Official Publication of the European Union: Luxembourg, 2014.
5. Santamouris, M.; Ding, L.; Fiorito, F.; Oldfield, P.; Osmond, P.; Paolini, R.; Prasad, D.; Synnefa, A. Passive and active cooling for the outdoor built environment—Analysis and assessment of the cooling potential of mitigation technologies using performance data from 220 large scale projects. *Sol. Energy* **2017**, *154*, 14–33.
6. Barradas, V.L. Air temperature and humidity and human comfort index of some city parks of Mexico City. *Int. J. Biometeorol.* **1991**, *35*, 24–28.
7. Jauregui, E. Influence of a large urban park on temperature and convective precipitation in a tropical city. *Energy Build.* **1991**, *15*, 457–463.
8. Chang, C.R.; Li, M.H. Effects of urban parks on the local urban thermal environment. *Urban For. Urban Green.* **2014**, *13*, 672–681.
9. Spronken-Smith, R.; Oke, T. Scale modelling of nocturnal cooling in urban parks. *Bound.-Layer Meteorol.* **1999**, *93*, 287–312.
10. Cao, X.; Onishi, A.; Chen, J.; Imura, H. Quantifying the cool island intensity of urban parks using ASTER and IKONOS data. *Landsc. Urban Plan.* **2010**, *96*, 224–231.
11. Ca, V.T.; Asaeda, T.; Abu, E.M. Reductions in air conditioning energy caused by a nearby park. *Energy Build.* **1998**, *29*, 83–92.
12. Shashua-Bar, L.; Hoffman, M.E. Vegetation as a climatic component in the design of an urban street: An empirical model for predicting the cooling effect of urban green areas with trees. *Energy Build.* **2000**, *31*, 221–235.
13. Doick, K.J.; Peace, A.; Hutchings, T.R. The role of one large greenspace in mitigating London's nocturnal urban heat island. *Sci. Total Environ.* **2014**, *493*, 662–671.
14. Chandler, T.J. *The Climate of London*; Hutchinson: London, UK, 1965.
15. Upmanis, H.; Eliasson, I.; Lindqvist, S. The influence of green areas on nocturnal temperatures in a high latitude city (Göteborg, Sweden). *Int. J. Climatol.* **1998**, *18*, 681–700.

16. Jansson, C.; Jansson, P.E.; Gustafsson, D. Near surface climate in an urban vegetated park and its surroundings. *Theor. Appl. Climatol.* **2007**, *89*, 185–193.

17. Skoulika, F.; Santamouris, M.; Kolokotsa, D.; Boemi, N. On the thermal characteristics and the mitigation potential of a medium size urban park in Athens, Greece. *Landsc. Urban Plan.* **2014**, *123*, 73–86.

18. Oke, T.R.; Crowther, J.; McNaughton, K.; Monteith, J.; Gardiner, B. The micrometeorology of the urban forest [and discussion]. *Philos. Trans. R. Soc. B Biol. Sci.* **1989**, *324*, 335–349.

19. Lacombe, M.; Bousri, D.; Leroy, M.; Mezred, M. *WMO Field Intercomparison of Thermometer Screens/Shields and Humidity Measuring Instruments, Ghardaia, Algeria, November 2008–October 2009*; Technical Report; World Meteorological Organization: Geneva, Switzerland, 2011.

20. Lacombe, M. Results of the WMO intercomparison of thermometer screens/shields and hygrometers in hot desert conditions. In Proceedings of the TECO-2010—WMO Technical Conference on Meteorological and Environmental Instruments and Methods of Observation, Helsinki, Finland, 30 August–1 September 2010.

21. Oke, T. *Initial Guidance to Obtain Representative Meteorological Observations at Urban Sites*; World Meteorological Organization: Geneva, Switzerland, 2004; Volume 81.

22. Perini, K.; Magliocco, A. Effects of vegetation, urban density, building height, and atmospheric conditions on local temperatures and thermal comfort. *Urban For. Urban Green.* **2014**, *13*, 495–506.

23. Rodriguez, F. FluxSAP—A collaborative experimental campaign on water and energy fluxes in urban areas and the relation with the vegetation: The case of a Nantes district. In Proceedings of the 9th International Conference on Urban Climate (ICUC9), Toulouse, France, 20–24 July 2015.

24. Peel, M.C.; Finlayson, B.L.; McMahon, T.A. Updated world map of the Köppen-Geiger climate classification. *Hydrol. Earth Syst. Sci. Discuss.* **2007**, *4*, 439–473.

25. McPherson, G.E.; Nowak, D.J.; Rowntree, R.A. *Chicago's Urban Forest Ecosystem: Results of the Chicago Urban Forest Climate Project*; US Department of Agriculture, Forest Service, Northeastern Forest Experiment Station: Radnor, PA, USA, 1994.

26. Morille, B.; Lauzet, N.; Musy, M. SOLENE-microclimate: A tool to evaluate envelopes efficiency on energy consumption at district scale. *Energy Procedia* **2015**, *78*, 1165–1170.

27. Musy, M.; Malys, L.; Morille, B.; Inard, C. The use of SOLENE-microclimat model to assess adaptation strategies at the district scale. *Urban Clim.* **2015**, *14*, 213–223.

28. Malys, L. Évaluation des Impacts Directs et Indirects des Façades et des Toitures Végétales sur le Comportement Thermique des Bâtiments. Ph.D. Thesis, École Nationale Supérieure d'Architecture Nantes, Nantes, France, 2012.

climate

MDPI

Article

The Solar Reflectance Index as a Tool to Forecast the Heat Released to the Urban Environment: Potentiality and Assessment Issues

Alberto Muscio *

Energy Efficiency Laboratory, University of Modena & Reggio Emilia, Modena 41121, Italy;
alberto.muscio@unimore.it; Tel.: +39-059-2056194

Received: 31 December 2017; Accepted: 13 February 2018; Published: 15 February 2018

Abstract: Overheating of buildings and urban areas is a more and more severe issue in view of global warming combined with increasing urbanization. The thermal behavior of urban surfaces in the hot seasons is the result of a complex balance of construction and environmental parameters such as insulation level, thermal mass, shielding, and solar reflective capability on one side, and ambient conditions on the other side. Regulations makers and the construction industry have favored the use of parameters that allow the forecasting of the interaction between different material properties without the need for complex analyses. Among these, the solar reflectance index (SRI) takes into account solar reflectance and thermal emittance to predict the thermal behavior of a surface subjected to solar radiation through a physically rigorous mathematical procedure that considers assigned air and sky temperatures, peak solar irradiance, and wind velocity. The correlation of SRI with the heat released to the urban environment is analyzed in this paper, as well as the sensitivity of its calculation procedure to variation of the input parameters, as possibly induced by the measurement methods used or by the material ageing.

Keywords: ageing; emissivity; measurement; solar reflectance; solar reflectance index; thermal emittance; urban heat island

---

1. Introduction

Overheating of buildings and urban areas is a more and more severe issue due to increasing urbanization combined with global warming. In fact, 54% of the world's population already lives in urban areas, and the percentage is expected to rise to 66% by 2050 [1]. In cities of any size the urban heat island (UHI) phenomenon shows up, which is the development of higher ambient temperatures compared to the surrounding rural and suburban areas: a recent study on one hundred Asian and Australian cities [2] reported average UHI intensities between 4.1 °C and 5.0 °C, and peaks of 11.0 °C. The UHI effect is also unfavorably superposed to the sharp increase of the mean global temperature observed in recent years [3]. In short, humanity is gathering in cities that are a lot warmer than the surrounding world, which is itself warming up!

A strong relationship exists between ambient temperature and mortality, with a threshold temperature above which mortality rates increase very rapidly and are considerably higher than the annual average [4]; such a threshold temperature is a function of local climate, architecture, and the physiological characteristics of the population. Moreover, for each degree of temperature increment, the increase of the peak electricity load was shown to vary between 0.45% and 4.6%, and the increase of total electricity consumption between 0.5% and 8.5% [5]. On the other hand, a proper design of urban surfaces based on the use of solar reflective materials for roofs and pavements has been known for decades to have great potential for the mitigation of building and urban overheating [6,7].

The thermal behavior of buildings in the hot seasons is the result of a complex balance of construction and environmental parameters such as insulation level, thermal mass, shielding, and solar reflective capability on one side, temperature, wind, and solar irradiance on the other side. Advanced software tools for dynamic simulation are generally needed to accurately predict thermal comfort and cooling energy needs, but their complexity of use makes it difficult to exploit them for preliminary selection of single building components. As a result, performance parameters of materials and building elements such as steady-state thermal transmittance, thermal mass, or solar reflectance, are often considered to operate a product comparison and, from that, a preliminary selection. Nonetheless, it is well known that making reference to distinct performance parameters independently of the installation context may be inadequate and sometimes misleading. For example, a low steady-state thermal transmittance may not prevent lightweight dark roofs from overheating and yielding a strong cooling load due to lack of inertia. Analogously, a relatively high solar reflectance may not prevent a roof surface from overheating if it is coupled with a low thermal emittance of the external surface, unless the solar reflectance is really high or a strong wind is blowing. Therefore, regulation makers and the construction industry have often favored the use of parameters such as the periodic thermal transmittance—which combines thermal insulation and thermal inertia [8]—and the solar reflectance index (SRI), in which all relevant surface properties are contemporarily taken into account [9]. Such composite parameters allow forecasting of the interaction between different material properties without the need of complex analyses.

This paper is focused on the SRI, which has raised significant interest thanks to its relative ease of calculation and, above all, its effective representation of the thermal behavior of opaque built surfaces subjected to solar radiation. In particular, it is contemplated by voluntary rating systems such as LEED [10], or regulations on energy efficiency such as Title 24 of California [11]. SRI is calculated from the solar reflectance and the thermal emittance of the analyzed surface through a physically rigorous procedure that considers assigned and highly demanding ambient conditions such as air and sky temperatures, wind velocity, and peak solar irradiance [9]. More specifically, this paper aims to analyze the correlation of SRI with the heat released to the near ground air, as well as the sensitivity of SRI calculation to variation in the input parameters, as possibly induced by the measurement methods or material weathering and soiling. The objective is to assess the potential of SRI as a tool to easily compare the performance of built surfaces in terms of heat released to the urban environment.

2. The Solar Reflectance Index (SRI) and Its Calculation

2.1. Surface Radiative Properties and SRI

The solar reflectance, or albedo, of a surface is the reflected fraction of incident solar radiation. It ranges from 0 to 1 (or 100%). Its value ρ_{sol} can be calculated by averaging over the range from 300 nm to 2500 nm, in which about 99% of total solar irradiance falls; the measured spectral reflectivity ρ_λ, defined as the ratio of reflected part and total amount of incident radiation at the considered wavelength λ (nm); weighted by the spectral irradiance of the sun at the earth surface, $I_{sol,\lambda}$ (W/(m^2 nm));

$$\rho_{sol} = \frac{\int_{300}^{2500} \rho_\lambda \cdot I_{sol,\lambda} \cdot d\lambda}{\int_{300}^{2500} I_{sol,\lambda} \cdot d\lambda} \tag{1}$$

Minimum values of solar reflectance are specified in many countries by regulations on building energy efficiency such as Title 24 of California [11], as well as by voluntary programs incentivizing energy efficiency like the Energy Star of EPA [12]; those values are usually differentiated for low-sloped and steep-sloped roofs and, possibly, for building use (e.g., residential/nonresidential). Minimum values are often set also for another surface property, the thermal emittance (also called infrared emittance, or emissivity). This is the ratio of the energy emitted in the far infrared (i.e., in the range from 4 to 40 μm) toward the sky by the considered surface and the maximum theoretical emission at the same surface temperature; this likewise ranges from 0 to 1 (or 100%). A low thermal emittance may cause a surface to overheat even if this is highly reflective because the fraction of solar radiation that is absorbed, however small it is, cannot be effectively released to the atmosphere. In many cases, however, only the solar reflectance or its complement to 1, the solar absorptance, are considered in regulations for opaque building elements.

The combined effect of different surface properties can be expressed through the "solar reflectance index" (SRI), a parameter calculated by the relationship [9]

$$SRI = 100 \cdot \frac{T_{sb} - T_{se}}{T_{sb} - T_{sw}}, \qquad (2)$$

where T_{se} (K) is the temperature that the considered surface would steadily reach when irradiated by a reference solar flux $I_{sol,max} = 1000$ W/m^2 at atmospheric air temperature $T_{air} = 310$ K; sky temperature $T_{sky} = 300$ K; and convection heat transfer coefficient h_{ce} to which three different values are assigned, equal to 5, 12, and 30 W/(m^2 K) for, respectively, low ($v_{wind} < 2$ m/s), intermediate (2 m/s $< v_{wind} < 6$ m/s), and high (6 m/s $< v_{wind} < 10$ m/s) wind speed. Intermediate wind speed is generally taken into account for product comparison. T_{sb} (K) and T_{sw} (K) are the temperatures that would be steadily reached by two reference surfaces, respectively, a black one ($\rho_{sol,b} = 0.05$) and a white one ($\rho_{sol,w} = 0.80$), with both surfaces having high thermal emittance ($\varepsilon_e = 0.90$) (see Figures 1 and 2).

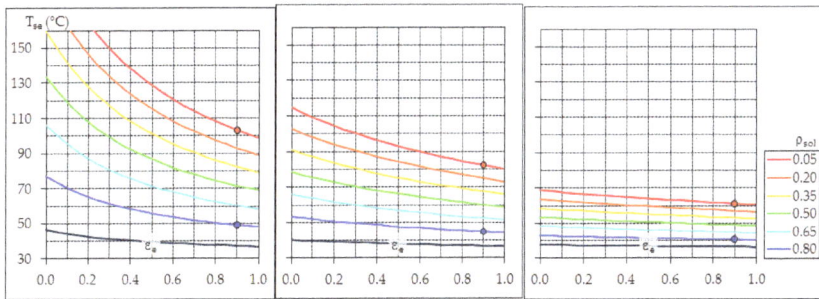

Figure 1. Surface temperature T_{se} for reference ambient conditions and (from left to right) low ($v_{wind} < 2$ m/s, $h_{ce} = 5$ W/(m^2 K)), intermediate (2 m/s $< v_{wind} < 6$ m/s, $h_{ce} = 12$ W/(m^2 K)), and high (6 m/s $< v_{wind} < 10$ m/s, $h_{ce} = 30$ W/(m^2 K)) wind speed; blue and red dots are for the reference white and black surfaces, respectively.

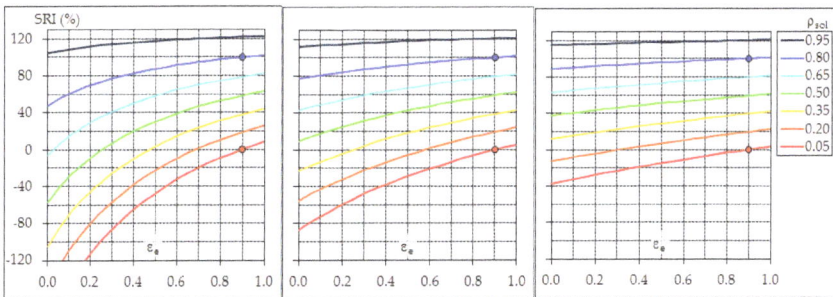

Figure 2. Solar reflectance index (SRI) for reference ambient conditions and (from left to right) low ($v_{wind} < 2$ m/s, $h_{ce} = 5$ W/(m^2 K)), intermediate (2 m/s $< v_{wind} < 6$ m/s, $h_{ce} = 12$ W/(m^2 K)), and high (6 m/s $< v_{wind} < 10$ m/s, $h_{ce} = 30$ W/(m^2 K)) wind speed; blue and red dots are for the reference white and black surfaces, respectively.

SRI represents the temperature decrement that the analyzed surface would allow with respect to the reference black surface in the reference ambient conditions, divided by the analogous decrement allowed by the reference white surface, eventually given in percentage terms. The surface temperature T_{se} (as well as T_{sb} and T_{sw}) is determined by iteratively solving the following heat balance, based on the hypothesis of an adiabatic irradiated surface:

$$(1 - \rho_{sol}) \cdot I_{sol} = \varepsilon_e \cdot \sigma_0 \cdot \left(T_{se}^4 - T_{sky}^4 \right) + h_{ce} \cdot (T_{se} - T_{air}) \tag{3}$$

SRI takes into account either the solar reflectance ρ_{sol} or the thermal emittance ε_e, as well as the reference wind velocity through the convection coefficient h_{ce}. Even if expressed in percent, its definition allows for values lower than zero or higher than 100 since materials worse than the black reference surface, or better than the white reference surface, can exist. The plots in Figures 1 and 2 show that high values of both solar reflectance and thermal emittance are contemporarily needed to obtain a low overheating of the involved surface in all wind conditions, unless its solar reflectance is really high.

2.2. Potentiality and Limitations of SRI

SRI matches the need of a single performance parameter as it allows to easy comparison of the performance of different solar reflective solutions. When applied to roofs and pavements, it is a clear indicator of the capability of their surface to return the incident solar radiation to the atmosphere by direct reflection and far infrared radiation. SRI does not take into account the sky view factor of urban surfaces nor the canyon effect, so it is mostly significant with regard to pavements and horizontal or low-slope roofs with a sky view factor close to 1. In such cases, the reflected part of solar radiation and the far-infrared thermal radiation emitted by roof and pavement surfaces can reach the high atmosphere and, therefore, the contribution to ambient warming at ground level is mostly due to convection heat transfer to the near ground air, whose rate per unit surface can be evaluated as follows:

$$q_{ce}'' = h_{ce} \cdot (T_{se} - T_{air}) \tag{4}$$

Given the temperatures of the black and white reference surfaces, T_{sb} and T_{sw}, in the ambient conditions assigned for SRI calculation, the fraction of incident solar radiation that is transferred to the near ground air can be evaluated from the definition in Equation (2) combined with Equation (4) as follows:

$$\frac{q_{ce}''}{I_{sol}} = \frac{1}{I_{sol}} \cdot h_{ce} \cdot \left[T_{sb} - \frac{SRI}{100} \cdot (T_{sb} - T_{sw}) - T_{air} \right] \tag{5}$$

Figure 3 shows that, being assigned the local wind velocity, a linear correlation exists between SRI and both the surface temperature T_{se} and the fraction of incident solar heat that is transferred to the near ground air. Since a given SRI value can be the result of different pairs of solar reflectance and thermal emittance values, as shown in Figure 2, SRI seems a more effective indicator of the capability of a surface to limit urban warming than solar reflectance alone. In fact, while thermal emittance of urban surfaces is generally around 0.9, metal roof panels with bare metal finish or thin coating and, consequently, much lower thermal emittance are often used in tropical and warm areas.

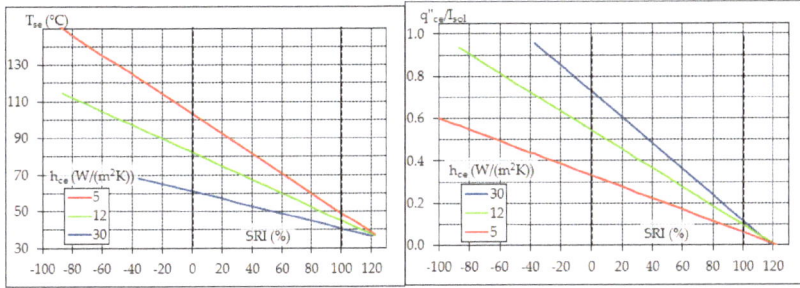

Figure 3. Surface temperature T_{se} (**left**) and fraction of incident solar heat transferred to the near ground air (**right**) vs SRI with convection coefficient h_{ce} for low, intermediate, or high wind speed and other standard conditions for SRI calculation (I_{sol} = 1000 W/m^2, T_{air} = 310 K, T_{sky} = 300 K).

SRI is based on the hypothesis of an adiabatic irradiated surface and does not consider the insulation or the inertia of the materials below, nonetheless it can again work well as an indicator of the heat transmitted by convection to the external near ground air because the heat flow rate conducted to the ground or through a roof can be lower by one or two orders of magnitude than solar irradiance. Regarding the heat flow rate entering an inhabited space below a roof, this one and the ceiling surface temperature, relevant to indoor thermal comfort, can again be correlated to T_{se}. In particular, the heat transmitted into the inhabited space is the result of the dynamic behavior of the roof structure with its insulation and thermal mass while subjected to the solar cycle, so in principle it must be calculated through mathematical tools such as the quadrupole method specified in EN ISO 13786 [8], or numerical simulation. This issue has been addressed with the recently proposed Solar Transmittance Index (STI) [13], which extrapolates the SRI approach by comparing the peak of oncoming heat flow rate (as resulting from the surface and bulk properties of the roof) with that obtained for two low- and high-performance reference cases; STI is currently under development in the specific aspect of identifying the reference cases. Nevertheless, for many roofing solutions with very low thermal inertia such as the commonly used corrugated metal panels or insulated sandwich panels, quasi-steady state conditions can be assumed throughout the day and, in particular, when the peak of solar irradiance occurs. In particular, given the indoor temperature T_i ($^\circ$C), the R factor (m^2 K/W) of the roof, and the inner surface resistance R_{si} (m^2 K/W) as resulting from convection and far-infrared radiation heat transfer, the transmitted heat flux q_i (W/m^2) can be estimated as follows:

$$q_i = \frac{T_{se} - T_i}{R + R_{si}} \tag{6}$$

Consequently, a heat balance can be extrapolated for the roof from that developed for adiabatic surfaces as presented in Equation (3):

$$(1 - \rho_{sol}) \cdot I_{sol} = \varepsilon_e \cdot \sigma_0 \cdot \left(T_{se}^4 - T_{sky}^4 \right) + h_{ce} \cdot (T_{se} - T_{air}) + \frac{T_{se} - T_i}{R + R_{si}} \tag{7}$$

Iteratively solving the heat balance in Equation (7) with respect to T_{se} for common values of the R factor yields a very low change of SRI from that of an adiabatic surface (i.e., ΔSRI = SRI$_{nonadiabatic}$ − SRI$_{adiabatic}$), especially for common surfaces with high thermal emittance (ΔSRI < 0.01 for ε_e > 0.75) (Figure 4). A confirmation is thus obtained that the SRI can be an excellent parameter for product comparison even for non-adiabatic surfaces. Moreover, a verification is provided to the hypothesis that the peak of transmitted heat flow rate is much lower than the absorbed solar irradiance and, consequently, much lower than the total irradiance (Figure 5).

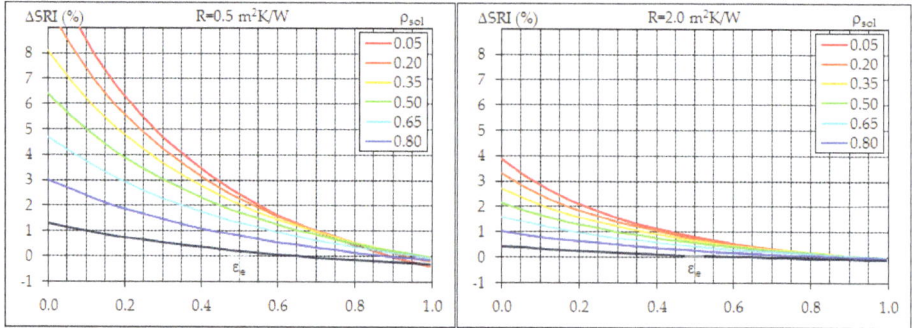

Figure 4. Change of SRI for a non-adiabatic roof surface with respect to an adiabatic one for $T_i = 27\,°C \cong 300$ K, $R_{si} = 0.17$ m^2 K/W, $h_{ce} = 12$ W/(m^2 K) and (from left to right) uninsulated (R ≈ 0.5 m^2 K/W) and well insulated (R ≈ 2.0 m^2 K/W) roofs, both with negligible thermal mass.

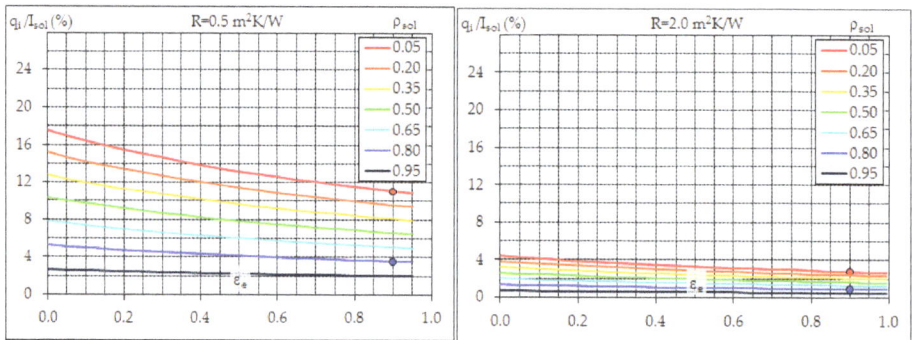

Figure 5. Ratio of peak transmitted heat flow rate and solar irradiance for $T_i = 27\,°C \cong 300$ K, $R_{si} = 0.17$ m^2 K/W, $h_{ce} = 12$ W/(m^2 K) and (from left to right) uninsulated (R ≈ 0.5 m^2 K/W) and well insulated (R ≈ 2.0 m^2 K/W) roofs, both with negligible thermal mass.

In brief, SRI has found a strong interest in the construction sector in view of its effectiveness of representation of the surface behavior. As already stated, it is contemplated by voluntary rating systems such as LEED when dealing with the summer performance of opaque roof elements and pavements: initial values of SRI $\geq$ 82 for low-sloped roofs (slope $\leq$ 2:12 or 9.5°), SRI $\geq$ 39 for steep-sloped roofs, or SRI $\geq$ 33 for pavements and other non-roof surfaces are required in the U.S. [10]. 2016 Title 24 of California [11] permits the SRI instead of separately considering solar reflectance and thermal emittance when complying with requirements on building energy efficiency.

The SRI calculation procedure is currently specified by an ASTM Standard, but a derivative European Standard is also under development. Radiative properties from which SRI is calculated, and the SRI itself, can be rated through bodies like the Cool Roof Rating Council of the U.S. [14] or the European Cool Roof Council [15]. Figure 6 shows the number of samples tested since 2014 by the Energy Efficiency Laboratory (EELab) at Modena, an accredited ECRC test laboratory; since 2015 the tests have been performed mostly under ISO/IEC 17025 accreditation [16] and some of them for ECRC rating. The sharp increase in the total number of tested samples testifies to the increasing interest for solar reflective solutions in Europe, which may rapidly approach that in the U.S. Moreover, the high fraction of tests aimed at SRI assessment testifies for the interest of the construction sector in this specific parameter.

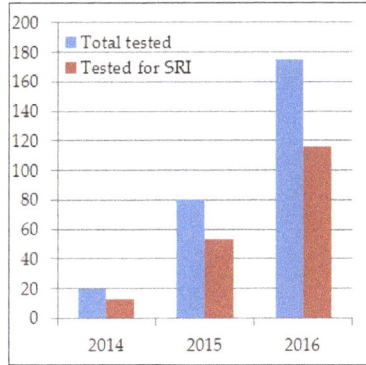

Figure 6. Samples tested at EELab for assessment of surface properties.

An issue regarding the determination of SRI may concern the arbitrary identification of the convection heat transfer coefficient at the external surface. Different measurement methods can also be used to determine solar reflectance and thermal emittance, providing slightly different results and thus affecting the SRI value. Moreover, aged values of the surface properties, the solar reflectance above all, may be significantly lower than initial values due to weathering and soiling. All these issues and their potential impact on SRI are analyzed in the following section.

3. Sensitivity of SRI Calculation to Input Parameters, and Measurement Issues

3.1. External Convection Coefficient

ASTM E1980 [9] proposes a kind of step function to estimate the convection coefficient h_{ce} to be used for SRI calculation (Figure 7). Since h_{ce} has a significant impact on the value of T_{se} and, from that, on SRI, a comparison has been made here between ASTM E1980 specifications and the literature. In this regard, EN ISO 6946 [17], a standard calculation method commonly used in the construction sector of Europe, correlates h_{ce} to the wind velocity v_{wind} (m/s) through the following empirical relationship:

$$h_{ce} = 4 + 4 \cdot v_{wind} \tag{8}$$

The formula yields higher values than those in ASTM E1980 (Figure 7), but EN ISO 6946 is presumably focused onto winter conditions, relevant to building heating, so it is supposed to adopt a precautionary approach that overestimates heat loss. More specialized literature on single phase convection heat transfer, summarized in [18] and other reference handbooks [19,20], proposes the following relationship for natural convection as induced by buoyancy forces on a horizontal plate with hot top:

$$Nu_n = 0.15 \cdot Ra^{1/3} \tag{9}$$

The Nusselt number Nu_n is the dimensionless form of the natural convection heat transfer coefficient $h_{ce,n}$

$$Nu_n = \frac{h_{ce,n} \cdot L_n}{\kappa} \tag{10}$$

and Ra is the Rayleigh number, defined as follows:

$$Ra = \frac{g \cdot b \cdot \Delta T \cdot L_n^3}{\nu \cdot \alpha} \tag{11}$$

The formula in Equation (9) is valid for $10^7 < Ra < 10^{11}$. Thermal conductivity κ (W/(m·K)), cinematic viscosity ν (m^2/s) and thermal diffusivity α (m^2/s) of the air are evaluated at the film temperature—that is, the average of surface and air temperature—whereas the coefficient of thermal expansion b (K^{-1}) is the inverse of the absolute temperature of the air T_{air}, in kelvin. The temperature difference $\Delta T = T_{se} - T_{air}$ (°C) is that between the plate surface and the air, the reference length L_n (m) is the ratio of plate area and perimeter, and g is gravity.

The following relationship is also proposed in the literature [18–20] for forced convection heat transfer on a horizontal flat plate with a turbulent boundary layer that begins at the leading edge:

$$Nu_f = 0.037 \cdot Re^{4/5} \cdot Pr^{1/3} \tag{12}$$

The Nusselt number Nu_f is the dimensionless form of the forced convection heat transfer coefficient $h_{ce,f}$ averaged over the plate

$$Nu_f = \frac{h_{ce,f} \cdot L_f}{\kappa} \tag{13}$$

and the Reynolds number Re includes the undisturbed wind velocity:

$$Re = \frac{v_{wind} \cdot L_f}{\nu} \tag{14}$$

The formula in Equation (12) is valid for $5 \cdot 10^5 < Re < 10^7$. The Prandtl number Pr and the other thermal properties of the air are again evaluated at the film temperature. The reference length L_f (m) is the length of the plate in the direction of fluid flow.

When natural convection and forced convection are superposed and have the same order of magnitude, a mixed convection Nusselt number can be evaluated as follows:

$$Nu^n = Nu_f^n \pm Nu_n^n \tag{15}$$

If buoyancy-induced flow and forced flow have the same or transverse direction, a case of assisting or transverse flow occurs, respectively, and the + applies, otherwise the − applies for opposing flow. For the exponent n the most used value is 3, however for the case of transverse flow over flat plates n = 3.5 may provide a better correlation of data [19,20]. In principle, the reference length should be the same ($L_n = L_f = L$), but for Equation (9) and Equation (12) it is not. In order to overcome such issues, the following relationship can be extrapolated from Equation (15) to combine the natural and forced convection coefficients (for transverse flow):

$$h_{ce}^n \approx h_{ce,f}^n + h_{ce,n}^n \tag{16}$$

In Figure 7 the convection coefficient calculated according to the different estimate methods is plotted against the wind velocity at ambient reference conditions for SRI calculation and sea level air properties. For calculation of h_{ce} by means of Equation (16) a horizontal roof surface 10 m × 10 m or 20 m × 20 m is considered, as well n = 3.5 and two different values of the temperature difference $\Delta T = T_{se} - T_{air}$, equal to 5 °C and 20 °C. For intermediate and high wind conditions, forced convection seems to dominate natural convection and the influence of ΔT on h_{ce} is very low or negligible. The reference values of h_{ce} specified by ASTM E1980 are halfway between those calculated according to EN ISO 6946 and the empirical formulas from the literature. Indeed the choice of h_{ce} is arbitrary since its actual value is a complex function of building shape, wind velocity and direction, and local ambient conditions; thus, one can conclude that ASTM E1980 provides arbitrary, but reasonable, values. Moreover, for those values, SRI permits an effective comparison of roofing and pavement solutions based on their radiative properties since variations of such properties have a negligible effect on h_{ce} for products which have similar performance.

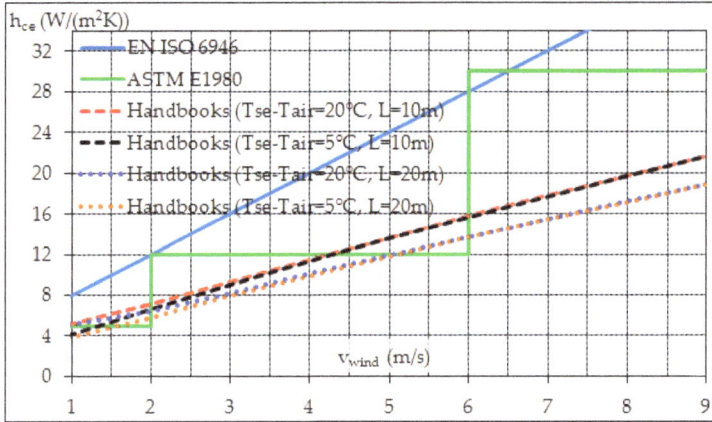

Figure 7. Convection coefficient vs. wind velocity according to different estimation methods.

3.2. Assessment of Solar Reflectance and Ageing Issues

Solar reflectance can be measured through Equation (1) by means of several methods, based on spectrophotometers [21–24] or solar reflectometers [23–25]. Independently of the method and instrument used, a main issue is the choice of the reference solar spectrum with which the average of the reflectivity spectrum is weighted. The correlation pattern of solar irradiance versus wavelength depends on parameters such as air mass (ratio of the mass of atmosphere in the actual observer-sun path to the mass that would exist if the observer were at sea level, at standard barometric pressure, and the sun were directly overhead), orientation and inclination of the irradiated surface, inclusion of direct, circumsolar or global (i.e., direct and diffuse) irradiance, position of the sun in the sky, sky conditions, etc. Among the many standard spectra available (Figure 8), the one initially recommended by the ASTM E903 Standard [21] in its 1996 version and thus considered by the Cool Roof Rating Council of U.S. [14] for product rating, also permitted by the European Cool Roofs Council [15], is that specified by the ASTM E891 Standard [26] for air mass 1.5 and beam normal solar irradiance (i.e., solar flux coming from the solid angle of the sun's disk on a surface perpendicular to the axis of that solid angle), designated 'E891BN' in the following. Such a spectrum was presumably intended for sun-tracking photovoltaic panels at the latitude of the U.S. and it may not be the most proper choice when the thermal behavior of a built surface subjected to solar radiation is considered. Moreover, ASTM E891 has been withdrawn and substituted by ASTM G173 [27], nevertheless E891BN is still largely use for product rating, most likely for ease and fairness of comparison with already rated products. The current version of ASTM E903 recommends the couple of spectra reported in ASTM G173, again for air mass 1.5 and for direct circumsolar irradiance (i.e., solar flux coming from a solid angle with aperture half-angle of 2.9° centered on the sun's disk) or global irradiance (i.e., direct and diffuse solar flux coming from the whole hemisphere overhead) on a sun-facing 37° tilted surface, in the following designated 'G173DN' and 'G173GT', respectively. Also these spectra are probably intended for use with sun tracking or fixed photovoltaic panels at the latitude of the U.S. In recent research on solar reflective built surfaces [23,24], the air mass 1 global horizontal spectrum, considering direct and diffuse radiation on a horizontal surface with the sun directly overhead in clear sky, here designated 'AM1GH', has been recommended as the spectrum of choice. Using AM1GH allows the measurement of solar reflectance under conditions that best predict annual peak solar heat gain, which is helpful because air conditioning systems are typically sized to meet annual peak cooling load. Moreover, electric grid peak load and health issues mostly arise for annual peak cooling load. While optimal for horizontal surfaces at mainland U.S. latitudes, AM1GH was shown to also apply well to moderately pitched roofs with a slope up to 23°, moreover it is expected to work well from 49° S to 49° N [23]. In the European Union a similar spectrum of global irradiance for air mass 1 is specified in the EN 410 Standard [22], intended for testing of window glass and designated 'EN410G' in the following. The use of such types of spectrum seems to be confirmed in the draft of a standard test method for solar reflectance measurement that is currently under development. In Table 1 the percent energy content of the different spectra in the UV (ultraviolet), visible (Vis) and near infrared (NIR) ranges is also summarized. Of course further spectra may be considered [23].

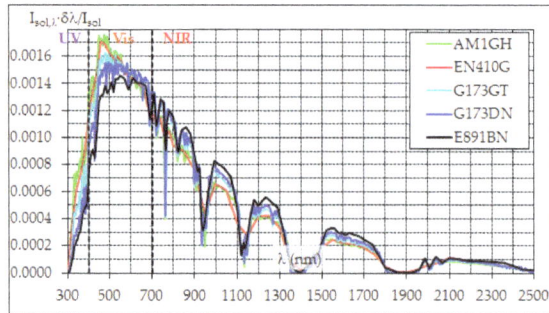

Figure 8. Standard solar irradiance spectra, normalized.

Table 1. Energy content (%) of solar spectra in the UV-Vis-NIR ranges.

	AM1GH	EN410G	G173GT	G173DN	E891BN
UV (300–400 nm)	6.8	6.4	4.7	3.5	3.0
Vis (400–700 nm)	44.7	44.1	43.3	42.0	39.0
NIR (700–2500 nm)	48.5	49.5	52.0	54.5	58.0

The solar spectra presented here, shown in Figure 5, clearly have different spectral contents. One can integrate them from 300 nm to a given wavelength λ in order to calculate the fraction of total irradiance included in the spectral range from 300 nm to λ, $F_{300 \to \lambda}$, thus obtaining the plots in Figure 9. These show that, at the boundary between visible and NIR (commonly set at 700 nm for analyses on solar reflective materials) there is up to 10% of difference, in terms of cumulated energy content, between air mass 1 global horizontal spectra such as AM1GH and EN410G, and direct/beam normal spectra such as E891BN or G173DN. One can also see in Table 1 that the NIR content of E891BN (evaluated as $1 - F_{300 \to 700}$) is as high as 58%, whereas it is slightly lower than 50% for air mass 1 spectra such as AM1GH and EN410G. This means that a selective material highly absorbent in the visible range, for example due to a dark color mandatory for the considered built surface, but at the same time highly reflective in the NIR range, may be rated differently as a result of the selected reference spectrum.

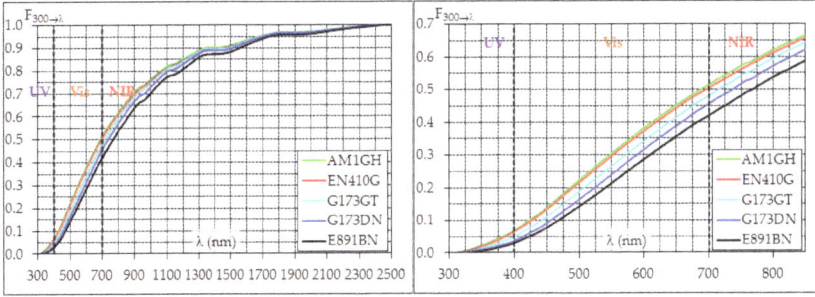

Figure 9. Cumulative energy content vs wavelength (full spectra and close-up on UV-VIS).

An experimental campaign was carried out at EELab on a set of samples representative of different product types and colors, summarized in Figure 10 and Table 2. In agreement with other research [23], selective surfaces such as samples 0218 and 0219, having NIR reflectance clearly higher than visible reflectance (Figure 11), yielded a difference as large as 0.04 between the lowest reflectance values returned by using EN410G (similar to AMIGH) and the highest ones returned by using E891BN. The same occurs, at a lesser extent, for samples 0220 and 0004, which show a highly variable reflectivity spectrum over the range relevant to solar radiation. The lowest difference occurs for a flat spectrum such as that of sample 0011. A lower difference exists, but still in favor of E891BN, between this and G173GT, whereas the situation with G173DN is about halfway. It is also worth noting that measurements made according to ASTM C1549 [25], which are carried out by an instrument with broadband sensors that fundamentally implement Equation (1) and allows selection of a weighting solar spectrum among the most commonly used ones, returns similar but slightly lower and thus precautionary values than truly spectrophotometric measurements (Table 2). In the authors' knowledge, only one instrument compliant with ASTM C1549 is commercially available.

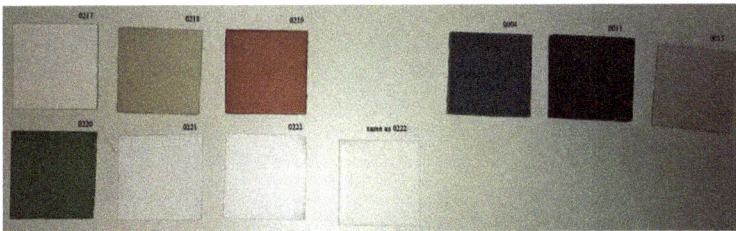

Figure 10. Sample set tested at EELab.

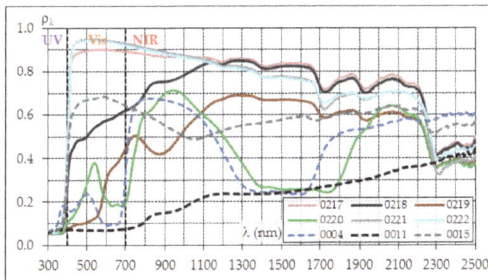

Figure 11. Measured reflectivity spectra.

Table 2. Measured solar reflectance values.

Sample		Reference Solar Spectrum			
Lab. Code	Description	EN410G	G173GT	G173DN	E891BN
0217	Elastomeric water based paint on aluminum	0.816 -	0.826 0.799	0.836 0.810	0.839 0.815
0218	Elastomeric water based paint on aluminum	0.612 -	0.625 0.606	0.637 0.623	0.650 0.635
0219	Elastomeric water based paint on aluminum	0.359 -	0.372 0.367	0.385 0.380	0.401 0.389
0220	Elastomeric water based paint on aluminum	0.357 -	0.367 0.358	0.377 0.367	0.388 0.378
0221	Elastomeric water based paint on aluminum	0.835 -	0.845 0.817	0.853 0.826	0.854 0.828
0222	Elastomeric water based paint on aluminum	0.828 -	0.838 0.815	0.846 0.824	0.848 0.830
0004	Engobed ceramic til	0.347 -	0.356 0.348	0.365 0.354	0.377 0.364
0011	Engobed ceramic til	0.125 -	0.128 0.125	0.132 0.128	0.138 0.133
0015	Engobed ceramic til	0.577 -	0.583 0.564	0.588 0.570	0.589 0.570

Note: data in italic were measured by a solar reflectometer compliant with ASTM C1549 [25].

When dealing with rating of commercial products, a few percentage points of difference generated by the selection of the reference spectrum rather than the true product performance may unfairly influence the choice of designers and end users. This is presumably the reason why the E891BN spectrum is still in use though ASTM E891 [26] was withdrawn in 1999. On the other hand, it seems reasonable to measure solar reflectance under conditions that best predict the annual peak of solar heat gain, that is when the sun is high in a cloudless sky and irradiates horizontal or low-pitched surfaces, so spectra such as AM1GH or EN410G should preferred. Their adoption, however, would artificially lower the measured performance of newly rated products, so a correction would be needed to allow a fair comparison with previously rated ones. This could be accomplished in terms of a clear separation of values rated with the two different types of spectrum in two different sections of the databases of the qualification bodies, to be contemporarily implemented by all relevant bodies in the world to take care of globalization, also showing in the databases and/or the test reports/commercial specifications of newly rated products a clear statement on the maximum expected penalization with respect to the previous rating approach, of course to be quantified. Best wishes to people in charge of that!

An issue even more relevant than that regarding the reference solar spectrum is probably the effect of ageing of solar reflective products due to weathering and soiling. Significant studies [28,29] have shown that many products suffer significant loss of solar reflectance after a few years of ageing, especially those with high initial values of the reflectance. In the U.S. aged samples can be obtained by exposition in three different weathering sites accredited by CRRC [14], representative of three different climates. Aged solar reflectance, thermal emittance and SRI values can eventually be calculated as the mean of those obtained for samples weathered in the three sites. Indeed, the LEED rating system requires in the U.S. [10] not only initial values as previously mentioned, but also three-years aged values such that SRI ≥ 64 for low-sloped roofs, SRI ≥ 32 for steep-sloped roofs, SRI ≥ 28 for pavements and other non-roof surfaces. 2016 Title 24 of California [11] does not consider initial properties at all, but it requires minimum values of aged properties; more specifically, three-years aged values of SRI ≥ 64 or 75 (depending on the building use) are required for low-sloped roofs and SRI ≥ 16 for steep-sloped roofs.

In case three-years aged values of surface properties are not available, the following provisional formula has been proposed to predict the aged solar reflectance $\rho_{sol,aged}$ from the initial value $\rho_{sol,init}$ [29] by the formula

$$\rho_{sol,aged} = \rho_{sol,0} + \beta \cdot (\rho_{sol,init} - \rho_{sol,0}) \tag{17}$$

where $\rho_{sol,0}$ is the solar reflectance of an opaque soil layer, and β is the resistance to soiling of the considered surface. With $\beta = 1$ the surface does not change its reflectance after ageing, whereas with $\beta = 0$ the aged reflectance becomes equal to that of the soil. The higher the initial reflectance, the larger is its expected decay. Values of $\rho_{sol,0} = 0.20$ and $\beta = 0.70$ were initially proposed in the 2008 Title 24 of California, but subsequent analyses on a large set of aged products rated in the CRRC and EPA databases [29] showed that actual values generally over-predict the aged reflectance returned by Equation (17) and thus suggested using product specific values of β, ranging from 0.76 for field-applied coatings to close to unity for factory-applied coatings. 2016 Title 24 [11] specifies $\beta = 0.65$ for a field-applied coating and $\beta = 0.70$ for not a field-applied coating in order to have a precautionary (but arbitrary) over-prediction. The effects of the latter values on SRI decay are summarized in Figure 12, where initial and aged SRI values are also shown for a surface such as the white reference one for SRI calculation, for which SRI = 100; a change ±0.05 is also considered for thermal emittance, more extensively explained in Section 3.3.

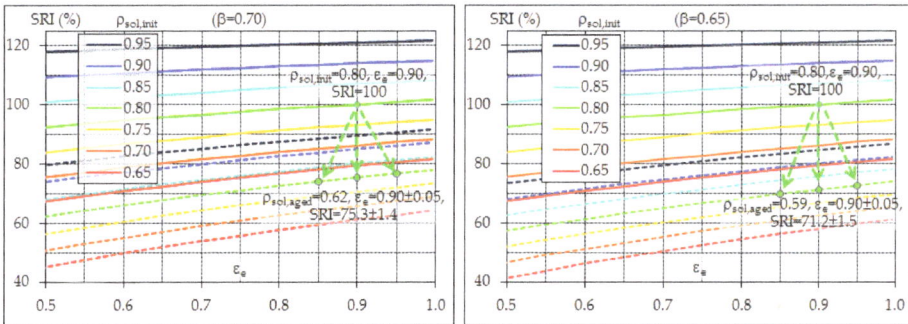

Figure 12. Initial (continuous lines) and aged (dashed lines) SRI values of highly reflective surfaces ($\rho_{sol,init} \geq 0.65$) for intermediate wind speed (2 m/s < v_{wind} < 6 m/s, h_{ce} = 12 W/(m² K)) and (from left to right) resistance to soiling $\beta = 0.70$ and $\beta = 0.65$, with soil solar reflectance $\rho_{sol,0} = 0.20$.

Indeed, the significant scattering of data obtained for products naturally aged in the weathering sites of CRRC [29] suggests that measurements on aged samples are necessary to obtain the aged solar reflectance of each specific product. Nonetheless at least three years of natural ageing are needed, a requisite that is not easily accepted by the industry. Moreover, randomly variable conditions are always possible in the weathering sites due to variations in meteorology and air pollution, and this may affect significantly the results. The solution seems to be provided by a recently developed laboratory method for accelerated ageing [30,31]—already a standard test method as ASTM D7897 [32]—which can condense into three days the three-year long process of natural ageing used by the CRRC to rate roofing products sold in the U.S. Entering into details, a calibrated aqueous soiling mixture of dust minerals, black carbon, humic acid, and salts is sprayed onto 10 cm × 10 cm preconditioned coupons of the tested materials, which are then subjected to cycles of ultraviolet radiation, heat, and water in a commercial weatherometer. The method proved to be easy and fast to perform, repeatable, and above all able to reproduce the reflectance obtained in a wide range of naturally exposed roofing products [30,31].

In many countries different from the U.S. the regulations on solar reflective materials are still under development and only initial values of surface properties are generally considered. Moreover, only producers selling in the U.S. already have three-years aged samples, which are commonly unavailable elsewhere for locally built and/or country specific products. On the other hand, a fast improvement of regulations and commercial products may be impeded by the long time required for natural ageing, moreover heavily polluted areas of Europe are probably not perfectly represented by the weathering sites of the U.S. The accelerated soiling method of ASTM D7897, possibly verified against local climate and pollution levels by means of validation studies, can thus provide a powerful tool to support the improvement of both regulations and product quality.

A short note can be made on ASTM E1918 Standard [24,33], a measurement method that compares the direct and reflected irradiances measured by a couple of pyranometers facing upwards and downwards, respectively, in order to retrieve the solar reflectance value. It is an effective method, especially for field measurements and if these are needed to calibrate mathematical models where the reflectance relevant to the current solar spectrum is concerned. Nonetheless it uses the randomly variable spectrum of the place and time of measurement, therefore it may not be the ideal choice for product rating and certification of the SRI.

3.3. Assessment of Thermal Emittance

It has already been evidenced that high values of thermal emittance allow rejection of solar energy absorbed by irradiated opaque surfaces [34], especially in low wind conditions. Solar reflectance is the key parameter to limit overheating, but a low thermal emittance may also affect re-emission of the absorbed solar energy and, therefore, the SRI. This is the case of uncoated metal surfaces, which can warm up as much as black roofing materials [35–38]. In this regard, regulations such as 2016 Title 24 of California [11] require minimum values of both solar reflectance and thermal emittance—higher than 0.75 for the latter—unless compliance is verified for their combination through the SRI.

Several test methods are available to measure thermal emittance, but most of them can be used only in the laboratory, often on small specimens made of pure materials, therefore they are of low practical usefulness in the construction industry. Only two methods seem available for measurement on actual building elements, usable with relative ease either in the laboratory or in the field. These are specified in the ASTM C1371 Standard [39] and the EN 15976 Standard [40]. To the authors' knowledge, only one instrument compliant with each test method is commercially available.

The emissometer based on ASTM C1371, called 'thermal emissometer' in the following, is probably the most used one, endorsed for performance assessment of solar reflective materials by both CRRC of the U.S. and ECRC (which, however, allows using also the alternative emissometer based on EN 15976). The total hemispherical emittance of the sample surface is evaluated through the following relationship [39]:

$$\Delta V = k \cdot \frac{\sigma_0 \cdot \left(T_d^4 - T_s^4\right)}{1/\varepsilon_s + 1/\varepsilon_d - 1} \tag{18}$$

In the above formula, the voltage signal ΔV (V) is that returned by a differential thermopile sensor embedded in the instrument head, which is placed onto the sample and left there until a steady output is reached. ΔV is proportional by a calibration constant k to the radiative heat flux exchanged between the bottom surface of the head and the sample surface. The former has assigned thermal emittance ε_d and absolute thermodynamic temperature stabilized at an assigned value T_d (K), significantly higher than that of the analyzed surface ($T_d > T_s$) and the ambient. The latter surface has thermal emittance ε_s unknown and absolute thermodynamic temperature stabilized at a value T_s (K) as close as possible to the ambient. The calibration constant k multiplies the heat flux exchanged by thermal radiation between the two surfaces, which are assumed to be flat, parallel, virtually infinite and facing each other, as well as gray and diffusive. The emissometer is calibrated before each test by measuring two reference samples with known thermal emittance, respectively equal to ε_{low} (for which a voltage signal ΔV_{low} is yielded) and ε_{high} (for which ΔV_{high} is yielded). Linearity of the instrument and uncertainty ± 0.01 are ensured in the range $0.03 < \varepsilon_s < 0.93$ by the producer of the emissometer (which provided two reference samples with declared emittances $\varepsilon_{low} = 0.06$ and $\varepsilon_{high} = 0.87$), so that:

$$\varepsilon = \varepsilon_{low} + \left(\varepsilon_{high} - \varepsilon_{low}\right) \cdot \frac{\Delta V - \Delta V_{low}}{\Delta V_{high} - \Delta V_{low}} \tag{19}$$

The instrument indeed measures something between normal and hemispherical emittance, nonetheless it was shown to yield the hemispherical emittance value when the hemispherical emittances of the reference samples are interpolated [41,42]. If the tested sample has a non-negligible resistance to heat transfer, due to a low thermal conductivity of the support material, the heat input applied by the hot emissometer head to the sample surface induces a thermal gradient across the thickness of the sample itself. As a result, the temperature T_s of the sample surface rises to a value significantly higher than that of the ambient. In such cases, the actual value of thermal emittance can be recovered by using one among the modifications of the standard method suggested by the producer of the emissometer, among which the most commonly used one is the so-called 'slide method' [43,44]. Entering into details, the head of the emissometer is allowed to slide above the sample in order to prevent the measured surface from warming up during the test session. The sliding operation is carried out by hand and time is needed to achieve a stabilized output of the instrument, therefore the measurement may be time-consuming; moreover, it may be affected by the operator's expertise. An approach was recently proposed [45] to solve both problems, based on automating the sliding operation by means of a robotized arm, and acquiring the voltage output returned by the emissometer by means of a computerized data acquisition system that allows visualization of its time-evolution pattern and may also interact with the robot. The approach has eventually provided encouraging results, with measurements in very good agreement with manual operation and also excellent repeatability. It is worth mentioning that the slide method is not specified in ASTM C1371 [39], but only in the technical notes of the instrument's producer [43,44].

The emissometer based on EN 15976 Standard [40], also known as 'TIR emissometer', is again based on a hot head that embeds a hemispherical cavity kept at a temperature significantly higher than that of the sample, which must instead be as close as possible to the ambient temperature. The head is placed onto the sample and the infrared radiation emitted by the cavity and reflected by the sample surface is measured by a fast response sensor viewing the surface from the bottom of the cavity, with near-normal orientation. The acquired signal is clearly correlated to the infrared reflectance of the sample surface, which is, for opaque surfaces, the complement to 1 of the emittance at the same temperature. It is acquired immediately after the head is placed onto the sample, so there is no time for the sample surface to significantly warm up and a low thermal conductivity of the support material does not disturb the measurement. A calibration is again performed before each test, automatically managed by the instrument, by measuring two reference samples with known emittances, a polished metal plate and a black anodized finned surface. The producer of the TIR emissometer (which provided two reference samples with emittance $\varepsilon_{low} = 0.011$ and $\varepsilon_{high} = 0.964$) ensures uncertainty ± 0.01 or better in the range $0.02 < \varepsilon_s < 0.98$.

While the term 'emissivity' is generically mentioned in the EN 15976 Standard, without further specification, the TIR emissometer measures the near-normal thermal emittance. For the most common case of non-metallic surfaces, the near-normal thermal emittance is slightly higher than the hemispherical emittance, the one relevant for surface overheating and to be used for SRI calculation. Data from [46] are plotted in Figure 13, showing that the overestimation may be as high as 0.045–0.055 for dielectric samples with high emittance, so the use of near-normal emittance values for cool roofing or cool pavement products may result in an unfair overestimation with respect to hemispherical emittance values measured by the alternative method.

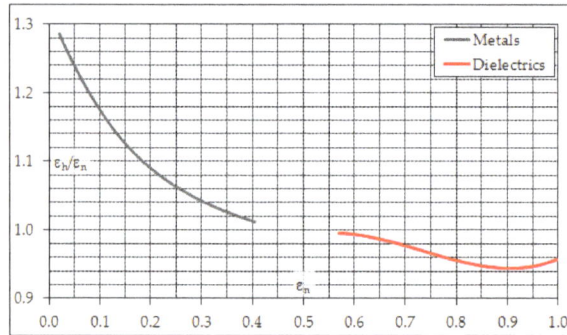

Figure 13. Hemispherical vs. near-normal emittance.

Near-normal emittance ε_n measured according to EN 15976 can be converted in hemispherical emittance ε_h by interpolating the data plotted in Figure 13 through the following formulas [46,47], respectively for dielectrics

$$\frac{\varepsilon_h}{\varepsilon_n} = 0.1569 + 3.7669 \cdot \varepsilon_n - 5.4398 \cdot \varepsilon_n^2 + 2.4733 \cdot \varepsilon_n^3 \tag{20}$$

and for metals

$$\frac{\varepsilon_h}{\varepsilon_n} = 1.3217 - 1.8766 \cdot \varepsilon_n + 4.6586 \cdot \varepsilon_n^2 - 5.8349 \cdot \varepsilon_n^3 + 2.7406 \cdot \varepsilon_n^4 \tag{21}$$

An advantage of the TIR emissometer on the thermal emissometer is that it does not require a time-consuming procedure such as the slide method with samples having a low thermal conductivity, as with all pavement products and most of roofing products apart from coated metals. On the other hand, the minimum size of the sample, imposed by the size of the hot head, is about 100 mm for the TIR emissometer versus some 60 mm for the thermal emissometer, even less when using optional port adapters [48]. Nevertheless, application of the slide method requires a size of the sample surface multiple of that of the emissometer head in order to allow space enough for the head itself to slide above the surface without progressively warming it up. The abovementioned port adapter helps in lowering the sample size, in principle making it possible to test the 10 cm × 10 cm samples required for accelerated ageing by ASTM D7897 [32].

Thermal emittance of high-emittance products is marginally affected by soiling, ±0.05 at most according to data in the CRRC database [49], but the plots in Figures 2 and 12 show that the effect on the SRI value is generally lower than ±1. Moreover, thermal emittance generally increases, up to 0.07, in low-emittance products [49]. Therefore, it may be reasonable, even if not necessarily precautionary, to assume the thermal emittance is constant after ageing, unless values measured on samples subjected to natural or accelerated ageing are available.

3.4. Combined Effects of Measurement Issues

As explained in Section 3.2, using the E891BN spectrum may yield an apparent increase of solar reflectance with respect to AM1GH or EN410G as high as 0.02 for highly reflective surfaces with $\rho_{sol} \approx 0.8$ (see Table 2). The increase may be 0.01 with respect to G173GT, but negligible with respect to G173DN. For more absorbing surfaces with $\rho_{sol} \approx 0.4$ use of the E891BN spectrum may yield an even higher apparent increase with respect to AM1GH or AN410G, up to 0.03–0.04, especially if those surfaces show selective behavior, i.e., the reflectivity spectrum is sharply different between visible and near infrared ranges. In Section 3.3 it is also explained that for high emittance surfaces with $\varepsilon_e \approx 0.9$ using the direct normal thermal emittance instead of the hemispherical one yields an apparent increase in the emittance as high as 0.055. Combining the two increases may yield significant discrepancies in the calculated SRI values, exemplified in Figure 14 for two categories of material: (a) highly reflective materials such as those used for flat roofs; and (b) moderately reflective materials such as those used for pitched roofs and pavements. With highly reflective surfaces, the SRI values calculated with AM1GH/EN410G spectra and hemispherical emittance may easily be lower by 3.5 points than those calculated with E891BN/G171DN spectra and near-normal emittance. The discrepancy may even rise to 7.5 points for moderately reflective surfaces. All this clearly applies to both new and aged samples.

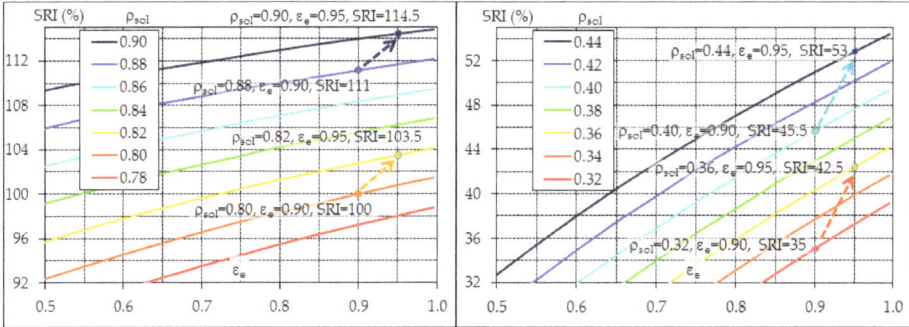

Figure 14. Theoretical discrepancies of calculated SRI values for intermediate wind speed ($2 \text{ m/s} < v_{wind} < 6 \text{ m/s}$, $h_{ce} = 12 \text{ W/(m}^2 \text{ K)}$) and (from left to right) highly reflective and moderately reflective surfaces.

All analyses of Sections 3.2–3.4 do not concern the uncertainty of measurement, which in the authors' experience can be low provided that the test methods are thoroughly implemented and calibration standards are properly chosen and maintained. This requisite should be automatically fulfilled when a laboratory operates under ISO/IEC 17025 accreditation.

4. Conclusive Remarks

- The SRI has raised significant interest in the construction sector thanks to its relative ease of calculation and, above all, its effective representation of the thermal behavior of a built surface subjected to solar radiation.
- A linear correlation exists between SRI and both the surface temperature and the fraction of incident solar heat that is transferred to the near ground air. Since a given SRI value can be the result of different pairs of solar reflectance and thermal emittance values, SRI seems a more effective indicator than solar reflectance alone for comparative evaluation of the capability of built surfaces to limit urban warming.
- SRI calculation is based on the hypothesis of an adiabatic irradiated surface and does not consider the insulation or the inertia of the materials below; nonetheless it can work well as an indicator of the heat transmitted to the external near ground air, and therefore of the contribution to urban warming, since the heat flow rate conducted through a roof or into the ground can be lower by one or two orders of magnitude than solar irradiance.
- Values of the convection heat transfer coefficient specified for SRI calculation are arbitrary; nonetheless they are in good agreement with the literature and are therefore reasonable from the perspective of product comparison.
- The standard solar spectra currently recommended in the U.S. by CRRC and allowed in Europe by ECRC to calculate the weighted average of surface reflectivity, considering air mass 1.5 and direct normal radiation, are probably improper for the prediction of annual peak solar heat gain; thus, spectra for air mass 1 global radiation on a horizontal surface have been recommended in

recent studies. On the other hand, the adoption of such spectra was shown to lower the measured performance by a non-negligible amount and to potentially inhibit a fair comparison between newly rated and already rated products. The issue still needs to be properly addressed.

- Two test methods of practical relevance are available to measure thermal emittance, one (ASTM C1371) returning a hemispherical value and the other (EN 15976) a near-normal value. Unless properly corrected, near-normal emittance may represent a non-negligible overestimation of hemispherical emittance, the one relevant to infrared heat transfer between a built surface and the sky.

- The combined use of air mass 1.5 direct normal radiation spectra and near-normal emittance can lead to a significant overestimation of SRI with respect to using air mass 1 global horizontal radiation spectra and hemispherical emittance, especially for surfaces with high or intermediate solar reflectance. The freedom of choice apparently allowed by standard test methods and rating systems may induce such an unfair situation.

- The surface performance achieved after ageing is relevant to the long-term behavior of built surfaces and therefore this should be considered in regulations rather than that initially measured. Nonetheless at least three years are needed for natural ageing—a requisite that is not easily accepted by the industry—and randomly variable conditions are also possible in the weathering sites due to variations in meteorology and air pollution.

- In many countries different from the U.S. the regulations on solar reflective materials are still under development and only initial values of surface properties are considered, or can be provided by product manufacturers. On the other hand, a fast improvement of commercial products and regulations may be impeded by the long time required for natural ageing.

- A recently developed laboratory test method for accelerated ageing—already a standard test method—makes it possible to condense into three days the three-year long process of natural ageing required by CRRC. It proved to be easy and fast to perform, repeatable, and, above all, able to reproduce the reflectance obtained in a wide range of naturally exposed roofing products.

- In order to create a fair global market for solar reflective products, as well as to favor their use where this is still undeveloped, a general alignment is required on measurement methods. The use of a common solar spectrum is desirable for solar reflectance measurement, possibly representative of peak heat load conditions. The use of hemispherical thermal emittance should also be clearly specified. Under these conditions, the effects of both reflectance and emittance can be effectively summarized by the SRI. In order to consider aged values of SRI—that is, those relevant to the long term performance of built surfaces—worldwide use of an accelerated soiling method such as that specified by ASTM D7897 may greatly speed up the development and qualification of durable products, thus favoring the diffusion of specific performance limits intended to improve building energy efficiency and limit urban warming.

Acknowledgments: The research was partially funded by Fondazione Cassa di Risparmio di Modena, which supported the development of the Energy Efficiency Laboratory (EELab) of the University of Modena and Reggio Emilia. The author also wishes to acknowledge personnel of EELab and members of the Technical Committee of the European Cool Roof Council for the useful discussions that provided hints and support to the development of the present work.

Conflicts of Interest: The author declares no conflict of interest. The founding sponsor had no role in the design of the study; in the collection, analyses, or interpretation of data; in the writing of the manuscript, and in the decision to publish the results.

References

1. United Nations. *World Urbanization Prospects: 2014 Revision*; United Nations: New York, NY, USA, 2014.
2. Santamouris, M. Analyzing the heat island magnitude and characteristics in one hundred Asian and Australian cities and regions. *Sci. Total Environ.* **2015**, *512–513*, 582–598. [CrossRef] [PubMed]

3. NASA—Goddard Institute for Space Studies. *NASA-GISS, 2017. GISS Surface Temperature Analysis (GISTEMP)*; NASA—Goddard Institute for Space Studies: New York, NY, USA, 2017.
4. Santamouris, M. Regulating the damaged thermostat of the cities—Status, impacts and mitigation challenges. *Energy Build.* **2015**, *91*, 43–56. [CrossRef]
5. Santamouris, M.; Cartalis, C.; Synnefa, A.; Kolokotsa, D. On the impact of urban heat island and global warming on the power demand and electricity consumption of buildings—A review. *Energy Build.* **2015**, *98*, 119–124. [CrossRef]
6. Akbari, H.; Pomerantz, M.; Taha, H. Cool surfaces and shade trees to reduce energy use and improve air quality in urban areas. *Sol. Energy* **2001**, *70*, 295–310. [CrossRef]
7. Akbari, H.; Cartalis, C.; Kolokotsa, D.; Muscio, A.; Pisello, A.L.; Rossi, F.; Santamouris, M.; Synnefa, A.; Wong, N.H.; Zinzi, M. Local climate change and urban heat island mitigation techniques—The state of the art. *J. Civ. Eng. Manag.* **2016**, *22*, 1–16. [CrossRef]
8. European Committee for Standardization (CEN). *EN ISO 13786:2007—Thermal Performance of Building Components—Dynamic Thermal Characteristics—Calculation Methods*; European Committee for Standardization (CEN): Brussels, Belgium, 2007.
9. ASTM International. *ASTM E1980-11—Standard Practice for Calculating Solar Reflectance Index of Horizontal and Low Sloped Opaque Surfaces*; ASTM International: West Conshohocken, PA, USA, 2011.
10. U.S. Green Building Council. *LEED v4 for Building Design and Construction*; (updated July 8, 2017); U.S. Green Building Council: Washington, DC, USA.
11. California Energy Commission. *2016 Building Energy Efficiency Standards for Residential and Nonresidential Buildings—Title 24, Part 6, and Associated Administrative Regulations in Part 1*; California Energy Commission: Sacramento, CA, USA, 2015.
12. Environmental Protection Agency. *ENERGY STAR®Program Requirements—Product Specification for Roof Products—Eligibility Criteria—Version 3.0*; Environmental Protection Agency: Washington, DC, USA, 2013.
13. Muscio, A.; Akbari, H. An index for the overall performance of opaque building elements subjected to solar radiation. *Energy Build.* **2017**, *157*, 184–194. [CrossRef]
14. Cool Roof Rating Council. *ANSI/CRRC S100 (2016)—Standard Test Methods for Determining Radiative Properties of Materials, Properties of Materials*; Cool Roof Rating Council: Portland, OR, USA, 2016.
15. European Cool Roof Council. *Product Rating Manual*; (November 2014); European Cool Roof Council: Brussels, Belgium, 2014.
16. European Committee for standardization (CEN). *EN ISO/IEC 17025:2005—General Requirements for the Competence of Testing and Calibration Laboratories*; European Committee for standardization (CEN): Brussels, Belgium, 2005.
17. European Committee for standardization (CEN). *EN ISO 6946:2007—Building Components and Building Elements—Thermal Resistance and Thermal Transmittance—Calculation Method*; European Committee for standardization (CEN): Brussels, Belgium, 2007.
18. American Society of Heating, Refrigerating and Air-Conditioning Engineers. *ASHRAE Handbook Fundamentals 2009*; (SI edition); American Society of Heating, Refrigerating and Air-Conditioning Engineers: Atlanta, GA, USA, 2009.
19. Kakaç, S.; Shah, R.K.; Aung, W. (Eds.) *Handbook of Single-Phase Convective Heat Transfer*; Wiley-Interscience: New York, NY, USA, 1987; ISBN 978-0-47-181702-4.
20. Incropera, F.P.; De Witt, D.P. *Fundamentals of Heat and Mass Transfer*, 5th ed.; John Wiley and Sons: Hoboken, NJ, USA, 2005; ISBN 978-0-47-138650-6.
21. ASTM International. *ASTM E903-12—Standard Test Method for Solar Absorptance, Reflectance, and Transmittance of Materials Using Integrating Spheres*; ASTM International: West Conshohocken, PA, USA, 2012.
22. European Committee for Standardization (CEN). *EN 410:2011—Glass in Building—Determination of Luminous and Solar Characteristics of Glazing*; European Committee for Standardization (CEN): Brussels, Belgium, 2011.
23. Levinson, R.; Akbari, H.; Berdahl, P. Measuring solar reflectance—Part I: Defining a metric that accurately predicts solar heat gain. *Sol. Energy* **2010**, *84*, 1717–1744. [CrossRef]
24. Levinson, R.; Akbari, H.; Berdahl, P. Measuring solar reflectance—Part II: Review of practical methods. *Sol. Energy* **2010**, *84*, 1745–1759. [CrossRef]

25. ASTM International. *ASTM C1549-09 (Reapproved 2014)—Standard Test Method for Determination of Solar Reflectance Near Ambient Temperature Using a Portable Solar Reflectometer*; ASTM International: West Conshohocken, PA, USA, 2014.

26. ASTM International. *ASTM E891-87(1992)—Tables for Terrestrial Direct Normal Solar Spectral Irradiance Tables for Air Mass 1.5 (Withdrawn 1999)*; ASTM International: West Conshohocken, PA, USA, 1992.

27. ASTM International. *ASTM G173-03(2012)—Standard Tables for Reference Solar Spectral Irradiances: Direct Normal and Hemispherical on 37° Tilted Surface*; ASTM International: West Conshohocken, PA, USA, 2012.

28. Berdahl, P.; Akbari, H.; Levinson, R.; Miller, W.A. Weathering of roofing materials—An overview. *Constr. Build. Mater.* **2008**, *22*, 423–433. [CrossRef]

29. Sleiman, M.; Ban-Weiss, G.; Gilbert, H.E.; Francois, D.; Berdahl, P.; Kirchstetter, T.W.; Destaillats, H.; Levinson, R. Soiling of building envelope surfaces and its effect on solar reflectance—Part I: Analysis of roofing product databases. *Sol. Energy Mater. Sol. Cells* **2011**, *95*, 3385–3399. [CrossRef]

30. Sleiman, M.; Kirchstetter, T.W.; Berdahl, P.; Gilbert, H.E.; Quelen, S.; Marlot, L.; Preble, C.V.; Chen, S.; Montalbano, A.; Rosseler, O.; et al. Soiling of building envelope surfaces and its effect on solar reflectance—Part II: Development of an accelerated aging method for roofing materials. *Sol. Energy Mater. Sol. Cells* **2014**, *122*, 271–281. [CrossRef]

31. Sleiman, M.; Chen, S.; Gilbert, H.E.; Kirchstetter, T.W.; Berdahl, P.; Bibian, E.; Bruckman, L.S.; Cremona, D.; French, L.H.; Gordon, D.A.; et al. Soiling of building envelope surfaces and its effect on solar reflectance—Part III: Interlaboratory study of an accelerated aging method for roofing materials. *Sol. Energy Mater. Sol. Cells* **2015**, *143*, 581–590. [CrossRef]

32. ASTM International. *ASTM D7897-15—Standard Practice for Laboratory Soiling and Weathering of Roofing Materials to Simulate Effects of Natural Exposure on Solar Reflectance and Thermal Emittance*; ASTM International: West Conshohocken, PA, USA, 2015.

33. ASTM International. *ASTM E1918-16—Standard Test Method for Measuring Solar Reflectance of Horizontal and Low-Sloped Surfaces in the Field*; ASTM International: West Conshohocken, PA, USA, 2016.

34. Levinson, R.; Berdahl, P.; Akbari, H.; Miller, W.A.; Joedicke, I.; Reilly, J.; Suzuki, Y.; Vondran, M. Methods of creating solar-reflective nonwhite surfaces and their application to residential roofing materials. *Sol. Energy Mater. Sol. Cells* **2007**, *91*, 304–314. [CrossRef]

35. Rosenfeld, A.H.; Akbari, H.; Bretz, S.; Fishman, B.L.; Kurn, D.M.; Sailor, D.; Taha, H. Mitigation of urban heat islands: Materials, utility programs, updates. *Energy Build.* **1995**, *22*, 255–265. [CrossRef]

36. Bretz, S.; Akbari, H.; Rosenfeld, A.H. Practical issues for using solar-reflective materials to mitigate urban heat islands. *Atmos. Environ.* **1998**, *32*, 95–101. [CrossRef]

37. Libbra, A.; Muscio, A.; Siligardi, C.; Tartarini, P. Assessment and improvement of the performance of antisolar surfaces and coatings. *Prog. Org. Coat.* **2011**, *72*, 73–80. [CrossRef]

38. Libbra, A.; Muscio, A.; Siligardi, C. Energy performance of opaque building elements in summer: Analysis of a simplified calculation method in force in Italy. *Energy Build.* **2013**, *64*, 384–394. [CrossRef]

39. ASTM International. *ASTM C1371-15—Standard Test. Method for Determination of Emittance of Materials Near Room Temperature Using Portable Emissometers*; ASTM International: West Conshohocken, PA, USA, 2015.

40. European Committee for standardization (CEN). *EN 15976—Flexible sheets for waterproofing—Determination of emissivity*; European Committee for standardization (CEN): Brussels, Belgium, 2011.

41. Kollie, T.G.; Weaver, F.J.; McElroy, D.L. Evaluation of a commercial, portable, ambient temperature emissometer. *Rev. Sci. Instrum.* **1990**, *61*, 1509–1517. [CrossRef]

42. Devices & Services Co. *D&S Technical Note 92-11—Emissometer Model. AE—Hemispherical vs Normal Emittance*; Devices & Services Co.: Dallas, TX, USA, 1992.

43. Devices & Services Co. *D&S Technical Note 04-1—Emissometer Model AE—Slide Method for AE Measurements*; Devices & Services Co.: Dallas, TX, USA, 2004.

44. Devices & Services Co. *D&S Technical Note 10-2—Emissometer Model AE1—Slide Method for High. Emittance Materials with Low Thermal Conductivity*; Devices & Services Co.: Dallas, TX, USA, 2010.

45. Pini, F.; Ferrari, C.; Libbra, A.; Leali, F.; Muscio, A. Robotic implementation of the slide method for measurement of thermal emissivity of building elements. *Energy Build.* **2016**, *114*, 241–246. [CrossRef]

46. Rubin, M.; Arasteh, D.; Hartmann, J. A correlation between normal and hemispherical emissivity coatings on glass. *Int. Commun. Heat Mass Transf.* **1997**, *14*, 561–565. [CrossRef]

47. National Fenestration Rating Council Inc. *NFRC 301-2014[E0A0]—Standard Test Method for Emittance of Specular Surfaces Using Spectrometric Measurements*; National Fenestration Rating Council Inc.: Greenbelt, MD, USA.

48. Devices & Services Co. *D&S Technical Note 11-2—Model. AE1 Emittance Measurements Using a Port. Adapter, Model. AE-ADP*; Devices & Services Co.: Dallas, TX, USA, 2011.

49. Paolini, R.; Zinzi, M.; Poli, T.; Carnielo, E.; Mainini, A.G. Effect of ageing on solar spectral reflectance of roofing membranes: Natural exposure in Roma and Milano and the impact on the energy needs of commercial buildings. *Energy Build.* **2014**, *84*, 333–343. [CrossRef]

climate

MDPI

Article

Recognition of Thermal Hot and Cold Spots in Urban Areas in Support of Mitigation Plans to Counteract Overheating: Application for Athens

Thaleia Mavrakou [1],*, Anastasios Polydoros [1], Constantinos Cartalis [1] and Mat Santamouris [2]

[1] Department of Physics, National and Kapodistrian University of Athens, 15772 Athens, Greece;
apoly@phys.uoa.gr (A.P.); ckartali@phys.uoa.gr (C.C.)

[2] Faculty of Built Environment, University of New South Wales, Sydney 2052, Australia;
m.santamouris@unsw.edu.au

* Correspondence: thmavrakou@phys.uoa.gr; Tel.: +30-210-727-6843

Received: 30 January 2018; Accepted: 5 March 2018; Published: 9 March 2018

Abstract: Mitigation plans to counteract overheating in urban areas need to be based on a thorough knowledge of the state of the thermal environment, most importantly on the presence of areas which consistently demonstrate higher or lower urban land surface temperatures (hereinafter referred to as "hot spots" or "cold spots", respectively). The main objective of this research study is to develop a methodological approach for the recognition of thermal "hot spots" and "cold spots" in urban areas during summer; this is accomplished with (a) the combined use of high and medium spatial resolution satellite data (Landsat 8 and Terra-MODIS, respectively); (b) the downscaling of the Terra-MODIS satellite data so as to acquire spatial resolution similar to the Landsat one and at the same time take advantage of the high revisit time as compared to the respective one of Landsat (16 days); and (c) the application of a statistical clustering technique to recognize "hot spots" and "cold spots". The methodological approach was applied as a case study for the urban area of Athens, Greece for a summer period. Results demonstrated the capacity of the methodological approach to recognize "hot spots" and "cold spots", revealed a strong relationship between land use and "hot spots" and "cold spots", and showed that the average land surface temperature (LST) difference between the "hot spots" and "cold spots" can reach 9.1 °K.

Keywords: land surface temperature; "hot spots"; "cold spots"; MODIS downscaling

1. Introduction

The thermal environment in several urban areas worldwide has experienced a rapid deterioration, leading to the overheating of cities. The increase of temperature may be attributed to the influence of the radiative properties of urban surfaces and the three-dimensional configuration and heat capacity of buildings [1–6]. In response to the increased urban (air and land surface) temperatures and the strong relationship between high urban air temperatures and adverse heat-related health outcomes (e.g., thermal discomfort, mortality, and heat-related illness) [7–12], urban planners and decision makers are recognizing the growing need to create more attractive, thermally comfortable, and sustainable cities, especially as urban populations expand and climatic variability and extremes increase [13–15]. Results of specific building adaptation studies for Athens, Greece showed that this is possible when appropriate and energy-efficient building technologies are used [16].

In support of mitigation plans, satellite imagery in the thermal infrared (TIR) can be used as a tool to identify thermal "hot spots" and "cold spots" and correlate the state of the thermal environment to land use composition [17–20]. "Hot spots" and "cold spots" are defined as areas which consistently demonstrate higher or lower land surface temperature (LST), respectively, as compared to their surrounding areas. Taken that areas of high air temperature coincide with areas of LST when coarse spatial resolution data is used (with a resolution >30 m) [13], the recognition of "hot spots" and "cold spots" can also support the extraction of information on the spatial distribution of air temperatures. As a matter of fact, several studies have successfully correlated air and land surface temperatures, while several researchers have successfully estimated air temperature from remote sensing data [21–26]. Coutts et al. [13] used very high resolution airborne thermal infrared data in conjunction with Landsat and MODIS data in order to assess the adequacy of satellite TIR data to "hot spot" recognition. They found that Landsat data are suitable for "hot spot" recognition and supportive for mitigation plans to counteract overheating due to the adequate spatial coverage, the free data accessibility, the routine capture, and the well-documented processes and corrections. A disadvantage for the use of Landsat TIR data is the low temporal resolution, as the period between two successive passes of the satellite is 16 days [27,28]. This may result to an unavailability of cloud free images for a long period if only Landsat data are used. To overcome this problem, TIR data from the MODIS instrument on board the Terra satellite are used; such data have a daily temporal resolution, but fall back in spatial resolution (1 km) [29–32]. To this end, disaggregation methods (i.e., the improvement of the coarser spatial resolution of TIR images by introducing spatial detail from a finer resolution image) for downscaling MODIS TIR need to be applied, so as to match the respective spatial resolution of the Landsat data.

Several disaggregation methods have been suggested in order to enhance the spatial resolution of the TIR data by linking TIR and visible and near infrared (VNIR) data [33–35]. These methods utilize the relation between the land cover and the LST; at first the relation between the coarser resolution TIR and VNIR bands is estimated and is thereafter applied at the finer resolution VNIR data in order to obtain finer resolution TIR data [33]. Different disaggregation methods in support of the MODIS LST product have been proposed in literature [36,37], while Bisquert et al. [33] applied and compared many of these methods to MODIS and Landsat sensors. Results showed that the linear regression between the normalized difference vegetation index (NDVI) and LST lead to better results, with RMSE values around ±2 K, compared to other regressions or to more sophisticated approaches (Neural Networks and Data Mining).

The main objective of the current research is to develop a methodological approach for the recognition of "hot spots" and "cold spots" in urban areas during summer; this is accomplished with the combined use of high and medium spatial resolution satellite data in the thermal infrared, from the Landsat 8 and Terra satellites, respectively. A secondary objective is to investigate the relationship of "hot spots" and "cold spots" with land use/land cover. Both objectives support the drafting of mitigation plans for counteracting the overheating of urban areas. The methodological approach for "hot spots" and "cold spots" recognition is presented in Section 2. In Section 3, the methodological approach is applied for the urban area of Athens, and results are discussed accordingly.

2. Methodological Approach

The adopted methodology for the identification of "hot spot" and "cold spot" areas is based on the use of satellite data. These areas are characterized by very high ("hot spots") and very low ("cold spots") LST values in multi-temporal satellite images, assuming stable meteorological conditions and very limited, if any, land cover changes. Consequently, the satellite-based approach needs to be replicated for more than one day in order to assess the spatial consistency of the hot and cold spot areas. To this end, the analysis cannot be based merely on Landsat, taken that it overpasses every 16 days. MODIS provides LST data on a daily basis, whereas the application of downscaling allows for LST products at the same spatial resolution as Landsat. The multi-temporal analysis is even more necessary for the implementation of the methodology in other regions that may be characterized by unstable weather conditions, in order to investigate any large spatial variations of the hot/cold spot areas. In the following sections the required steps are described in detail.

2.1. Land Surface Temperature

The estimation of *LST* was based on the algorithm developed by Jiménez-Muñoz and Sobrino [38,39]. *LST* is retrieved from only one thermal channel, i.e., TIRS-10 (Equation (1)).

$$LST = \gamma \left[\varepsilon^{-1} (\psi_1 L_{sen} + \psi_2) + \psi_3 \right] + \delta \tag{1}$$

where ε is the surface emissivity, L_{sen} is the radiance of the thermal channel TIRS-10, and γ, δ are two parameters given by

$$\gamma = \frac{T_{sen}^2}{b_\gamma L_{sen}}; \delta = T_{sen} - \frac{T_{sen}^2}{b_\gamma}$$

where T_{sen} is the brightness temperature of the thermal channel TIRS 10, $b_\gamma = 1324$ for TIRS-10, and ψ_1, ψ_2, and ψ_3 are atmospheric functions given by

$$\psi_1 = \frac{1}{\tau}; \psi_2 = -L_d - \frac{L_u}{\tau}; \psi_3 = L_d$$

where τ is the atmospheric transmission, L_d is the downwelling radiance, and L_u is the upwelling radiance. These parameters are calculated from the Atmospheric Correction Parameter Calculator provided by NASA (http://atmcorr.gsfc.nasa.gov) [40]. Land surface emissivity (ε) was estimated using the NDVI Thresholds Method as proposed by Sobrino et al. [41].

2.2. Downscaling Modis Data

In this study, the adopted disaggregation method for downscaling MODIS TIR data is based on the linear regression approach that leads to better results, as stated in [33]. Linear regression between Terra MODIS and Landsat 8 TIRS is applied in the event that no Landsat overpass coincides with a MODIS one, with low viewing angle, thus a close Landsat NDVI image is used under the assumption that NDVI remains constant.

Initially, a normalization process needs to be applied to Landsat data in order to minimize the discrepancies that result from the differences in spectral resolution, atmospheric correction, viewing angle, pixel footprint, time acquisition, etc., between the two sensors [33]. Subsequently, according to the disaggregation procedure followed in this study and presented schematically in Figure 1, the relationship between the coarser MODIS NDVI and LST at 960 m is obtained and subsequently applied to the MODIS NDVI at 960 m, resulting in the LST' 960 m. The residuals as calculated by subtracting the LST' 960 m from the LST 960 m, were added to the finer LST' 30 m, which resulted from the application of the initial relationship to the finer Landsat NDVI 30 m, giving eventually the downscaled MODIS LST 30 m.

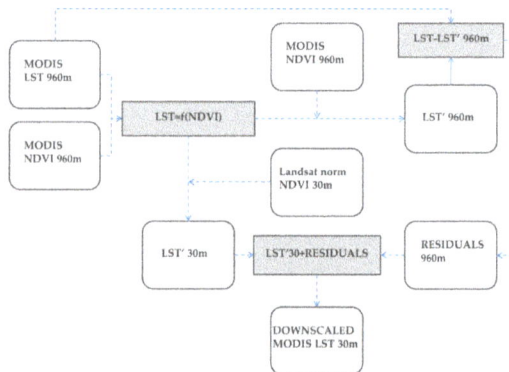

Figure 1. Disaggregation procedure for downscaling MODIS Terra land surface temperature (LST) product.

2.3. "Hot Spot" Recognition

A statistical approach is applied so as to detect—and cluster—the presence of "hot spot" or "cold spot" areas over a study area by calculating G statistics [42] for each pixel in the image, as shown in Equations (2)–(4). The resulting z-scores show where the pixels with either high or low LST values cluster spatially, while they are also measures of statistical significance (Table 1). It is important to note that a pixel with a high LST value may not be a statistically significant "hot spot" if its neighbor pixels have much lower LST values. To be a statistically significant "hot spot", a pixel must have a high value and be surrounded by other pixels with high values as well. The methodological approach suggests the use of all eight neighbor pixels in addition to the central one, in order to calculate the local sum and subsequently obtain the local average (i.e., local sum/9) of the main pixel (Figure 2). This procedure was selected in order to ensure that LST "hot spots" are related to air temperature "hot spots" given that a larger spatial domain is expected to relate better with the near-surface air temperature source area. This pixel value is then compared to the respective average value of the image (sum of all pixels in the study area/number of pixels in the study area). When the local average is very different from the image average, and the difference is too large to be the result of random chance, a statistically significant z-score is found. For statistically significant positive z-scores, the larger the z-score, the more intense is the clustering of high values ("hot spot"). For statistically significant negative z-scores, the smaller the z-score, the more intense is the clustering of low values ("cold spot").

$$G_i^* = \frac{\sum_{j=1}^n w_{i,j} x_j - \overline{X} \sum_{j=1}^n w_{i,j}}{S \sqrt{\frac{\left[n \sum_{j=1}^n w_{i,j}^2 - (\sum_{j=1}^n w_{i,j})^2 \right]}{n-1}}} \tag{2}$$

$$\overline{X} = \frac{\sum_{j=1}^n x_j}{n} \tag{3}$$

$$S = \sqrt{\frac{\sum_{j=1}^n x_j^2}{n} - (\overline{X})^2} \tag{4}$$

where G_i^* is the resultant G statistics (z-scores and p-values) for pixel i, x_j is the LST value for pixel j, $w_{i,j}$ is the spatial weight between pixel i and neighboring pixel j, n is equal to the total number of pixels, $\overline{X}$ is the mean LST of all pixels, and S the variance.

Table 1. Critical p-values and z-scores for different confidence levels.

Significance Level (p Value)	Critical Value (z Score)	Confidence Level
−0.01	z < −3.3	99.9%
−0.1	−3.30 < z < −2.58	99%
0	−2.58 < z < 2.58	-
0.1	2.58 < z < 3.30	99%
0.01	z > 3.3	99.9%

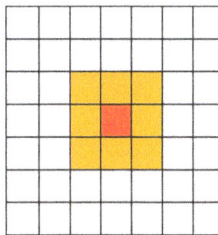

Figure 2. The calculation of the local LST sum for a pixel under consideration (red) includes all of its neighbors (orange).

3. Application of the Methodological Approach for the Urban Area of Athens, Greece

3.1. Application Area

In this section, the methodological approach is applied for the area denoted in Figure 3 with the red polygon representing the Municipality of Athens. The area has coverage of about 39 km^2 and includes high-density urban residential areas, commercial and industrial areas, transport modes, and the associated road network. Urban space is characterized in particular by a high degree of mixed land-use, limited green spaces or open public spaces, and limited pathways that allow for the influx of airflow from the surrounding countryside [43,44].

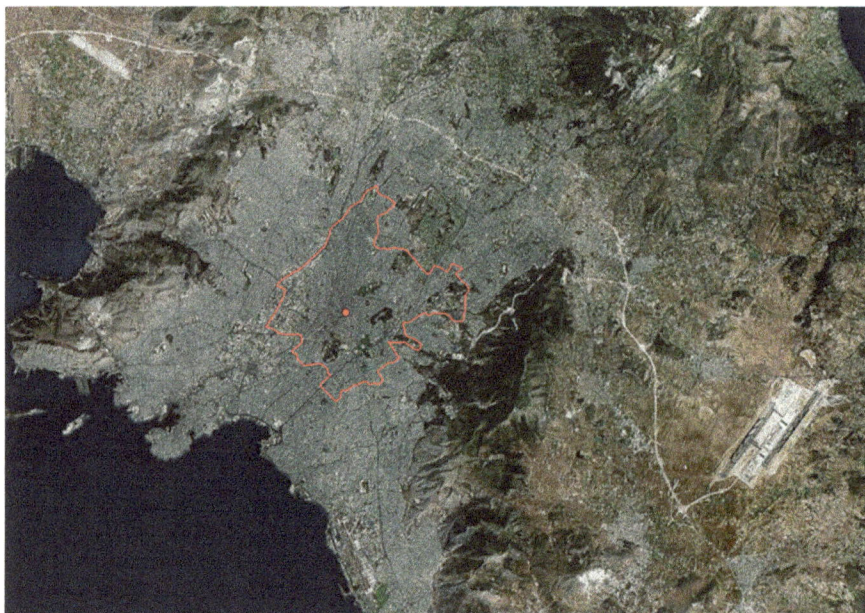

Figure 3. The urban agglomeration of Athens and the administrative boundaries of the municipality of Athens (red polygon).

3.2. Data

Landsat 8 images (morning pass) were used in order to calculate the LST of the urban agglomeration of Athens. The acquisition dates were 27 July 2016 and 12 August 2016, and the selection of these dates were in compliance with the need for similar meteorological conditions. As a matter of fact, during this period no precipitation occurred over the Athens basin, while the wind and the temperature regime were similar for all days under study. In particular, clear sky conditions were observed from 26 July 2016 until 12 August 2016, the mean temperature was 30.5 °C, the maximum air temperature was 36.9 °C, the minimum air temperature was 23.4 °C, and the mean wind speed was 5.5 km/h southwards—typical Mediterranean summer conditions.

MODIS LST data from Terra satellite (MOD11A1 product [45]) were selected so as to refer to the acquisition time of the Landsat 8 images. They were used for four days with low MODIS viewing angle (<±30°), as high angles lead to higher errors, between the two Landsat 8 overpasses (27 July and 12 August). These days are the 26th and 28th of July as well as the 2nd and 11th of August. The selected time period along with the acquisition time of the satellite data and the satellite viewing angle are presented in Table 2. In addition, the MODIS Terra VNIR data (MOD13Q1) were used for the NDVI calculation, while the Urban Atlas dataset [46] was employed for the land use analysis.

Table 2. Selected dates and characteristics of the LST data.

Day	Satellite	Sensor	Acquisition Time (UTC)	View Angle
26/7/2016	Terra	MODIS	09:37	15°
27/7/2016	Landsat 8	TIRS	09:05	0°
28/7/2016	Terra	MODIS	09:25	−9°
2/8/2016	Terra	MODIS	09:43	25°
11/8/2016	Terra	MODIS	09:42	14°
12/8/2016	Landsat 8	TIRS	09:05	0°

4. Results and Discussion

4.1. Land Surface Temperature

The LST evolution during the period under study is presented in Figure 4. The surface temperature patterns are similar in all images, whilst the mean LST values range from 312.505 °K to 314.076 °K, indicating rather stable conditions from the 26th of July to the 12th of August. The latter is in accordance with the stable meteorological conditions characterising the period under study. Furthermore, all images clearly depict that the western part of the Municipality of Athens exhibits higher LST than the eastern part, mainly due to the presence of industrial buildings and the high-density urban areas. On the contrary, the southeastern part shows lower LST as it is less densely built and consists of many urban green areas. Interestingly, small areas inside the boundaries of the municipality of Athens exhibit very high LST values that exceed 319 °K in all cases. This temperature value is much higher than the average LST of the municipality, indicating the presence of "hot spots". Intra-urban variations of LST are clearly depicted in the Landsat images (Figure 4b,f) and in the downscaled images too (Figure 4a,c–e).

It should be mentioned that the Landsat LST error is typically less than 1.5 K [39] and the accuracy of the MODIS LST product is better than 1 K [47]. Furthermore, a discrepancy of ±1.9 K is observed between MODIS downscaled LST and Landsat LST [33], when Landsat overpass coincided with a MODIS one, the latter with low viewing angle. The summation of the above leads to the worst case scenario error of the downscaled MODIS LST. The above errors do not affect the recognition of the "hot spots" and "cold spots" patterns in the urban area under investigation; this is due to the fact that the recognition is based on differences between local and image averages, rather than to absolute LST values.

4.2. Recognition of "Hot Spots"

The calculation of the G statistics for each LST image resulted in the assignment of a z-value to each pixel. In order to define the "hot spot" areas, the suitable threshold for the z-values (Table 1) was determined as follows. Usually a z-value over 2.58 (99% confidence level) indicates a statistically significant result, but a more rigorous threshold of a z-value over 3.3 (99.9% confidence level) can be used in order to narrow down the number and the extent of the "hot spots". Comparing the results using both thresholds, it was found out that the less rigorous threshold of 2.58 increases the extent of the "hot spot" areas by 29% and the number of "hot spots" by 27%. Additionally, this threshold increases the extent and the number of "cold spots" by 68% and 109% respectively. Taken the above, the threshold 3.3 of z-value was chosen as the most suitable for this study.

Subsequently, "hot spots" and "cold spots" were determined when a pixel was characterized as a "hot spot" or "cold spot", respectively, in at least four images (i.e., over 50% of the available data). Based on the above, three categories of "hot spots" and "cold spots" emerged according to the number of times they were recognized. Table 3 provides the statistical analysis for each "hot spot" and "cold spot" category. A gradual change in the minimum, maximum, and mean values is evident.

The average LST of a 6-day "hot spot" is 4.43 °K higher than the average LST of the areas that are neither "hot spots" nor "cold spots". Even the 4-day "hot spots" have a substantial difference from these areas, reaching 2.16 °K. In addition, the average LST difference between the 6-day "hot spots" and the 6-day "cold spots" reaches 9.1 °K.

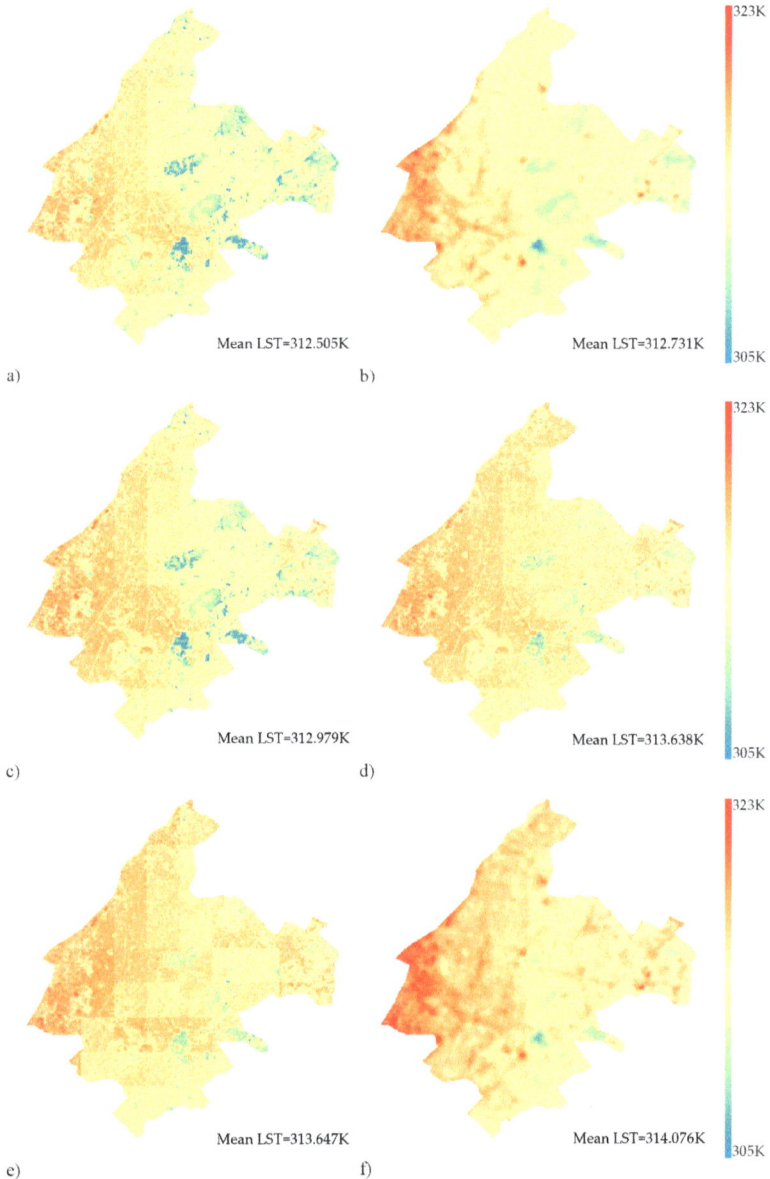

Figure 4. Land surface temperatures of the municipality of Athens for (**a**) 26/7/2016 (downscaled)—no available Landsat image, (**b**) 27/7/2016 (Landsat), (**c**) 28/7/2016 (downscaled)—no available Landsat image (**d**) 2/8/2016 (downscaled)—no available Landsat image (**e**) 11/8/2016 (downscaled)—no available Landsat image and (**f**) 12/8/2016 (Landsat).

Table 3. LST statistics for the "hot spots" and "cold spots" categories (in °K).

	"Hot Spots"			"Cold Spots"		
	Min.	Max.	Mean	Min.	Max.	Mean
6-day	313.94	319.24	317.75	305.57	313.04	308.65
5-day	313.86	318.69	316.19	306.69	312.24	309.78
4-day	313.11	318.37	315.48	306.19	313.56	310.38

Figure 5 depicts the "hot spots" and "cold spots" of the Municipality of Athens. The majority of the "hot spots" are located in the western part of the city, which has greater LST values. The old industrial area (denoted as A in Figure 5) in the western boundaries of the municipality appears as a large "hot spot" due to the homogeneous land cover of impervious materials and the lack of vegetation. Some of the "hot spots" are located in the historic center of Athens (denoted as B in Figure 5), where major sightseeing places are located. On the contrary, the majority of "cold spots" are located at the eastern part of the municipality in the vicinity of urban green areas.

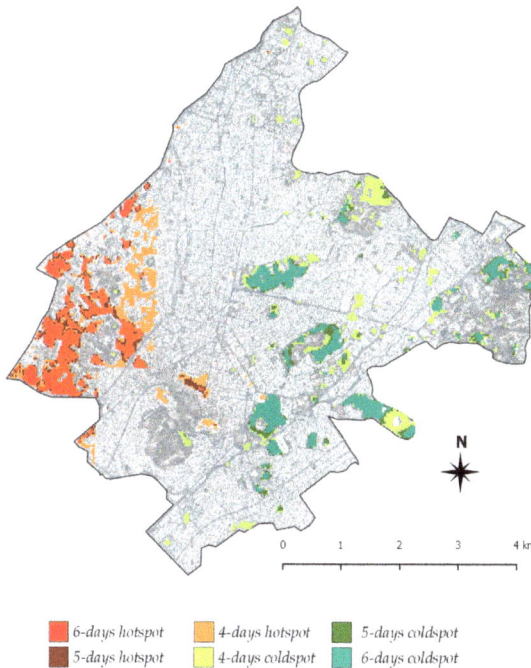

Figure 5. Location of the hot/cold spots within Athens Municipality. (A: old industrial area of Athens, B: the historic center of Athens).

The extracted map of "hot spots" and "cold spots" in the Municipality of Athens (Figure 5) can be easily processed in Geographic Information Systems (GIS) environment by urban planners and decision makers as it contains geospatial information. Using GIS software, it was calculated that the "hot spots" occupy an area of 2.75 km² (7% of the total municipality area) and the "cold spots" occupy an area of 2.78 km². Moreover, various data, such as census, energy consumption, greenery, age of buildings, etc., can be overlaid, thus enabling the design of mitigation plans to counteract overheating.

The above results indicate that further investigation of the "hot spot" and "cold spot" areas is needed in order to examine the relation of land use/land cover with the presence of "hot spots" or "cold spots". To this end, land use data from the Urban Atlas were used; Table 4 provides the land use information for the municipality of Athens. High-density urban fabric covers more than 45% of the area, while proportions of land are covered by roads (26.73%), industrial and commercial structures (19.83%), and urban parks (15.53%).

Table 4. Land use percentages for the municipality of Athens based on the Urban Atlas data.

Land Use	Percentage
Continuous urban fabric (S.L.: >80%)	45.61
Discontinuous dense urban fabric (S.L.: 50–80%)	6.22
Discontinuous medium-density urban fabric (S.L.: 30–50%)	0.69
Discontinuous low-density urban fabric (S.L.: 10–30%)	0.15
Industrial, commercial, public, military, and private units	19.83
Other roads and associated land	26.73
Railways and associated land	0.56
Construction sites	0.04
Land without current use	0.17
Green urban areas	15.53
Sports and leisure facilities	1.44

The above results in conjunction with the LST differences as derived between the 6-day and the 4-day "hot spots" (2.27 °K) highlight the need to prioritize the interventions to the 6-day "hot spot" areas. Furthermore, land use analysis (Figure 6) lead to the recognition of various land use patterns in each "hot spot" or "cold spot" category, indicating the need for customized interventions.

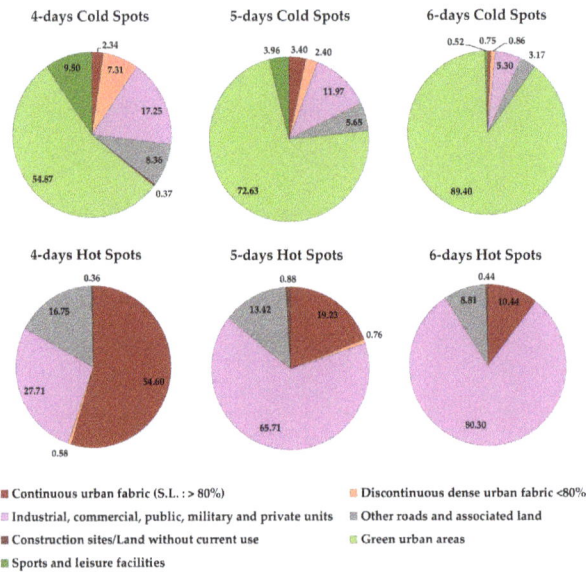

Figure 6. Land use percentages of the "hot spot" and "cold spot" categories.

5. Conclusions

In this study, a methodological approach was developed in order to support mitigation plans for counteracting overheating in urban areas. The methodological approach recognizes—with the use of satellite remote sensing data in the thermal infrared—areas that consistently demonstrate high or low land surface temperatures during summer time. A critical criterion in the methodological approach is that a pixel (an area) with a high LST value may not be a "hot spot" if its neighbor pixels (areas) have much lower LST values. To be considered as a "hot spot", a pixel (area) must have high LST value and must be surrounded by pixels (areas) with high LST values as well. This criterion overcomes trivial approaches to represent the thermal environment of urban areas, based merely on the LST value on a pixel by pixel basis.

The methodological approach is based on: (a) the combined use of various satellite data, in order to circumvent—through a downscaling technique—the inherent difficulty which characterizes such data, namely the inverse relationship between spatial and temporal resolutions; (b) the estimation of LST on a daily basis and at a spatial resolution of 30 meters; (c) the statistical analysis of the land surface temperatures so as to recognize and cluster "hot spots" and "cold spots"; and (d) the correlation of the "hot spots" and "cold spots" temporal endurance and spatial extent to land use classes.

The methodological approach was applied for the Municipality of Athens on the basis of six—either original or downscaled—high-resolution LST images during a 16-day summer period; "hot spots" were recognized when a pixel was characterized as a "hot spot" in at least in four images (i.e., over the 50% of the available data) using G statistics.

The average LST of a 6-day "hot spot" is 4.43 °K higher than the average LST of the areas that are neither "hot spots" nor "cold spots". Even the 4-day "hot spots" have a substantial difference from these areas, reaching 2.16 °K. In addition, the average LST of the 6-day "hot spot" is 2.27 °C higher than the 4-day "hot spots", whereas the average LST difference between the 6-day "hot spots" and the 6-day "cold spots" reaches 9.1 °K.

Furthermore, land use analysis revealed the strong relationship between land use on the one hand and "hot spots" and "cold spots" on the other. "Hot spots" are almost exclusively covered by industrial/commercial, high-density urban fabric, and street type surface materials. Conversely, "cold spots" have, as expected, an increased presence of urban greenery.

The methodological approach, as developed in this study, is replicable to all urban areas and can be highly supportive to mitigation plans to counteract overheating as it can detect "hot spot" or "cold spot" areas in fine resolution (30 × 30 m). Moreover, various data, such as census, energy consumption, greenery, age of buildings, etc., can be overlaid, thus enabling a thorough design of mitigation plans to counteract overheating. Further work to be done refers to the application of the methodological approach for night time satellite images from MODIS as Landsat night time images are very limited. The successful identification of nighttime "hot spots" will lead to an improved understanding of the thermal environment of the city.

Author Contributions: All authors designed the methodology. Thaleia Mavrakou and Anastasios Polydoros implemented the methodology and performed the necessary statistical analysis; all authors contributed to the analysis of the data and to the drafting of the paper.

Conflicts of Interest: The authors declare no conflicts of interest

References

1. Oke, T.R. City size and the urban heat island. *Atmos. Environ.* **1973**, *7*, 769–779. [CrossRef]
2. Oke, T.R. The energetic basis of the urban heat island. *Q. J. R. Meteorol. Soc.* **1982**, *108*, 1–24. [CrossRef]
3. Jenerette, G.D.; Harlan, S.L.; Stefanov, W.L.; Martin, C.A. Ecosystem services and urban heat riskscape moderation: Water, green spaces, and social inequality in Phoenix, USA. *Ecol. Appl.* **2011**, *21*, 2637–2651. [CrossRef] [PubMed]
4. Tong, S.; Wong, N.H.; Jusuf, S.K.; Tan, C.L.; Wong, H.F.; Ignatius, M.; Tan, E. Study on correlation between air temperature and urban morphology parameters in built environment in northern China. *Build. Environ.* **2018**, *127*, 239–249. [CrossRef]
5. Akbari, H.; Kolokotsa, D. Three decades of urban heat islands and mitigation technologies research. *Energy Build.* **2016**, *133*, 834–842. [CrossRef]
6. Santamouris, M. *Energy and Climate in the Urban Built Environment*; James & James: London, UK, 2001; ISBN 978-1-873936-90-0.

7. Gabriel, K.M.; Endlicher, W.R. Urban and rural mortality rates during heat waves in Berlin and Brandenburg, Germany. *Environ. Pollut.* **2011**, *159*, 2044–2050. [CrossRef] [PubMed]

8. Loughnan, M.E.; Tapper, N.; Phan, T.; Lynch, K.; McIn, J. *A Spatial Vulnerability Analysis of Urban Populations during Extreme Heat Events in Australian Capital Cities*; National Climate Change Adaptation Research Facility (Australia): Southport, Australia; Monash University: Clayton, Australia, 2013; ISBN 978-1-921609-73-2.

9. Dousset, B.; Gourmelon, F.; Laaidi, K.; Zeghnoun, A.; Giraudet, E.; Bretin, P.; Vandentorren, S. Satellite monitoring of summer heat waves in the Paris metropolitan area. *Int. J. Climatol.* **2011**, *31*, 313–323. [CrossRef]

10. Stathopoulou, M.I.; Cartalis, C.; Keramitsoglou, I.; Santamouris, M. Thermal remote sensing of Thom's discomfort index (DI): Comparison with in-situ measurements. *SPIE Remote Sens.* **2005**, *5983*. [CrossRef]

11. Giannopoulou, K.; Livada, I.; Santamouris, M.; Saliari, M.; Assimakopoulos, M.; Caouris, Y. The influence of air temperature and humidity on human thermal comfort over the greater Athens area. *Sustain. Cities Soc.* **2014**, *10*, 184–194. [CrossRef]

12. Paravantis, J.; Santamouris, M.; Cartalis, C.; Efthymiou, C.; Kontoulis, N. Mortality Associated with High Ambient Temperatures, Heatwaves, and the Urban Heat Island in Athens, Greece. *Sustainability* **2017**, *9*, 606. [CrossRef]

13. Coutts, A.M.; Harris, R.J.; Phan, T.; Livesley, S.J.; Williams, N.S.; Tapper, N.J. Thermal infrared remote sensing of urban heat: Hotspots, vegetation, and an assessment of techniques for use in urban planning. *Remote Sens. Environ.* **2016**, *186*, 637–651. [CrossRef]

14. IPCC. *Contribution of Working Group III to the Fourth Assessment Report of the Intergovernmental Panel on Climate Change*; Metz, B., Davidson, O.R., Bosch, P.R., Dave, R., Meyer, L.A., Eds.; Cambridge University Press: Cambridge, UK; New York, NY, USA, 2007; p. 852. ISBN 978-0-521-88011-4.

15. Jabareen, Y.R. Sustainable Urban Forms: Their Typologies, Models, and Concepts. *J. Plan. Educ. Res.* **2006**, *26*, 38–52. [CrossRef]

16. Asimakopoulos, D.A.; Santamouris, M.; Farrou, I.; Laskari, M.; Saliari, M.; Zanis, G.; Zerefos, S.C. Modelling the energy demand projection of the building sector in Greece in the 21st century. *Energy Build.* **2012**, *49*, 488–498. [CrossRef]

17. Larsen, L. Urban climate and adaptation strategies. *Front. Ecol. Environ.* **2015**, *13*, 486–492. [CrossRef]

18. Wang, J.; Ouyang, W. Attenuating the surface urban heat island within the local thermal zones through land surface modification. *J. Environ. Manag.* **2017**, *187*, 239–252. [CrossRef] [PubMed]

19. Estoque, R.C.; Murayama, Y.; Myint, S.W. Effects of landscape composition and pattern on land surface temperature: An urban heat island study in the megacities of Southeast Asia. *Sci. Total Environ.* **2017**, *577*, 349–359. [CrossRef] [PubMed]

20. Bonafoni, S.; Baldinelli, G.; Verducci, P. Sustainable strategies for smart cities: Analysis of the town development effect on surface urban heat island through remote sensing methodologies. *Sustain. Cities Soc.* **2017**, *29*, 211–218. [CrossRef]

21. Gallo, K.; Hale, R.; Tarpley, D.; Yu, Y. Evaluation of the Relationship between Air and Land Surface Temperature under Clear- and Cloudy-Sky Conditions. *J. Appl. Meteorol. Clim.* **2011**, *50*, 767–775. [CrossRef]

22. Nichol, J.E.; Fung, W.Y.; Lam, K.S.; Wong, M.S. Urban heat island diagnosis using ASTER satellite images and "in situ" air temperature. *Atmos. Res.* **2009**, *94*, 276–284. [CrossRef]

23. Agathangelidis, I.; Cartalis, C.; Santamouris, M. Estimation of Air Temperatures for the Urban Agglomeration of Athens with the Use of Satellite Data. *Geoinf. Geostat. Overv.* **2016**, *4*. [CrossRef]

24. Chen, E.; Allen, L.H., Jr.; Bartholic, J.F.; Gerber, J.F. Comparison of winter-nocturnal geostationary satellite infrared-surface temperature with shelter—Height temperature in Florida. *Remote Sens. Environ.* **1983**, *13*, 313–327. [CrossRef]

25. Green, R.M.; Hay, S.I. The potential of Pathfinder AVHRR data for providing surrogate climatic variables across Africa and Europe for epidemiological applications. *Remote Sens. Environ.* **2002**, *79*, 166–175. [CrossRef]

26. Stathopoulou, M.; Cartalis, C. Downscaling AVHRR land surface temperatures for improved surface urban heat island intensity estimation. *Remote Sens. Environ.* **2009**, *113*, 2592–2605. [CrossRef]

27. Gao, F.; Masek, J.; Schwaller, M.; Hall, F. On the blending of the Landsat and MODIS surface reflectance: Predicting daily Landsat surface reflectance. *IEEE Trans. Geosci. Remote Sens.* **2006**, *44*, 2207–2218. [CrossRef]

28. Hilker, T.; Wulder, M.A.; Coops, N.C.; Seitz, N.; White, J.C.; Gao, F.; Masek, J.G.; Stenhouse, G. Generation of dense time series synthetic Landsat data through data blending with MODIS using a spatial and temporal adaptive reflectance fusion model. *Remote Sens. Environ.* **2009**, *113*, 1988–1999. [CrossRef]

29. Weng, Q.; Fu, P.; Gao, F. Generating daily land surface temperature at Landsat resolution by fusing Landsat and MODIS data. *Remote Sens. Environ.* **2014**, *145*, 55–67. [CrossRef]

30. Kim, J.; Hogue, T.S. Evaluation and sensitivity testing of a coupled Landsat-MODIS downscaling method for land surface temperature and vegetation indices in semi-arid regions. *J. Appl. Remote Sens.* **2012**, *6*, 063569. [CrossRef]

31. Bindhu, V.M.; Narasimhan, B.; Sudheer, K.P. Development and verification of a non-linear disaggregation method (NL-DisTrad) to downscale MODIS land surface temperature to the spatial scale of Landsat thermal data to estimate evapotranspiration. *Remote Sens. Environ.* **2013**, *135*, 118–129. [CrossRef]

32. Mukherjee, S.; Joshi, P.K.; Garg, R.D. A comparison of different regression models for downscaling Landsat and MODIS land surface temperature images over heterogeneous landscape. *Adv. Space Res.* **2014**, *54*, 655–669. [CrossRef]

33. Bisquert, M.; Sánchez, J.M.; Caselles, V. Evaluation of disaggregation methods for downscaling MODIS land surface temperature to Landsat spatial resolution in Barrax test site. *IEEE J. Sel. Top. Appl.* **2016**, *9*, 1430–1438. [CrossRef]

34. Kustas, W.P.; Norman, J.M.; Anderson, M.C.; French, A.N. Estimating subpixel surface temperatures and energy fluxes from the vegetation index–radiometric temperature relationship. *Remote Sens. Environ.* **2003**, *85*, 429–440. [CrossRef]

35. Jeganathan, C.; Hamm, N.A.S.; Mukherjee, S.; Atkinson, P.M.; Raju, P.L.N.; Dadhwal, V.K. Evaluating a thermal image sharpening model over a mixed agricultural landscape in India. *Int. J. Appl. Earth Observ. Geoinf.* **2011**, *13*, 178–191. [CrossRef]

36. Ha, W.; Gowda, P.H.; Howell, T.A. A review of downscaling methods for remote sensing-based irrigation management: Part I. *Irrig. Sci.* **2013**, *31*, 831–850. [CrossRef]

37. Agam, N.; Kustas, W.P.; Anderson, M.C.; Li, F.; Neale, C.M. A vegetation index based technique for spatial sharpening of thermal imagery. *Remote Sens. Environ.* **2007**, *107*, 545–558. [CrossRef]

38. Jiménez-Muñoz, J.C.; Sobrino, J.A. A generalized single-channel method for retrieving land surface temperature from remote sensing data. *J. Geophys. Res.-Atmos.* **2003**, *108*. [CrossRef]

39. Jiménez-Muñoz, J.C.; Sobrino, J.A.; Skoković, D.; Mattar, C.; Cristóbal, J. Land surface temperature retrieval methods from Landsat-8 thermal infrared sensor data. *IEEE Trans. Geosci. Remote Sens.* **2014**, *11*, 1840–1843. [CrossRef]

40. Barsi, J.A.; Schott, J.R.; Palluconi, F.D.; Hook, S.J. Validation of a web-based atmospheric correction tool for single thermal band instruments. *Opt. Photonics* **2005**, *5882*. [CrossRef]

41. Sobrino, J.A.; Raissouni, N.; Li, Z.L. A comparative study of land surface emissivity retrieval from NOAA data. *Remote Sens. Environ.* **2001**, *75*, 256–266. [CrossRef]

42. Getis, A.; Ord, J.K. The analysis of spatial association by use of distance statistics. *Geogr. Anal.* **1992**, *24*, 189–206. [CrossRef]

43. Chorianopoulos, I.; Pagonis, T.; Koukoulas, S.; Drymoniti, S. Planning, competitiveness and sprawl in the Mediterranean city: The case of Athens. *Cities* **2010**, *27*, 249–259. [CrossRef]

44. Papamanolis, N. The main characteristics of the urban climate and the air quality in Greek cities. *Urban Clim.* **2015**, *12*, 49–64. [CrossRef]

45. Wan, Z. MODIS Land Surface Temperature Products Users' Guide. Available online: http://www.icess.ucsb.edu/modis/LstUsrGuide/MODIS_LST_products_Users_guide_C5.pdf (accessed on 19 January 2018).

46. European Union. *Copernicus Land Monitoring Service*; European Environment Agency (EEA): Copenhagen, Denmark, 2018.

47. Wan, Z. New refinements and validation of the MODIS land-surface temperature/emissivity products. *Remote Sens. Environ.* **2008**, *112*, 59–74. [CrossRef]

climate

MDPI

Article

Assessment and Mitigation Strategies to Counteract Overheating in Urban Historical Areas in Rome

Flavia Laureti *, Letizia Martinelli and Alessandra Battisti

Department of Planning, Design and Technology of Architecture, La Sapienza University, Via Flaminia 72, 00196 Rome, Italy; letizia.martinelli@gmail.com (L.M.); alessandra.battisti@uniroma1.it (A.B.)
* Correspondence: flavia.laureti@gmail.com; Tel.: +39-329-9666516

Received: 24 January 2018; Accepted: 14 March 2018; Published: 18 March 2018

Abstract: As urban overheating is increasing, there is a strong public interest towards mitigation strategies to enhance comfortable urban spaces, for their role in supporting urban metabolism and social life. The study presents an assessment of the existing thermal comfort and usage of San Silvestro Square in Rome during the summer, and performs the simulation of cooling strategies scenarios, to understand their mitigation potential for renovation projects. The first stage concerns a field analysis of the thermal and radiative environment on the 1st and 2nd of August 2014, including meteorological measurements and unobtrusive observations, to understand how people experience and respond to extreme microclimate conditions. In the second stage, the research proposes scenario simulations on the same day to examine the influence of cool colored materials, trees and vegetative surfaces on thermal comfort. The thermal comfort assessment was based on Physiologically Equivalent Temperature (PET), whereas microclimatic simulations were conducted with CFD calculations (ENVImet v.4.3.1). The first stage shows a strong relationship between lower PET values and attendance rate, depending on daily shading patterns. The second stage shows a relevant improvement of thermal comfort, with PET values of $-12\ ^\circ$C comparing to the no-intervention scenario, associated with a combination of cool materials and trees.

Keywords: overheating; summer heat stress; urban open space; shading; thermal comfort; Physiologically Equivalent Temperature; mitigation strategies; cooling technologies; cool materials

1. Introduction

There is strong public interest towards the sustainable and environmental regeneration of urban open spaces, particularly in the light of the increasing urban overheating phenomena due to multiple factors, among which the most relevant are: loss of green spaces, impervious paving surfaces, materials' characteristics of the built environment, combined with a very high increase of released anthropogenic heat [1,2]. The expression "urban open space" has been used largely and in many different thematic areas: architects and urban planners refer to it from a spatial perspective as all the outdoor areas of cities that are physically accessible to the public with few restrictions, that compromises several activities, planned or spontaneous, promoting the interaction of citizens and their feeling of identity and security, and ultimately improving indirectly local economy and people's health [3,4]. As stated by Katzschner [5], it is acknowledged that urban open spaces can contribute consistently to the quality of life within cities or contrarily enhance isolation and social exclusion [6], depending on the thermo-physical as well as social environment [7]. Thus, a good urban open space must address not only the "imageability" and aesthetic value of the city [8,9], but also support the daily activities and sociability of citizens and therefore it should represent a comfortable and livable place [4,10]. Its environmental quality and success depends on outdoor thermal comfort and microclimate, as stated by Gehl [10]: «few topics have greater relevance for comfort and well-being in city space than the actual climate where one is sitting, walking or biking» (p.168). Responses to microclimate may be unconscious, but they often result in a different use of open space in different climatic conditions [11].

Regarding thermal comfort, ASHRAE 55 defines it as «that condition of mind that expresses satisfaction with the thermal environment and is assessed by subjective evaluation» [12]. Thus, human thermal comfort is both physical and perceptual, it comprises interactions with the local microclimate, as well as physiological and psychological mechanisms [13]. Outdoor, these mechanisms mainly depend on four factors, variable in quantity, distribution and duration: air temperature (T_a), air velocity (v), vapor pressure (VP) and Mean Radiant Temperature (T_{mrt}), which parameterizes the impact of all short- and long-wave radiation fluxes on human body [14] and has a paramount daily impact [15]. Within the vertical scale of Urban Canopy Layer (UCL), which constitutes the lower part of the roughness sub-level, approximately from the ground to the rooftops of buildings, the effect of urban design on microclimate factors becomes substantial and it strongly affects thermal comfort [16], with the most relevant parameters being: topography, urban morphology, soil structure, the paving surfaces and the building materials [17], comprehending vegetated surfaces, green spaces and "cool sinks" [18], such as the ground and water [19,20].

The high density of buildings and urban structures, the use of urban materials with low albedo, that is the ratio of global radiation reflected to the global radiation received by a surface [21], the lack of green spaces and pervious surfaces and the anthropogenic heat contribute to the general higher temperatures measured within the built environment comparing to the relatively cooler suburban and rural surroundings [22]. Unlike what happens in many rural areas, where plant cover and evaporation of soil moisture may moderate the increase of surface temperature that occurs when solar radiation is absorbed by the earth, a large proportion of urban areas consists of dry impervious surfaces, such as pavement and buildings, absorbing a great amount of solar radiation (short wave) and releasing it into the urban atmosphere in the thermal infrared (long wave) of the electromagnetic radiation spectrum. If we consider the historical city centre of Rome, in particular the area surrounding the present case study, the rate between built environment and green areas is around 70%. The temperature of these surfaces, after receiving the solar infrared radiation, gets warmer than that of the overlying near-surface air, resulting in a sensible heat flux directed upward leading to an increase in the UCL air temperature. The phenomenon, known as Urban Heat Island (UHI), occurs chiefly during night-time [23], contributes significantly to the urban overheating phenomenon and it worsen the impact of climate change within cities because it intensifies energy consumption, deteriorates comfort conditions, put in danger vulnerable populations and amplify pollution problems [18,24–26]. To counteract these effects, many different solutions have been investigated and studied, among which the most significant ones are the mitigation strategies. Mitigation strategies aim to reduce greenhouse gas sources and emissions and prevent urban overheating [27], applied on the local scale, mitigation techniques aim to balance the thermal budget of cities by increasing thermal losses and decreasing the corresponding gains [18]. Effective mitigating approaches are the increase of vegetated surfaces and green spaces and the raise of urban materials' albedo, that is the reflection coefficient, measured from 0, for a blackbody theoretically absorbing 100% of the incident solar radiation, to 1 corresponding to a total white surface, an ideal reflective surface [28].

Regarding the first class of mitigation strategies, vegetated surfaces and green spaces lower the dry-bulb air temperatures and increase latent cooling through evapotranspiration processes, which are the sum of evaporation and plant transpiration [29–31]. In addition, vertical elements such as trees also reduce direct solar radiation by shading, depending on the shape, dimension, Leaf Area Density (LAD) of the single plant, which is defined as the portion of leaf surface in m^2 within a m^3 of air, and Leaf Area Index (LAI) which is defined as a dimensionless value of the leaf area per unit of ground area. If we consider the interaction between solar radiation and a tree during summer, between 10% and 30% of solar radiation generally reaches the base of a tree [32]. Thus, types, sizes and arrangements of leaves play an important role in improving efficiency in radiation absorption and reflection. The most used greening techniques include: green pavements, ranging from traditional green lawn to the hybrid permeable pavements characterized by tiles mixed with greenery (i.e. grass pavers); trees, vertical greening and green roofs.

As underlined by Upreti et al. [33] and Saaroni et al. [34] urban trees represent a feasible and advantageous form of green infrastructures for mitigating urban overheating and in addition to their microclimate benefits, trees present other benefits ranging from social (e.g., enhancing quality of urban life), economic (e.g., increase of property value), health (e.g., air quality improvement, reducing stress, improving physical health), visual (e.g., creating seasonal interest by highlighting seasonal changes) and aesthetic (e.g., improving scenic quality) advantages to humans. Whereas, the problems and hazards associated to the extensive use of trees within cities are related to a reduction of the solar access and wind speed, carbon pollution through landscape and tree management practices, maintenance problems and pollen diffusion, as underlined by Roy et al. who [35] conducted a systematic quantitative review of urban tree benefits, costs and assessment methods across cities in different climatic zones.

Regarding their mitigation potential, Shashua et al. have investigated the effect of trees and cool pavements on urban streets in Athens during hot summer conditions [36]. Among the different findings, they highlight how the effect of trees represents the dominant factor affecting comfort, since the increase in vegetation street coverage from 8% to 50% leads to a T_{mrt} value decrease of 13.6 K and to a PET value decrease about 8.3 K at noontime. As stated by the authors, «the tree option is the most beneficial agent in improving the microclimate, as well as the comfort situation». Another study conducted by Ketterer and Matzarakis in Stuttgart [37] underlines how the maximal value of PET over a green unshaded area is 35 °C, whereas it can exceed 58 °C over a sealed and asphaltic surface. The authors also suggest that a green surface can reduce PET by up to 7 K in the neighboring streets and areas, especially on the lee side, thus helping to establish an «ideal urban climate» [38]. In another recent study conducted by Cheung and Jim [39] the cooling effects of a tree and a concrete shelter have been compared using PET and UTCI. The research underlines how the cooling effect related to a tree shade was significant but nevertheless higher than the one provided by the concrete artificial shading. In fact, as stated by the authors, the mean daytime cooling effects generated by the tree were 0.6 °C (air temperature), 3.9 °C (PET) and 2.5 °C (UTCI), which were higher than the shelter at 0.2 °C, 3.8 °C and 2.0 °C respectively. Nevertheless, in order to inform the urban designers and planners on the strategic use of natural and artificial shading devices, an overview of the general benefits provided by the green infrastructure solutions, such as trees in this case, must be done as exposed in the aforementioned researches.

Regarding the second class of mitigation techniques called "cool materials", the present study focuses on specific innovative materials presenting a high solar reflectance and high thermal emittance compared to traditional ones, which results in a lower surface temperature, thus the name "cool materials", affecting the radiative environment and T_{mrt}. At building scale, this means that the heat penetrating into the building decreases, whereas at city scale this contributes to decreased air temperatures as the heat transfer from warmer surfaces is lower [40].

Nowadays, the cool materials class comprises: the first generation, including natural materials with high reflectivity [41], the second generation, based on the development of artificial white materials design to have a high albedo equal or higher than 0.85 [42]. These first high-reflective materials, known traditionally as cool materials, presented visibility problems for the reflected UV radiation toward the pedestrian; therefore, they lead to the development of the third generation of cool materials, known as "cool colored materials", comprehending colored high reflective materials that present a high reflectivity in the infrared spectrum of solar radiation (NIR) [43], thus appearing like traditional dark materials but with low surface temperature, and finally the fourth generation, based on nanotechnological additives like thermochromic paints [44] or PCM (phase change materials) doped cool materials [45]. Moreover, recent studies are now focusing on the advances of cool colored coatings to be applied both on roof and on building façades in historic context, due to their appearance similar to the traditional colors of the consolidated city. Among these cool coatings, a recent interest has emerged towards mineral-based coatings that present a similar behavior to high-reflective materials, as demonstrated by Kolokotsa (2012) [42], and directionally reflective materials, these materials present multifaceted surfaces whose color and reflectance change accordingly to the surface orientation and the direction of incident solar radiation, reflecting sunlight during the summer and absorbing light during winter [46].

The interest towards these type of cooling measures related to the built surfaces, depends on the massive presence of roofs and pavements representing a major part of the urban fabric: in fact, roofing surfaces correspond to about 20–40% of the total area exposed to solar radiation and pavements amount approximately to 29–44% of the total urban ground [47], thus their implementation could be massively handled with promising local and global benefits energy and environmental benefits [48,49].

There is significant body of research on cool materials and coatings, among which Santamouris et al. [50], Santamouris [18] and Pisello [49] offer a consistent review of the main cool materials and coatings, investigated in laboratory or on field, developed firstly for pavements and roofs and then for building façades. In particular, regarding the benefits related to the cool coatings class, as stated by Pisello [49], when applied over roof and wall, they provide a very effective solution for passive cooling of buildings indoors and of local outdoor microclimate. Studies conducted on different cities, such as New York (USA), Los Angeles (USA), Athens (Greece), show that the expected mean decrease of the average ambient temperature is close to 0.3 K per 0.1 rise of the albedo, while the corresponding average decrease of the peak ambient temperature is close to 0.9 K [18]. Regarding the effect of cool roof on the city scale, a study on 14 cool reflective coatings, selected from the international market [51], demonstrated that the use of reflective coatings can reduce a white concrete tiles surface temperature under hot summer conditions by 4 °C and during the night by 2 °C.

Although cool materials can be applied for both pavements, roof and façade surfaces, the development of cool pavements for outdoor spaces has remained at laboratory test level, only recently they have been applied in outdoor spaces [52]. In fact, pavements' thermal behavior is affected by convection due to traffic movement and by shadings due to people, traffic and urban structures, and a pavement's coat is subject to rapid dirtying and wearing. However, in the last decades, cool pavements have been developed and analyzed following the same principles used for the researches on cool roofs [21,53,54]. One of the first and relevant field measurement regarding cool pavements, conducted in Flisvos urban park in the greater Athens area, shows that extensive application of reflective pavements may reduce the peak daily ambient temperature during a typical summer day up to 1.9 °C, while surface temperatures were reduced up to 12 °C [52]. Regarding the mitigation potential of surface temperature values, a study of Kinouchi [55] conducted in Japan in 2003, highlighted the lower Ts values of cool pavements. The study considers the application on an asphalt road of a cool colored coating, with low reflectivity in the visible part of sunlight spectrum (23%) and high reflectivity of near infra-red (87%). The cool colored coating was applied over two sample of street pavement in two different sites, respectively in Tsukuba and in Okinawa, and resulted in an average surface temperature of 15 °C lower than a conventional asphalt pavement.

Most of the researches on cool materials have been focusing on their mitigation potential towards ambient air temperatures, surface temperatures or mean radiant temperatures. A recent study conducted by Piselli et al. [56] highlights the different benefits related to cool pavements and trees, in terms not only T_a, T_s but also of comfort (PET) by means of in situ monitoring campaigns, questionnaires to the moving pedestrians, and microclimate simulations. From the analysis and comparison of the air temperature, T_{mrt} and PET values among the five scenarios object of study, the most beneficial scenarios were the one which comprised an overall upgrade of the green and trees areas (scenario 2) and the mixed scenario (scenario 5) which included a combination of cool pavements, trees and photovoltaic pavement. Regarding this last scenario, the air temperature at 12:00 was found to be 1 °C lower than the initial condition, the T_{mrt} was found to be about 20 °C lower and the related comfort conditions were improved up to 15 °C, underlying not only an effective mitigation potential of cool pavements but also a good performance regarding its effects on thermal comfort when coupled with green technologies.

Nevertheless, as aforementioned, until recently the majority of existing studies concentrated mainly on the reduction of surface temperature and air temperature, which depends on the spatial scale of the area in question [31,37] or on the related energy saving for indoor spaces, neglecting the general effect of cool materials on outdoor thermal comfort through its influence on T_{mrt}. However, as the human body is non-selective in its absorption of solar energy, the effect of albedo modification on thermal sensation may actually increase thermal stress in warm environments [57]. Regarding this aspect, Chatzidimitriou and Yannas [58] verified that lower surface temperatures, by use of reflective materials, does not compensate for the higher amounts of reflected solar radiation released in the urban environment and directly effecting the thermal comfort of the pedestrian.

Therefore, given the relevance of these technologies in counteracting UHI and in mitigating overheating, the present research aims to assess the effect of urban greenery and cool colored materials, in different combinations on outdoor thermal comfort. It also presents a different perspective for the renovation of historical urban squares, attentive to the microclimatic effect of urban morphology and materials and based on field analysis of existing thermal comfort conditions and visitors' consequent behavior. Mediterranean historical city centers are often characterized by a minimal use of greenery, since they are often perceived as paved space; conversely, the proposed approach goes beyond the typical example of Mediterranean paved "placas duras" [3], incorporating urban greenery and cool colored materials as minimal strategies to upgrade the sustainability of historical urban open spaces, while still maintaining the image of the city in the preservation of its traditional color patterns.

Following this assumption, the current study:

1. analyzes the existing relationship between attendance, shading patterns depending on urban morphology and thermal comfort in a public square in the Mediterranean climate during overheating periods [3];
2. evaluates the effects on outdoor thermal comfort of renovation scenarios of the square, defined according to the field analysis and based on urban greenery and cool colored materials, single or combined;
3. compares the thermal comfort conditions of different points of the square, corresponding to different morphological and material conditions, for the renovation scenarios.

2. Materials and Methods

2.1. Case Study: Site Location and Microclimatic Characteristics

The study was carried out on San Silvestro square (Figure 1), a square located in the historic city center of Rome (coordinates 41°54' N 12°30' E, elevation 20 m), the capital of Italy and its largest and most populated city. The principal river, the Tiber, branches through the core of the city north to south and the center of the city is about 24 km from the Mediterranean Sea; its climate is dry-subtropic Csa, according to Köppen-Geigen classification [59].

The square was selected for its social and cultural relevance and for its location, since it is near one of the main street of the historic centre (Via del Corso), close to relevant tourist attractions, shopping boulevards and offices and the related public transport hub. The square is a quadrangular paved open space measuring approximately 80×60 m^2 and has undergone a renovation in 2011, proposing a complete pedestrianization of the square for recreational use, furnishing it with long, marble benches. The renovation divided the area in two different squares: one on the West side with the benches outlining a rectangular area, and the other one, wider, located in the centre-East of the square and characterized by benches following an elliptical pattern.

Even though the initial design proposed an integration of trees and green lawn, as well as the insertion of a fountain in the elliptical square, the final construction didn't include those elements and proved to have no concerns for the microclimatic conditions and the thermal behavior of the place.

Figure 1. A picture of san Silvestro square taken in March 2015 (this is the reason for many visitors lingering on the sunny benches).

The surrounding buildings, representing the urban interface of the open space and its main interacting surfaces, are about 4–5 floor height, with a height/width ratio (H/W) equal to 0.36. Thus, the square does not offer a natural or shaded environment other than the buildings' shade.

In addition, it must be underlined that San Silvestro square has been selected in this study because it represents a typical urban open space of the historic tissue of Rome for its dimensions, the morphology of the surrounding buildings, the relation between its sides and especially for the height and width ratio (H/W = 0.30 ca.). In fact, if we consider the evolution of the urban open space in the historic centre of Rome, San Silvestro square recalls the *Centuratio* from the ancient Roman period, that is to say the quadrangular open space resulting from the *Cardo* and *Decumanus* grid system, which measured about 70 m per side and the typical height of the surroundings *Insulae* which were about 20 m, resulting in a H/W ratio of about 0.3. Similar proportions remain through the urban planning evolution of the historical tissues in Rome, and are visible in other historical squares such as the Immacolata square (beginning of the 20th century) in the San Lorenzo neighbourhood which presents a H/W ratio of 0.35 ca. or Madonna dei Monti square (Renaissance period) in the Monti neighbourhood which presents a H/W ratio of 0.4 ca.

2.2. Field Data Collection and Meteorological Measurements

The first stage of the study involved a field collection of meteorological data and unobtrusive observations of the visitors of the square, acquired on Friday 1 and Saturday 2 August 2014, to explore people behavior and usage of the square on work days and on the weekend, each 10 min from 8:00 to 20:00 [3]. The observations recorded the people lingering or sitting for more than 10 min, omitting passersby and separating the people in shaded or unshaded locations.

Following the definition given by Kántor and Unger [32], data about the number of visitors every 10 min are referred to as momentary attendance. Visitors' behavior depends on local time during summer, so all results refer to Local Standard Time (LST) + 1.

Air temperature and relative humidity was measured using a humidity Temperature Meter, HH314A, with a measurement resolution of 0.1 K and 0.1%, with an accuracy of 2.5% for relative humidity and 0.7 K for air temperature.

The instrument was mounted on a lightweight aluminum tripod camera stands at a height of 1.1 m, corresponding to the average height of the gravity centre of adults [60], and was positioned in an open point (point A on Figure 2) and sheltered from direct solar radiation, as recommended by Oke [16]. The Collegio Romano meteorological station, situated approximately 500 m from San Silvestro square, provided data for air velocity with 10 min resolution and for cloud cover (*cc*). Since the station is above roof level (40 m above ground), air velocity was reduced to the reference height of 1.1 m:

$$v_{1.1} = v_h \cdot \left(\frac{1.1}{h}\right)^\alpha, \tag{1}$$

$$\alpha = 0.12 \cdot z_0 + 0.18, \tag{2}$$

where $v_{1.1}$ is the air velocity at 1.1 m height, v_h is the air velocity at the meteorological station's height, h is the meteorological station's height, α is an empirical exponent depending on the surface roughness and z_0 is the roughness length. In this study, z_0 was set at 1.5 m, which represents a densely built-up urban area (high floor area ratio) according to the classification of Davenport [61] and several studies on thermal comfort in urban environments [37,62,63]. Finally, the Mean Radiant Temperature (T_{mrt}) has been calculated throughout the day using the Rayman software [15], by importing cloud cover, time of year and surrounding obstacles.

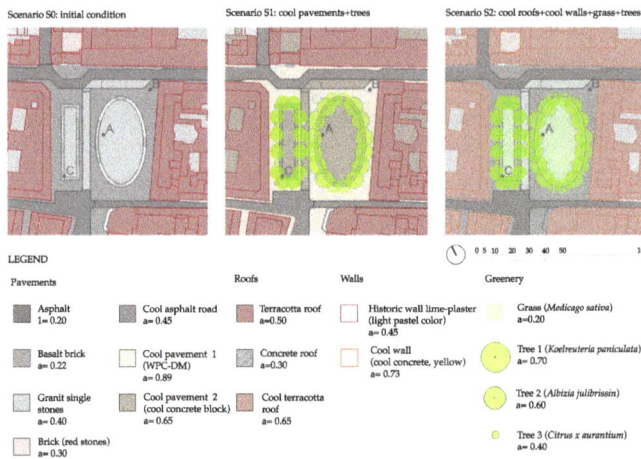

Figure 2. Initial condition scenario S0 (**a**), first renovation scenario S1 (**b**), second renovation scenario S2 (**c**). Point A, B and C represent the selected location for the comparisons among the scenarios.

The microclimate parameters measured and collected in the first stage analysis, on the first day (Friday 1 August 2014), are used as input data for the microclimate simulations of the renovation scenarios in ENVI-met, modelled and analyzed in the second stage of the study.

2.3. Scenario Simulation Definition

The second stage of study focuses on the simulation and comparison of two different renovation scenarios applied to the square, involving the use of cool materials for pavements, façades and roofs and the use of greenery. Each scenario has been compared to the initial condition used as a test reference model (S0) to assess the renovation strategies performance.

The percentage and distribution of applied materials depends on morphological and typological limitations of the area. In particular, the artificial materials were selected and applied in order to satisfy the following criteria:

1. to present the higher possible non-specular reflectivity to solar radiation for each surfaces' common range,
2. to present the highest possible emissivity factor,
3. to present the highest possible aesthetic value and integration within the image of the historical context.

The natural materials and elements such as grass and trees, were selected in order to satisfy the following criteria:

1. for the trees to present a medium-high Leaf Area Density (LAD),
2. to be applicable and in compliance within the Mediterranean context and traditional urban greenery of the city.

Moreover, the position of greenery is based on the results of the field analysis of the first stage of the study, which indicates a correspondence between shading patterns and attendance of sitting places.

The first renovation scenario (S1) focuses on the minimal intervention and on the principal surfaces interacting with the pedestrian level: the urban pavement. It aims to analyze the combined cooling effect of trees and cool pavements and their related outdoor comfort. The second renovation scenario (S2) focuses on the combination of different materials, applying cool colored materials on roofs and walls and increasing trees and the vegetated surfaces of urban pavements.

Figure 2 represents the initial condition scenario S0 (initial condition) and the two renovation scenarios S1 (cool pavements + trees) and S2 (cool roofs + cool walls + trees + grass), with the specific renovation strategies. There are different institutions and organizations that are creating and implementing their digital database of cool materials, such as the Energy Star Roof Products, the Lawrence Berkley Institute, the US Cool Roof Rating Council and the EU Project Cool Roofs [50]. In this study, the cool materials were selected from two studies carried out by Santamouris et al. and Gobakis et al. [50,64].

Table 1 summarizes the materials database used in the initial scenario, S0; in the first renovation scenario S1 (cool pavements + trees) and in the second renovation scenario S2 (cool roofs + cool walls + trees + grass); whereas Table 2 summarizes the albedo and LAD coefficients of the tree' species modelled in ENVI-met V.4.3.1.

2.4. Thermal Comfort Analysis

The thermal comfort analysis is based on PET (Physiological Equivalent Temperature) index, defined as the equivalent temperature at any given place (outdoors or indoors) to the air temperature at which, in a typical indoor setting, the heat balance of the human body, with light activity (80 W) and heat resistance of clothing of 0.9 clo, is maintained with equal core and skin temperatures to those under the conditions being assessed [65].

The PET index belongs to a set of detailed thermal indices based on the human energy balance which, as underlined by Matzarakis et al. [66], give us information on the effects of complex thermal environments on humans, taking into account the human thermoregulatory mechanisms and the human circulatory system like the constriction or dilation of peripheral blood vessels and the physiological sweat rate, in order to produce a comprehensive, reliable representation of human body's actual physiological sensations [3]. As a thermal index, PET requires basic meteorological input parameters: air temperature, air humidity, wind speed, short and long wave radiation fluxes and in order to calculate the human energy balance, thermo-physiological parameters, such as heat resistance of clothing (clo) and activity of humans (in W), are required. The following assumptions are made for the indoor reference climate: $T_{mrt} = T_a$; $v = 0.1$ ms^{-1}; $VP = 12$ hPa. PET depends on four main physical parameters, which characterize the thermal environment: air temperature (T_a), water vapor pressure (VP), air velocity (v) and Mean Radiant Temperature (T_{mrt}).

Table 1. Materials' database of the analyzed scenarios: S0 initial condition, S1 first renovation scenario (cool pavements + trees), second renovation scenario (cool roofs + cool walls + trees + grass).

Surface	S0 Name	S0 Albedo (a)	S1 Name	S1 Albedo (a)	S2 Name	S2 Albedo (a)
Roofs	clay tiles (* terracotta roof)	0.50	clay tiles (* terracotta roof)	0.50	cool red brick tiles (* cool terracotta roof)	0.65
	concrete tiles	0.30	concrete tiles	0.30		
Walls	areated brick block with lime plaster (* historical wall)	0.45	areated brick block with lime plaster (* historical wall)	0.45	yellow cool coating with TiO_2 (* cool wall)	0.73
Pavements	asphalt	0.20	cool colored thin layer asphalt	0.45	asphalt	0.20
	basalt tiles	0.22	white Portland cement with dolomitic marbles plaster	0.89	basalt tiles	0.22
	flint blocks	0.40	cool colored pigmented concrete tiles	0.65	flint blocks	0.40
Greenery	—	—	*Citrus x aurantium* (deciduous)	0.40	*Citrus x aurantium* (deciduous)	0.40
			Albizia julibrissin (deciduous)	0.70	*Albizia julibrissin* (deciduous)	0.70
			Koelreuteria paniculata (deciduous)	0.60	*Koelreuteria paniculata* (deciduous)	0.60
					Medicago sativa (evergreen)	0.20

* abbreviated classification as illustrated in the plans' legend.

Table 2. Greenery characteristics.

Name	Albedo (a)	Leaf Area Density (LAD) m^2/m^3
Citrus x aurantium (deciduous)	0.40	0.50–0.60
Albizia julibrissin (deciduous)	0.70	0.60–0.70
Koelreuteria paniculata (deciduous)	0.60	0.70–0.80

Regarding the current study, the PET index has been selected among other thermal indexes, because of its extensive application to analyze thermal comfort in different climates [67–69] and its measurement unit (°C), making the results easy to understand for urban planners or decision makers, who are generally not so familiar with modern human biometereological terminology [70]. However, similarly to other thermal indices, the PET shows some limitations in its applicability as underlined by Chen and Matzarakis [71]: in particular variations in air humidity and clothing insulation show weak influence on PET. Nevertheless, for the aforementioned characteristics, specifically for the extensive scientific literature and because it has been used in urban built-up area with complex shading patterns and generated accurate predictions of thermal environments [72] the PET index was selected as the primary thermal index in this study.

The main focus of the second stage of the study is to assess the impact of combination of tree density, green surfaces and albedo of materials; therefore plant, surface and air interaction are essential. Micro-scale model ENVI-met v.4.3.1 [73,74] was selected to analyze the human-biometeorological conditions and their changes due to the renovation scenarios, as the software calculates micro-scale surface-air-plant interactions in a three-dimensional non-hydrostatic way. In the latest updated version (28 November 2017) solar radiation analyses, vegetation modeling, advanced plant simulation, detailed building physics and air pollution are included. It is also possible to assess the energy budget at ground surfaces and to visualize the surface temperatures in 3D mode on roofs, walls as well as pavements. Therefore, its spatial and temporal resolution provides a good basis for the quantification of changes and microclimatic interactions in the urban environment.

We approximated the scenario solutions in terms of spatial configurations within the specific constraints of the software, thus:

1. the grid cell dimension used for the ENVI-met models measures 2 m × 2 m, in order to optimize the time needed for the calculation as well as to maintain a proper level of detail;
2. walls and roofs were modelled in the Database Manager specifying their three characteristic layers, in order to take into account, the thermal characteristic of the surrounding building volumes;
3. for the characterization of the exterior layer of walls and roofs, in order to balance the uncertain application of materials in 3Dmode within the ENVI-met SPACE and to approximate the real conditions of building materials, we chose to apply the prevalent material, that is clay tiles for the roofs and aerated brick block with lime plaster for the walls.

The ENVI-met simulations were carried out from 8:00 to 20:00 of 1st August 2014, plus a stabilizing period. The input data used in ENVI-met derived from the measurements and calculations made during the first stage of the study, specifically they include: air temperature (T_a) and relative humidity (RH) measured at San Silvestro square, air velocity (v) and cloud cover (cc) from the Collegio Romano meteorological station. In order to calculate PET, we used the ENVImet package's PET index calculator BioMet. As for the thermo-physiological parameter of the human body, we assumed a "typical European male" (35 years old, 1.75 m tall, weight 75 kg), with a clothing index of 0.6 clo (corresponding to a summer business suit) and an activity rate of 80 W.

3. Results

The following data represent the average value between the two days of analysis conducted in San Silvestro square, on the 1st and 2nd of August 2014.

During the day of analysis in the 1st stage of the study, the minimum value of air temperature of 22.6 °C was recorded at 8:00, while the maximum value of 32.1 °C occurred at 13:00. The average air temperature in daytime was 29.2 °C. The minimum value of water vapor pressure of 19.9 hPa was reached at 18:10, while the maximum value was 25.9 hPa at 9:20 and 9:40. The average of water vapor pressure was 23.1 hPa. Table 3 shows a systematic overview of the principal meteorological parameters measured in the square, in particular: air temperature (T_a), air velocity (v) relative humidity (RH) and cloud cover (*cc*), and finally the Mean Radiant Temperature (T_{mrt}) calculated with Rayman on the 1 August 2014, from 8:00 to 19:00.

Table 3. Summary of the meteorological parameters measured in the square and collected from the Collegio Romano meteorological station on the 1st and 2nd of August 2014: air temperature (T_a), air velocity (*v*), relative humidity (RH) and cloud cover (*cc*), and the Mean Radiant Temperature (T_{mrt}) calculated with the Rayman software.

Day	Hour	Air Temperature (T_a)	Relative Humidity (RH)	Air Velocity (*v*)	Cloud Cover (*cc*)	Mean Radiant Temperature (T_{mrt})
1 August 2014	08:00	22.6	77.2	0.4	0	23.6
1 August 2014	09:00	24	73	0.4	0	25.2
1 August 2014	10:00	25.7	65.2	0.8	0	27
1 August 2014	11:00	28.6	58.9	0.4	0	44.2
1 August 2014	12:00	30.2	53.4	0.4	1	48
1 August 2014	13:00	32.1	47	0.8	1	50.2
1 August 2014	14:00	30.6	55.8	0.8	1	45.7
1 August 2014	15:00	30.3	55.4	1.2	1	41.7
1 August 2014	16:00	30.9	51.5	2.5	1	43.4
1 August 2014	17:00	30.4	53.8	1.7	1	42.6
1 August 2014	18:00	29.1	57	2.1	1	32.8
1 August 2014	19:00	29.8	60.5	1.7	1	39

3.1. Relationship between Attendance, Shading Patterns and Thermal Comfort

The first stage of the study analysed the relationship between the attendance of the square, daily shading patterns and thermal comfort conditions during the day. Shade in the square derives entirely from the surrounding buildings, therefore is higher during the morning and late afternoon, with a percentage of shaded area of 99% at 8:00 and 100% at 20:00 and decreases towards midday, when in reaches the minimum of 13% at 12:00 and 14:00, 7% at 13:00.

The majority of visitors concentrated in specific time periods (Figure 3), corresponding to the main usages of the square by workers of the nearby offices and tourists: early morning (8:00 to 9:00), presumably when the workers have breakfast or read newspapers at the square before going to work around 9:00; lunchtime (13:00 to 14:30) and late afternoon (17:00 to 19:00).

Momentary attendance in unshaded locations is significantly lower than in shaded locations: visitors massively preferred shaded locations, throughout the day, whenever possible. The momentary attendance in unshaded locations changed from 0–4 people during the morning from 8:00 to 11:00, to reach the peak of 9–8 people at 12:00–13:00, to lower again to 2–3 in the afternoon after 17:00. On the contrary, in shaded locations the number of visitors during the morning extended from 8–20 people at 10:00 to 11–25 at 11:00; then, it drastically lowered to 0–4 people at 12:00 and 0–9 at 13:00, to increase in the afternoon, reaching 52–94 at 18:00. The low attendance in shaded locations at 12:00 and 13:00 corresponds with the almost complete absence of sitting possibilities in shade, either formal on the benches or informal on building's steps or flowerpots. This preference is in accordance with thermal comfort analysis, as PET in shaded locations was consistently lower than PET in unshaded locations, due to the reduction of direct solar radiation [3].

The boxplot in Figure 4 divides the median momentary attendance into classes of PET values for four-time periods: it indicates that visitors generally preferred locations corresponding to PET below 35 °C, which coincided with shade. In the morning interval (from 8:00 to 10:50), PET of 26 °C–30 °C and PET of 30 °C–35 °C matched with the highest median of momentary attendance of 12; from 14:00 to 19:50, the highest median of momentary attendance reached 40 people for PET between 30 °C and 35 °C, while in the late afternoon (from 17:00 to 19:50), the highest median of momentary attendance of 60 people corresponds to PET between 30 °C and 35 °C. In the interval 11:00–13:50 there was an exception, with the highest median of attendance of 5.5 for PET between 35 °C and 40 °C; however, as depicted by Figure 3, during that time period there was the lowest percentage of shaded area and shaded sitting locations.

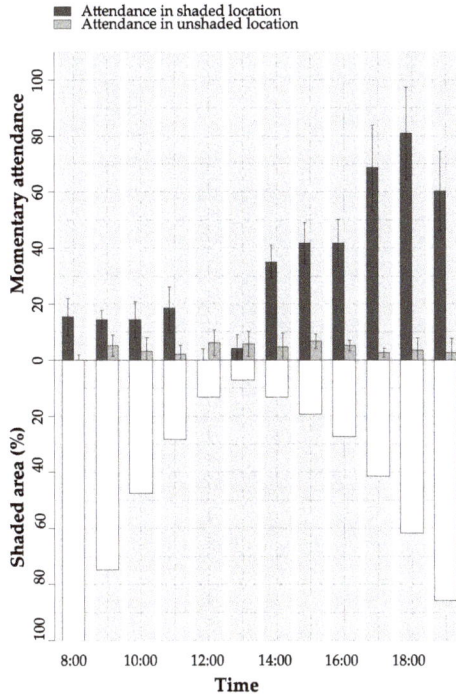

Figure 3. Hourly median and standard deviation of the attendance for shaded (dark grey) and unshaded (light grey) locations, on 1st and 2nd of August 2014, from 8:00 to 19:00, compared with the hourly percentage of shaded area (white bars) in San Silvestro Square [3].

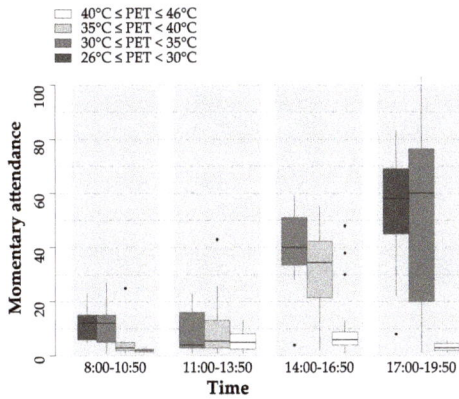

Figure 4. Attendance for classes of PET for four-time intervals during the day: morning (8:00, 10:50), midday (11:00, 13:50), afternoon (14:00, 16:50), late afternoon (17:00, 19:50) in San Silvestro Square [3].

3.2. Spatial Analysis of Thermal Comfort Distribution for the Simulation Scenarios

The second stage of the study compares the thermal comfort of the initial condition scenario S0 and the two renovation scenarios S1 and S2. Figure 5 shows the spatial distribution of PET values among the square for S0, S1 and S2 at three selected hours: 8:00, 13:00, 17:00, which corresponds to the highest attendance of the day, according to the observed momentary attendance shown in Figure 3.

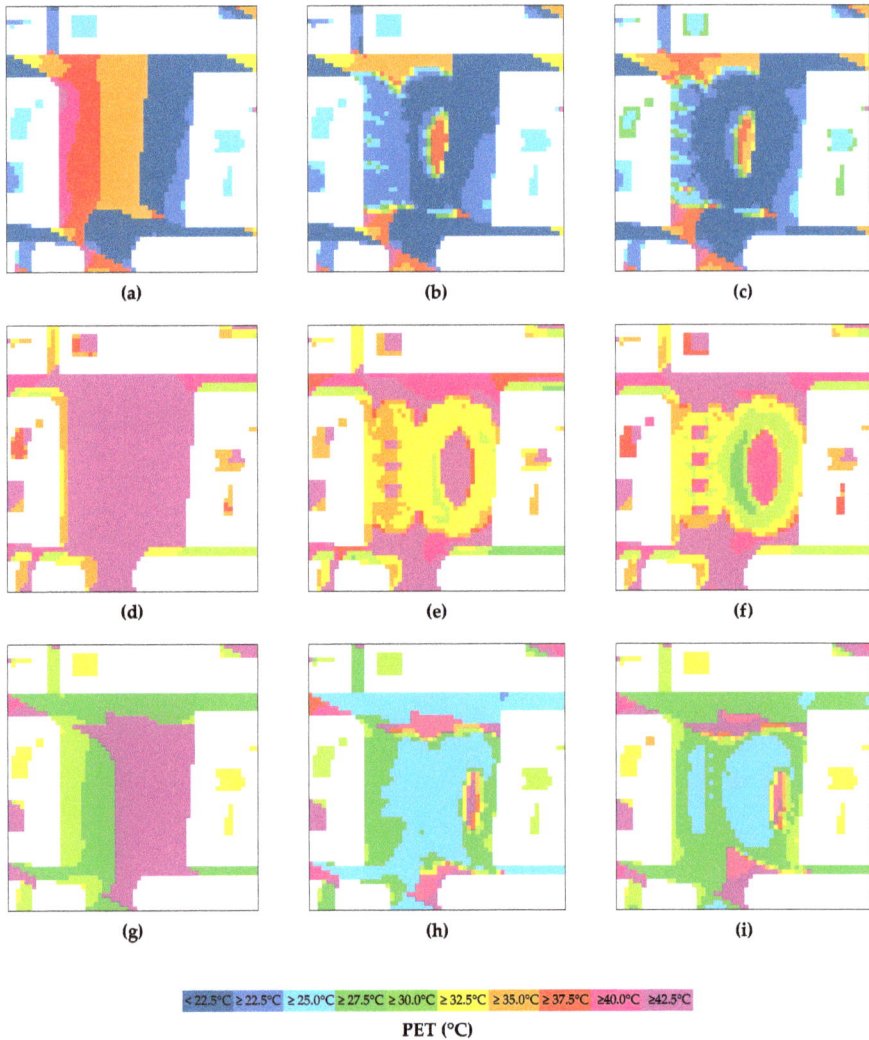

PET (°C)

Figure 5. PET maps at 8:00, 13:00 and 17:00, on the 1 August 2014, of S0 (initial condition), S1 (cool pavements + trees) and S2 (cool roofs + cool walls + trees + grass), in detail: S0 PET at 8:00 (**a**), S1 PET at 8:00 (**b**), S2 PET at 8:00 (**c**), S0 PET at 13:00 (**d**), S1 PET at 13:00 (**e**), S2 PET at 13:00 (**f**), S0 PET at 17:00 (**g**), S1 PET at 17:00 (**h**), S2 PET at 17:00 (**i**).

At 8:00, PET values for scenario S0 (a) range from a minimum value below 22.5 °C in the SE area of the square, shaded by the surrounding buildings, to a maximum value above 42.5 °C alongside the western façades of the square, which receives direct solar radiation. Whereas, the central area of the square presents a uniform distribution of PET values, between 35 °C and 37.5 °C. For scenario S1, PET values (b) are consistently lower throughout the square, not only in the shaded areas near the buildings, but also in the areas corresponding to the cool pavements and trees. A limited area on the SW angle of the square and the central area not shaded by trees exhibit PET above 37.5 °C, while the minimum value below 22.5 °C is located in the same SE area of scenario S0, affected by buildings' shading patterns, and in the central area underneath the *Koelreuteria paniculata* trees. On NW, under the other trees, *Citrus x aurantium* and *Albizia julibrissin*, characterized by with a smaller crown (LAD respectively of 0.55 m^2/m^3 and 0.65 m^2/m^3), PET is comprised between 22.5 °C and 25.0 °C. Scenario S2 (c) presents a distribution of PET values considerably similar to S1, though with minimum values below 22.5 °C spreading on a larger area, corresponding to the external portion of the central area shaded by *Koelreuteria paniculata* trees. Consistently with S0, along the northern façade, the unshaded asphalt area displays PET values between 35.0 °C and 37.5 °C, while the PET of corresponding area of S1, covered with cool materials, varies between 32.5 °C and 35.0 °C. At 13:00, S0 (d) presents minimum PET values between 32.5 °C and 35 °C in the shaded area near the northern façades in the adjacent urban canyons, while maximum PET values above 42.5 °C comprehend the 90% of the square exposed to direct solar radiation. The distribution of PET values for S1 (e) mainly coincides with S0 in the periphery of the square, but it is significantly lower in the shaded areas below the trees, where the PET values vary between 32.5 °C and 35.5 °C, with a reduction of approximately −8.7 °C compared to the corresponding unshaded areas in S0. PET values of S2 (f) have a distribution similar to S1, but the absolute values are lower under the *Koelreuteria paniculata* trees (between 32.5 °C and 35 °C for S1, between 27.5 °C and 30.0 °C for S2) and in the unshaded centre of the square (above 42.5 °C for S1, between 40.0 °C and 42.5 °C for S2) and higher in the unshaded asphalt area along the northern façade, which in S1 is covered with cool materials (above 42.5 °C for S2, between 40.0 °C and 42.5 °C for S1). At 17:00, the distribution of PET values for S0 (g) is symmetrical to 8:00, following the symmetrical shading patterns, while absolute values are generally higher: the minimum PET values between 27.5 °C and 32.5 °C concentrated on the NW areas shaded by buildings, while maximum PET values exceeding 42.5 °C distributed on the SE areas directly exposed to solar radiation. S1 (h) displays an overall decrease of PET throughout the square, with a minimum of PET between 25.0 °C and 27.5 °C in the center of the square, affected by the shade of both groups of trees and the effect of cool materials, and PET between 27.5 °C and 30.0 along the SE and NW façades. The PET values for S2 (i) mostly correspond to S1, but they are higher along the northern façades, showing the mitigation potential of cool asphalt (S1) in relation to a traditional asphalt (S2), and are lower in the center of the square corresponding to the grass surface and in the inner circle of trees (S2) comparing to a cool pavement (S1).

3.3. Comparison between Surface Temperature and Thermal Comfort of Different Points of the Square for the Simulation Scenarios

The percentage and distribution of greenery and cool materials vary for the different areas of the square, whereas the shading patterns, depending on the urban morphology of the area and on the trees arrangement, vary as well. A direct comparison of the surface temperature and PET of 3 points of the square throughout the day for scenario S0, S1 and S2 can stress the coupled effects of these parameters on microclimate and thermal comfort. The selected points depicted in Figure 2, are:

1. point A, corresponding to the measurement point, central and exposed to direct radiation in S0, displays the effect of trees shading and cool and green pavements in S1 and S2;
2. point B, located in the upper NE area of the square, not affected by trees and greenery in any scenario, shows the effect of different artificial pavements on thermal comfort with or without buildings' shade;
3. point C, located in the SW area of the square, under two trees and in proximity with the building façade, depicts the interactions between building shading patterns, artificial pavement and greenery (grass and trees).

Figure 6a shows the daily trend of surface temperature (T_s) among the scenarios at point A. If we consider the general tendency, T_s for S0 starts from a minimum value of 23.4 °C registered at 8:00, increasing until the peak of 40.5 °C between 14:00 and 15:00, then decreasing till 20:00 with a of 29.0 °C. At 8:00, T_s for S1 presents a value of 23.8 °C, and an increasing trend consistent with S0 up to 29.5 °C at 10:00; then the trend changes considerably, with a constant decrease up to 24.0 °C at 20:00. S2, on the other hand, displays a T_s at 8:00 of 25.2 °C, slightly higher than scenario S0 and then follows a trend similar to scenario S1, but lower, with a peak of 29.4 °C at 10:00 and a minimum of 21.4 °C at 20:00. Focusing the attention on 8:00, 13:00, 17:00, which correspond to the highest momentary attendance (Figure 3), at 8.00 T_s for S0 is 23.4 °C, for S1 is slightly higher, presenting a value of 23.8 °C, whereas it is 25.2 °C for S2, higher than S0 and S1. At 13:00, T_s for S0 is 39.1 °C, whereas it is 26.5 °C for S1 and 24.6 °C for S2, with a variation from S0 of −12.6 °C and −14.5 °C respectively. At 17:00 T_s for S0 is 37.3 °C, for S1 26.1 °C, 24.2 °C for S2, with a significant decrease of −11.2 °C and −13.1 °C comparing to scenario S0.

Figure 6b shows the daily trend of T_s among the scenarios for point B. T_s for S0 slightly increases from the initial value of 22.1 °C at 8:00 up to 27.4 °C at 11:00, then it abruptly increases to 39.6 °C at 12:00, when the shade of the surrounding buildings recedes, with a peak of 44.6 °C at 14:00 and a decrease up to 29 °C at 20.00. S2 shows a considerably similar trend, overlapping and, in some hours, also exceeding the value of S0, whereas T_s for S1 is drastically lower, with a minimum of 22.5 °C at 8:00 and a maximum of 27.6 °C at 16:00. Comparing T_s at 8:00, 13:00, 17:00 for the three scenarios, at 8:00, S0 presents a T_s for 22.1 °C, while S1 of 20.7 °C and S2 of 22.5 °C, with a difference with S0 of −1.4 °C and +0.4 °C. At 13:00, S0 displays a T_s of 43.1 °C, for S1 the corresponding value of 26.5 °C is significantly lower, whereas S2 appears to have a similar T_s as S0, with a value of 43.6 °C. The differences between S1 and S2 with S0 are, respectively, −16.6 °C and +0.5 °C. At 17:00, T_s for S0 is 40.5 °C, whereas for S1 it is 27.3 °C, with a notable decrease of 26.8 °C, and 40.4 °C for S2, only 0.1 °C lower.

Figure 6c shows the daily trend of T_s among the scenarios for point C. T_s for S0 rises from 24.9 °C at 8.00 to 41.4 °C at 14:00, then it significantly decreases to 35.4 °C at 15:00, when the point starts being shaded, until 20:00 when T_s is around 25.0 °C. S1 and S2 present a daily trend lower than S0: S1 has a slight rise from 22 °C at 8:00 to 27.0 °C at 14:00, then it decreases till 24.5 at 20:00. On the other hand, S2 displays values lower than S0 but slightly higher than S1 from 22. 2°C at 8:00 to 26.8 °C at 14:00, then it constantly decreases until 22.3 °C at 20:00. Taking into account the three selected hours, T_s at 8:00 for S0 is 24.9 °C, whereas S1 has a T_s of 22 °C and S2 a similar value of 22.2 °C, both with a difference towards S0 of −3.0 °C approximately. At 13:00, the T_s for S0 is 40.5 °C, whereas S1 presents a value of 26.4 °C and scenario S2 a value of 28 °C. The relative difference between S0 and the other two scenarios is −14 °C for S1 and −12.5 °C for S2. At 17.00, T_s for S0 is 32.1 °C, whereas for S1 is 26.4 °C and for S2 is 25.4 °C. In this case, the difference between S0 and S1 is around −6 °C, between S0 and S2 is around −7.0 °C.

As shown in Figure 7a, which represents the daily trends of PET values for point A, S0 displays a PET value of 36.8 °C at 8:00, it increases until a peak of 47.8 °C at 15:00 and then shows an abrupt decrease between 16:00 and 17:00, with a value of 25.1 °C, until the evening at 20:00, when it reaches 20.1 °C. PET trends for S1 and S2 are similar and appear to have lower values corresponding to S0 throughout the day, except in the morning at 8:00 and 9:00, when both scenarios have slightly higher values than S0. In general, PET for S1 is higher than for S2: in fact, S0 has a mean PET difference with S1 and S2 of −7.2 °C and −8.4 °C respectively, while the difference of mean PET between S1 and S2 corresponds to +1.2 °C.

(a)

(b)

(c)

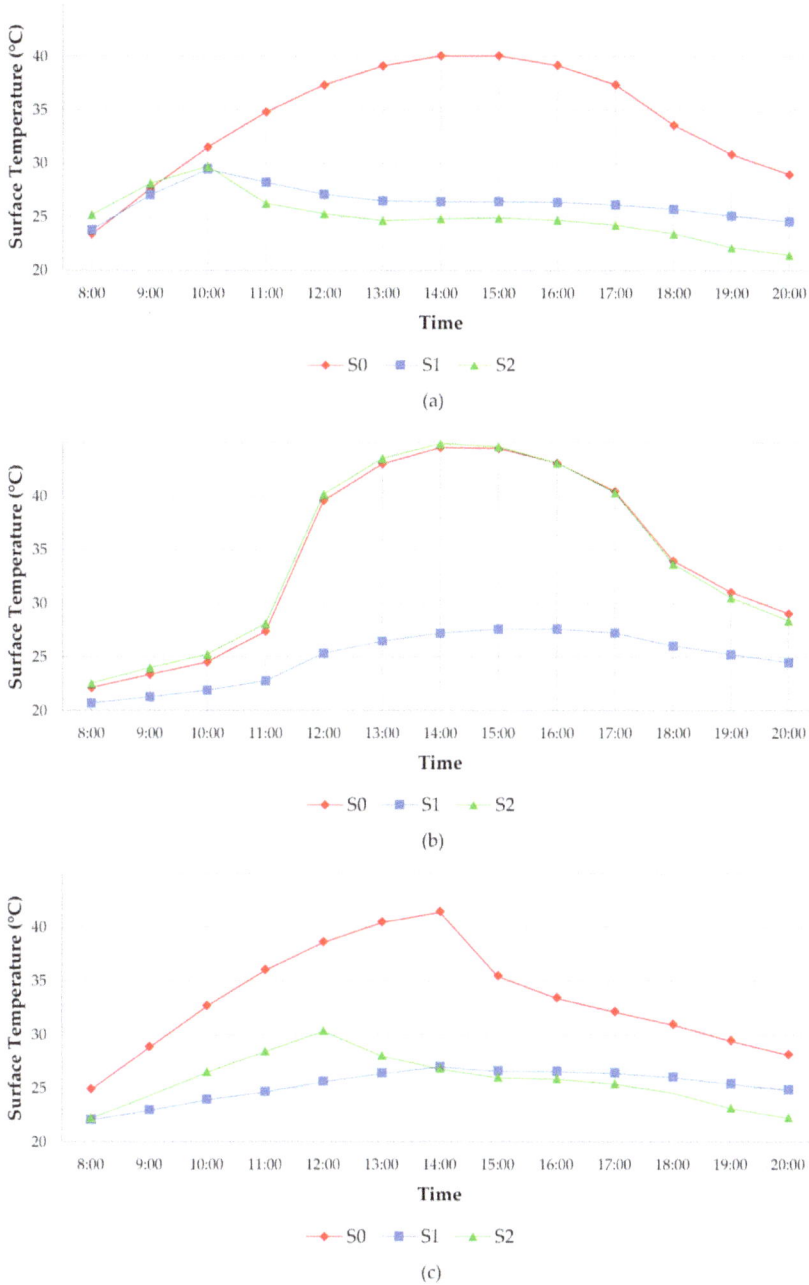

Figure 6. (**a**) Daily trend, on the 1st August 2014, of surface temperature (T_s) in point A among the three scenarios: S0 (initial condition), S1 (cool pavements + trees) and S2 (cool roofs + cool walls + trees + grass); (**b**) daily trend of surface temperature (T_s) in point B among the three scenarios: S0, S1, S2; (**c**) daily trend of surface temperature (T_s) in point C among the three scenarios: S0, S1, S2.

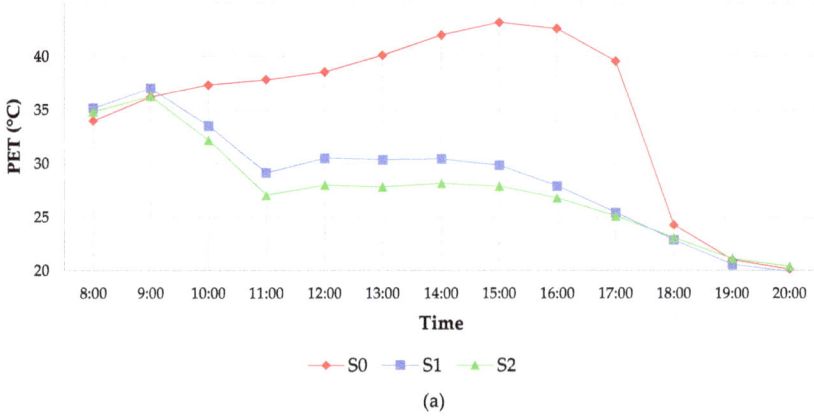

(a)

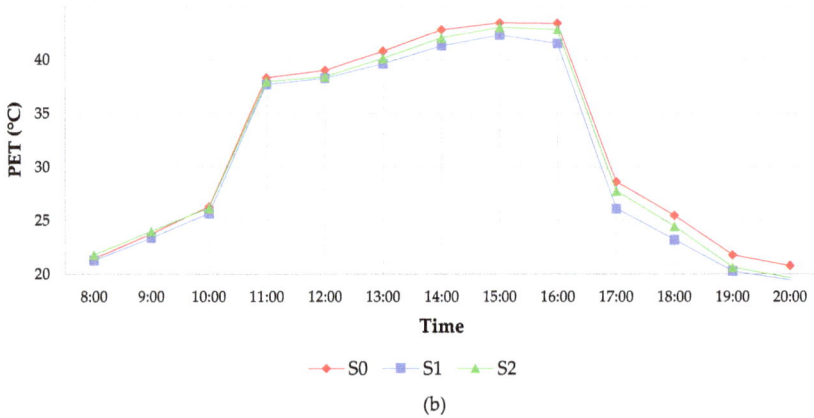

(b)

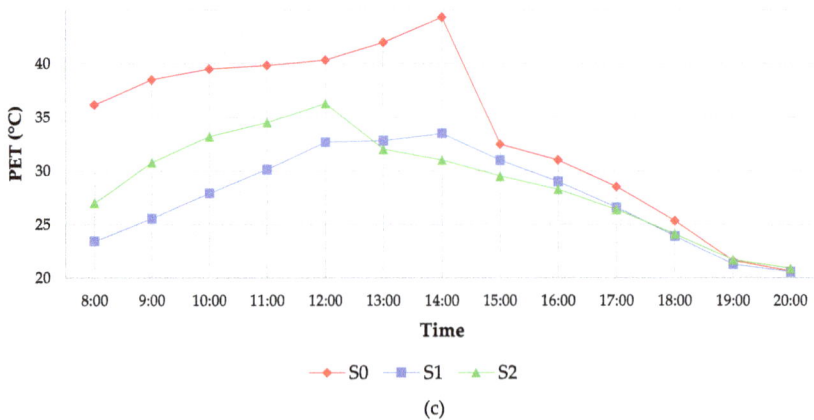

(c)

Figure 7. (**a**) Daily trend, on 1st August 2014, of PET values in point A among the three scenarios: S0 (initial condition), S1 (cool pavements + trees) and S2 (cool roofs + cool walls + trees + grass); (**b**) daily trend of PET values in point B among the three scenarios: S0, S1, S2; (**c**) daily trend of PET values in point C among the three scenarios: S0, S1, S2.

If we consider PET values for the scenario at 8:00, 13:00 and 17:00, at 8:00 S0 presents a PET value of 36.8 °C, whereas S1 a PET of 38.2 °C and S2 a PET of 37.8 °C. The difference between S0 with S1 and S2 are of +1.4 °C for S1 and of −1 °C for S2. At 13:00, the PET value for S0 is 44.1 °C, S1 presents a PET of 32.4 °C and for S2 the PET value is 29.4 °C, with a variation of −12 °C for S1 and −15 °C for S2 against S0. At 17:00, S0 presents a PET value of 43.5 °C, whereas S1 depicts a value of 26.5 °C and similarly S2 a value of 26.1 °C.

Figure 7b shows the daily trends of PET for point B. In this case, all the three scenarios display a very similar tendency in terms of intensity and daily distribution of PET, since point B is far from the direct effects of trees and vegetation and, compared to S0, the most relevant changed urban design parameter for S1 is the cool pavement, whereas for S2 it is the cool material of the near building façades. The scenarios present an PET value at 8:00 between 21.0 °C and 22.0 °C, then they all increase consistently after 10:00, when shade from the buildings moves, until they reach the peak at 15:00 with PET values between 47.0 °C and 48.0 °C, and then they decrease suddenly 15:00 and 16:00, when the point is shaded again, with PET values around 29.0 °C, continuing to decrease until 20:00 with PET values of about 20.0 °C. The mean PET difference for S1 and S2 in relation to scenario S0 is, respectively, −1.3 °C and −0.6 °C. At 8:00, S0 presents a PET value of 21.7 °C, whereas S1 registers a PET of 21.5 °C and S2 a PET of 22.1 °C. The difference between S1 and S2 to S0 are of −0.2 °C for S1 and of +0.4 °C for S2. At 13:00, the PET value of S0 is 44.1 °C, S1 presents a PET of 43.5 °C and S2 a PET value of 44.1 °C, registering a variation of −0.6 °C for S1 in relation to S1, and no variation for S2. At 17:00, S0 presents a PET value of 30.3 °C, whereas S1 of 27.3 °C and S2 of 29.2 °C, decreasing of −3 °C and −1.1 °C from S0, respectively.

Figure 7c shows the daily trend of PET values among the three scenarios for point C. In this case, the PET of S0 presents generally a higher tendency in relation to S1 and S2. It is rather high at 8:00, with a value of 39.4 °C, then it increases until it reaches the peak at 14:00, when it measures 49.2 °C and then it decreases suddenly at 15:00, with a value of 34.5 °C, until it reaches the minimum at 20:00, with a value of 20. 8 °C. Regarding S1, PET presents a steady rise from 24.1 °C at 8:00 to 33.2 °C at 14:00 and then a decrease until 20:00, when the PET value is 20.5 °C. S2 presents higher PET values than S1 from 8:00 to the maximum of 34.4 °C at 12:00, then it decreases, with values lower than the corresponding of S2, until 19:00 and 20:00 when it measures 22.0 °C and 21.6 °C, again slightly higher than S1. The mean PET differences of S1 and S2 in relation to S0 is, respectively, around −7.0 °C and −5.5 °C, whereas the difference in terms of mean PET value for S1 in relation to S2 is around −1.5 °C. Considering the three selected hours, at 8:00 the PET value for S0 is 39.4 °C, for S1 is 24.1 °C and for S2 is 28.4 °C. At 13:00, the PET for S0 reaches 46.4 °C, whereas for S1 the PET is equal to 35.4 °C and for S2 to 34.4 °C, displaying a difference in relation to S0 of −11 °C and of −12 °C, respectively. At 17:00, S0 displays a PET value of 30.2 °C, S1 shows a PET value of 27.9 °C and S2 a PET value of 27.6 °C, with a difference in relation to S0 of about −2.3 °C for S1 and −2.6 °C for S2.

The following tables, Tables 4 and 5, offer a systematic overview of the surface temperature and PET values' trends throughout the day in the investigated scenarios, in order to underline the minimum and maximum values as well as the general trends of each scenarios.

Table 4. Principal surface temperature values in point A, B and C for scenarios S0 (initial condition), S1 (cool pavements + trees) and S2 (cool roofs + cool walls + tress + grass) at 8:00, 14:00 and 17:00, on 1 August 2014.

Time	T_S (A)			T_S (B)			T_S (C)		
	S0	S1	S2	S0	S1	S2	S0	S1	S2
8:00	23.4	23.8	25.2	22.1	20.7	22.6	24.9	22.1	22.2
14:00	40.0	26.4	24.8	44.6	27.3	44.9	41.4	27.0	26.8
17:00	37.4	26.1	24.2	40.5	27.3	40.4	32.1	26.4	25.4

Table 5. Principal PET values in point A, B and C for scenarios S0 (initial condition), S1 (cool pavements + trees) and S2 (cool roofs + cool walls + tress + grass) at 8:00, 14:00 and 17:00, on 1 August 2014.

Time	PET (A)			PET (B)			PET (C)		
	S0	**S1**	**S2**	**S0**	**S1**	**S2**	**S0**	**S1**	**S2**
8:00	36.8	38.2	37.8	21.7	21.5	22.1	39.4	24.1	28.4
14:00	46.4	32.5	29.8	47.3	45.5	46.4	49.2	36.2	33.2
17:00	43.5	26.5	26.1	30.3	27.3	29.2	30.2	27.9	27.7
19:00	21.2	20.7	21.3	22.1	20.3	20.7	21.9	21.5	22.0

4. Discussion

The field analysis of the microclimatic conditions and attendance on 1st August 2014 in San Silvestro Square has highlighted the close connection between the usage of square and outdoor thermal comfort, which is influenced by the morphological features of the urban open space, especially by the shading patterns: in fact, the significantly lower PET values of shaded locations consistently correspond to higher attendance rates.

These results are visible in Figure 3 that compares shading patterns and attendance rate and in Figure 4 that illustrates the PET values in the shaded and unshaded areas of the square, the T_{mrt} in this case has been calculated as shown in Table 3 but not illustrated in a graph since it represents only a part of the more complex thermal comfort sensation; thus, it is more accurate and easy to understand to compare the PET index with shading patterns and attendance rate. In fact, if we take a closer look at the data exposed in the graph (Figure 3) and consider as reference hours the early morning and evening: 8:00 and 18:00, excluding the interval between 12:00 and 13:00 because of the extended solar radiation in the square, it appears clearly that at 8:00 when the shaded areas comprises all of the square surfaces (100% of shaded area), the registered average momentary attendance is 18 people, corresponding to a PET value of 26 °C (Figure 4); the correlation between shaded areas, attendance rate and thermal comfort is confirmed at 18:00 when the shaded area percentage is about 60% of total area of the square and the registered average momentary attendance in shaded locations is 80 people corresponding to a PET value of 30 °C (Figure 4), whereas at the same hour in the evening (18:00) in unshaded locations (about 40% of the square) the average momentary attendance is 5 people corresponding to a PET value of 32 °C (Figure 4).

Based on this finding, the second stage of the study proposes a simulation analysis of two renovation scenarios, S1 and S2, against the existing conditions scenario S0, taking into account the behavior of cool materials and vegetation, in terms of trees and grass, on outdoor thermal comfort focusing the attention on the pedestrian level.

If we consider the PET variations displayed by the two renovation scenarios, both S1 and S2 show a larger distribution of lower PET values during the day comparing to S0. Especially at 13:00, S1 (Figure 5e) depicts a larger distribution of PET values between 32.5 °C and 35 °C while S2, (Figure 5f) displays a larger distribution of PET values between 27.5 °C and 30 °C, demonstrating their mitigation potential of the outdoor thermal comfort during extreme microclimatic conditions.

Regarding the effect of vegetation, the study highlights the significant mitigation potential of trees, which is in line with Ketterer [37] findings. This potential mainly depends on the obstruction of direct solar radiation, which is determined by the trees' geometry, characteristics of the crown and LAD (leaf area density, m^2/m^3). In fact, in S1, Figure 5e, as well as in S2, Figure 5f, the minimum PET values between 27.5 °C and 30 °C, are concentrated underneath the *Koelreuteria paniculata* tree crowns, that presents the highest value of LAD in the square (0.75 m^2/m^3), whereas underneath the *Citrus x aurantium* and the *Albizia julibbrissin*, that present lower LAD values (respectively of 0.55 m^2/m^3 and 0.65 m^2/m^3), the PET values vary between 32.5 °C and 35 °C. This impact appears to be more relevant than the "artificial shade" casted by buildings, probably on the account of their evapotranspiration processes. An example of higher PET values under building shade is visible in

Figure 5e where, in those area underneath the shade of surrounding buildings, for example between point C and the near building's façade or in the courtyard located at the western side of the square, the PET values are slightly higher ranging between 35 °C and 37.5 °C.

Concerning the cool materials, it is important to underline how the present study does not attempt to model all aspects of the complex human response to high-albedo environments which, as stated by Erell et al. [57], is likely to be affected by other considerations in addition to thermal comfort, such as visual comfort; this concerns especially S1 that models and analyzes a cool material characterized by a white appearance and thus a high albedo, whereas S2 models and analyzes a cool colored materials, that presents a light orange colored appearance.

Whether cool materials are applied on pavements (cool pavements) or on façades (cool walls), they exhibit, as expected, a significant cooling effect on surface temperature, which correspond to a variable mitigation effect on thermal comfort, especially when they are directly exposed to solar radiation, without the contribution of natural or artificial shadings. Their performance in lowering the surface temperature can be observed in Figure 6. If we consider a point near a building façade and directly exposed to the solar radiation, as in Figure 6b (point B), the difference between the T_s of S1 in relation to S0 and S2 is significant: the cool pavement modelled in S1 ($\alpha = 0.89$), when exposed directly to solar radiation, registers lower T_s values both compared to S0 and to S2 throughout the day and especially at the peak hour at 14:00 (T_s (S1) = 27.3 °C), when the difference between T_s of S1 in relation to T_s of S0 and S2 is around -17 °C (T_s (S0) = 44.6 °C; T_s (S2) = 45 °C). Regarding the difference between T_s of S1 and T_s of S2 in point B, it is related most likely to the nearby cool wall effect, thus leading to a possible contribution of the cool wall in slightly increasing the T_s of the pavement nearby, around +2 °C at 14:00. On the other hand, if the cool pavement is shaded by a tree, as shown in Figure 6a at point A, its T_s values are lower than the T_s values of S0 (T_s (S1) = 26.4 °C; T_s (S0) = 40 °C), around -13 °C at 14:00, but they are slightly higher than those registered by the combined effect of grass and tree in S2 (24.8 °C), with a difference at 14:00 of +1.6 °C. This is most likely due to the coupled effect of the tree' shading and the evapotranspiration process between tree and grass, contributing to the lowering the T_s. However, cool pavements display a better performance in lowering T_s than grass during the early hours of the morning as visible in Figure 6a, from 8:00 to 13:00. Cool materials generally have a mitigating effect on thermal comfort, but their performance is variable, due to their complex effect on the radiative environment: in fact, Figure 7a shows higher PET values at 8:00 and 9:00 for S1 (38.2 °C and 40.4 °C, respectively) in comparison to S0 (36.7 °C and 39.4 °C). Moreover, Figure 7b, shows that cool pavements of S1 presents lower PET values than S0 at 14:00, around -2 °C, but slightly higher than S2, around +1 °C, due most likely to the contribution of the cool wall on a cobblestone pavement. This finding is in line with Erell et al. [57], and it presumably depends on the reflected solar radiation, which can either be absorbed by the surrounding surfaces increasing their surface temperature [21] or directly by the human body, increasing its energy balance.

Considering shaded cool materials, the coupled effect of tree and an impervious high reflective cool colored pavement of S1 results in lower PET values than the existing impervious low albedo pavement of S0 (Figure 5e), but conversely in higher PET values than the combination of tree and grass of S2 depicted in Figure 5f. This is most likely due to the evapotranspiration and metabolic processes of both tree and grass.

5. Conclusions

The current study presents a two-step analysis, centered first on the evaluation of the connection between shade and thermal comfort, through a field analysis of the thermal and radiative environment on the 1st and 2nd August 2014, highlighting the effect of daily shading patterns on thermal comfort in urban open spaces under extreme microclimate conditions, by means of attendance studies in San Silvestro square, in Rome. In the second stage, the study, by means of a single day microclimate simulation on 1st August 2014, focuses the attention on the evaluation of the mitigation effect that two selected renovation scenarios based on urban greenery and advanced materials, such as cool colored materials, have on outdoor thermal comfort conditions in existing open spaces, in the light of increasing urban overheating.

Thus, it must be underlined that the study represents a very specific analysis on outdoor thermal comfort in a square of the historical Mediterranean tissue. Firstly, because the results analyzed applied under a limited set of meteorological conditions, collected at one site and over two days during the summer period. Second, because of the typology of the space investigated: in fact, all the renovation scenarios have been modelled and studied for a type of square considered typical of the historical tissue of the city of Rome, in terms of size, height of surrounding buildings, H/W ratio and SVF. Consequently, this also implies that the technologies and scenarios analyzed can be replicated in other urban historical squares characterized by a similar H/W ratio and SVF, in which there is a direct effect on the outdoor microclimate from the surrounding built environment, particularly from the urban façades and pavements. Third, because the type of materials used and inquired, in fact even though characterized by high albedo coefficients, on the other hand they constitute a small percentage of the expanding high reflective materials class, thus more investigations are required for other combinations and solutions in this regard.

From the point of view of urban design and urban planning, the research emphasizes the importance of outdoor thermal comfort concerns within the renovation of open spaces and demonstrates the potential of well-chosen "minimum intervention" solutions, respectful of the traditional image of the city, as a driving principle for the sustainable renovation and adaptation of the historic tissues to urban overheating.

The results of this study suggest two main microclimatic aspects relevant for informing urban designers in the renovation processes of the historical tissues. Firstly, the correlation between shading patterns and thermal comfort highlighted by the attendance rate in the square conducted in the first stage of the study; promotes the widespread use of shading devices, integrated with local urban morphology and usage patterns. Second, the different upgrade in terms of pedestrian thermal comfort offered by trees and advanced materials technologies, such as cool pavements, in urban open spaces. In fact, the second stage of analysis highlights the effective result that innovative surface materials can have on urban overheating, especially when coupled with natural shading devices, such as trees.

In general, the findings from this study show how greenery appears to have a better effect on outdoor thermal comfort than traditional artificial materials, because of the influence of evapotranspiration on the cooling loads [37]: in fact, the second renovation scenario (S2) with an extensive use of trees and grass pavements, shows the largest distribution of lower PET values in hours of the day with the highest attendance (Figure 5).

Nevertheless, cool pavements can also bring a tangible improvement on urban environment, as they generally have a positive mitigating effect on outdoor thermal comfort and are less expensive and with low maintenance rate than vegetated surfaces [49]. However, their interaction with the radiative environment and the higher reflection of solar radiation could have a counteracting effect [21], especially when coupled with high reflective walls, thus their use requires an attentive evaluation going beyond surface temperature and air temperature analysis, but focused on overall comfort parameters such as PET index. In addition, it could be relevant to investigate the effects that high reflective materials have, not only on the microscale but on the large scale as well, if applied on pavements and façades as well.

Particularly, if we consider the different results among the three case studies scenarios at 13:00, selected because of the high solar irradiance, the performance of the cooling technologies and strategies investigated is evident. Both the renovation scenarios, the first scenario (cool pavements + trees) and the second scenario (cool roofs + cool walls + trees + grass), show, as aforementioned, a general upgrade of the thermal conditions regarding the T_s values and specifically the PET values.

T_s was found to be 16.6 °C lower in correspondence of a cool pavement characterized by a high albedo (a = 0.89) and exposed to direct solar radiation (point B, first scenario) as visible in Figure 6b, whereas it was found to be 0.5 °C higher, in the same location, for the second scenario in correspondence of grass (a = 0.2). *Ts* was found to be 12.6 °C lower (point A, first scenario) in correspondence of cool pavement (a = 0.89) and of a high crown density tree such as the *Koeulreuteria paniculata* tree (a = 0.70, LAD = 0.75), whereas it was found to be 14.5 °C lower in correspondence with grass (a = 0.2) and the *Koeulreuteria paniculata* tree (point A, second scenario).

PET was found to be 14.8 °C lower in correspondence of the *Koeulreuteria paniculata* tree (a = 0.70, LAD = 0.75) and grass (point A, second scenario; see Figure 7a), a similar performance of the PET value is visible in correspondence of a high crown density tree such as the *Koeulreuteria paniculata* tree and a high albedo pavement (a = 0.89), where the PET is 11.8 °C lower (point a; first scenario). However, if we consider a point without tree (e.g., point B) the PET value trends are different, in fact: in correspondence of a cool pavement with a high albedo (a = 0.89) the PET is 1.4 °C lower, whereas it is 0.8 °C lower in correspondence of grass only, showing a better performance of the cool pavement.

Thus, if we consider surface temperature at 13:00 in those areas of the square mostly affected by solar radiation, the most efficient strategies in terms of cooling the pavement surfaces is the cool pavement used alone (point B, first scenario; see Figure 6b), followed by the combination of tree and grass (point A, second scenario; see Figure 6a).

On the other hand, if we consider the thermal comfort conditions in terms of PET at 13:00, in a point exposed to solar radiation excluded by the shades of surrounding buildings but under natural shading devices, such as the *Koeulreuteria paniculata* tree, the most effective design combination is represented by tree and grass, in the second scenario, with a corresponding cooling effect of −14.8 °C, followed by cool pavement and tree, first scenario, with a corresponding cooling effect of −11.8 °C.

From a design point of view, the present study underlines the importance of shaded areas and the positive effect of combined green and cool technologies, such as trees and cool pavements, on outdoor thermal comfort. Therefore, taking into account these observations, the findings of the study are intended for the urban planners and designers not as detailed quantitative guidelines but as designing considerations for the configuration of the urban open space. In this regard, the microclimatic analysis is relevant not in the way it offers specific quantities to apply thoroughly in a determined urban area, but in the way it informs the urban designer of the processes and behaviors related to morphological and materials factors.

Regarding the shading factor, it is important to guarantee a certain amount of shaded spaces, which depends on the climate and the use of the square. In fact, the effect of the shadow on the microclimate is local and site specific; in other words, shadow, whether derived by an artificial or natural shading object, has a very high effect in the specific local area, as visible from the ENVI-met thermal maps and from the graphs, and as previously underlined in an interesting study conducted in Stuttgart by Ketterer and Matzarakis [37].

Regarding the cool materials technologies, cool pavements showed a general upgrade of the outdoor thermal comfort conditions and a mitigation effect towards the urban surfaces temperature, albeit when applied to building walls they showed a different performance, especially regarding the related comfort conditions. Thus, further investigations are needed in order to better understand their behavior in different orientation (the surface orientation) and configurations (in relation to other materials or natural element): in fact, if we consider the solution of cool pavements and trees, the combination showed a good performance for the mitigation potential and for the thermal comfort upgrade, but the same technology when applied on a vertical surface such as a wall showed different effects on the outdoor microclimate and nearby thermal comfort.

These findings suggest different themes and directions, especially in the field of innovative and advanced materials. Particularly, it could be interesting to investigate the effect of a wall characterized by a combination of technologies: such as green wall in the lower section, at the pedestrian level, and cool materials applied on the upper part of the façades, thus influencing the microclimate and avoiding the direct negative effects the pedestrians. Regarding the vegetation, further simulations are needed in order to evaluate the performance of different trees, since the cooling effect on the outdoor environment vary considerably in green typology, location and orientation. From a methodological point of view, it could be interesting to test in situ the designed renovations scenarios and to carry out a field survey of the pedestrian thermal comfort. In conclusion, concerning the future developments, the observations made in this study have identified some key issues, in particular the different thermal performance of cool materials according to the surface orientation and to the combination with other cooling strategies, which require attention at the level of urban design, in addition the findings emphasize the importance to proceed with further research on the influence of urban design parameters on microclimate and comfort in a wider and more complete range of climatic conditions (eg., winter, spring, autumn), possibly in other squares in the city of Rome in order to analyze open spaces belonging to the same climate and typology of tissue but with different SVFs, and then to carry out further investigations in a wider range of locations characterized by other climates.

Acknowledgments: The authors are also grateful to Ms Silvia Cimini, Ms Arianna Marino and Ms Irene Trombetta for their help during the field data collection and to Ms Ioanna Skoufali for her help in the ENVI-met PET calculation.

Author Contributions: Letizia Martinelli conceived and conducted the experiments and analysis on the summer attendance in San Silvestro square, Flavia Laureti and Alessandra Battisti conceived the renovation scenarios, Flavia Laureti designed and performed the microclimate simulations, Flavia Laureti and Letizia Martinelli analyzed the data and wrote the paper, Alessandra Battisti supervised the research and reviewed the paper.

Conflicts of Interest: The authors declare no conflict of interest.

References

1. Santamouris, M.; Kolokotsa, D. *Urban Climate Mitigation Techniques*; Routledge-Earthscan: London, UK, 2016.
2. Stocker, T. *Climate Change 2013: The Physical Science Basis: Working Group I Contribution to the Fifth Assessment Report of the Intergovernmental Panel on Climate Change*; Cambridge University Press: Cambridge, UK; New York, NY, USA, 2014.
3. Martinelli, L.; Lin, T.-P.; Matzarakis, A. Assessment of the influence of daily shadings pattern on human thermal comfort and attendance in Rome during summer period. *Build. Environ.* **2015**, *92*, 30–38. [CrossRef]
4. Gehl, J.; Koch, J. *Life Between Buildings: Using Public Space*; Van Nostrand Reinhold: New York, NY, USA, 1987.
5. Katzschner, L.; Bosch, U.; Röttgen, M. Behaviour of people in open spaces in dependence of thermal comfort conditions. In Proceedings of the 23rd Conference on Passive and Low Energy Architecture (PLEA), Geneva, Switzerland, 6–8 September 2006.
6. Hwang, R.-L.; Lin, T.-P.; Matzarakis, A. Outdoor thermal comfort in university campus in hot-humid regions. In Proceedings of the Seventh International Conference on Urban Climate, Yokohama, Japan, 29 June–3 July 2009.
7. Nikolopoulou, M. *Designing Open Spaces in the Urban Environment: A Bioclimatic Approach: Ruros–Rediscrovering the Urban Realm and Open Spaces*; Center for Renewable Energy Sources: Attiki, Greece, 2004.
8. Lynch, K. *The Image of The City*; MIT Pr.: Cambridge, MA, USA, 1979.
9. Krier, R. *Lo Spazio Della Città*; Clup: Milano, Italy, 1982.
10. Gehl, J. *Cities for People*; Island Press: Washington, DC, USA, 2010.
11. Thorsson, S.; Lindqvist, M.; Lindqvist, S. Thermal bioclimatic conditions and patterns of behaviour in an urban park in Göteborg, Sweden. *Int. J. Biometeorol.* **2004**, *48*, 149–156. [CrossRef] [PubMed]
12. American Society of Heating, Refrigerating and Air-Conditioning Engineers. *Thermal Environmental Conditions for Human Occupancy: ANSI/ASHRAE Standard 55-2004*; ANSI: Atlanta, GA, USA, 2004.
13. Nikolopoulou, M.; Steemers, K. Thermal comfort and psychological adaptation as a guide for designing urban spaces. *Energy Build.* **2003**, *35*, 95–101. [CrossRef]
14. Kántor, N.; Unger, J. The most problematic variable in the course of human-biometeorological comfort assessment: The mean radiant temperature. *Cent. Eur. J. Geosci.* **2011**, *3*, 90–100. [CrossRef]
15. Matzarakis, A.; Rutz, F.; Mayer, H. Modelling radiation fluxes in simple and complex environments: Basics of the RayMan model. *Int. J. Biometeorol.* **2010**, *54*, 131–139. [CrossRef] [PubMed]
16. Oke, T.R. *Initial Guidance to Obtain Representative Meteorological Observations at Urban Sites*; World Meteorological Organization: Geneva, Switzerland, 2004.
17. Oke, T.R. *Boundary Layer Climates*; Routledge: London, UK; New York, NY, USA, 1987; ISBN 978-0-415-04319-9.

18. Santamouris, M. Cooling the cities—A review of reflective and green roof mitigation technologies to fight heat island and improve comfort in urban environments. *Sol. Energy* **2014**, *103*, 682–703. [CrossRef]
19. Mihalakakou, G.; Santamouris, M.; Papanikolaou, N.; Cartalis, C.; Tsangrassoulis, A. Simulation of the Urban Heat Island Phenomenon in Mediterranean Climates. *Pure Appl. Geophys.* **2004**, *161*, 429–451. [CrossRef]
20. Kolokotsa, D.; Santamouris, M.; Zerefos, S. Green and cool roofs' urban heat island mitigation potential in European climates for office buildings under free floating conditions. *SE Sol. Energy* **2013**, *95*, 118–130. [CrossRef]
21. Yang, J.; Wang, Z.-H.; Kaloush, K.E. Environmental impacts of reflective materials: Is high albedo a 'silver bullet' for mitigating urban heat island? *Renew. Sustain. Energy Rev.* **2015**, *47*, 830–843. [CrossRef]
22. Santamouris, M. *Energy and Climate in the Urban Built Environment*; James & James: London, England, 2001; ISBN 978-1-873936-90-0.
23. Erell, E.; Boneh, D.; Pearlmutter, D.; Bar-Kutiel, P. Effect of high-albedo materials on pedestrian thermal sensation in urban street canyons in hot climates. In Proceedings of the 29th Conference, Sustainable Architecture for a Renewable Future (PLEA2013), Munich, Germany, 10–12 September 2013.
24. Cartalis, C.; Synodinou, A.; Proedrou, M.; Tsangrassoulis, A.; Santamouris, M. Modifications in energy demand in urban areas as a result of climate changes: an assessment for the southeast Mediterranean region. *Energy Convers. Manag.* **2001**, *42*, 1647–1656. [CrossRef]
25. Akbari, H.; Konopacki, S. Energy effects of heat-island reduction strategies in Toronto, Canada. *Energy* **2004**, *29*, 191–210. [CrossRef]
26. Vallati, A.; Vollaro, A.D.L.; Golasi, I.; Barchiesi, E.; Caranese, C. On the Impact of Urban Micro Climate on the Energy Consumption of Buildings. *Energy Procedia* **2015**, *82*, 506–511. [CrossRef]
27. Organisation de Coopération et de Développement Économiques. *Climate Change Mitigation What Do We Do?*; OECD: Paris, France, 2008.
28. Shashua-Bar, L.; Hoffman, M.E. Vegetation as a climatic component in the design of an urban street: An empirical model for predicting the cooling effect of urban green areas with trees. *Energy Build.* **2000**, *31*, 221–235. [CrossRef]
29. Lindberg, F.; Grimmond, C.S.B. The influence of vegetation and building morphology on shadow patterns and mean radiant temperatures in urban areas: model development and evaluation. *Theor. Appl. Climatol.* **2011**, *105*, 311–323. [CrossRef]
30. Akbari, H. Shade trees reduce building energy use and CO2 emissions from power plants. *Environ. Pollut.* **2002**, *116*, S119–S126. [CrossRef]
31. Berry, R.; Livesley, S.J.; Aye, L. Tree canopy shade impacts on solar irradiance received by building walls and their surface temperature. *Build. Environ.* **2013**, *69*, 91–100. [CrossRef]
32. Srivanit, M.; Hokao, K. Evaluating the cooling effects of greening for improving the outdoor thermal environment at an institutional campus in the summer. *Build. Environ.* **2013**, *66*, 158–172. [CrossRef]
33. Upreti, R.; Wang, Z.-H.; Yang, J. Radiative shading effect of urban trees on cooling the regional built environment. *UFUG Urban For. Urban Green.* **2017**, *26*, 18–24. [CrossRef]
34. Saaroni, H.; Amorim, J.; Hiemstra, J.; Pearlmutter, D. Urban Green Infrastructure as a tool for urban heat mitigation: Survey of research methodologies and findings across different climatic regions. *UCLIM Urban Clim.* **2018**, *24*, 94–110. [CrossRef]
35. Roy, S.; Byrne, J.; Pickering, C. A systematic quantitative review of urban tree benefits, costs, and assessment methods across cities in different climatic zones. *UFUG Urban For. Urban Green.* **2012**, *11*, 351–363. [CrossRef]
36. Shashua-Bar, L.; Tsiros, I.X.; Hoffman, M. Passive cooling design options to ameliorate thermal comfort in urban streets of a Mediterranean climate (Athens) under hot summer conditions. *BAE Build. Environ.* **2012**, *57*, 110–119. [CrossRef]
37. Ketterer, C.; Matzarakis, A. Human-biometeorological assessment of heat stress reduction by replanning measures in Stuttgart, Germany. *Landsc. Urban Plan.* **2014**, *122*, 78–88. [CrossRef]
38. Mayer, H. Workshop "Ideales Stadtklima" am 26. Oktober 1988 in München. *DMG Mitteilungen 3/89*; 1989, pp. 52–54. Available online: http://www.geographie.uni-freiburg.de/publikationen/fgh-index (accessed on 10 December 2017).
39. Cheung, P.K.; Jim, C. Comparing the cooling effects of a tree and a concrete shelter using PET and UTCI. *Build. Environ.* **2018**, *130*, 49–61. [CrossRef]

40. Levinson, R.; Akbari, H.; Konopacki, S.; Bretz, S. Inclusion of cool roofs in nonresidential Title 24 prescriptive requirements. *Energy Policy* **2005**, *33*, 151–170. [CrossRef]

41. Bretz, S.; Akbari, H.; Rosenfeld, A.; Taha, H. *Implementation of Solar-Reflective Surfaces: Materials and Utility Programs*; Lawrence Berkeley Lab.: Berkeley, CA, USA, 1992.

42. Kolokotsa, D.; Maravelaki-Kalaitzaki, P.; Papantoniou, S.; Vangeloglou, E.; Saliari, M.; Karlessi, T.; Santamouris, M. Development and analysis of mineral based coatings for buildings and urban structures. *SE Sol. Energy* **2012**, *86*, 1648–1659. [CrossRef]

43. Levinson, R.; Berdahl, P.; Akbari, H. Solar spectral optical properties of pigments—Part I: model for deriving scattering and absorption coefficients from transmittance and reflectance measurements. *Sol. Energy Mater. Sol. Cells* **2005**, *89*, 319–349. [CrossRef]

44. Karlessi, T.; Santamouris, M.; Apostolakis, K.; Synnefa, A.; Livada, I. Development and testing of thermochromic coatings for buildings and urban structures. *Sol. Energy* **2009**, *83*, 538–551. [CrossRef]

45. Karlessi, T.; Santamouris, M.; Synnefa, A.; Assimakopoulos, D.; Didaskalopoulos, P.; Apostolakis, K. Development and testing of PCM doped cool colored coatings to mitigate urban heat island and cool buildings. *Build. Environ.* **2011**, *46*, 570–576. [CrossRef]

46. Akbari, H.; Cartalis, C.; Santamouris, M.; Synnefa, A.; Kolokotsa, D.; Muscio, A.; Pisello, A.L.; Rossi, F.; Wong, N.H.; Zinzi, M. Local climate change and urban heat island mitigation techniques–The state of the art. *J. Civ. Eng. Manag.* **2016**, *22*, 1–16. [CrossRef]

47. Akbari, H.; Matthews, H.D.; Seto, D. The long-term effect of increasing the albedo of urban areas. *Environ. Res. Lett.* **2012**, *7*. [CrossRef]

48. Rossi, F.; Bonamente, E.; Nicolini, A.; Anderini, E.; Cotana, F. A carbon footprint and energy consumption assessment methodology for UHI-affected lighting systems in built areas. *Energy Build.* **2016**, *114*, 96–103. [CrossRef]

49. Pisello, A.L. State of the art on the development of cool coatings for buildings and cities. *Sol. Energy* **2017**, *144*, 660–680. [CrossRef]

50. Santamouris, M.; Synnefa, A.; Karlessi, T. Using advanced cool materials in the urban built environment to mitigate heat islands and improve thermal comfort conditions. *Sol. Energy* **2011**, *85*, 3085–3102. [CrossRef]

51. Synnefa, A.; Santamouris, M.; Livada, I. A study of the thermal performance of reflective coatings for the urban environment. *SE Sol. Energy* **2006**, *80*, 968–981. [CrossRef]

52. Santamouris, M.; Gaitani, N.; Spanou, A.; Saliari, M.; Giannopoulou, K.; Vasilakopoulou, K.; Kardomateas, T. Using cool paving materials to improve microclimate of urban areas–Design realization and results of the flisvos project. *BAE Build. Environ.* **2012**, *53*, 128–136. [CrossRef]

53. Kolokotsa, D.-D.; Santamouris, M.; Akbari, H. *Advances in the Development of Cool Materials for the Built Environment*; Bentham Science Publishers: Sharjah, United Arab Emirates, 2013.

54. Georgakis, C.; Zoras, S.; Santamouris, M. Studying the effect of "cool" coatings in street urban canyons and its potential as a heat island mitigation technique. *SCS Sustain. Cities Soc.* **2014**, *13*, 20–31. [CrossRef]

55. Development of Cool Pavement with Dark Colored High Albedo Coating. Available online: https://ams. confex.com/ams/AFAPURBBIO/techprogram/paper_79804.htm (accessed on 10 January 2018).

56. Piselli, C.; Castaldo, V.; Pigliautile, I.; Pisello, A.; Cotana, F. Outdoor comfort conditions in urban areas: on citizens' perspective about microclimate mitigation of urban transit areas. *Sustain. Cities Soc.* **2018**. [CrossRef]

57. Erell, E.; Pearlmutter, D.; Boneh, D.; Kutiel, P.B. Effect of high-albedo materials on pedestrian heat stress in urban street canyons. *UCLIM Urban Clim. Part 2* **2014**, *10*, 367–386. [CrossRef]

58. Chatzidimitriou, A.; Yannas, S. Microclimate development in open urban spaces: The influence of form and materials. *ENB Energy Build.* **2015**, *108*, 156–174. [CrossRef]

59. Köppen, W. The thermal zones of the Earth according to the duration of hot, moderate and cold periods and to the impact of heat on the organic world. *Meteorol. Z.* **2011**, *20*, 351–360. [CrossRef] [PubMed]

60. Mayer, H.; Höppe, P. Thermal comfort of man in different urban environments. *Theor. Appl. Climatol.* **1987**, *38*, 43–49. [CrossRef]

61. Davenport, A.G.; Oke, T.R.; Grimmond, C.S.B. Estimating the roughness of cities and sheltered country. In Proceedings of the 12th Conference on Applied Climatology, Asheville, North Carolina, Boston, MA, USA, 8–11 May 2000; pp. 96–99.

62. Nikolopoulou, M.; Lykoudis, S. Use of outdoor spaces and microclimate in a Mediterranean urban area. *Build. Environ.* **2007**, *42*, 3691–3707. [CrossRef]

63. Spagnolo, J.; de Dear, R.J. A field study of thermal comfort in outdoor and semi-outdoor environments in subtropical Sydney Australia. *Build. Environ.* **2003**, *38*, 721–738. [CrossRef]

64. Gobakis, K.; Kolokotsa, D.; Maravelaki-Kalaitzaki, N.; Perdikatsis, V.; Santamouris, M. Development and analysis of advanced inorganic coatings for buildings and urban structures. *ENB Energy Build.* **2015**, *89*, 196–205. [CrossRef]

65. Höppe, P. The physiological equivalent temperature: A universal index for the biometeorological assessment of the thermal environment. *Int. J. Biometeorol.* **1999**, *43*, 71–75. [CrossRef] [PubMed]

66. Matzarakis, A.; Martinelli, L.; Ketterer, C. Relevance of Thermal Indices for the Assessment of the Urban Heat Island. In *Counteracting Urban Heat Island Effects in a Global Climate Change Scenario*; Springer: Cham, Switzerland, 2016; pp. 93–107; ISBN 978-3-319-10424-9.

67. Cohen, P.; Potchter, O.; Matzarakis, A. Human thermal perception of Coastal Mediterranean outdoor urban environments. *JAPG Appl. Geogr.* **2013**, *37*, 1–10. [CrossRef]

68. Kantor, N.; Unger, J. Benefits and opportunities of adopting GIS in thermal comfort studies in resting places: An urban park as an example. *Landsc. URBAN Plan.* **2010**, *98*, 36–46. [CrossRef]

69. Gulyás, Á.; Unger, J.; Matzarakis, A. Assessment of the microclimatic and human comfort conditions in a complex urban environment: Modelling and measurements. *Build. Environ.* **2006**, *41*, 1713–1722. [CrossRef]

70. Matzarakis, A.; Mayer, H.; Iziomon, M.G. Applications of a universal thermal index: Physiological equivalent temperature. *Int. J. Biometeorol.* **1999**, *43*, 76–84. [CrossRef] [PubMed]

71. Chen, Y.-C.; Matzarakis, A. Modification of physiologically equivalent temperature. *J. Heat Isl. Inst. Int.* **2014**, *9*, 2. [CrossRef]

72. Lin, T.-P. Thermal perception, adaptation and attendance in a public square in hot and humid regions. *BAE Build. Environ.* **2009**, *44*, 2017–2026. [CrossRef]

73. Huttner, S.; Bruse, M.; Dostal, P. Using ENVI-met to simulate the impact of global warming on the microclimate in central European cities. In Proceedings of the 5th Japanese-German Meeting on Urban Climatology, Freiburg, Germany, 6–8 October 2008; pp. 307–312.

74. Winter Release 2017—New Features. *ENVI_MET*. Available online: http://www.envi-met.com/winter-release-2017-new-features/ (accessed on 10 December 2017).

climate

MDPI

Article

The Effect of Increasing Surface Albedo on Urban Climate and Air Quality: A Detailed Study for Sacramento, Houston, and Chicago

Zahra Jandaghian and Hashem Akbari *

Heat Island Group, Building, Civil and Environmental Engineering Department, Concordia University, Montreal, QC H3G 1M8, Canada; z_janda@encs.concordia.ca
* Correspondence: hashem.akbari@concordia.ca

Received: 28 January 2018; Accepted: 15 March 2018; Published: 21 March 2018

Abstract: Increasing surface reflectivity in urban areas can decrease ambient temperature, resulting in reducing photochemical reaction rates, reducing cooling energy demands and thus improving air quality and human health. The weather research and forecasting model with chemistry (WRF-Chem) is coupled with the multi-layer of the urban canopy model (ML-UCM) to investigate the effects of surface modification on urban climate in a two-way nested approach over North America focusing on Sacramento, Houston, and Chicago during the 2011 heat wave period. This approach decreases the uncertainties associated with scale separation and grid resolution and equip us with an integrated simulation setup to capture the full impacts of meteorological and photochemical reactions. WRF-ChemV3.6.1 simulated the diurnal variation of air temperature reasonably well, overpredicted wind speed and dew point temperature, underpredicted relative humidity, overpredicted ozone and nitrogen dioxide concentrations, and underpredicted fine particular matters ($PM_{2.5}$). The performance of $PM_{2.5}$ is a combination of overprediction of particulate sulfate and underprediction of particulate nitrate and organic carbon. Increasing the surface albedo of roofs, walls, and pavements from 0.2 to 0.65, 0.60, and 0.45, respectively, resulted in a decrease in air temperature by 2.3 °C in urban areas and 0.7 °C in suburban areas; a slight increase in wind speed; an increase in relative humidity (3%) and dew point temperature (0.3 °C); a decrease of $PM_{2.5}$ and O_3 concentrations by 2.7 μg/m^3 and 6.3 ppb in urban areas and 1.4 μg/m^3 and 2.5 ppb in suburban areas, respectively; minimal changes in $PM_{2.5}$ subspecies; and a decrease of nitrogen dioxide (1 ppb) in urban areas.

Keywords: WRF-Chem; urban climate; air quality; urban heat island; surface albedo

1. Introduction

Increasing urban albedo is a verifiable and repeatable heat island mitigation strategy to decrease urban temperatures, photochemical reaction rates, and pollution, thus improving human health and comfort [1,2]. Land use changes and anthropogenic heat emissions contribute to the urban heat island effect by increasing air temperature by 0.42 °C and 0.22 °C, respectively [3]. The reflectivity of surfaces in urban areas range from 0.1 to 0.2 and can be increased by use of high reflective materials on roofs, walls, and pavements [1,4]. The effects of surface modifications on urban climate and atmospheric conditions have been investigated in various regions and episodes. Salamanca and Martilli [5] have shown that a higher albedo decreases urban temperature by 1.5–2 °C during hot summer days in Madrid. Fallmann et al. [6,7] showed that increasing surface albedo lead to a decrease in 2-m air temperature and ozone concentrations by nearly 0.5 °C and 5–8% in urban areas of Stuttgart during the 2003 heat wave period. Touchaei et al. [8] showed a decrease in maximum air temperature, ozone (O_3), and fine particulate matters ($PM_{2.5}$) concentrations by up to 0.7 °C, 0.2 ppb, and 1.8 μg/m^3, respectively by increasing surface albedo in Montreal during the 2005 heat wave period. Jandaghian et al. [9] showed that by increasing surface reflectivity during a rainy episode in summer 2009, the air temperature decreases by 0.2 °C, wind speed slightly increases, and relative humidity and precipitation decrease by 2.8% and 0.2 mm, respectively over Montreal.

Air-quality prediction models have been developed in response to the increased concerns regarding the effects of air quality on human health. The interaction between chemistry and meteorology is complicated. Meteorology has its effects on chemistry through temperature (increasing the chemical reactions, photolytic rates and biogenic emissions), cloud formation (affecting mixing, transformation, and scavenging of chemical compounds), precipitation (increasing the removal of trace gases and aerosols), radiation (affecting photolysis rates, isoprene emissions, and chemical reaction rates), wind speed and direction (affecting horizontal transport and vertical mixing of chemical species and aerosol emissions), planetary boundary layer (PBL) height (affecting pollutant concentrations). In turn, chemistry influences meteorology through aerosol (affecting radiation transfer, cloud life time and optical depth, boundary layer meteorology, precipitation, and scattering/absorption); ozone (affecting radiation and temperature); NO_x, CO, and VOCs (as the precursor of O_3 contributions to ozone radiative effects). Land surface properties also affect natural emission and dry deposition [10]. The online coupled weather research and forecasting model with chemistry (WRF-Chem) is developed to simulate meteorological quantities and air pollution concentrations simultaneously [11]. The model is a fully compressible, non-hydrostatic mesoscale numerical weather prediction (NWP) system. WRF-Chem has several physical and chemical parameterizations [12]. The component of air quality is consistent with the meteorological ones within the same transport scheme, grid and physics schemes, and time steps. The spatial and temporal aspects of WRF-Chem application have been analyzed in many studies through one-way approaches, in local, regional or global scales [13–18]. However, a two-way nested approach in WRF-Chem needs to be applied over a larger geographical area through regional and local scales to equip us with an integrated simulation setup to capture the full impacts of meteorological and photochemical reactions. This approach reduces the uncertainties associated with scale separation and grid resolution to investigate the effects of UHI and surface modifications on urban climate and air quality. The morphological, thermal, and micro-scale properties of the urban canopy are considered by coupling of the multi-layer of the urban canopy model (ML-UCM) [19] within WRF-ChemV3.6.1. Applying the multi-layer of the UCM is necessary when analyzing the urbanization impacts on regional climate [20].

The intent of this research is to investigate the effects of urban heat island and increasing surface albedo on meteorological parameters, such as 2-m air temperature, 10-m wind speed, dew point temperature and 2-m relative humidity, and air quality parameters; namely ozone, fine particulate matters, nitrogen dioxide (NO_2), $PM_{2.5}$ subspecies (particulate sulfate ($SO4_{2.5}$), particulate nitrate ($NO3_{2.5}$), and organic carbon ($OC_{2.5}$) concentrations. The simulation domain covers North America with focus on three populated cities in the United States of America: Sacramento in California, Houston in Texas, and Chicago in Illinois. The simulation episode is during the 2011 heat wave period. On 17 July 2011, the heat wave event started over eastern Ontario, southern Quebec and northern New England. The intense heat moved eastwards and peaked on 22 July with Central Park in New York City breaking the record for the day at 40 °C, which was the hottest temperature the city had experienced in over three decades. The heat wave continued to intensify, with temperatures in north Texas exceeding 38 °C, most days, beginning in mid-June. By 23 July, the heat had intensified and reached to 45 °C. For the entire U.S., the 2011 heat wave period was the hottest in 75 years. The duration of the heat event was at the minimum of 7 days to a maximum of 69 days [21]. The paper structure is as follows: the methodology includes model description and simulation setup and surface modification approach, the results and analyses include model performance evaluation and estimating the effects of increasing surface reflectivity on urban climate and regional air quality, the discussion includes the limitations and assumptions of the study, the conclusion includes a summary of the study and future steps.

2. Methodology

2.1. Model Description and Simulation Setup

We used the online weather research and forecasting model with chemistry (WRF-Chem) and coupled it with the multi-layer of the urban canopy model (ML-UCM) to investigate the effects of urban heat island and its mitigation strategy over continental scale through urban scales during the 2011 heat wave period. The first domain covers North America (NA) including Canada, the United States, and the Northern part of Mexico with 445 grids in west–east direction and 338 grids in south–north direction. The horizontal resolution is 12 km. The second, third, and fourth domains cover the Sacramento area (36 × 31 grids), Houston area (41 × 31 grids), and Chicago area (36 × 31 grids) with the horizontal resolution of 2.4 km. The vertical resolution includes 35 vertical layers from the surface to a fixed pressure of ~100 mb (~16 km AGL). Figure 1 shows the simulation domains and land use/land cover. The simulation period extended over the seven consecutive hottest days in 2011, from the 17 to 23 July. We disregarded the first 72 h of the simulation as spin-up period.

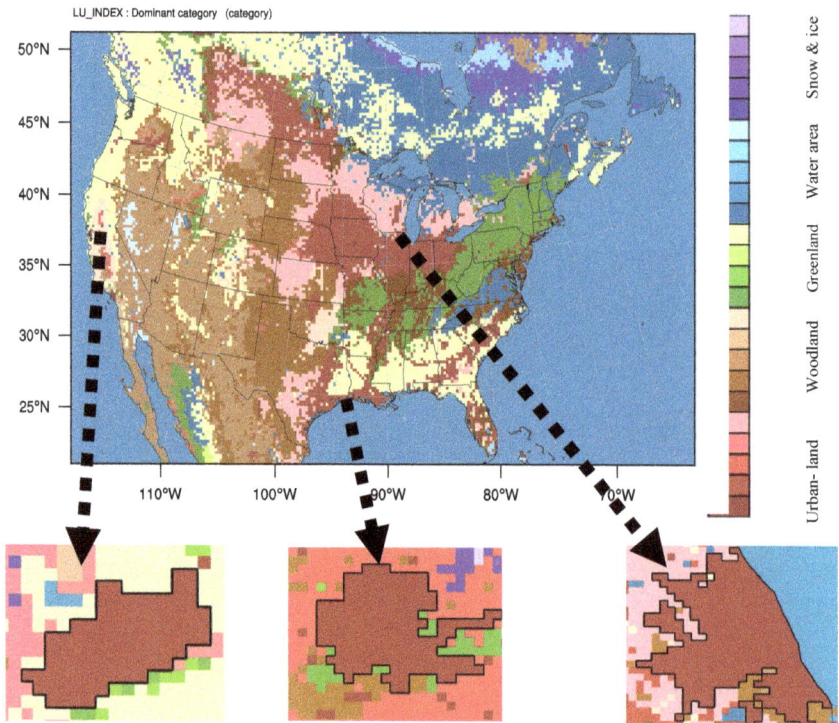

Figure 1. Simulation domains and land use/land cover over North America (mother domain) Sacramento, Houston, and Chicago (inner domains).

The simulation is conducted with the initial and boundary conditions obtained from the North American Regional Reanalysis (NARR) [22,23] with a grid resolution of 32 km and a time resolution of 3-h. Land use was derived from the USGS 24-category data set. We used the Lin scheme as microphysics parameters to evaluate six classes of hydrometeors [24]. Goddard scheme [25] and rapid radiative transfer model (RRTMG; [26]) are respectively selected for shortwave and longwave radiations. The planetary boundary layer is simulated by the Mellor–Yamada–Janjic scheme using Eta similarity theory [27]. The unified Noah land surface model is applied as the land surface scheme. For cumulus parameterization, the Grell–Devenyi ensemble scheme [28] is used. We activated the positive-definite advections of moisture, scalars, and turbulent kinetic energy for model stability [29,30].

Modelling of the chemical compositions in the atmosphere, requires preliminary information concerning the emissions of chemical compounds within the modelling domain. We used the United States National Emission Inventory for 2011 (US-NEI11). The US-NEI11 contains the anthropogenic emissions of the U.S., southern Canada, and northern Mexico in 4-km spatial resolution, divided into point and area sources [31]. Guenther et al. [32] incorporated the quantitative understanding of biogenic emissions into a numerical model named Model of Emissions of Gases and Aerosols from Nature (MEGAN). MEGAN estimates the time resolved gridded BVOC emission estimation in mole/km^2/h and is designed for regional and global emission simulations. The Modal Aerosol Dynamics Model for Europe (MADE) [33] is coupled with the chemistry package to evaluate the radiation and clouds interactions. Within WRF_Chem, the Regional Atmospheric Chemistry Mechanism (RACM) [34] is used for gas-phase reactions. The secondary organic aerosols (SOA) have been incorporated into MADE by means of the Secondary ORGanic Aerosol Model (SORGAM; [35]) Photolysis frequencies are calculated by the Fast_J model scheme [11,36–39]. Table 1 presents the physical and chemical parameterizations applied in WRF-ChemV3.6.1.

Table 1. Physical and chemical parameterizations in WRF_ChemV3.6.1.

Category	Option Used
Microphysics	Lin scheme
Shortwave radiation	Goddard
Longwave radiation	RRTMG
Land surface model	NOAH
Planetary boundary layer scheme	Mellor–Yamada–Janjic Scheme
Cumulus parameterization	Grell Devenyi
Chemistry option	RACM
Photolysis scheme	Fast_J
Aerosol option	MADE/SORGAM
Advection scheme	Runge–Kutta third order
LULC data	USGS 24-class
Anthropogenic emissions	US-NEI11
Biogenic emissions	MEGAN
Urban canopy model	ML-UCM

2.2. Increasing Surface Reflectivity

We conducted two sets of simulations with different scenarios concerning urban surface modifications: the base case condition that the albedo of roofs, walls, and pavements are assumed to be 0.2 (hereafter referred to as CTRL); and the increasing surface albedo scenario where the reflectivity of roofs, walls, and pavements is increased to 0.65, 0.60, and 0.45, respectively (hereafter referred to as ALBEDO). We selected three cities: Sacramento (CA), Houston (TX), and Chicago (IL) based on Akbari et al. [4,40–42] and Rose et al. [43] findings on the urban fabric of these cities. Using high-resolution orthophotography, they found that roofs cover 20–25% and pavements cover 30–40% of urban surfaces. Table 2 presents the urban fabric of Sacramento, Chicago, and Houston [43].

Here, we applied these fractions to characterize the fabric of the selected cities. Hence, the changes of surface albedo modification from the CTRL case as 0.2 to full adoption of roofs and pavements can be calculated as: $0.20 \times 0.65 + 0.45 \times 0.45 = 0.33$ (as an example for Sacramento; an increase of 0.13 of roof albedo and 0.20 of pavement albedo on the urban scale). The change to gridded ALBEDO can be calculated as: (Surface albedo enhancement (roofs, walls, and pavements) × Fraction of urban areas per grid cell) [44].

Table 2. Urban fabric of three cities in NA [43].

Metropolitan Areas	Roofs (%)	Pavements (%)
Sacramento	20	45
Chicago	25	37
Houston	22	30

3. Results and Analyses

3.1. Model Performance Evaluation

The local meteorological patterns affect the transport, transmission, advection, and diffusion of pollutants over the regional scale and further continental scales, which cannot be fully captured by the model to some extent [18]. We evaluated the model performance of WRF_ChemV3.6.1 by comparing the simulation results with observations obtained from weather and air-quality stations in Sacramento, Houston, and Chicago. The weather and air quality monitoring stations were chosen based on their locations close to the downtown of the selected cities (hereafter referred to as urban) and their surroundings (hereafter referred to as suburb). The hourly 2-m air temperature (T2), 10-m wind speed (WS10), 2-m relative humidity (RH2), and dew point temperature (Td) simulation results are compared with the measurements obtained from the U.S. Environmental Protection Agency (EPA) Clean Air Status and Trend Network (CASTNET). The daily averaged modelled fine particular matters ($PM_{2.5}$), ozone (O_3), nitrogen dioxide (NO_2), $PM_{2.5}$ subspecies (particulate sulfate ($SO4_{2.5}$), particulate nitrate ($NO3_{2.5}$), and organic carbon ($OC_{2.5}$)) concentrations are compared with the EPA Air Quality System (AQS) observations using 24-h average data. [45–49]

Here, the time series of simulation results changed to the local time for each specific location: Sacramento: LST = UTC − 7 h; Houston and Chicago: LST = UTC − 5 h. The performance and accuracy of the simulation results are quantitatively based on a series of metrics estimations [50]. Here, we followed the Zhang et al. [51] calculations for the mean bias error (MBE), mean absolute error (MAE), and the root mean square error (RMSE) estimations of the meteorological and chemical parameters.

In terms of meteorological components of the model, the WRF-ChemV3.6.1 effectively captures the diurnal variations of 2-m air temperature, overpredicts 10-m wind speed, overpredicts dew point temperature, and underpredicts 2-m relative humidity. The MBA of T2 (−0.07 °C) shows that the model is capable in predicting air temperature. A small underprediction can be seen in urban areas (~−0.3 °C) that indicates the model deficiency in calculating the heat emission from anthropogenic sources in urban areas accurately. The MAE and RMSE of T2 are approximately 1 °C. Wind speed plays an important role in calculation of air temperature from skin temperature in the land surface model. The 10-m wind speed comparisons show small to large overpredictions (0.3 to 3.15 m/s). The MBA is 1.65 m/s, that shows the model is unable to capture the effects of micro scales and wind patterns. The MAE and RMSE of WS10 are almost 2 m/s. Relative humidity is a function of moisture content, air temperature, and surface pressure. The spatial distribution of RH2 represents an underestimation with the MBE of −1.42%. This underestimation shows that the microphysics scheme miscalculated the processes of transforming water (rain, vapor, cloud, etc.) and moisture fluxes. It also shows the model limitation in capturing the sea surface temperature, wind speed, and their impacts on water mixing ratio and water content of the air properly. The MAE and RMSE of RH2 are nearly 10% and 13%, respectively. Figure 2 shows the time series (hourly) of the observed vs. simulated T2 (°C), WS10 (m/s), and RH2 (%) in the urban areas of Sacramento, Houston, and Chicago. We also calculated the dew point temperature to avoid the dependency on air temperature for the moisture variable. The MBA, MAE, and RMSE of dew point temperature (0.39, 0.53, and 0.65 °C, respectively) show that the model overpredicts the moisture content in the atmosphere especially in urban areas (~0.5 °C).

In terms of the chemical component of the model, the WRF_ChemV3.6.1, as configured here, tends to underpredict 24-h fine particular matters ($PM_{2.5}$) and over-predict the 24-h O_3 concentrations during the 2011 heat wave period. The MBE of the 24-h avg. $PM_{2.5}$ is -1.42 $\mu g/m^3$. The MAE and RMSE of $PM_{2.5}$ are approximately 4 $\mu g/m^3$. This is because the accuracy in fine particular matters concentrations is to some extent a function of its subspecies estimations as particulate sulfate, particulate nitrate, and organic carbons. Thus, we also compared the simulation results of $SO4_{2.5}$ ($\mu g/m^3$), $NO3_{2.5}$ ($\mu g/m^3$), and $OC_{2.5}$ ($\mu g/m^3$) with observations at urban areas of aforementioned cities. We observed that the performance of $PM_{2.5}$ subspecies is a combination of overprediction of particulate sulfate (MBE ~5 $\mu g/m^3$) and underprediction of particulate nitrate (MBE ~-4 $\mu g/m^3$) and organic carbon (MBE ~-3 $\mu g/m^3$). The MAE and RMSE of $SO4_{2.5}$, $NO3_{2.5}$, and $OC_{2.5}$ are approximately 5, 4, and 3 $\mu g/m^3$, respectively. The comparison between simulated ozone and measurements indicated an overestimation of O_3 across the domains (MBE ~5 ppb). The O_3 concentrations is overestimated due to the NO_x and VOCs overestimation in emission inventories and their calculations in chemistry packages (US-NEI11 and MEGAN). The average MAE and RMSE of O_3 are around 7 ppb and 8 ppb, respectively. We also calculated the NO_2 concentrations as one of the precursor in ozone formation. The MBE of NO_2 in urban areas (~2.5 ppb) show that the model tends to overpredict the nitrogen dioxide. The MAE and RMSE of NO_2 is around 4 ppb.

Figure 3 shows the observed vs. simulated $PM_{2.5}$ ($\mu g/m^3$) and O_3 (ppb) concentrations in the urban areas of Sacramento, Houston, and Chicago. Tables 3–5 respectively represent the mean bias error (MBE), mean absolute error (MAE), and the root mean square error (RMSE) of T2 ($^\circ$C), WS10 (m/s), RH2 (%), $PM_{2.5}$ ($\mu g/m^3$), and O_3 (ppb) for aforementioned cities. There are several limitations and assumptions in these comparisons. The simulation results are extracted hourly for all variables, whereas the observation in terms of $PM_{2.5}$ and O_3 are reported as a 24-h average. Figure 4 shows the overall comparison between observed vs. simulated aforementioned parameters in terms of MBA, MAE, and RMSE. Despite the model biases in simulating meteorological and chemical variables, the performance of WRF-ChemV3.6.1 is generally consistent with most air quality models. For comparison of thermal components, the fifth-generation NCAR/Penn State Mesoscale Model (MM5) presented the MBE of T2 as 0.4 $^\circ$C to -3.8 $^\circ$C during a year [52–55]. For comparison of chemical components, the CMAQ model was run during a year and indicated an under estimation of $PM_{2.5}$ as the MBE of -0.6 $\mu g/m^3$ and an overprediction of seasonal O_3 as the MBE of 4.4 ppb [56]. However, given the various differences in physical and chemical parameterizations and input data (different simulation year and observations), the online coupled WRF-Chem is mostly suited for application of simulating and investigating the effects of urban heat island and its mitigation strategies.

Table 3. Mean bias error (MBE) of T2 ($^\circ$C), WS10 (m/s), Td ($^\circ$C), RH2 (%), O_3 (ppb), $PM_{2.5}$ ($\mu g/m^3$), $SO4_{2.5}$ ($\mu g/m^3$), $NO3_{2.5}$ ($\mu g/m^3$), $OC_{2.5}$ ($\mu g/m^3$), and NO_2 (ppb) at selected monitoring stations across Sacramento, Houston, and Chicago.

	Mean Bias Error (MBE)						
Variables	Sacramento		Houston		Chicago		Average
	Suburb	Urban	Suburb	Urban	Suburb	Urban	
T2 ($^\circ$C)	0.15	-0.34	0.22	-0.34	-0.30	0.19	-0.07
WS10 (m/s)	1.90	3.15	0.87	0.34	1.28	1.05	1.43
Td ($^\circ$C)	0.21	0.61	0.24	0.47	0.33	0.47	0.39
RH2 (%)	-5.43	-5.63	-1.03	8.16	1.88	-6.45	-1.42
24-h avg. O_3 (ppb)	9.72	4.68	3.17	3.85	2.31	4.38	4.68
24-h avg. $PM_{2.5}$ ($\mu g/m^3$)	-5.94	2.30	-3.26	2.07	-3.86	-2.33	-1.84
24-h avg. $SO4_{2.5}$ ($\mu g/m^3$)	-	4.20	-	5.30	-	3.89	4.46
24-h avg. $NO3_{2.5}$ ($\mu g/m^3$)	-	-3.75	-	-4.40	-	-3.52	-3.91
24-h avg. $OC_{2.5}$ ($\mu g/m^3$)	-	-1.80	-	-2.33	-	-3.68	-2.60
24-h avg. NO_2 (ppb)	-	2.61	-	3.40	-	1.25	2.42

Note: The definitions of statistical measurements are as follows Zhang et al. (2006) [51]: MBE $= \frac{1}{N}\sum_1^N (C_M - C_O)$, C_M and C_O are modeled and observed concentrations, respectively and N is the total number of model and observation pairs.

Table 4. Mean absolute error (MAE) of T2 (°C), WS10 (m/s), Td (°C), RH2 (%), O_3 (ppb), $PM_{2.5}$ ($\mu g/m^3$), $SO4_{2.5}$ ($\mu g/m^3$), $NO3_{2.5}$ ($\mu g/m^3$), $OC_{2.5}$ ($\mu g/m^3$), and NO_2 (ppb) at selected monitoring stations across Sacramento, Houston, and Chicago.

Variables	Mean Absolute Error (MAE)						Average
	Sacramento		Houston		Chicago		
	Suburb	Urban	Suburb	Urban	Suburb	Urban	
T2 (°C)	1.05	0.88	1.20	0.88	0.77	1.12	0.99
WS10 (m/s)	1.96	3.33	1.26	1.20	2.31	1.79	1.97
Td (°C)	0.56	0.30	0.49	0.63	0.56	0.66	0.53
RH2 (%)	15.32	9.45	5.38	9.54	8.98	10.11	9.80
24-h avg. O_3 (ppb)	9.72	9.90	6.92	6.01	2.56	5.88	6.83
24-h avg. $PM_{2.5}$ ($\mu g/m^3$)	6.24	3.05	3.26	3.70	3.86	2.33	3.74
24-h avg. $SO4_{2.5}$ ($\mu g/m^3$)	-	4.20	-	5.30	-	3.89	4.46
24-h avg. $NO3_{2.5}$ ($\mu g/m^3$)	-	3.75	-	4.45	-	3.52	3.91
24-h avg. $OC_{2.5}$ ($\mu g/m^3$)	-	1.80	-	2.33	-	3.68	2.60
24-h avg. NO_2 (ppb)	-	4.71	-	3.40	-	2.54	3.55

Note: The definitions of statistical measurements are as follows Zhang et al. (2006) [51]: MAE $= \frac{1}{N}\sum_1^N |C_M - C_O|$, C_M and C_O are modeled and observed concentrations, respectively and N is the total number of model and observation pairs.

Table 5. Root mean square error (RMSE) of T2 (°C), WS10 (m/s), Td (°C), RH2 (%), O_3 (ppb), $PM_{2.5}$ ($\mu g/m^3$), $SO4_{2.5}$ ($\mu g/m^3$), $NO3_{2.5}$ ($\mu g/m^3$), $OC_{2.5}$ ($\mu g/m^3$), and NO_2 (ppb) at selected monitoring stations across Sacramento, Houston, and Chicago.

Variables	Root Mean Square Error (RMSE)						Average
	Sacramento		Houston		Chicago		
	Suburb	Urban	Suburb	Urban	Suburb	Urban	
T2 (°C)	1.35	1.13	1.44	1.13	1.01	1.32	1.23
WS10 (m/s)	2.29	3.68	1.58	1.47	2.86	2.22	2.35
Td (°C)	0.68	0.37	0.58	0.77	0.67	0.80	0.65
RH2 (%)	18.94	12.32	7.52	12.26	11.36	13.48	12.65
24-h avg. O_3 (ppb)	10.51	14.21	7.89	6.71	3.09	8.21	8.44
24-h avg. $PM_{2.5}$ ($\mu g/m^3$)	7.74	3.25	4.04	4.30	4.82	2.81	4.49
24-h avg. $SO4_{2.5}$ ($\mu g/m^3$)	-	4.44	-	6.24	-	3.93	4.87
24-h avg. $NO3_{2.5}$ ($\mu g/m^3$)	-	3.96	-	4.87	-	4.28	4.37
24-h avg. $OC_{2.5}$ ($\mu g/m^3$)	-	1.93	-	2.39	-	3.76	2.69
24-h avg. NO_2 (ppb)	-	5.74	-	4.10	-	2.89	4.24

Note: The definitions of statistical measurements are as follows Zhang et al. (2006) [51]: RMSE $= \left[\frac{1}{N}\sum_1^N (C_M - C_O)^2\right]^{1/2}$, C_M and C_O are modeled and observed concentrations, respectively and N is the total number of model and observation pairs.

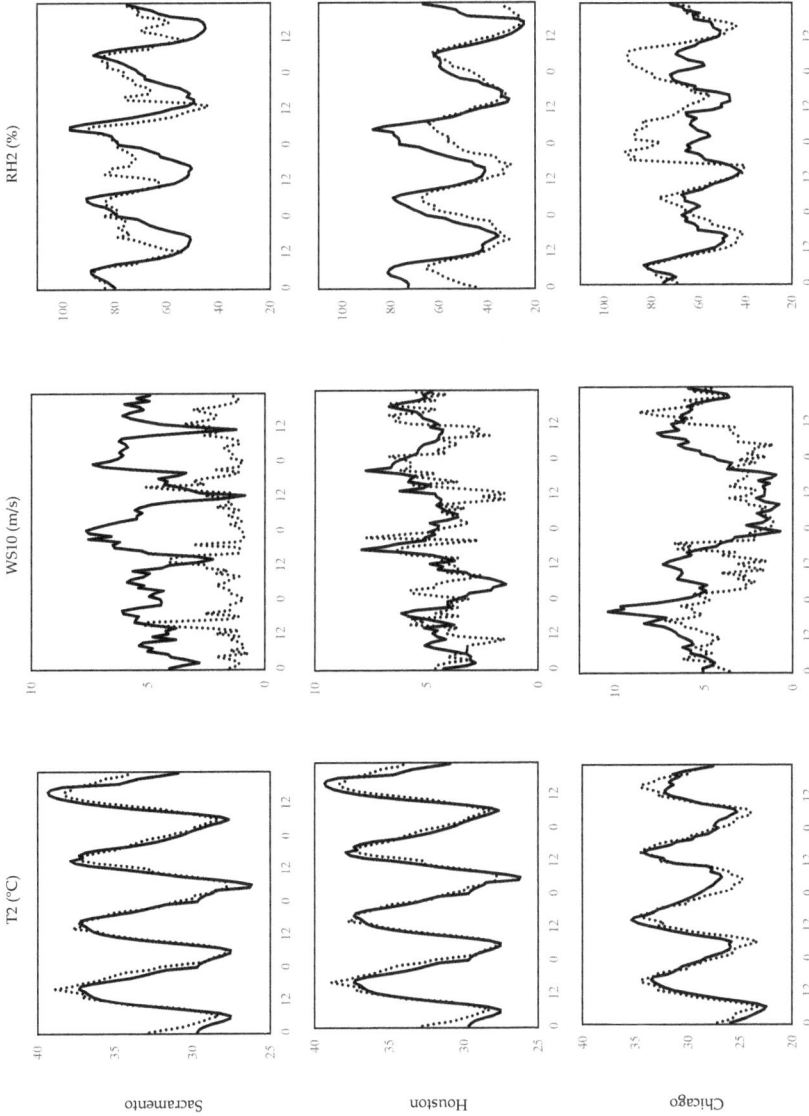

Figure 2. The time series (hourly) of the simulated (solid line) *vs.* measurements (dashed line) T2 (°C), WS10 (m/s), and RH2 (%) at urban monitoring stations across Sacramento, Houston, and Chicago.

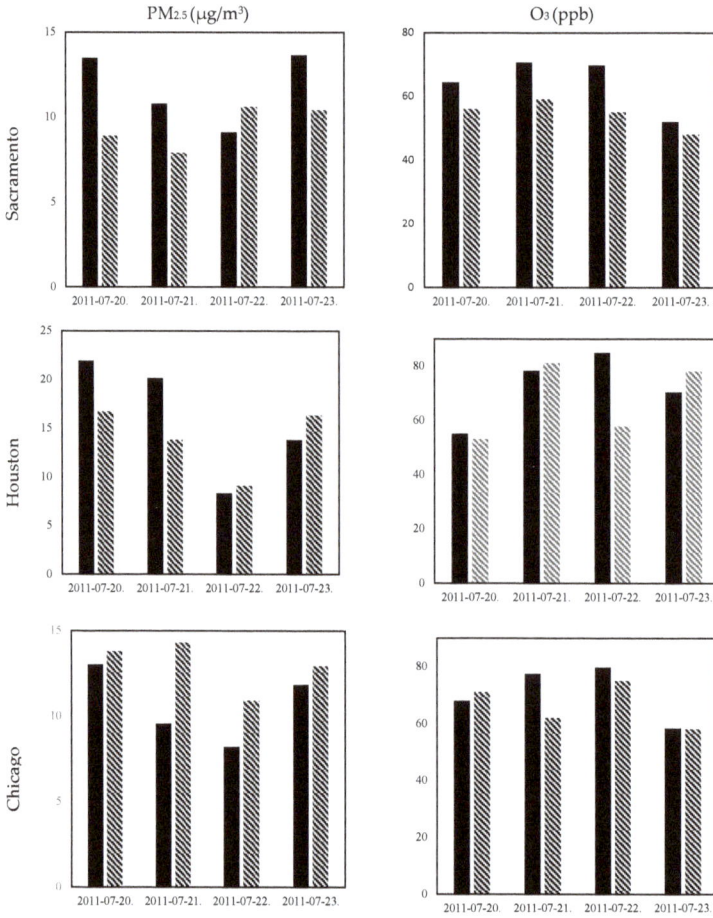

Figure 3. The time series (averaged 24-h) of simulated (black bar chart) vs. measurements (patterned downward diagonal bar chart) of PM$_{2.5}$ (μg/m^3) and O$_3$ (ppb) concentrations at urban monitoring stations across Sacramento, Houston, and Chicago.

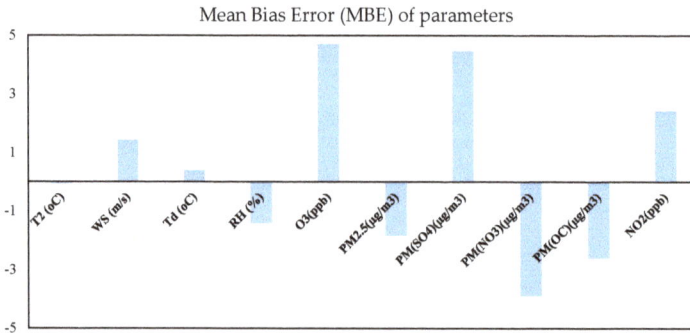

Figure 4. *Cont.*

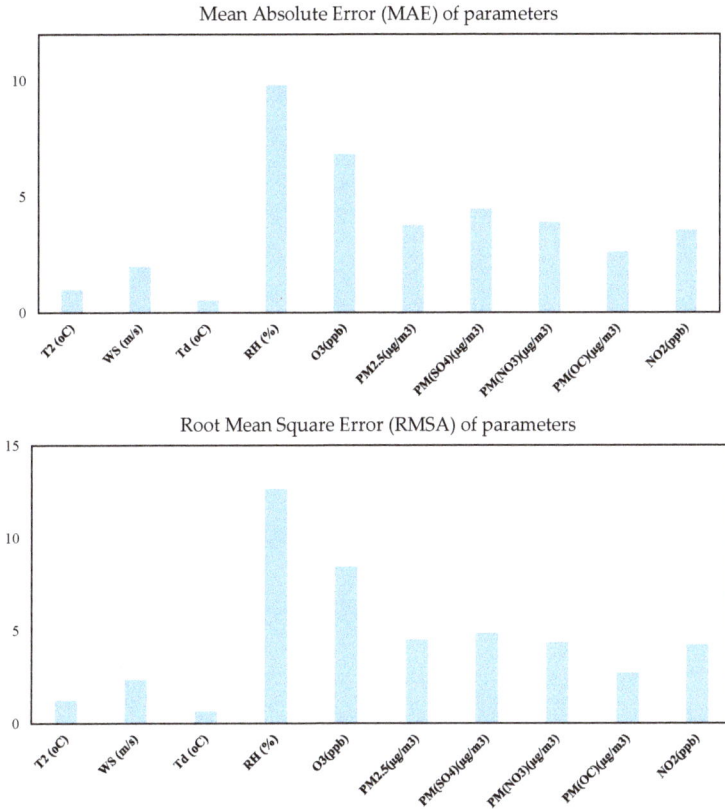

Figure 4. The overall mean bias error (MBE), mean absolute error (MAE), and root mean square error (RMSA) of T2 (°C), WS10 (m/s), Td (°C), RH2 (%), O_3 (ppb), $PM_{2.5}$ ($\mu g/m^3$), $SO4_{2.5}$ ($\mu g/m^3$), $NO3_{2.5}$ ($\mu g/m^3$), $OC_{2.5}$ ($\mu g/m^3$), and NO_2 (ppb) during the 2011 heat wave period.

3.2. Effects of Increasing Urban Albedo

Results discussed here are based on the comparison between the ALBEDO and CTRL scenarios for each city. Table 6 and Figure 5 represent the average differences in T2 (°C), WS10 (m/s), Td (°C), RH2 (%), $PM_{2.5}$ ($\mu g/m^3$), O_3 (ppb), $SO4_{2.5}$ ($\mu g/m^3$), $NO3_{2.5}$ ($\mu g/m^3$), $OC_{2.5}$ ($\mu g/m^3$), and NO_2 (ppb) during the 2011 heat wave period across the second, third, and fourth domains: Sacramento (CA), Houston (TX), and Chicago (IL). Figure 6 shows the averaged differences of T2 (°C) and O_3 (ppb) concentrations in suburb and urban areas of the aforementioned cities.

Sacramento, California is located in the central valley near the Sierra foothills. It is at the confluence of the Sacramento River and the American River and is known as the Sacramento Valley. The city has a population of approximately 500,000 people and covers over 253 km^2 [57,58]. Its climate is characterized by mild year-round temperature. It has a hot-dry-summer Mediterranean climate with little humidity and an abundance of sunshine. Based on the National Oceanic and Atmospheric Administration (NOAA) Online Weather Data [22], Sacramento has the summer temperature exceeding 32 °C on 73 days and 38 °C on 15 days. The State of the Air 2017 report, by American Lung Association [59], ranks the metropolitan areas based on ozone and particular pollutions during 2013, 2014, and 2015 period. They used the official data from the U.S. Environmental Protection Agency (EPA). Sacramento ranks eighth because of its high ozone concentration.

Table 6. The differences between CTRL and ALBEDO scenarios of T2 (°C), WS10 (m/s), RH2 (%), O_3 (ppb), $PM_{2.5}$ (μg/m^3), $SO4_{2.5}$ (μg/m^3), $NO3_{2.5}$ (μg/m^3), $OC_{2.5}$ (μg/m^3), and NO_2 (ppb) during the 2011 heat wave period across Sacramento, Houston, and Chicago.

	Δ ALBEDO						
CTRL-ALBEDO	**Sacramento**		**Houston**		**Chicago**		**Average**
	Suburb	**Urban**	**Suburb**	**Urban**	**Suburb**	**Urban**	
Δ T2 (°C)	0.72	2.37	0.81	2.68	0.75	1.76	1.52
Δ WS10 (m/s)	0.03	0.02	0.33	0.02	−0.08	0.00	0.05
Δ Td (°C)	−0.26	−0.39	−0.27	−0.46	−0.21	−0.34	−0.32
Δ RH2 (%)	−2.99	−6.88	−2.44	−6.89	−0.81	0.21	−3.30
24-h avg. O_3 (ppb)	2.98	7.52	2.85	7.23	1.77	4.23	4.43
24-h avg. $PM_{2.5}$ (μg/m^3)	0.98	2.36	2.59	3.49	0.61	2.48	2.08
24-h avg. $SO4_{2.5}$ (μg/m^3)	-	0.02	-	0.01	-	0.06	0.03
24-h avg. $NO3_{2.5}$ (μg/m^3)	-	0.01	-	0.05	-	0.23	0.09
24-h avg. $OC_{2.5}$ (μg/m^3)	-	0.00	-	0.00	-	0.00	0.00
24-h avg. NO_2 (ppb)	-	0.82	-	1.21	-	0.91	0.98

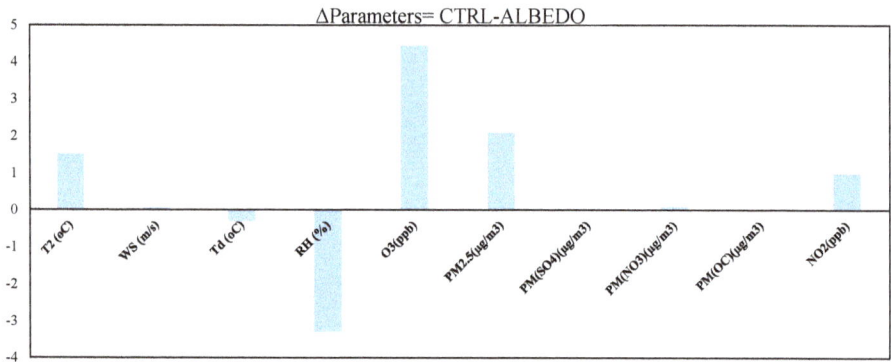

Figure 5. The average differences between CTRL and ALBEDO scenarios in T2 (°C), WS10 (m/s), RH2 (%), O_3 (ppb), $PM_{2.5}$ (μg/m^3), $SO4_{2.5}$ (μg/m^3), $NO3_{2.5}$ (μg/m^3), $OC_{2.5}$ (μg/m^3), and NO_2 (ppb) during the 2011 heat wave period.

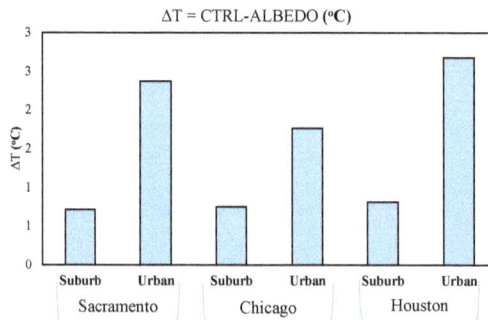

Figure 6. *Cont.*

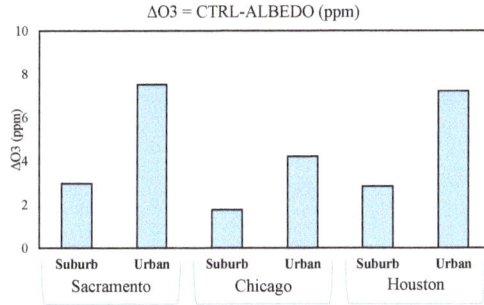

Figure 6. The average differences between CTRL and ALBEDO scenarios of T2 (°C) and O$_3$ (ppb) during the 2011 heat wave period in suburb and urban areas of Sacramento, Chicago, and Houston.

In a study performed at the Lawrence Berkeley National Laboratory (LBNL), Taha et al. [60] applied the Colorado State Urban Meteorological Model (CSUMM) and the Urban Airshed Model (UAM-IV) to estimate the impacts of heat island mitigation strategies in Sacramento on the area's local meteorology and ozone air quality in 2000. The albedo level and vegetative cover increased by approximately 0.11 and 0.14, respectively. Using 11–13 July 1990 as the modeling period, the ozone and temperature decreased by up to 10 ppb and 1.6 °C, respectively. In a more recent study, Taha et al. [61] applied WRF with CMAQ in Sacramento Valley with the inner domain of 1 km resolution. The albedo of roofs, walls, and pavements increased by 0.4, 0.1, and 0.2, respectively. The surface temperature and air temperature were reduced by up to 7 °C and 2–3 °C, respectively. The ozone concentrations also decreased by up to 5–11 ppb during the daytime.

Our simulation results for Sacramento show that albedo enhancement leads to a net decrease in 2-m air temperature by up to 2.5 °C and 0.7 °C in urban and suburban areas, respectively. Most of the decreases occur between 1200 and 1600 LST as shown in Figure 7-S. Figure 8-Sa (CTRL) shows the maximum air temperature across the simulation domain in the heat wave period. By increasing surface reflectivity, the maximum temperature reduction is around 3 °C almost in all parts of the city (Figure 8-Sb-ALBEDO) and this reduction is more obvious in the western part of the domain. The wind speed slightly decreased over the entire domain. The relative humidity increased by 7% and 3% in urban and suburban areas, respectively. Increasing surface reflectivity affords a decrease of nearly 2.4 µg/m^3 in PM$_{2.5}$ concentrations in urban area (Figure 7-S) and 1 µg/m^3 in suburb. Figure 8-Sc shows the maximum PM$_{2.5}$ concentrations across the domain. The maximum is around 12 µg/m^3 in urban area that decreases by 2–3 µg/m^3 as the results of albedo enhancement (Figure 8-Sd). The heat island mitigation strategy causes a decline in O$_3$ by almost 8 ppb in urban (Figure 7-S) and 3 ppb in suburb of the Sacramento area. Figure 8-Se shows the maximum O$_3$ concentrations as nearly 80 ppb across the simulation domain that decreases to nearly 70 ppb by UHI mitigation strategy (Figure 8-Sf). Our results resemble those of previous studies [60–63]. We have also compared the CTRL and ALBEDO simulations results of particulate sulfate, particulate nitrate, organic carbon, and nitrogen dioxide. Albedo enhancement causes no changes (OC) to minimal changes to particular matters subspecies (~0.01 reduction) but decreases the NO$_2$ concentration by 0.82 ppb.

Houston is the fourth most populous city in the U.S. with a population of 2.3 million within a land area of 1700 km^2 [57,58]. It is located in the Southeast Texas near the Gulf of Mexico. Houston's climate is classified as humid subtropical. During the summer, the temperature commonly reaches 34 °C, and some days it reaches to even 40 °C. The wind comes from the south and southeast and brings heat and moisture from the Gulf of Mexico. The highest temperature recorded in Houston is 43 °C, which occurred during the 2011 heat wave period [57,58]. Houston also suffers from excessive ozone levels and the American Lung Association [59] named the city as the 12th most polluted city in the U.S., based on EPA 2013, 2014, and 2015 data base.

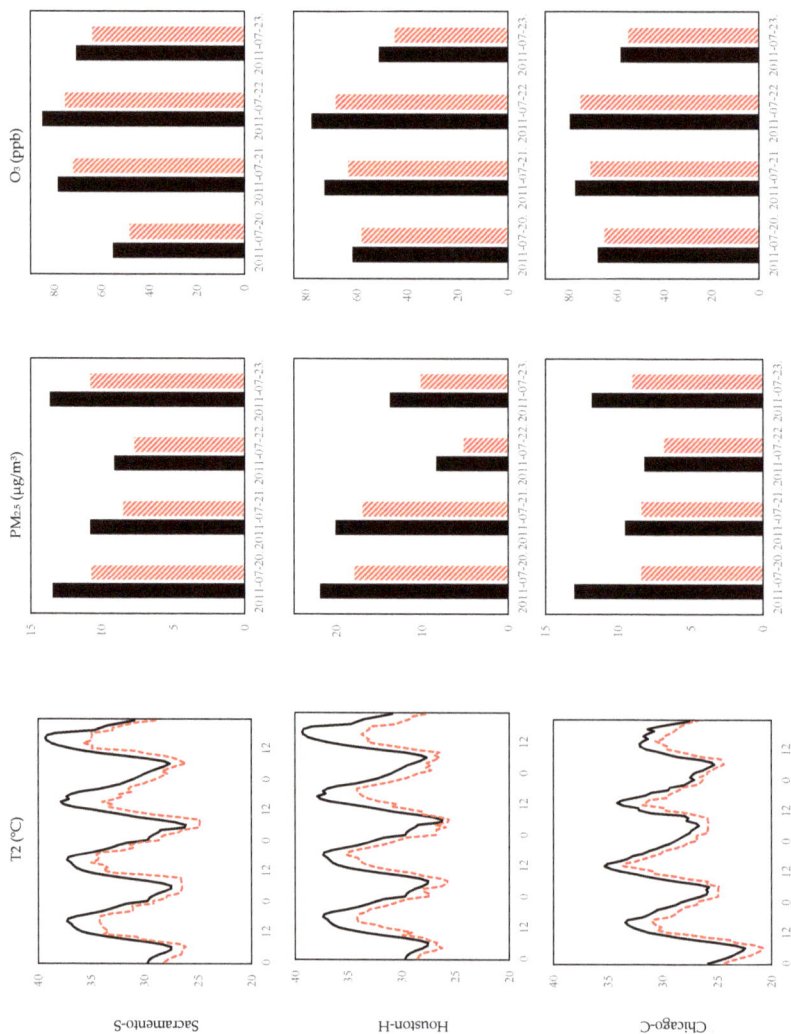

Figure 7. The differences between CTRL (solid line and black bar chart) and ALBEDO (red dashed line and patterned downward diagonal bar chart) scenarios in hourly T2 (°C) and 24-h avg. PM$_{2.5}$ (µg/m^3) and O$_3$ (ppb) concentrations during the 2011 heat wave period across the urban areas of Sacramento-S, Houston-H, and Chicago-C.

Taha [64] used MM5 to evaluate the model's episode performance and its response to increasing surface albedo and vegetation in Houston during several days in August 2000. In ALBEDO scenario, the roof albedo was increased from an average of 0.1 to an average of 0.3; wall albedo was increased from an average of 0.25 to an average of 0.3; pavement albedo was increased from an average of 0.08 to 0.2. The results indicated a reduction in temperature by up to 3.5 °C, and also caused warming in some areas by up to 1.5 °C. Results indicated that cooling usually occurs during daytime, while heating occurs at night. The other simulations show the same results [64–66].

Our simulation results for Houston show that albedo enhancement leads to a net decrease in 2-m air temperature by up to 3 °C and 0.8 °C in urban (Figure 7-H) and suburban areas, respectively. We witness no heating effect in our simulation. The reason is due to the sea breeze consideration in the solver of WRF-Chem. Figure 8-Ha illustrates the maximum air temperature across the Houston in the heat wave period. The maximum temperature reduction is above 3 °C almost in all parts of the city (Figure 8-Hb). Our model tends to perform relatively better in urban rather than in suburb areas. With albedo enhancement, the wind speed slightly decreased, and the relative humidity increased by up to 7% in urban and 3% in suburb. Increasing surface reflectivity affords a decrease of $PM_{2.5}$ concentrations by up to 3.5 $\mu g/m^3$ and 2.6 $\mu g/m^3$ in urban and suburban areas, respectively. Figure 8-Hc shows the maximum $PM_{2.5}$ concentrations across Houston. The maximum is above 20 $\mu g/m^3$ in urban area that decreases to 16 $\mu g/m^3$ as the results of albedo enhancement (Figure 8-Hd). The O_3 concentrations also decrease by up to 7.2 ppb and 3 ppb in urban and suburban areas, respectively. Figure 8-He shows the maximum O_3 concentrations as above 80 ppb across the simulation domain that decreases to nearly 70 ppb all over the domain (Figure 8-Hf). Our results resemble to previous studies [64–66]. Increasing surface albedo in the urban area of Houston causes no changes in particular matters subspecies and a decrease of 1.2 ppb in NO_2 concentration.

Chicago is the third most populous city in the U.S. with over 2.7 million residents. The city area is 606 km^2 [57,58]. The city lies on the southwestern shores of Lake Michigan and has two rivers: the Chicago River and the Calumet River. Chicago has a humid continental climate. Summer temperatures can reach up to 32 °C. Taha et al. [67] used a three-dimensional, Eulerian, mesoscale meteorological model (CSUMM) to simulate the effects of large scale surface modifications on meteorological conditions in 10 cities across the U.S. Surface modifications included increasing albedo by 0.03 ± 0.05 and increasing vegetative fraction by 0.03 ± 0.04. The results indicated that the air temperature was reduced by up to 1 °C in the Chicago area.

Our simulation results for Chicago show that albedo enhancement leads to a net decrease in 2-m air temperature by up to nearly 2 °C and 0.8 °C in urban (Figure 7-C) and suburban areas, respectively. Figure 8-Ca shows the maximum air temperature across the simulation domain. With albedo enhancement, the air temperature reduced over the domain (Figure 8-Cb). The wind speed slightly reduces in suburbs, with no changes in urban areas. The results show a slight decrease in relative humidity by up to 0.2% in Chicago's urban areas. The reason is due to the wind speed direction that is north to west (passing the bodies of water) and the city's location that is along one of the Great Lakes, Lake Michigan, and has the Mississippi River Watershed and the Chicago River. The other reason is due to the increasing surface reflectivity that reduces the skin temperature and thus air temperature that might also decrease the chance of evaporation and thus decreases moisture content above the ground. This strategy also affords a decrease of $PM_{2.5}$ concentrations by up to 2.5 $\mu g/m^3$ and 0.6 $\mu g/m^3$ in urban and suburban areas, respectively. The maximum $PM_{2.5}$ concentrations across Chicago is nearly 12 $\mu g/m^3$, that decreases to nearly 9 $\mu g/m^3$ as the results of albedo enhancement (Figure 8-Cc and Cd). The O_3 concentrations decrease by up to 4.2 ppb in urban area and 1.7 ppb in suburb. Figure 8-Ce shows the maximum O_3 concentrations as nearly 70 ppb across the simulation domain that decreases to almost 65 ppb all over the domain (Figure 8-Cf). Increasing urban albedo in Chicago leads to an increase of particulate nitrate by 3 ppb and a decrease of NO_2 concentration by 0.9 ppb.

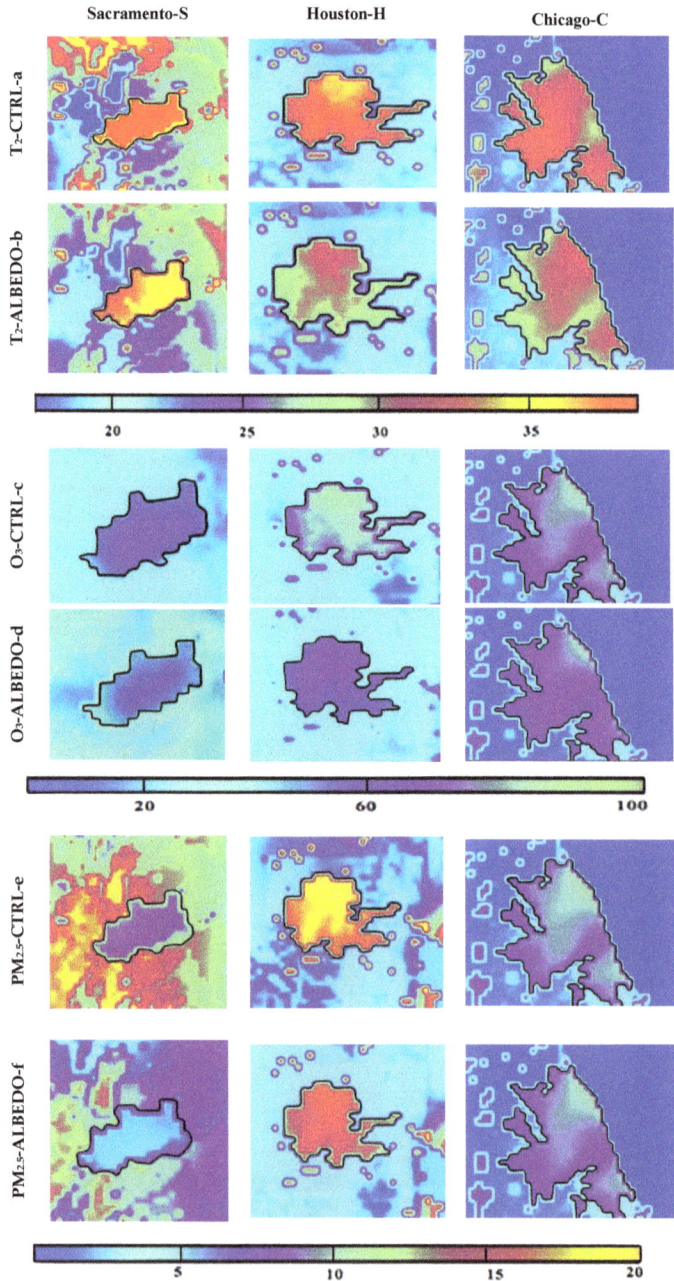

Figure 8. The maximum 2-m air temperature (°C), $PM_{2.5}$ ($\mu g/m^3$) and O_3 (ppb) concentrations in CTRL and ALBEDO scenarios across Sacramento, Houston, and Chicago during the 2011 heat wave period.

Overall, the results indicate that with albedo enhancement, the air temperature drops (~1.5 °C) and thus causes a decrease in ozone concentrations (~5 ppb) and nitrogen dioxide (~1 ppb). Increasing surface solar reflectance lead to a minimal decrease in particular matters (~2 $\mu g/m^3$) and no significant changes in its subspecies. The $SO4_{2.5}$ and $NO3_{2.5}$ concentrations reduced slightly in urban areas (~0.1 µg/m^3) due to the decrease in air temperature and thus photochemical reaction rates, but there is no change in $OC_{2.5}$ (µg/m^3). The UHI mitigation strategy increased the relative humidity and dew point temperature. Our results show that there are no significant changes in the wind speed over the domain and the differences between two scenarios is 0.05 m/s. This minimal change can be due to the WRF-Chem configurations and it does not reflect any changes in momentum transport from the shallow boundary layer.

4. Discussion

By comparing the simulation results with the observations, we acknowledge that WRF-Chem generally reproduces well the hourly variations of meteorological variables, but overpredicts or underpredicts the air pollutant concentrations during the 2011 heat wave period. One of the reasons is due to the method of comparison; the simulation outputs are extracted at the start of each hour, whereas the measurements are reported as hourly or daily averages. Another reason is because of the anthropogenic and the biogenic estimations by US-NEI11 (spatial resolution of 1 km) and MEGAN (spatial resolution of 4 km). The other issue is concerning the simulation episode which is seven consecutive days during the heat wave period, whereas the pollutants dispersion and formation may occur during a longer time period. We suggest a study to assess the effects of increasing surface albedo for a whole year to see its effects during the winter season and during a year as well. The WRF-Chem meteorological and chemical settings configured here are reasonable, but different settings may also be reasonable. We suggest alternative parameterizations to see the effects of surface albedo in a two-way nested approach such as ours for further studies. We also suggest assessing the effects of other UHI mitigation strategies (as increasing vegetative fraction) on urban climate and air quality within a two-way nested approach.

At the inception of our study, the WRF-Chem version 3.6.1 was the most recent release. Since our objectives are concerning the effects of urban heat island and its mitigation strategies on urban climate and there were no modifications in the multilayer of the urban canopy model, we decided to continue using this version. We run the model on 48 cores, 96 Gb for 14 days. Each actual day took 48 h to run on a supercomputer. The data assimilation in WRF can be used to update WRF model's initial conditions. The WRF-DA is based on an incremental vibrational data assimilation technique and has both 3D-Var and 4D-Var capabilities. It also includes the capability of hybrid data assimilation (vibrational + ensemble). The conjugate gradient method is utilized to minimize the cost function in the analysis control variable space. Analysis increments are also interpolated to staggered grids and it gets added to the background (first guess) to get the final analysis of the WRF-model grid. The WRF has the capability to maintain data assimilation in its solver [29,30]. That is why our results show a good agreement with the observations. We believe our results and analyses are trustable and can be used for further investigations.

The 2011 heat wave period is selected for our simulations to investigate the effects of increasing albedo on the worst-case scenarios in each city. However, in order to specify the effects of increasing albedo, another simulation should be carried out in a normal condition during summer. Then the results need to be compared with the heat wave period to see the typical effects of albedo enhancement in each location. Simulation of the entire year can reveal more information on the annual effects of the mitigation strategy. To gain better results of the effectiveness of high-albedo strategy in improving the regional ozone air quality, other episodes and locations with more reliable emission inventories should be further investigated, modeled, and analyzed in a more detailed modeling approach. The information on an area's local climate can help to focus on heat island mitigation strategies that best suit their region. For example, cities with dry climates may achieve greater benefits from increasing vegetative fraction of urban areas (more evapotranspiration) than would cities in humid climates. However, dry-climate cities also need to consider the availability of water to maintain vegetation. We suggest a more detailed analysis of simulation results to investigate the effects of surface modifications on decreasing the temperature-dependent photochemical reaction rates, as well as decreasing evaporation losses of organic compounds from industrial sectors, and mobile and stationary sources.

5. Summary and Conclusions

We applied a two-way nested approach in WRF-ChemV3.6.1 coupled with the ML-UCM to evaluate the surface modification consequences on air temperature, wind speed, relative humidity, dew point temperature, ozone, nitrogen dioxide, fine particulate matters, $PM_{2.5}$ subspecies (particulate sulfate ($SO4_{2.5}$), particulate nitrate ($NO3_{2.5}$) and organic carbon ($OC_{2.5}$)) concentrations in a unified continental scale through regional scales (North America through Sacramento, Houston, and Chicago) during the 2011 heat wave period. The two-way nested approach with fine-resolution modelling framework can equip us with an integrated simulation setup to capture the full impacts of meteorological and photochemical reactions. The applied method would serve as a basis for future model improvements and parameterization development, fine-resolution dispersion, and photochemical modelling for other geographical locations.

The model performance is evaluated by comparing the simulation results with the observations. Despite the model biases in simulating meteorological and chemical variables, the performance of WRF-ChemV3.6.1 is generally consistent with the most air quality models. [52–55,60–67], thus is mostly suited for application of simulating and investigating the effects of urban heat island and its mitigation strategies. The MBA, MAE, and RMSE estimations confirmed the model capabilities. For meteorological components, the WRF-ChemV3.6.1, as configured here, captures well the diurnal variations of 2-m air temperature (MBA ~-0.07 °C), overpredicts 10-m wind speed (MBA $\sim$1.65 m/s), overpredicts dew point temperature (MBA $\sim$0.4 °C), underpredicts 2-m relative humidity (MBA ~-1.4%). For chemical component, the model underpredicts the daily fine particular matters ($PM_{2.5}$) (MBA ~-1.5 $\mu g/m^3$) and overpredicts the O_3 concentrations (MBA $\sim$5 ppb). The model underpredicts the NO_2 ($\sim$2.5 ppb) and overpredicts particulate sulfate (MBE $\sim$5 $\mu g/m^3$) and underpredicts particulate nitrate (MBE ~-4 $\mu g/m^3$) and organic carbon (MBE ~-3 $\mu g/m^3$) in urban areas of aforementioned cities during the 2011 heat wave period. The model tends to perform relatively better in urban, rather than in suburban areas.

Two sets of simulations are conducted with regard to surface modifications: CTRL scenario and ALBEDO scenario. With albedo enhancement we observed: a decrease in air temperature by 2.3 °C in urban areas and 0.7 °C in suburban areas; a slight increase in wind speed; an increase in relative humidity (3%) and dew point temperature (0.3 °C); a decrease of $PM_{2.5}$ and O_3 concentrations by 2.7 $\mu g/m^3$ and 6.3 ppb in urban areas and 1.4 $\mu g/m^3$ and 2.5 ppb in suburban areas, respectively; minimal changes in $PM_{2.5}$ subspecies and a decrease of nitrogen dioxide (1 ppb) in urban areas. The results presented here are episode- and region-specific and thus may not provide a suitable basis for generalization to other circumstances. Overall, the results confirm that for Sacramento in California, Houston in Texas, and Chicago in Illinois, the albedo enhancement is an effective mitigation strategy to reduce the air temperature and improve air quality. The results show that Sacramento and Houston benefit more from increasing surface solar reflectance. These findings are an asset for policy makers and urban planning designers. However, we suggest that the effects of other UHI mitigation strategies on urban climate and air quality should also be investigated before making decisions on applying any surface modifications. We also suggest running the simulations with more accurate emission inventories. We recommend a simulation for the entire year that can reveal more information of the mitigation strategy impacts.

Acknowledgments: Funding for this research was provided by the National Science and Engineering Research Council of Canada (NSERC) under Discovery Grants Program. Calcul Quebec and Compute Canada provided the computational facilities for this research.

Author Contributions: Zahra Jandaghian performed the simulations, analyzed the data and wrote the paper. Professor Hashem Akbari supervised the research, advised on the analyses, reviewed various drafts of the paper.

Conflicts of Interest: The authors declare no conflict of interest.

References

1. Akbari, H.; Kolokotsa, D. Three decades of urban heat islands and mitigation technologies research. *Energy Build.* **2016**, *133*, 834–842. [CrossRef]
2. Taha, H. Urban surface modification as a potential ozone air-quality improvement strategy in California: A mesoscale modeling study. *Bound. Layer Meteorol.* **2008**, *127*, 219–239. [CrossRef]
3. Chen, F.; Yang, X.; Zhu, W. WRF simulations of urban heat island under hot-weather synoptic conditions: The case study of Hangzhou City, China. *Atmos. Res.* **2013**, *138*, 364–377. [CrossRef]
4. Akbari, H.; Pomerantz, M.; Taha, H. Cool surfaces and shade trees to reduce energy use and improve air quality in urban areas. *Sol. Energy* **2001**, *70*, 295–310. [CrossRef]
5. Salamanca, F.; Martilli, A. A numerical study of the Urban Heat Island over Madrid during the DESIREX (2008) campaign with WRF and an evaluation of simple mitigation strategies. *Int. J. Climatol.* **2012**, *32*, 2372–2386. [CrossRef]
6. Fallmann, J.; Emeis, S.; Suppan, P. Mitigation of urban heat stress—A modelling case study for the area of Stuttgart. *J. Geogr. Soc. Berl.* **2013**, *144*, 202–216.
7. Fallmann, J.; Forkel, R.; Emeis, S. Secondary effects of urban heat island mitigation measures on air quality. *Atmos Environ.* **2016**, *125*, 199–211.
8. Touchaei, A.G.; Akbari, H.; Tessum, C.W. Effects of increasing urban albedo on meteorology and air quality Montreal (Canada)—Episodic simulation of heat wave in 2005. *Atmos. Environ.* **2016**, *132*, 188–206. [CrossRef]
9. Jandaghian, Z.; Touchaei, G.A.; Akbari, H. Sensitivity analysis of physical parameterizations in WRF for urban climate simulations and heat island mitigation in Montreal. *Urban Clim.* **2017**. [CrossRef]
10. Seinfeld, J.H.; Pandis, S.N. *Atmospheric Chemistry and Physics: From Air Pollution to Climate Change*, 2nd ed.; John Wiley & Sons, Inc.: Hoboken, NJ, USA, 2012.
11. Grell, G.A.; Peckham, S.E.; Schmitz, R.; McKeen, S.A.; Frost, G.; Skamarock, W.C.; Eder, B. Fully coupled "online" chemistry within the WRF model. *Atmos. Environ.* **2005**, *39*, 6957–6975. [CrossRef]
12. Skamarock, W.C.; Klemp, J.B.; Dudhia, J.; Gill, D.O.; Barker, D.M.; Wang, W.; Powers, J.G. *A Description of the Advanced Research WRF Version 3*; National Center for Atmospheric Research: Boulder, CO, USA, 2008.
13. Ahmadov, R.; McKeen, S.A.; Robinson, A.L.; Bahreini, R.; Middlebrook, A.M.; de Gouw, J.A.; Meagher, J.; Hsie, E.-Y.; Edgerton, E.; Shaw, S.; et al. A volatility basis set model for summertime secondary organic aerosols over the eastern United States in 2006. *J. Geophys. Res.* **2012**, *117*, 06–31. [CrossRef]
14. Chuang, M.-T.; Zhang, Y.; Kang, D. Application of WRF/Chem-MADRID for real-time air quality forecasting over the southeastern United States. *Atmos. Environ.* **2011**, *45*, 6241–6250. [CrossRef]
15. Misenis, C.; Zhang, Y. An examination of sensitivity of WRF/Chem predictions to physical parameterizations, horizontal grid spacing, and nesting options. *Atmos. Res.* **2010**, *97*, 315–334. [CrossRef]
16. Zhang, Y.; Chen, Y.; Sarwar, G.; Schere, K. Impact of gas-phase mechanisms on Weather Research Forecasting Model with Chemistry (WRF/Chem) predictions: Mechanism implementation and comparative evaluation. *J. Geophys. Res.* **2012**, *117*, D01301. [CrossRef]
17. Yahya, K.; Wang, K.; Gudoshava, M.; Glotfelty, T.; Zhang, Y. Application of WRF/Chem over North America under the AQMEII Phase 2: Part I. Comprehensive evaluation of 2006 simulation. *Atmos. Environ.* **2015**, *155*, 733–755. [CrossRef]
18. Tessum, C.W.; Hill, J.D.; Marshall, J.D. Twelve-month, 12 km resolution North American WRF-Chem v3.4 air quality simulation: Performance evaluation. *Geosci. Model. Dev.* **2015**, *8*, 957–973. [CrossRef]
19. Martilli, A.; Clappier, A.; Rotach, M. An urban surface exchange parameterization for mesoscale models. *Bound. Layer Meteorol.* **2002**, *104*, 261–304. [CrossRef]
20. Liao, J.; Wang, T.; Wang, X.; Xie, M.; Jiang, Z.; Huang, X.; Zhu, J. Impacts of different urban canopy schemes in WRF/Chem on regional climate and air quality in Yangtze River Delta, China. *Atmos. Res.* **2014**, *146*, 226–243. [CrossRef]
21. US-National Climate Data Centre (NOAA). Available online: https://www.ncdc.noaa.gov (accessed on 28 January 2018).
22. NOAA. National Oceanic and Atmospheric Administration Changes to the NCEP Meso Eta Analysis and Forecast System: Increase in Resolution, New Cloud Microphysics, Modified Precipitation Assimilation, Modified 3DVAR Analysis. 2001. Available online: http://www.emc.ncep.noaa.gov/mmb/mmbpll/eta12tpb/ (accessed on 28 January 2018).

23. Mesinger, F.; DiMego, G.; Kalnay, E.; Mitchell, K.; Shafran, P.C.; Ebisuzaki, W.; Jović, D.; Woollen, J.; Rogers, E.; Berbery, E.H.; et al. North American regional reanalysis. *Bull. Am. Meteorol. Soc.* **2006**, *87*, 343–360. [CrossRef]
24. Lin, Y.; Farley, R.; Orville, H.D. Bulk parameterization of the snow field in a cloud model. *J. Clim. Appl. Meteorol.* **1983**, *22*, 1065–1092. [CrossRef]
25. Chou, M.-D.; Suarez, M.J. *A Solar Radiation Parameterization (CLIRAD-SW) Developed at Goddard Climate and Radiation Branch for Atmospheric Studies*; NASA Technical Memorandum NASA/Goddard Space Flight Center Greenbelt: Greenbelt, MD, USA, 1999.
26. Iacono, M.J.; Delamere, J.S.; Mlawer, E.J.; Shephard, M.W.; Clough, S.A.; Collins, W.D. Radiative forcing by longlived greenhouse gases: Calculations with the AER radiative transfer models. *J. Geophys. Res.* **2008**, *113*, 131–153. [CrossRef]
27. Janjic, Z.I. The stepmountain Eta coordinate model: further developments of the convection, viscous sublayer, and turbulence closure schemes. *Mon. Weather Rev.* **1994**, *122*, 927–945. [CrossRef]
28. Grell, G.A.; Devenyi, D. A generalized approach to parameterizing convection combining ensemble and data assimilation techniques. *Geophys. Res. Lett.* **2002**, *29*, 38–48. [CrossRef]
29. NCAR. *WRF User's Guide; Mesoscale & Microscale Meteorology Division*; National Center for Atmospheric Research (NCAR): Boulder, CO, USA, 2016.
30. ARW User Guide. *ARW Version 3 Modeling System User's Guide*; National Center for Atmospheric Research: Boulder, CO, USA, 2012.
31. US EPA (US Environmental Protection Agency). 2011 National Emissions Inventory (NEI). Available online: http://www.epa.gov/ttn/chief/emch/index.html (accessed on 7 March 2016).
32. Guenther, A.B.; Jiang, X.; Heald, C.L.; Sakulyanontvittaya, T.; Duhl, T.; Emmons, L.K.; Wang, X. The Model of Emissions of Gases and Aerosols from Nature Version 2.1 (MEGAN2.1): An extended and updated framework for modeling biogenic emissions. *Geosci. Model. Dev.* **2012**, *5*, 1471–1492. [CrossRef]
33. Ackermann, I.J.; Hass, H.; Memmesheimer, M.; Ebel, A.; Binkowski, F.S.; Shankar, U. Modal aerosol dynamics model for Europe: Development and first applications. *Atmos. Environ.* **1998**, *32*, 2981–2999. [CrossRef]
34. Stockwell, W.R.; Kirchner, F.; Kuhn, M.; Seefeld, S. A new mechanism for regional atmospheric chemistry modeling. *J. Geophys. Res. Atmos.* **1997**, *102*, 25847–25879. [CrossRef]
35. Schell, B.; Ackermann, I.J.; Hass, H.; Binkowski, F.S.; Ebel, A. Modeling the formation of secondary organic aerosol within a comprehensive air quality model system. *J. Geophys. Res. Atmos.* **2001**, *106*, 28275–28293. [CrossRef]
36. Fast, J.D.; Gustafson, W.I., Jr.; Easter, R.C.; Zaveri, R.A.; Barnard, J.C.; Chapman, E.G.; Grell, G.A.; Peckham, S.E. Evolution of ozone, particulates, and aerosol direct radiative forcing in the vicinity of Houston using a fully coupled meteorology-chemistry-aerosol model. *J. Geophys. Res.* **2006**, *111*, 203–213. [CrossRef]
37. Grell, G.A.; Freitas, S.R. A scale and aerosol aware stochastic convective parameterization for weather and air quality modeling. *Atmos. Chem. Phys.* **2014**, *14*, 5233–5250. [CrossRef]
38. NCAR. *WRF-CHEM Emission Guide*; National Center for Atmospheric Research (NCAR) and The University Corporation for Atmospheric Research (UCAR): Boulder, CO, USA, 2016.
39. NCAR. *WRF-CHEM User's Guide, WRF-Chem Emissions Guide*; National Center for Atmospheric Research (NCAR): Boulder, CO, USA, 2016.
40. Akbari, H.; Rose, L.S. *Characterizing the Fabric of the Urban Environment: A Case Study of Metropolitan Chicago, Illinois*; Report LBNL-49275; Lawrence Berkeley National Laboratory: Berkeley, CA, USA, 2001.
41. Akbari, H.; Rose, L.S. *Characterizing the Fabric of the Urban Environment: A Case Study of Salt Lake City, Utah*; Report No. LBNL-47851; Lawrence Berkeley National Laboratory: Berkeley, CA, USA, 2001.
42. Akbari, H.; Rose, L.S.; Taha, H. Analyzing the land cover of an urban environment using high-resolution orthophotos. *Landsc. Urban Plan.* **2003**, *63*, 1–14. [CrossRef]
43. Rose, L.S.; Akbari, H.; Taha, H. *Characterizing the Fabric of the Urban Environment: A Case Study of Greater Houston, Texas*; Report LBNL-51448; Lawrence Berkeley National Laboratory: Berkeley, CA, USA, 2003.
44. Millstein, D.; Menon, S. Regional climate consequences of large-scale cool roof and photovoltaic array deployment. *Environ. Res. Lett.* **2011**, *6*, 34–44. [CrossRef]
45. EPA. 2005. Available online: http://www.epa.gov/ttnchie1/net/2005inventory.html (accessed on 28 January 2017).

46. US EPA (Environmental Protection Agency). Technology Transfer Network (TTN) Air Quality System (AQS). 2005. Available online: http://www.epa.gov/ttn/airs/airsaqs/detaildata/downloadaqsdata.htm (access on 6 March 2016).

47. US EPA (US Environmental Protection Agency). Air Quality Modeling Technical Support Document for the Regulatory Impact Analysis for the Revisions to the National Ambient Air Quality Standards for Particulate Matter, Research Triangle Park, NC 27711. 2012. Available online: http://www.regulations.gov/ (accessed on 28 January 2018).

48. UCAR (University Corporation for Atmospheric Research). GCIP NCEP Eta Model Output. 2005. Available online: http://rda.ucar.edu/datasets/ds609.2/ (accessed on 15 January 2017).

49. University of California Davis. *IMPROVE Data Guide: A Guide to Interpret Data*; Prepared for National Park Service, Air Quality Research Division: Fort Collins, CO, USA, 1995; Available online: http://vista.cira.colostate.edu/improve/publications/OtherDocs/IMPROVEDataGuide/IMPROVEdataguide.htm (accessed on 18 September 2017).

50. Boylan, J.W.; Russell, A.G. PM and light extinction model performance metrics, goals, and criteria for three-dimensional air quality models. *Atmos. Environ.* **2006**, *40*, 4946–4959. [CrossRef]

51. Zhang, Y.; Liu, P.; Pun, B.; Seigneur, C. A comprehensive performance evaluation of MM5-CMANQ for the summer 1999 southern oxidant study episode—Part I. Evaluation protocols, databases and meteorological predictions. *Atmos. Environ.* **2006**, *40*, 4825–4838. [CrossRef]

52. Gilliam, R.C.; Hogrefe, C.; Rao, S.T. New methods for evaluating meteorological models used in air quality applications. *Atmos. Environ.* **2006**, *40*, 5073–5086. [CrossRef]

53. Wu, S.-Y.; Krishnan, S.; Zhang, Y.; Aneja, V. Modelling atmospheric transport and fate of ammonia in North Carolina, part I. Evaluation of meteorological and chemical predictions. *Atmos. Environ.* **2008**, *42*, 3419–3436. [CrossRef]

54. Wang, K.; Zhang, Y.; Jang, C.J.; Phillips, S.; Wang, B.-Y. Modelling study of Intercontinental air pollution transport over the Trans-Pacific region in 2001 using the community multiscale air quality (CMAQ) modelling system. *J. Geophys. Res.* **2009**, *114*, 4–19. [CrossRef]

55. Liu, X.-H.; Zhang, Y.; Olsen, K.; Wang, W.-X.; Do, B.; Bridgers, G. Responses of future air quality to emission controls over North Carolina—Part I: Model evaluation for current-year simulations. *Atmos. Environ.* **2010**, *44*, 2443–2456. [CrossRef]

56. Appel, K.W.; Chemel, C.; Roselle, S.J.; Francis, X.V.; Hu, R.-M.; Sokhi, R.S.; Rao, S.T.; Galmarini, S. Examination of the Community Multiscale Air Quality (CMAQ) model perfor-mance over the North American and European domains. *Atmos. Environ.* **2012**, *53*, 142–155. [CrossRef]

57. US Census Bureau. Cartographic Boundary Shapefiles Regions. 2013. Available online: https://www.census.gov/geo/maps-data/data/cbf/cbf_region.html (accessed on 10 February 2017).

58. US Census Bureau. Year-2014 US Urban Areas and Clusters. 2014. Available online: ftp://ftp2.census.gov/geo/tiger/TIGER2014/UAC/ (accessed on 10 February 2017).

59. American Lung Association. 2017. Available online: http://www.lung.org/our-initiatives/healthy-air/sota/city-rankings/most-polluted-cities.html (accessed on 28 January 2018).

60. Taha, H. Meso-urban meteorological and photochemical modeling of heat island mitigation. *Atmos. Environ.* **2008**, *42*, 8795–8809. [CrossRef]

61. Taha, H.; Wilkinson, J.; Bornstein, R.; Xiao, Q.; McPherson, G.; Simpson, J.; Anderson, C.; Lau, S.; Lam, J.; Blain, C. An urban–forest control measure for ozone in the Sacramento, CA Federal Non-Attainment Area (SFNA). *Sustain. Cities Soc.* **2015**, *21*, 51–65. [CrossRef]

62. Taha, H. Meteorological, emissions, and air-quality modeling of heat-island mitigation: Recent findings for California, USA. *Int. J. Low Carbon Technol.* **2013**, *10*, 3–14. [CrossRef]

63. Taha, H. Ranking and Prioritizing the Deployment of Community-Scale Energy Measures Based on Their Indirect Effects in California's Climate Zones. 2013. Available online: http://www.energy.ca.gov/2011publications/CEC-500-2011-FS/CEC-500-2011-FS-021.pdf (accessed on 28 January 2018).

64. Taha, H. *Potential Meteorological and Air-Quality Implications of Heat-Island Reduction Strategies in the Houston-Galveston TX Region*; Technical Note HIG-12-2002-01; Lawrence Berkeley National Laboratory: Berkeley, CA, USA, 2003.

65. Taha, H. Episodic Performance and Sensitivity of the Urbanized MM5 (uMM5) to Perturbations in Surface Properties in Houston Texas. *Bound. Layer Meteor.* **2008**, *127*, 193–218. [CrossRef]

66. Taha, H. *Evaluating Meteorological Impacts of Urban Forest and Albedo Changes in the Houston-Galveston Region: A Fine-Resolution (UCP) Meso-Urban Modeling Study of the August–September 2000 Episode*; Report Prepared for the Houston Advanced Research Center by Altostratus Inc.; Altostratus Inc.: Martinez, CA, USA, 2005.

67. Taha, H.; Konopacki, S.; Gabersek, S. Impacts of Large-Scale Surface Modifications on Meteorological Conditions and Energy Use: A 10-Region Modeling Study. *Theor. Appl. Climatol.* **1999**, *62*, 175–185. [CrossRef]

climate

MDPI

Article

Subjective Human Perception of Open Urban Spaces in the Brazilian Subtropical Climate: A First Approach

João Paulo Assis Gobo *, Emerson Galvani * and Cássio Arthur Wollmann *

Department of Geography, University of São Paulo (USP), 338 Prof. Lineu Prestes Avenue,
São Paulo 05508-000, Brazil
* Correspondence: jpgobo@usp.br (J.P.A.G.); egalvani@usp.br (E.G.); cassio_geo@yahoo.com.br (C.A.W.)

Received: 1 March 2018; Accepted: 26 March 2018; Published: 3 April 2018

Abstract: This research concerns a first approach to adapt the thermal comfort bands of the Physiological Equivalent Temperature (PET), New Standard Effective Temperature (SET), and Predicted Mean Vote (PMV) indices to Santa Maria's population, Rio Grande do Sul, Brazil, on the basis of the application of perception/sensation questionnaires to inhabitants while, at the same time, recording meteorological attribute data. Meteorological and thermal sensation data were collected from an automatic weather station installed on paved ground in the downtown area, which contained the following sensors: a scale gauge; a global radiation sensor; a temperature and humidity sensor; a speed and wind direction sensor; a gray globe thermometer. First of all, air temperature, gray globe temperature, relative air humidity, wind speed, wind gust, global solar radiation and precipitation were collected. People were interviewed using a questionnaire adapted from the model established by ISO 10551. The results demonstrated the efficiency of the linear regression model and the adequacy of the interpretive indexes, presenting results different from those analyzed by other authors in different climatic zones. These differences meet the analyzed literature and attest to the effectiveness of the calibration method of the PET, SET, and PMV indices for the Brazilian subtropical climate. After calibration, the PET index hit rate increased from 32.8% to 69.3%. The SET index, which had an initial hit rate of 34.6% before calibration, reached a hit-rate of 64.9%, while the PMV index increased from 35.9% to 58.7%.

Keywords: climatic perception; urban areas; thermal comfort; subtropical climate

---

1. Introduction

Models for the prediction of thermal comfort employ seven- or nine-point thermal sensitivity scales for the assessment of the average person's perception of atmospheric weather conditions in open spaces. Human thermal comfort ranges have been proposed or modified in recent years [1,2]. For decades, researchers have investigated ways to predict the thermal sensation of individuals in their typical environments on the basis of personal, environmental, and physiological variables. As a result, mathematical models that simulate the thermal responses of individuals in their environments were developed.

As a rule, the instruments used to construct this type of scale consist of questionnaires with items investigating respondents' personal impressions of the atmospheric weather, while a meteorological station simultaneously measures several climatic attributes, such as air temperature, relative humidity, and wind speed [3–5].

Studies and human comfort indexes are being developed to measure thermal comfort in open spaces under uncontrolled conditions [6–9]. Some such studies focus on modeling and assessment methods from the thermophysiological perspective, e.g., those by Gulyas et al. [10] and Hoppe [1], whereas others are based on the relationships between the climate parameters that determine the thermal comfort level of humans outdoors [11,12].

A survey of cross-sectional studies that have investigated the thermal comfort patterns and preferences of people at different times of the year have detected a large number of studies that employed the indexes of the New Standard Effective Temperature (SET) [13], Predicted Mean Vote (PMV) [14], and Physiological Equivalent Temperature (PET) [15], relative to temperate [2–18] and hot and wet [19–21] climates.

In Brazil, Hirashima et al. [22] described interpretive ranges for the PET calibrated for two different climatic regions: Belo Horizonte, Brazil (tropical climate) and Kassel/Freiburg, Germany (temperate climate in both cases). In turn, Lucchese et al. [23] analyzed the thermal comfort of visitors to a public square in Campo Grande, Mato Grosso do Sul, Brazil, during the hot and cold seasons, and compared the predictive capacities of the PET, Universal Thermal Climate Index (UTCI), Perceived Equivalent Temperature (TEP), Sense of Thermal Comfort (YDS), and PMV indexes.

Krüger, Rossi, and Drach [24] described a preliminary procedure for the calibration of the PET for three different climatic regions: Curitiba, Brazil (subtropical climate), Rio de Janeiro, Brazil (tropical climate), and Glasgow, United Kingdom (high-latitude climate).

Several studies have detected differences in outdoor thermal comfort across different climatic regions [24–27] and have pointed to the need for additional fieldwork research on subjective human perception in different climatic regions [20], given that most of those surveys currently available have mainly been conducted in areas with temperate or tropical climates.

Lin's study [28], developed in Taichung, Taiwan, showed that, unlike a temperate climate, in a subtropical climate, mild temperatures and weak sunshine are generally desirable during the warm season. In his field research, Lin [28] found that more than 90% of people visiting the public square in the summer chose to stand under shade trees or in constructed shelters, indicating the importance of shade in outdoor environments in a subtropical climate.

Cheng et al. [29], in his studies about outdoor thermal comfort under the influence of Hong Kong's subtropical climate, revealed that air temperature, wind speed, and solar radiation intensity are the most influential factors in determining people's thermal sensation. Based on the data collected, a predictive formula for estimating the subjective outdoor temperature sensation was also developed, highlighting the fact that wind speed change has a negative influence on thermal sensation, especially in the summer in Hong Kong [29,30].

For studies in which the PET index in subtropical climates was used, different patterns of definition of the thermal comfort zone for different locations were observed, even when these places are in the same climatic zone [12–32]. The same behavior can be observed for the SET and PMV indices as well [17–35].

A literature review showed that no substantial study using some of the main thermal comfort indexes has yet to be conducted, relative to the Brazilian subtropical region.

A survey of studies conducted in Brazil showed that the SET, PMV, and PET are some of the indexes more widely used for open spaces. However, they have not been applied relative to the Brazilian subtropical region, except for the PET, which was used by Rossi et al. [7] to define the thermal comfort range for Curitiba, Paraná.

Therefore, the aim of the present study was to adapt the thermal comfort range for the PET, SET, and PMV indexes relative to the population of Santa Maria, Rio Grande do Sul, Brazil, on the basis of the application of questionnaires investigating the inhabitants' perceptions/sensations and simultaneously recording the local meteorological attributes.

2. Materials and Methods

The meteorological and thermal sensation data required for the present study were collected. For this purpose, a Campbell CR100 Automatic Weather Station (AWS) was used, at a maximum height of 2.0 m, with a mobile aluminum tripod containing the following sensors: Rain gage model TE525 Tipping Bucket Rain Gage, with a resolution of 0.10 mm; global radiation sensor model LI200X Pyranometer, with a resolution of 0.2 kW·m^{-2}; temperature and humidity sensor model HMP35C Temperature and Relative Humidity Probe, with resolutions of 0.1 °C and 0.6%; speed and wind direction sensor models 03001 R.M. Young Wind Setry Set 03101 R.M. Young Wind Sentry Anemometer 03301 R. M. Young Wind Sentry Vane, with resolutions of 0.5 m·s^{-1} and 5°; TMCx-HD gray globe thermometer, with a resolution of 0.03 °C. Primary data on air temperature, gray globe temperature—because the station was set up in an open space exposed to direct solar radiation [4]—relative humidity, wind speed, wind gusts, global solar radiation, and rain were collected. The station was placed on a paved area in Saldanha Marinho Square in downtown Santa Maria, where the flow of people is intense (Figure 1).

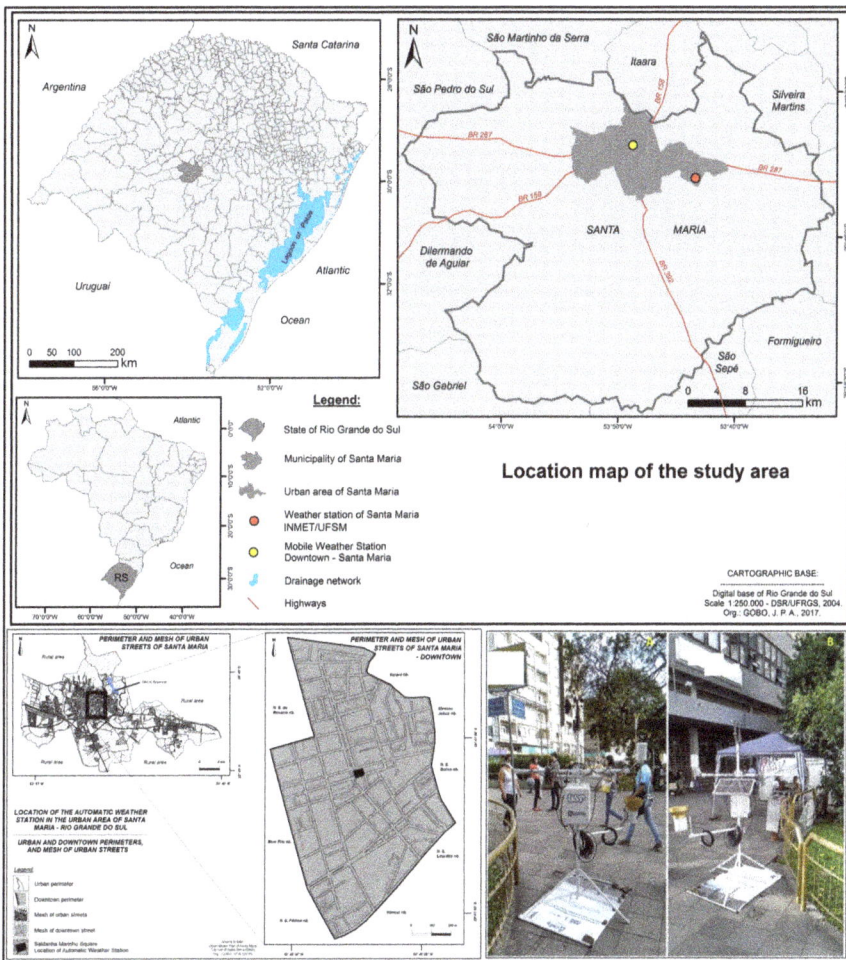

Figure 1. Location of the study area and automatic weather station.

Field data collection was performed 5–7 August 2015, 17–19 January 2016 and 6–8 July 2016. Collection of meteorological data and interviews with the local population were performed from 9:00 a.m. to 5:00 p.m. solar time, on each of the aforementioned days and periods.

On the days of field research, an atypical climatic situation, popularly known as "little summer inside winter" and due to an atmospheric block, was observed during 5–7 August 2015, which is characterized by a persistent high pressure anomaly (anticyclone belt around 30 degrees of latitude), with relatively slow displacement of high pressures, and may persist for several days. During the January 2016 field research days, it was possible to identify a pattern compatible with Santa Maria's normal climatological averages for this month, presenting high temperatures with maximums above 32 °C. In the next winter analysis, in July 2016, above-average temperatures for Santa Maria during this month were observed, mainly in the first day of analysis, but with periods of temperatures within the range expected in this season.

In the present study, only those individuals who had resided in the town for more than one year were interviewed to derive a function of the individual thermal history and environmental memory, as observed by Nikolopoulou [36], in a total of 1728 interviews. Interviewees also had to exhibit 0.3 to 1.5 of clothing insulation, which corresponds to wearing jeans and T-shirt or a suit [37] and 300 W of physical activity, because only people in motion (walking) were included [38]. The questionnaire administered was an adaptation of the one included in standard ISO 10551 [3] (Figure 2).

Data of the interviewee:

age (___)

Sex: (___) M (___) F

Weight (___)

Height (___)

With regard to the individual's dress, he is wearing:

0,3 clo (___); 0,5 clo (___); 0,8 clo (___); 1,0 clo (___); 1,5 clo (___)

1. At this very moment, I'm feeling:
( ) Cold -3
( ) Cool -2
( ) Slightly cool -1
( ) Neither cold nor hot 0
( ) Slightly warm 1
( ) Warm 2
( ) Hot 3

2. At this very moment, regarding the weather, I am:
( ) comfortable 0
( ) a little uncomfortable 1
( ) uncomfortable 2
( ) very uncomfortable 3

3. Right now, I'd rather be feeling:
( ) Much colder -3
( ) Colder -2
( ) A little more cold -1
( ) No changes 0
( ) A little more heat 1
( ) More heat 2
( ) Much more heat 3

4. With regard to air temperature, I would prefer it to be:
( ) lowest -1
( ) as is 0
( ) highest 1
( ) I do not know how to say X.

5. With regard to air humidity, I would prefer the air to be:
( ) drier -1
( ) as is 0
( ) wettest 1
( ) I do not know how to say X.

6. Regarding the wind, I would prefer it to be:
( ) weaker -1
( ) as is 0
( ) stronger 1
( ) I do not know how to say X.

7. With regard to solar radiation, I would prefer it to be:
( ) softer -1
( ) as is 0
( ) more intense 1
( ) I can not say X.

Figure 2. Questionnaire [3].

The AWS meteorological data were used to calculate the PET, SET, and PMV according to the RayMan model [39,40].

Following the calculation of the aforementioned indexes for the three analyzed periods (August 2015, June 2016, and July 2016), two multiple-linear regression models were fitted [41] to adjust the thermal comfort range of each index to the average thermal preference pattern of the interviewees. For this purpose, a code to optimize the proportion of hits of the models was formulated through changes in the cutoff points (index classes).

The multiple-linear regression method is described by the following equation:

$$Y = \alpha_1 x_1 + \alpha_2 x_2 + \alpha_3 x_3 + \ldots \alpha_n x_n + c \tag{1}$$

where n is the number of terms in the equation; αn corresponds to the (constant) parameters obtained by the regression; c is the constant that corresponds to the point where the fitted line intersects the y-axis; Y is the independent variable; x_n are the dependent variables.

What we called the Index Mean is actually the mean of the responses obtained during fieldwork for a given value of the climatic variables, during a 20 min interval. Because Y is a discrete random variable with a finite number of possible values, we considered that an equation for the expected values of discrete random variables (finite case) afforded the best technique by which to calculate the expected value of the Index Mean for a given interval of time, as the responses were limited to the ANSI/ASHRAE 55 [42] 7-point thermal sensation scale:

$$E[y] = \sum_{k=1}^{n} y_k p_k \tag{2}$$

where y_k is the possible value, and p_k is its corresponding probability in an independent assay.

In the case of the present study:

$$I_{mean} = \sum_{n=1}^{n} I_k p_k \tag{3}$$

where I_k is the Index Mean, and p_k is the corresponding probability estimated by dividing the number of responses for a given index by the number of interviews conducted over a given interval of time. That is, $I_1 = -3$, $I_2 = -2$, $I_3 = -1$, $I_4 = 0$, $I_5 = 1$, $I_6 = 2$, and $I_7 = 3$, and p_1, p_2, p_3, p_4, p_5, and p_6 are the probabilities of each respective I, obtained as $\frac{number\ of\ occurrences}{total\ number\ of\ interviews\ along\ the\ interval}$, resulting in:

$$I_{mean} = \sum_{n=1}^{n} I_k p_k = (-3)p_{-3} + (-2)p_{-2} + (-1)p_{-1} + (0)p_0 + (1)p_1 + (2)p_2 + (3)p_3 \tag{4}$$

This method is more appropriate for the obtained responses than calculating the weighted mean or the median.

New cutoff points were established for the values of the PMV, PET, and SET obtained on the assessment days. For this purpose, an optimization algorithm was used that had the original cutoff points of the indexes as its point of departure.

However, in the case of the indexes based on equivalent temperature analogies (i.e., PET and SET), we could not obtain the originally intended interpretive ranges, because, by principle, the interpretation of values is obtained from analogies between temperatures instead of interpretive scales. Nevertheless, we chose to use Hoppe's [43] classes for the PET, and Jendritzky's [44] classes for the PMV. In turn, the SET was described without the original ranges, but with the calibration ranges alone.

3. Results

Considering that the hit rate obtained in the linear model was approximately 45%, the results of the logistic model showed an index rate similar to 45%.

By analyzing the distribution of the values of the thermal comfort indexes as a function of the meteorological variables at each instant, together with the interviewees' responses to question 1 ("At this exact moment, I'm feeling ___."), we sought to obtain an index that is able to satisfactorily describe the thermal comfort of Santa Maria's population.

Just as Lucchese et al. [23] found for Campo Grande, Mato Grosso do Sul, Brazil, the limited predictive ability of the analyzed indexes (less than 50% precision) did not efficiently represent the respondents' thermal sensation reports. Thus, new cutoff points were established for such indexes, which were based on two multiple-linear regression models [41] and the use of an optimization code of the proportion of hits of the models through changes in the cutoff points. The optimized values are described in Table 1, which illustrates the rationale underlying the procedure, namely, to find range sizes that maximize the number of hits.

Table 1. Classes corresponding to score intervals following the calculation of optimized cutoff points and hit rates of the analyzed indexes. PMV, Predicted Mean Vote; PET, Physiological Equivalent Temperature; SET, New Standard Effective Temperature (SET).

Thermal Comfort Level	Score	Index	Hit Rate
Cold	$(-\infty; -2.379]$	PMV	35.9%
Cool	$(-2.379; -1.335]$	PET	32.8%
Slightly cool	$(-1.335; -0.254]$	SET	34.6%
Neither cold nor hot	$(-0.254; 0.965]$		
Slightly warm	$(0.965; 1.412]$		
Warm	$(1.412; 2.040]$		
Hot	$(2.040; +\infty)$		

Table designed by the author.

3.1. Calibration of Interpretive Ranges for the Predicted Mean Vote (PMV)

Relative to the PMV, the cutoff point found for the thermal comfort range following calibration was within the interval -1 to 0.8, which thusly differed from that suggested by Jendritzky et al. [44], i.e., -0.5 to 0.5 (Table 2).

Table 2. Cutoff points for classes corresponding to variable Response to Question 1 (*) of the questionnaire, calibrated for PMV.

PMV Jendritzky et al. (1979)		Calibrated PMV	
>2.5	Hot	<−3.6	Cold
1.5 to 2.5	Warm	−3.6 to −2.3	Cool
0.5 to 1.5	Slightly warm	−2.3 to −1.0	Slightly cool
−0.5 to 0.5	Comfortable	−1.0 to 0.8	Neither cold nor hot
−1.5 to −0.5	Slightly cool	0.8 to 1.9	Slightly warm
−2.5 to −1.5	Cool	1.9 to 3.5	Warm
<−2.5	Cold	>3.5	Very warm

(*) "At this exact moment, I'm feeling ___". Table designed by the author.

The comparison of the cutoff points obtained for Santa Maria, following calibration of the PMV ranges, to those found by Monteiro [45] shows a subtle difference, with the interval reported by the latter being from -0.9 to 0.6. However, the maximum and minimum cutoff points were 4 and 3.1, respectively.

Having been developed in an interior setting under the assumption of comfortable average temperature and sweat rate, its applicability to an outdoor environment is limited. The same finding was observed by Lai et al. [2] who, using microclimatic monitoring and interviews with the population in a park of Tianjin, China, analyzed outdoor thermal comfort under different climatic conditions through UTCI, PMV, PET, and Thermal Sensation Vote (TSV), and they compared with Lin's [28], in Taiwan, and Pantavou's et al. [16] studies. The analysis indicated that the PMV overestimated the outdoor thermal sensation by 1.3, leading the authors to conclude that the wide fluctuations in the outer thermal parameters resulted in frequent deviations from the neutral state and, therefore, its use was not indicated.

3.2. Calibration of Interpretive Ranges for the New Standard Effective Temperature (SET)

In the case of the SET, we could not find the interpretive ranges suggested by the original authors. The cutoff points for the comfort range were 17 °C and 23 °C, whereas the maximum and minimum values of the scale calibrated for Santa Maria were 33 °C and 6 °C, respectively (Table 3).

Table 3. Cutoff points for classes corresponding to the variable Response to Question 1 (*) of the questionnaire, calibrated for the SET.

Calibrated SET	
<6	Cold
6–12	Cool
12–17	Slightly cool
17–23	Neither cold nor hot
23–29	Slightly warm
29–33	Warm
>33	Hot

(*) "At this exact moment, I'm feeling ___". Table designed by the author.

The comparison of the calibrated SET values for Santa Maria to those values for São Paulo [46] evidence a significant similarity between the cutoff points for the interpretive ranges, particularly with regard to the comfort range, which for São Paulo extended from 17 °C to 22 °C. The maximum and minimum cutoff points of the interpretive ranges for São Paulo were practically identical to those for Santa Maria, i.e., 33 °C and 5 °C, respectively.

Considering the differences in climate and thermal preference between the populations of Santa Maria and São Paulo (subtropical and tropical climates, respectively), one might infer that the SET did not perform satisfactorily in terms of calibration for the subtropical climate. This finding might be accounted for by the lack of a reference interpretive range for this procedure.

When the results of SET found for Santa Maria were compared with those of Xi et al. [46], who studied thermal comfort at the Guangzhou University of Technology campus in southern China's subtropical climate region, the results pointed out that the comfort zone for young students in the summer of Guangzhou would be around 24 °C, only 1 °C above the upper limit of the comfort zone observed in Santa Maria. The authors also noted that cities in the subtropical climate zone in which the sky-vision factor (SVF) is very high receive more short-wave radiation and long-wave reflections, which may create an external, thermal overheated environment during the summer [46].

To check the patterns of thermal comfort and population preference in different seasons, Lin et al. [17] conducted 1644 interviews with simultaneous outdoor micrometeorological measurements in central Taiwan, in a hot and humid climatic zone. The results indicated a SET deviation of 1.3 °C relative to neutral temperatures between hot and cold seasons and a SET of 1.8 °C at the preferred temperature between hot and cold seasons. The authors observed that the SET's comfort range for Taiwan was between 25.4 °C and 28.9 °C, much more than that observed for Santa Maria.

3.3. Calibration of Interpretive Ranges for the Physiological Equivalent Temperature (PET)

The calibration of the PET performed by Monteiro [45] for São Paulo found a comfort range similar to that obtained for Santa Maria, with cutoff points of 17 °C and 22 °C; the same was true for the cases of the maximum and minimum values of 33 °C and 5 °C, respectively.

Following calibration, the comfort range extended from 16 °C to 24 °C, which shows little difference with regard to the cutoff range reported by Jendritzky [47], i.e., 18 °C to 23 °C (Table 4).

Table 4. Cutoff points for classes corresponding to the variable Response to Question 1 (*) of the questionnaire, calibrated for the PET.

PET Jendritzky (1991)		Calibrated PET	
<4	Cold	<5	Cold
4–8	Cool	5–11	Cool
8–18	Moderately cool	11–16	Slightly cool
18–23	Comfortable	16–24	Neither cold nor hot
23–35	Moderately hot	24–30	Slightly warm
35–41	Warm	30–39	Warm
>41	Hot	>39	Hot

(*) "At this exact moment, I'm feeling ___?" Table designed by the author.

The SET, just as the PET, is an index based on equivalent temperatures; as with the PET, for neither of these indexes were we able to locate the original interpretive ranges. Thus, we used Jendritzky's [47] ranges, because the ranges found after calibration were lower, for instance, than those reported by Monteiro [45] for São Paulo, ranging from 18 °C to 26 °C.

In turn, Krüger, Rossi, and Drach [24] found an average of 25 °C as defining the comfort range for Curitiba through the use of a binary method, according to the values of the assessed index. These authors performed a preliminary calibration of the PET for three different climatic regions. Thus, any comparison between the results of calibration for Curitiba and Santa Maria would not be valid, because the studies differ from a methodological point of view. For this reason, the authors advocate a standardization of the protocols used for the assessment of comfort in open spaces [24].

The comparison of the thermal comfort range calibrated for the PET in Santa Maria to that calibrated by Lucchese et al. [23] for Campo Grande showed a difference, mainly with respect to the beginning of the range of discomfort due to cold, with the ranges of comfort being from 21 °C to 28 °C and from 16 °C to 24 °C for Campo Grande and Santa Maria, respectively. This difference might be partially attributed to the different climate characteristics of Campo Grande and Santa Maria, because the former is located in an area of transition between the humid subtropical (Cfa) and tropical wet (Aw) climates (Köppen–Geiger) [23], and the latter is located entirely within the humid, subtropical climatic zone.

Farajzadeh et al. [18] compared simple thermal comfort indexes and indices derived from energy balance models in northwestern Iran, using air temperature, solar radiation, relative humidity, cloud cover, and wind speed data from 13 weather stations during the period from 1986 to 2007, using the Bioklima and RayMan models. The indices based on the human energy balance showed a significant correlation with each other, with an R value above 90% and a lowest R value of 70%, related to the subjective temperature index (STI). In addition, the indices based on relatively simple formulas had low correlation with the UTC and the PET [24], probably because of the radiation factor in the equations. The results of this analysis indicated the UTCI and the PET as the most adequate indices for determining the conditions of thermal comfort [18].

When the PET results for Santa Maria are compared with those observed by Kántor, Kovács, and Takács [48], who calibrated the PET index from more than 5800 outdoor comfort questionnaires in the city of Szeged, Hungary, using different analysis procedures for 78 days of monitoring, the results found varied according to the season and the method of analysis used, and the average comfort range was from 17.1 °C to 21 °C. This is a threshold of only 4 °C, which is much less in comparison to Santa Maria, where the difference between the lower and upper limits of the comfort range was 8 °C.

3.4. Analysis–Synthesis of the Calibration Suggested for the Interpretive Ranges of the PET, SET, and PMV

Following calibration of the interpretive ranges of the analyzed indexes, as shown in Table 1, their percent hit rates were very low relative to the average votes of the investigated population.

After calibration, the hit rate for the PET shifted from 32.8% to 69.3%, that for the SET from 34.6% to 64.9%, and that for the PMV from 35.9% to 58.7%.

Figure 3 depicts a visual analysis of the percent distribution of the interpretive ranges of the investigated indexes vis-à-vis the number of occurrences relative to each such range across the entire analyzed period.

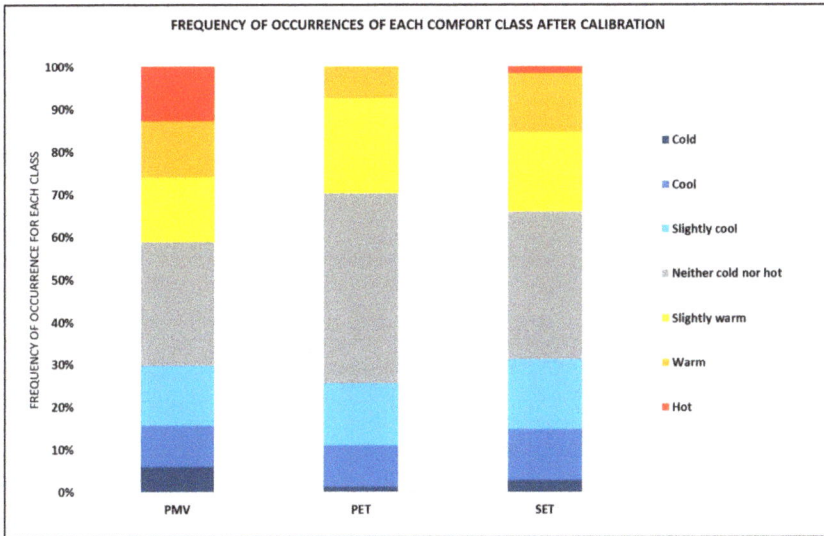

Figure 3. Frequency of occurrences relative to each interpretive range after calibration across the entire analyzed period.

As Figure 3 shows, class "neither cold nor hot" exhibited the highest percentage of occurrences for all three investigated indexes across the entire analyzed period. The PET was the single index not associated with an occurrence of the class "hot", following calibration.

4. Conclusions

We conclude that the analyzed scores allowed the estimation of the thermal comfort classes for an individual at a given time as a function of the values of the observed climatic conditions, which validates the linear regression method for determining the adequacy of the interpretive ranges of each index.

The results of the present study show the efficiency of the linear regression model for the purpose of adapting the interpretive ranges of the investigated indexes, because their values differed from those reported by other authors for other climatic regions. These differences agree with the surveyed literature and reinforce the validity of the calibration method for the PET, SET, and PMV relative to the Brazilian subtropical climate.

From the statistical point of view, there is still a need for a larger approach based on a more comprehensive daily analysis and a wider variety of climatic conditions. By this reasoning, the present study can be defined as "a first approach".

Further studies are needed on the effectiveness of the indices used in this research for open urban spaces in a subtropical climate, especially taking into account more detailed aspects of the characteristics and preferences of the individuals interviewed, such as the local demographic distribution, ethnicity, socioeconomic distribution, and thermal history, as well as the lifestyle and the frequency of the use of air conditioning, especially in places with well-defined climatic seasons, as the studied area. All such care, coupled with an appropriate standard method, is a suggestion for future studies that may fill the gaps still left.

Acknowledgments: The authors thank the National Counsel of Technological and Scientific Development (CNPq).

Author Contributions: This paper is one part of the doctoral thesis of the first author under the guidance of the second author.

Conflicts of Interest: The authors declare no conflict of interest.

References

1. Höppe, P. Different aspects of assessing indoor and outdoor thermal comfort. *Energy Build.* **2002**, *34*, 661–665. [CrossRef]
2. Lai, D.; Guo, D.; Hou, Y.; Lin, C.; Chen, Q. Studies of outdoor thermal comfort in northern China. *Build. Environ.* **2014**, *77*, 110–118. [CrossRef]
3. International Organization for Standardization (ISO). *Ergonomics of the Thermal Environment—Assessment of the Influence of the Thermal Environment Using Subjective Judgement Scales*; ISO 10551; ISO: Geneva, Switzerland, 1995.
4. International Organization for Standardization (ISO). *Ergonomics of the Thermal Environment—Instruments for Measuring Physical Quantities*; ISO 7726; ISO: Geneva, Switzerland, 1998.
5. Johansson, E.; Thorsson, S.; Emmanuel, R.; Krüger, E. Instruments and methods in outdoor thermal comfort studies e the need for standardization. *Urban Clim.* **2014**, *10*, 346–366. [CrossRef]
6. Bröde, P.; Krüger, E.L.; Rossi, F.A.; Faila, D. Predicting urban outdoor thermal comfort by the Universal Thermal climate index UTCI—A case study in Brazil. *Int. J. Biometeorol.* **2012**, *56*, 471–480. [CrossRef] [PubMed]
7. Rossi, F.A.; Krüger, E.L.; Bröde, P. Definição de faixas de conforto e desconforto térmico para espaços abertos em Curitiba, PR, com o índice UTCI. *Ambiente Construído* **2012**, *12*, 41–59. [CrossRef]
8. Zambrano, L.; Malafaia, C.; Bastos, L.E.G. Thermal Comfort Evaluation in Outdoor Space of Tropical Humid Climate. In Proceedings of the PLEA 23rd Conference on Passive and Low Energy Architecture, Geneva, Switzerland, 6–8 September 2006.
9. Rossi, F.A.; Krüger, E.L.; Guimarães, I.A. Modelo preditivo de sensação térmica em espaços abertos em Curitiba, PR. *Ra'e Ga.* **2013**, *29*, 209–238. [CrossRef]
10. Gulyas, A.; Unger, J.; Matzarakis, A. Assessment of the microclimatic and human comfort conditions in a complex urban environment: Modelling and measurements. *Build. Environ.* **2006**, *41*, 1713–1722. [CrossRef]
11. Cheng, V.; Ng, E. Thermal comfort in urban open spaces for Hong Kong. *Archit. Sci. Rev.* **2006**, *49*, 236–242. [CrossRef]
12. Spagnolo, J.; Dear, R. A field study of thermal comfort in outdoor and semi-outdoor environments in subtropical Sydney Australia. *Build. Environ.* **2003**, *38*, 721–738. [CrossRef]
13. Gagge, A.P.; Stolwijk, J.A.J.; Nish, Y. An effective temperature scale based on a simple model of human physiological regulatory response. *ASHRAE Trans.* **1971**, *77*, 247–263.
14. Fanger, P.O. *Thermal Comfort*; McGraw-Hill Book Company: New York, NY, USA, 1970.
15. Höppe, P. The physiological equivalent temperature: A universal index for the assessment of the thermal environment. *Int. J. Biometeorol.* **1999**, *43*, 71–75. [CrossRef] [PubMed]
16. Pantavou, K.; Theoharatos, G.; Santamouris, M.; Asimakopoulos, D. Outdoor thermal sensation in a Mediterranean climate and a comparison with UTCI. *Build. Environ.* **2013**, *66*, 82–95. [CrossRef]
17. Lin, T.P.; Dear, R.; Hwang, R.L. Effect of thermal adaptation on seasonal outdoor thermal comfort. *Int. J. Climatol.* **2011**, *31*, 302–312. [CrossRef]
18. Farajzadeh, H.; Saligheh, M.; Alijani, B.; Matzarakis, A. Comparison of selected thermal indices in the northwest of Iran. *Nat. Environ. Chang.* **2015**, *1*, 1–20.

19. Nasira, R.A.; Ahmada, S.S.; Ahmedb, A.Z. Psychological Adaptation of Outdoor Thermal Comfort in Shaded Green Spaces in Malaysia. *Procedia Soc. Behav. Sci.* **2012**, *68*, 865–878. [CrossRef]
20. Yang, W.; Wong, N.H.; Jusuf, S.K. Thermal comfort in outdoor urban spaces in Singapore. *Build. Environ.* **2013**, *59*, 426–435. [CrossRef]
21. Villadiego, K.; Velay-Dabat, M.A. Outdoor thermal comfort in a hot and humid climate of Colombia: A field study in Barranquilla. *Build. Environ.* **2014**, *75*, 142–152. [CrossRef]
22. Hirashima, S.; Katzschner, A.; Ferreira, D.; Assis, E.S.; Katzschner, L. Thermal comfort comparison and evaluation in different climates Toulouse, Françe. In Proceedings of the 9th International Conference on Urban Climate (ICUC), Toulouse, France, 20–24 July 2015.
23. Lucchese, J.R.; Mikuri, L.P.; De Freitas, N.V.S.; Andreasi, W.A. Application of selected indices on outdoor thermal comfort assessment in Midwest Brazil. *Int. J. Energy Environ.* **2016**, *7*, 291–302.
24. Krüger, E.L.; Rossi, F.A.; Drach, P. Calibration of the physiological equivalent temperature index for three different climatic regions. *Int. J. Biometeorol.* **2017**, *61*, 1323–1336. [CrossRef] [PubMed]
25. Hirashima, S.Q.S.; Assis, E.S.; Nikolopoulou, M. Daytime thermal comfort in urban spaces: A field study in Brazil. *Build. Environ.* **2016**, *107*, 245–253. [CrossRef]
26. Zeng, Y.; Dong, L. Thermal human biometeorological conditions and subjective thermal sensation in pedestrian streets in Chengdu, China. *Int. J. Biometeorol.* **2015**, *59*, 99–108. [CrossRef] [PubMed]
27. Golasi, I.; Salata, F.; de Lieto Vollaro, E.; Coppi, M. Complying with the demand of standardization in outdoor thermal comfort: A first approach to the Global Outdoor Comfort Index (GOCI). *Build. Environ.* **2018**, *130*, 104–119. [CrossRef]
28. Lin, T.P. Thermal perception, adaptation and attendance in a public square in hot and humid regions. *Build. Environ.* **2009**, *44*, 2017–2026. [CrossRef]
29. Cheng, V.; Ng, E.; Chan, C.; Givoni, B. Outdoor thermal comfort study in sub-tropical climate: A longitudinal study based in Hong Kong. *Int. J. Biometeorol.* **2012**, *56*, 43–56. [CrossRef] [PubMed]
30. Chen, L.; Wen, Y.; Zhang, L.; Xiang, W.N. Study of thermal comfort and space use in an urban park square in cool and cold seasons in Shanghai. *Build. Environ.* **2015**, *94*, 644–653. [CrossRef]
31. Li, K.; Zhang, Y.; Zhao, L. Outdoor thermal comfort and activities in the urban residential community in a humid subtropical area of China. *Energy Build.* **2016**, *133*, 498–511. [CrossRef]
32. Liu, W.; Zhang, Y.; Deng, Q. The effects of urban microclimate on outdoor thermal sensation and neutral temperature in hot-summer and cold-winter climate. *Energy Build.* **2016**, *128*, 190–197. [CrossRef]
33. Zhao, L.; Zhoua, X.; Li, L.; Heb, S.; Chena, R. Study on outdoor thermal comfort on a campus in a subtropical urban area in summer. *Sustain. Cities Soc.* **2016**, *22*, 164–170. [CrossRef]
34. Jeong, M.A.; Park, S.; Song, G.S. Comparison of human thermal responses between the urban forest area and the central building district in Seoul, Korea. *Urban For. Urban Green.* **2016**, *15*, 133–148. [CrossRef]
35. Song, G.S.; Jeong, M.A. Morphology of pedestrian roads and thermal responses during summer, in the Urban area of Bucheon city, Korea. *Int. J. Biometeorol.* **2016**, *60*, 999–1014. [CrossRef] [PubMed]
36. Nikolopoulou, M.; Baker, N.; Steemers, K. Thermal comfort in outdoor urban spaces: Understanding the human parameter. *Sol. Energy* **2001**, *70*, 227–235. [CrossRef]
37. International Organization for Standardization (ISO). *Ergonomics of the Thermal Environment—Estimation of Thermal Insulation and Water Vapour Resistance of a Clothing Ensemble*; ISO 9920; ISO: Geneva, Switzerland, 2007.
38. International Organization for Standardization (ISO). *Ergonomics of the Thermal Environment—Determination of Metabolic Rate*; ISO 8996; ISO: Geneva, Switzerland, 2004.
39. Matzarakis, A. Modelling radiation fluxes in simple and complex environments—Application of the RayMan model. *Int. J. Biometeorol.* **2007**, *51*, 323–334. [CrossRef] [PubMed]
40. Matzarakis, A. Climate Change: Temporal and spatial dimension of adaptation possibilities at regional and local scale. In *Tourism and the Implications of Climate Change: Issues and Actions*; Schott, C., Ed.; Emerald Group Publishing: Bingley, UK, 2010; Volume 3, pp. 237–259.
41. Neter, J.; Wasserman, W.; Kutner, M.H. *Applied Linear Statistical Models: Regression, Analysis of Variance, and Experimental Designs*; IRWIN: Burr Ridge, IL, USA, 1990.
42. Ansi/Ashrae. *Standard 55: Thermal Environmental Conditions for Human Occupancy*; American Society of Heating, Refrigerating and Air-conditioning Engineers: Atlanta, GA, USA, 2004.
43. Höppe, P. Die Physiologisch Äquivalente Temperatur PET. *Ann. Meteorol.* **1997**, *33*, 108–112.

44. Jendritzky, G.; Sönning, W.; Swantes, H.J. *Ein Objektives Bewertungsverfahren zur Beschreibung des Thermischen Milieus in der Stadt- und Landschaftsplanung ('Klima-Michel-Modell')*; Beiträge der Akademie für Raumforschung und Landesplanung, Bd. 28; Akademic für Raumforschung und Lancksplanung: Hannover, Germany, 1979.

45. Monteiro, L.M. Modelos Preditivos de Conforto Térmico: Quantificação de Relações Entre Variáveis Microclimáticas e de Sensação Térmica Para Avaliação e Projeto de Espaços Abertos. Ph.D. Thesis, Universidade de São Paulo, São Paulo, Brazil, 2008.

46. Xi, T.; Li, Q.; Mochida, A.; Meng, Q. Study on the outdoor thermal environment and thermal comfort around campus clusters in subtropical urban areas. *Build. Environ.* **2012**, *52*, 162–170. [CrossRef]

47. Jendritzky, G. Selected Questions of Topical Interest in Human Bioclimatology. *Int. J. Biometeorol.* **1991**, *35*, 139–150. [CrossRef] [PubMed]

48. Kántor, N.; Kovács, A.; Takács, Á. Seasonal differences in the subjective assessment of outdoor thermal conditions and the impact of analysis techniques on the obtained results. *Int. J. Biometeorol.* **2016**, *60*, 1615–1635. [CrossRef] [PubMed]

climate

MDPI

Article

Lighting Implications of Urban Mitigation Strategies through Cool Pavements: Energy Savings and Visual Comfort

Giuseppe Rossi [1], Paola Iacomussi [1,*] and Michele Zinzi [2]

[1] INRIM, Istituto Nazionale di Ricerca Metrologica, Strada delle Cacce 91, 10135 Torino, Italy; g.rossi@inrim.it
[2] ENEA, National Agency for New Technologies, Energy and Sustainable Economic Development,
    Via Anguillarese 301, 00123 Rome, Italy; michele.zinzi@enea.it
*   Correspondence: p.iacomussi@inrim.it; Tel.: +39-011-3919226

Received: 31 January 2018; Accepted: 26 March 2018; Published: 7 April 2018

Abstract: Cool materials with higher solar reflectance compared with conventional materials of the same color are widely used to maintain cooler urban fabrics when exposed to solar irradiation and to mitigate the Urban Heat Island (UHI). Photo-catalytic coatings are also useful to reduce air pollutants. Many studies related to these topics have been carried out during the past few years, although the lighting implication of reflective coatings have hardly been explored. To investigate these aspects, reflective coatings were applied on portions of a road and intensely analyzed in a laboratory and on the field. The applied cool coatings were found to have much higher solar and lighting reflectance than the existing road, which lead to lower surface temperatures up to 9 °C. Non-significant variations of chromaticity coordinates were measured under different lighting conditions. However, these materials showed a relevant variation of directional properties depending on the lighting and observation conditions with respect to conventional pavements. The optical behavior of these materials affects the uniformity of visions for drivers and requires ad-hoc installation of light sources. On the other hand, potential energy savings of up to 75% were calculated for the artificial lighting of a reference road.

Keywords: cool pavements; road lighting; urban heat island; road surface; material characterization; luminance coefficient; energy savings; Euramet; EMPIR 16NRM02

---

1. Introduction

Urban Heat Island (UHI) and urban overheating are two major environmental hazards arising from the combined force of climate change and unregulated growth of cities. The UHI is defined as the increase in urban temperatures in proportion to those of the countryside surrounding the city. The phenomenon is complex and there is still no complete explanation. However, the main causes are: the reduction of permeable surfaces, the use of high solar absorptive construction materials, the anthropogenic heat generated by energy uses in the civil and transport sectors and the urban texture geometry that creates favorable conditions for heat trapping and low heat release [1–3].

Extensive studies on the UHI phenomenon and associated mitigation strategies were started several decades ago [1–5], which have intensified in the new millennium due to experiences documented at all latitudes, which has been summarized in literature review studies [6–9]. The increase in urban ambient temperatures has been documented through pluriannual observations for many US and European cities [8,9] as result of the combined effect of global warming and urban heat islands. According to vast literature surveys carried out in recent years, it was found out that UHI intensity reached 15 °C in the most sever conditions [6].

This trend shows that these serious urban climate hazards have not been adequately tackled, especially considering that it has a relevant impact at the energy, environmental and social levels, as widely documented in several exemplary studies [10–15]. On the contrary, several mitigation strategies and technologies are available to counter this phenomenon. Some of these are consolidated enough to be considered as a trademark for vernacular architecture and urbanism in several areas of the world. Such solutions can be gathered in four large categories: urban and building greenery; natural and artificial water bodies; urban shading; and cool construction materials for pavements and buildings. The impact of such solutions, each separately and/or combined, has been proven in many studies. Exhaustive reviews have been already conducted [7,16–18].

The applications of white and light-colored materials for buildings have been carried out for centuries, becoming a trademark of vernacular architecture in the Mediterranean region [19]. This ancient concept gained attention with the so-called cool materials, which are products with higher solar reflectance compared with conventional materials of the same color. Due to this, they are able to remain cooler under solar irradiation. A higher solar reflectance results in lower material surface temperature under the sun, subsequently resulting in less heat released to the outdoor and indoor environments by convection and radiation. White and light-colored materials have the highest potential for thermal mitigation. Despite the effort exerted in maximizing the solar reflectance in the near-infrared region in the past few years so that even darker color could achieve good performances, several studies have demonstrated the stronger mitigation benefits of high albedo materials [20–23]. Generally, cool materials also have high thermal emissivity, although the impact of this property on UHI is less than that of their reflectance [24]. Comprehensive reviews about cool pavements and the impact on urban climate can be found in previous studies [25–28].

The potential impact of high solar reflectance materials in mitigating the urban environment was investigated in the past few years. Several studies are based on the preliminary monitoring of urban area and assessment of potential mitigation through Computational Fluid Dynamic (CFD) simulations. An application covering 25,000 m^2 was carried out in Tirana, Albania, where existing concrete and asphalt pavements were upgraded with cool paints. The albedo increased from 0.2 to above 0.65, with a 2.1-K predicted decrease in the peak ambient temperature [29]. A similar application was carried out in Athens, Greece, where the existing pavements were replaced with different products with albedo ranging between 0.35 and 0.78. The peak temperature reduction was estimated as being 1.2–2 K on average [30]. A relevant application was carried out in Putrajaya, Malaysia, in which a 420,000 m^2 urban area was upgraded with greenery and pavements, with 0.8 albedo. The ambient temperature reduction was estimated to be 1.5 K, of which only 0.1 K seemed to be related to the use reflective materials [31].

It can be inferred that a considerable amount of work has been done on the thermal aspects, although there has been little evidence related to environmental lighting conditions and on the lighting associated impact. The latter refers to artificial lighting energy savings as well as to glare and visual discomfort risks arising from materials with high reflectance in the visible range and enhanced directional properties [32].

Cool materials, which are based on photo-catalytic solutions with titanium dioxide or with dispersed silicates, are often used for road surfaces in urban zones due to their self-cleaning capability and their capability to reduce harmful substances (pollutants) in air. These products are applied on existing roads by spray guns using standard methods without considering their influence on the optical properties of the road surface and changes in lighting conditions and night/day vision.

The knowledge of the road surface luminance coefficient is important for the optimization of the energy performances of the lighting installations [33], glare evaluation [34] and improvement of safety, comfort during the night and energy savings. For example, this could occur through LED adaptive lighting systems (smart cities) [35]. The introduction of the European standards of energy performance indicators [33] and correlated requirements asks for improved optimization in the design of luminaries and selection of their luminous intensity distribution in addition to the installation layout and optical characteristics of the road surface. The use of brighter road surfaces can reduce the UHI effects during the day and improve energy saving and traffic safety during the night.

This study investigates the implications of applying high reflective coatings in design and performances of road lighting installations when these coatings are used in urban environments to rehabilitate existing pavements and road surfaces for thermal mitigation. For this purpose, cool coatings were tested for visual comfort and energy lighting assessment. To fully understand behavior and performances of such products, the following process was implemented:

- application of cool coatings in the real urban environment;
- optical, luminous and thermal characterization with laboratory measurements;
- luminous and thermal measurements in real urban conditions;
- simulation of artificial lighting energy saving and power performances of a standard road lighting installations.

All the analyses were carried out by considering reference asphalt pavements and cool pavements in order to compare differences in properties and performances for lighting related issues.

2. Materials and Methods

2.1. Theoretical Background

This section is dedicated to recall the basic theoretical concepts related to the directional properties of construction materials and the quality of vision at the road level.

For motorized traffic, the European standards for road lighting (EN 13201 series [33,34,36,37]) consider the road luminance level as the key parameter for obtaining adequate vision conditions and traffic safety. The technical report 115 [38] of International Lighting Commission (CIE) adopts the same concept, while the USA standard RP-8-14 [39] focuses on the road surface luminance requirements between the proposed design possibilities.

During night and at the lighting levels that road lighting standards specify, the eye is adapted under the mesopic conditions. Under these conditions, the spectral distribution of the radiation reflected by the road surface to reach the driver's eyes influences his/her perception of brightness. Photometric instruments, calibrated under photopic conditions (i.e., at the lighting level of daylight eye adaptation), are not able to correctly quantify the real conditions as they measure the photopic lighting level (luminance). Only some national standards (Italy and UK) [40,41] and international guidelines [38] acknowledge and reference mesopic conditions [42].

In some conditions, such as for the evaluation of glare due to the lighting installation, the standards consider that the road luminance is also important for pedestrians [34], although it needs to be evaluated in a different view, as explained hereafter.

The road luminance is directly related to the direction of the incident luminous flux, the direction of view and the reflection characteristics of the road surfaces.

Road lighting standards conventionally consider the angle of observation of the road surface α to be constant and equal to $1°$ [34], which is equivalent to a driver (with eyes at a height of 1.5 m above the road) seeing a portion of the road at a distance of about 87 m. Under this geometrical condition, luminance coefficient q (or the reduced luminance coefficient r).

According to the angular convention adopted in EU standards, the luminance coefficient can be written as:

$$q(\alpha, \varepsilon, \beta) = \frac{L\left(\alpha, \beta + \frac{\pi}{2}\right)}{E(\epsilon, \beta)}, \tag{1}$$

where $q(\alpha, \beta, \varepsilon)$ is the luminance coefficient of an elementary surface of the road considering the incident light path with angular coordinates (ε, β) and the viewing direction with angular coordinates $(\alpha, \beta + \pi/2)$, in reciprocal steradians; $L(\alpha, \beta + \pi/2)$ is the luminance of the lighted elementary road surface when the viewing direction has angular coordinates $(\alpha, \beta + \pi/2)$, in candelas per square meter;

$E(\varepsilon,\beta)$ is the illuminance on the lighted elementary road surface considering the incident light path with angular coordinates (ε,β), in lux; α is the angle of observation of the point on road surface measured from the road surface, in degrees; ε is the angle of incidence, which is the angle between the light path at the observed point on the road surface and the normal to this surface, in degrees; and β is the angle of deviation between the oriented vertical planes through the observer to the point of observation and from the point of observation through the luminaire, in degrees.

A scheme of the above angles is presented in Figure 1 for clarity of explanation. According to the assumption of the observation angle of the road surface α being equal to $1°$, the relevant standard introduces the reduced luminance coefficient $r(\varepsilon,\beta)$ for lighting installations, which is given by:

$$r(\varepsilon,\beta) = q(\varepsilon,\beta)\cos^3\varepsilon, \tag{2}$$

where $r(\varepsilon,\beta)$ is the luminance at the point, in candelas per square meter; and $q(\varepsilon,\beta)$ is the luminance coefficient of the elementary road surface for the incident light path with angular coordinates (ε,β) and for the view direction with angular coordinates $(\alpha = 1, \beta + \pi/2)$, in reciprocal steradians.

To compare the brightness of road surfaces when lighted under the same conditions and considering the typical layouts of road lighting installations, the average luminance coefficient Q_0 was introduced in a previous study [39]. It is defined by:

$$Q_0 = \frac{1}{\Omega} \int_\Omega q\,d\omega, \tag{3}$$

where Ω is the solid angle defined by a rectangle parallel to the road surface at a height h equal to the height of the installed luminaries. To the reference point of the road surface, the rectangle is $3\,h$ wide with a length of $4\,h$ in the direction of travel and $12\,h$ in the opposite direction (Figure 2).

The reference values of the luminance coefficient q and Q_0 are provided in CIE [43] and national standards, although they are more than 40 years old and do not consider coated road surfaces. For example, the Italian standard suggests the following values: $Q_0 = 0.07$ for asphaltic pavements and $Q_0 = 0.10$ for concrete pavements [40].

For pedestrians, the angles of observation of the road surface cannot be specified conventionally, but when α is greater than $10°$, the road surface works as a Lambertian surface and is characterized using the average diffuse reflectance. If measured data are not available, the European standard suggests $\rho = 0.2$ as the default value [34].

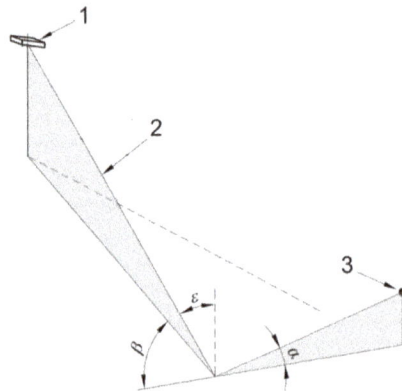

β	Deviation angle
ε	Incidence angle
α	Observation angle
1	Light Source
2	Light Direction
3	Observer

Figure 1. Relevant angles in the assessment of road lighting performances.

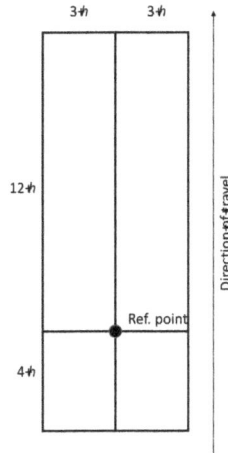

Figure 2. Dimension of the boundary area for Q_0 calculation where h is the installation height of the luminaire.

2.2. Sample and Test Field Preparation

Table 1 describes the sample identification and measurement conditions used in this research. Two products were identified and implemented for this study: a grey and an off-white elastomeric coating. The two coatings were manually applied on an old smooth asphaltic road (Figure 3) within the INRIM (Istituto Nazionale di Ricerca Metrologica) premises and compared with the untreated adjacent road surface. The test site was divided in three different parts (Figure 3):

- R_u: 4×10 m of road untreated surface (40 m^2);
- R_g: 4×10 m of road treated with grey coating (40 m^2);
- R_w: 4×4 m of road treated with off-white coating (16 m^2).

Two asphaltic cores with a diameter of 150 mm were partially painted with the two coatings to obtain samples for laboratory characterization (Figure 4). The main difference between the two cores, which were named C1 and C2, is the texture of their surfaces, which depends on the characteristics of the asphaltic substrate. The purpose of this research is the measurement of optical properties of the materials in real working environments. The samples cores were useful for carrying out peculiar characterizations that are not possible with the portable instruments on site.

Table 1. Samples and measurement conditions.

Sample		Type	Code	Description	Measurements	Figure
Road	R	Smooth old road	R_u	Untreated	On site	3
			R_g	With grey coating		
			R_w	With off-white coating		
Core 1	C1	Smooth asphaltic	C1_u	Untreated	Laboratory	4a
			C1_g	With grey coating		
			C1_w	With off-white coating		
Core 2	C2	Rough asphaltic	C2_u	Untreated	Laboratory	4b
			C2_g	With grey coating		
			C2_w	With off-white coating		

Figure 3. View of the experimental installation, taken in condition of glazing sun lighting quite parallel to the direction of view to emphasize the different behavior of the actual asphalt and coated pavements (from top to bottom: actual asphalt, grey and off-white elastomeric coatings).

(a)   (b)

Figure 4. The two asphaltic cores used for laboratory testing (**a**) core 1; and (**b**) core 2.

2.3. Measured Quantities

The selected samples have been characterized in the laboratory and on site to assess their thermal and optical properties under different environment conditions. With reference to Table 1, the following quantities were measured for the listed samples:

- spectral reflectance in the solar range ρ_e of 300–2500 nm (C1, C2);
- spectral reflectance in the visible range ρ_v of 380–780 nm (C1, C2, R);
- chromaticity coordinates in different geometries (C1, C2, R);
- q and r coefficients in standard required geometries (R); and
- thermal behavior (C1, C2, R).

2.3.1. Optical and Luminous Characterization

The spectral optical properties, spectral reflectance and chromaticity coordinates of the three pavement typologies were measured in the laboratory with a double beam spectrophotometer Lambda 950 by Perkin Elmer, which was equipped with a 150 mm diameter Spectralon coated integrating sphere by Labsphere. The sphere accessory is needed for measuring the optical properties of scattering materials. Measurements were carried out for the whole solar spectrum of 300–2500 nm with an angle of incidence of 8° and considering all the diffuse radiation (8/d geometry, specular component included). The spectral resolution was 1 nm and the absolute expanded measurement uncertainty was ±0.01. The slit aperture was set to 1 nm in the visible range and in servo mode in the near-infrared range. In this modality, the slit size changes according to the optimal energy input. Broadband values in the visible (v) and solar (e) range were calculated according to a previous study [44]. Visible quantities are calculated using the CIE standard Illuminant A as the reference incident source.

Chromaticity and reflection properties of the samples were also measured both in the laboratory and on site using a calibrated Hunter Lab MiniScan EZ4500 spectrophotometer with a spectral resolution and bandwidth of 10 nm. The instrument has a 45°/0° geometry with a directional annular 45° illumination and a 0° viewing (specular components excluded). The absolute expanded measurement uncertainty of reflectance was ±0.08 and the absolute expanded measurement uncertainty of the chromaticity coordinates in the CIE 1931 standard colorimetric system was ±0.005. The on-site measurement was carried out on a 1-m grid of measurement points to verify the homogeneity of the implementation carried out with hand tools.

2.3.2. Luminance Coefficient

Directional measurements were carried out with an incandescent light source to simulate the illuminant CIE A and a Photo Research PR650 spectroradiometer. This instrument has a spectral resolution of 4 nm and bandwidth of 8 nm. To fulfil requirement of the viewing angle of the road surface needing to be equal to 1°, the instrument was set at a height of about 30 cm and a measuring distance of about 17 m from the road measured surface. The measured area was an ellipse with the longest axis of about 3 m, which was centered in the middle of the testing area. In this way, the measured q was the average value that includes the inhomogeneity of the road surface coating. The relative expanded measurement uncertainty of the luminance coefficient was ±2% and the absolute expanded measurement uncertainty of the chromaticity coordinates in the CIE 1931 standard colorimetric system was ±0.005.

2.3.3. Thermal Characterization

Thermal measurements were carried out with a NEC G100 thermography camera calibrated at INRIM. A preliminary test was carried out in the laboratory to confirm that all samples have the same emissivity and to estimate the measurement uncertainty of field measurements.

The first field tests have been conducted in the morning (starting at 9:30 a.m.) of an autumn day with a clear sky (20 September) to consider the conditions of a low temperature (19 °C) of the road surface and the absence of direct solar radiation Under these conditions, the influence of the optical properties of the coating on the measured temperature is minimized. The second field tests were carried out in the afternoon (starting at 3:40 p.m.) to consider the conditions of a higher temperature (23 °C) of the road surface due to more than four hours of direct solar irradiation. Under these condition, the influence of the coating optical properties on the measured temperature is maximized with respect to the climatic conditions and the road surface position.

2.4. Calculations

To compare the performances of the coated surfaces and the untreated surfaces, a reference road lighting installation was simulated using the measured values of *q* and the measured luminous intensity distribution of a LED commercial luminaire design for road lighting. The luminaire has a luminous flux of 6150 ± 70 lm and a luminous efficacy of 55 ± 0.9 lm W^{-1}.

To evaluate and compare the power saving performances of road lighting installations, the European standard [33] defined the power density indicator D_P and the annual energy consumption indicator D_E. The first quantifies the ability of a luminaire to light the road at the given level with the minimum consumption of energy and as a first approximation, can be considered to not be correlated with the road surface characteristics. For our aim, the second one is more interesting because it depends on the road surface properties. For a simple installation, a single operation profile can be calculated with the following formula:

$$D_E = \frac{Pt}{A} \qquad (4)$$

where P is the system power of the lighting installation used to light the relevant area, in watts; t is the duration of the operation profile when the power P is consumed over a year, in hours; and A is the size of the area lit by the lighting installation, in meters.

Following the European standard [33], a two-lane road for motorized traffic was considered. The carriageway has a width of 8 m, the distance between luminaires was 35 m with a height of 8 m. The design target was to obtain a M3 lighting class [40] (minimum maintained average road luminance of 1 cd m^{-2}, minimum overall uniformity of 0.4 and minimum longitudinal uniformity of 0.60) when considering a road surface of type C2 with $Q_0 = 0.07$ [43] as the most common surface used by lighting engineers.

Considering the above described conditions, the European standard [33] provides the typical values of $D_P = 27$ mW lx^{-1} m^{-2} and $D_E = 1.6$ kWh m^{-2} when a full power operational profile is adopted with an annual operation time equal to 4000 h.

3. Results

3.1. Laboratory Characterisation

Figure 5 shows the spectral reflectance of the investigated materials, which was measured on the two cores (C1, C2 samples). Although the untreated surfaces show substantial differences both in the visible and in the IR ranges due to their different composition and mechanical structure (due to bitumen and different rocks dimensions as larger dimensions result in higher reflectance), the two coated surfaces show differences that are lower than the measurement uncertainty and surface homogeneity, with graphs that are similar to each other. From the spectral point of view, the coating masks the influence of optical properties of the substrate and of its mechanical characteristics. Thus, for a given coating, the main factor determining its on-site optical properties is the application technique.

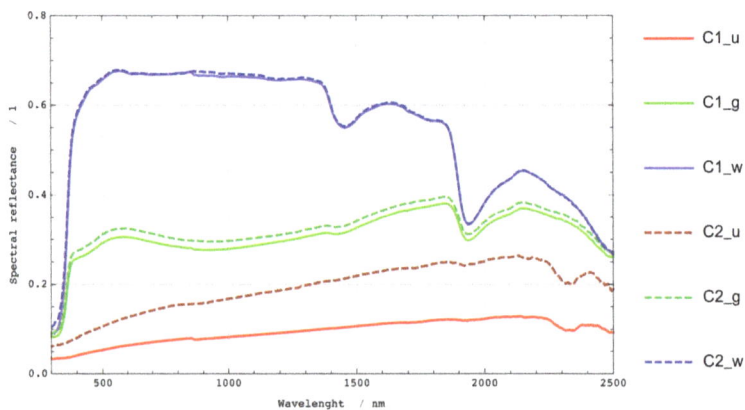

Figure 5. Spectral reflectance in the 300–2500 nm wavelength range of the cores (see Table 1) measured in laboratory conditions (8/d geometry).

With regards to the spectral response of the products, the grey coating (C1_g and C2_g) has a spectral reflectance of up to 1900 nm, while the reflectance of the off-white coating (C1_w and C2_w) starts decreasing at 1400 nm. The two coatings have practically the same behavior in the 1900–2500 nm range. The consequences of this behavior are highlighted in Figure 6 where the spectral distribution of the solar radiation reflected by the different surfaces of the cores is shown. The differences between the grey and off-white coatings are small over 1900 nm but they become remarkable in the near infrared and visible ranges. In Table 2, the solar reflectance [45] and the luminous reflectance considering the CIE Standard Illuminant A and CIE Standard Illuminant D65 as incident sources are presented. This shows that the spectral distribution of the incident light does not influence the reflectance values. This aspect is important in road lighting where sources with different correlated color temperature and color rendering properties are used.

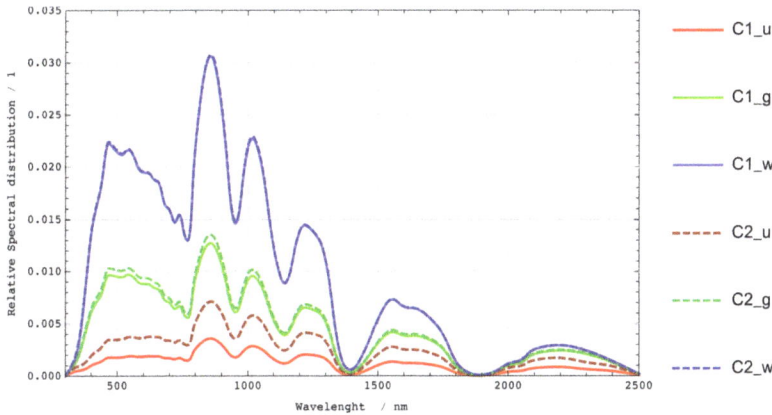

Figure 6. Spectral distribution in the 300–2500 nm wavelength range of the solar radiation reflected by the cores (see Table 1) considering the measured values of reflectance in laboratory conditions (8/d geometry).

Table 2. Solar reflectance ρ_e and luminous reflectance ρ_v of the two cores (see Table 1) measured in laboratory conditions (8/d geometry).

Sample	ρ_e	ρ_v Ill. A	ρ_v Ill. D65
C1_u	0.07	0.06	0.06
C1_g	0.29	0.30	0.30
C1_w	0.63	0.67	0.67
C2_u	0.14	0.12	0.12
C2_g	0.30	0.32	0.32
C2_w	0.63	0.67	0.67

The same behavior was found under different geometric measurement conditions, which is shown in Table 3. The measurements were carried out with two different instruments on a different measuring area. Considering the measurement uncertainty and the sample homogeneity, the CIE chromaticity coordinates and luminous reflectance are the same if the values of this quantity are slightly greater in the 45/0 geometry in the table. Colorimetric results refer to CIE Standard Illuminant A only for brevity.

Table 4 summarizes the influence of the azimuthal angle, which is essentially the effect of the surface texture and anisotropy with the incidence plane. In the geometrical conditions considered, similar to those usually observed for pedestrian traffic, the untreated surfaces show some moderate effects, while such effects are negligible for the cool coatings (off-white and grey). This difference depends on the influences of the mechanical structure and texture of the substrate, which are masked by the coatings.

Table 3. Luminous reflectance ρ_V and chromaticity coordinates x, y of the two cores considering different measurement geometries.

Sample	Measurement Conditions Incidence (°)	Observation (°)	ρ_V (−)	x (−)	y (−)
C1_u	45	0	0.07	0.47	0.41
	8	d	0.06	0.47	0.41
C1_g	45	0	0.30	0.46	0.42
	8	d	0.30	0.45	0.41
C1_w	0	45	0.76	0.45	0.41
	8	d	0.67	0.45	0.41
C2_u	45	0	0.14	0.46	0.41
	8	d	0.12	0.47	0.41
C2_g	45	0	0.32	0.45	0.41
	8	d	0.32	0.45	0.41
C2_w	0	45	0.69	0.45	0.41
	8	d	0.67	0.45	0.41

Table 4. Influence of the azimuthal angle on the reflectance of the analyzed products.

Sample	Measurement Conditions ε (°)	α (°)	ρ_V Relative Change (%) 0°	β 90°	180°	270°
C1_u	45	0	0	−1	5	3
C1_g	45	0	0	−2	−1	0
C1_w	45	0	0	0	0	−1
C2_u	45	0	0	2	5	4
C2_g	45	0	0	−1	0	0
C2_w	0	45	0	1	−1	−1

Considering the thermal properties, Figure 7 presents the false color thermal images of the two cores after 3 days of thermal stabilization in a temperature-controlled laboratory. These measurements show that treated and untreated cores have uniform surface temperatures. Negligible differences exist in the emissivity between the asphalt substrate and the two cool materials. In this sense, conventional and reflective materials have the same potential in cooling down by radiation at night.

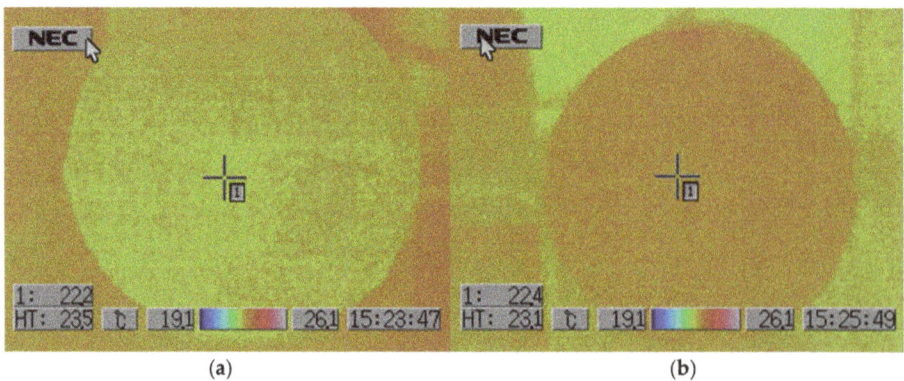

(a)   (b)

Figure 7. False color image of thermal measurement of (a) C1; and (b) C2.

3.2. On Field Characterisation

Considering the on-site measurements, the first step was to test the homogeneity of the coatings over the large treated surface area. The mean and the standard deviation of the measured values are reported in Table 5. The reflectance measurements have been carried out in the 45/0 geometry on several points in a grid with slots of 1×1 m. The low standard deviation demonstrated a good repeatability between the measured points and a good uniformity of the samples area. Visually, in typical motor traffic conditions (Figure 1), the homogeneity is not so good However, under these measurement conditions, the measured area is large enough to give an average value useful for the design of the road lighting installation.

Table 5. On-site measurement of the homogeneity of the untreated and coated road surfaces. The measurements have been carried out on a grid of points as described in the text.

Sample	Value	ρv (−)	x (−)	y (−)
R_u	Mean	0.12	0.468	0.415
	Standard deviation	0.01	0.006	0.001
R_g	Mean	0.33	0.461	0.414
	Standard deviation	0.03	0.009	0.002
R_w	Mean	0.64	0.453	0.412
	Standard deviation	0.01	0.002	0.001

Table 6 shows an example of the measured luminance coefficient q and the reduced luminance coefficient r for the standard measurement conditions in real urban environments. For comparison purposes, the tabulated values of a typical reference road (R_s), used at the designing stage of road lighting installations, are shown. By increasing the angle of incidence ε, the reduced luminance coefficient ($r\,10^4$) of the tested road (R_u) has a greater decrement compared to the standard road R_s. This can be explained by the conditions of the tested road, namely whether it is old and has deteriorated. On the other hand, it is important to note that the reduced luminance coefficient of the off-white surface (R_w) is 1.5−4 times that of R_s. The grey coating also significantly increases values of the design parameters. This aspect can be critical in retrofit works, since cool materials, characterized by a smoother surface, may severely affect the luminance uniformity of the lighted road. In practical terms, implementing this type of road surfaces needs the design and use of new luminaires, which are suitable for the peculiarities of these road coatings.

Table 6. Measured luminance coefficient and reduced luminance coefficient compared to the suggested values (R_s) of a reference road generally adopted in the design of road lighting installations [41,44].

ε	$\tan(\varepsilon)$	β	R_u		R_g		R_w		R_s	
			q	$r\,10^4$	q	$r\,10^4$	q	$r\,10^4$	q	$r\,10^4$
(°)	(−)	(°)	(sr^{-1})	(−)	(sr^{-1})	(−)	(sr^{-1})	(−)	(sr^{-1})	(−)
30	0.57	0	0.059	380	0.121	789	0.235	1527	0.058	379
35	0.70	0	0.064	350	0.107	589	0.215	1183	0.069	380
40	0.84	0	0.073	330	0.101	456	0.153	687	0.084	377
45	1.00	0	0.091	320	0.121	429	0.146	516	0.105	372

Table 7 presents the effect of the luminance coefficient q for different observation and radiation incidence angles on the selected materials. The table also shows the relative change Δq of the cool materials with respect to the untreated asphaltic road R_u. At high observation angles, the cool materials show Lambertian diffusing behavior for pedestrian traffic. At the same time, the reflected radiation increases with an increase in the α angle. This implies that the installed power of road lighting system should be adequately reduced in the case of using reflective coating to avoid excessive lighting levels and lighting pollution. In this sense, the ratio $Q_0\,(\alpha = 1°)/Q_0\,(\alpha = 45°)$ could be proposed as an indicator of the coating performances for the luminous pollution control.

Table 7. On-site characterization of the road surfaces considering geometric conditions of incidence and observation typical of pedestrian traffic.

Observation	Incidence	R_g		R_w		R_u
α	ε	q	Δq	q	Δq	q
(°)	(°)	(sr^{-1})	(%)	(sr^{-1})	(%)	(sr^{-1})
20		0.253	206	0.319	286	0.083
30	15	0.16	167	0.243	303	0.06
45		0.12	123	0.218	304	0.054
20		0.117	113	0.218	295	0.055
30	30	0.115	122	0.212	307	0.052
45		0.108	118	0.202	309	0.049
20		0.124	170	0.205	345	0.046
30	35	0.114	151	0.212	367	0.045
45		0.108	157	0.207	393	0.042
20		0.111	146	0.205	354	0.045
30	40	0.109	155	0.212	396	0.043
45		0.105	146	0.207	386	0.043
20		0.105	138	0.204	363	0.044
30	45	0.106	146	0.204	374	0.043
45		0.102	130	0.205	361	0.044

Thermal measurements were also carried out on site. Figure 8 presents the false color images acquired in the afternoon by the thermal camera of the actual asphalt, showing the grey coating (left) in addition to the off-white coating and the actual asphalt (right). The results, averaged across the whole area, are summarized in Table 8 and refer to the three pavement products. The results show the benefits of cool coating in reducing the surface temperature and thus, mitigating the urban environment. The strongest difference in thermal behavior between untreated road surface and off-white coated road surface was 9 °C during a day with clear sky and mild atmospheric temperatures. In the same conditions, the grey coating registered a surface temperature reduction of 3 °C.

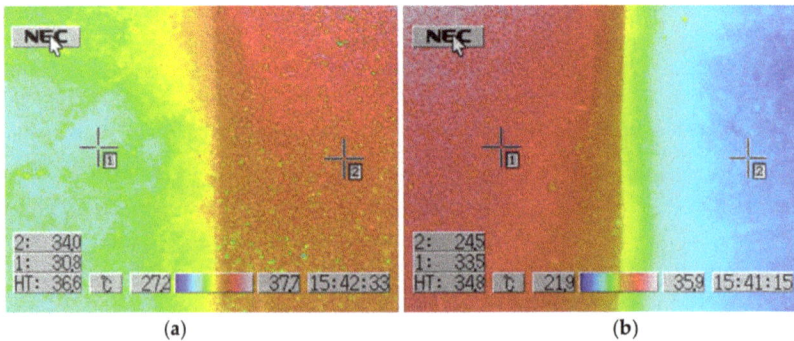

(a)   (b)

Figure 8. False color image of thermal measurement on site of (a) R1_u and R1_g; (b) R1_u and R1_w.

Table 8. On-site surface temperature measurements in the morning and in the afternoon.

	R_u	R_g		R_w		
Measurement	T	T	$T(u)-T(g)$	T	$T(u)-T(w)$	$T(g)-T(w)$
	(°C)	(°C)	(°C)	(°C)	(°C)	(°C)
Morning	19.8	19.9	−0.1	18.8	1	1.1
Afternoon	33.8	30.8	3	24.5	9.3	6.3

3.3. Energy Performances Results

The values of the annual energy use for road electric lighting D_E calculated for the three surfaces are presented in Table 9. In this table, ΔD_E is the relative percentage change of D_E between the road lighting installation with the given coated surface and the untreated road surface. The results in this table consider only the luminance level and not the uniformity because it would require optimization of installation layout and luminous intensity distribution of luminaires, which is out of the scope of this research. Energy consumption can be strongly reduced using cool materials, although such results could be achieved only using properly designed installation. Furthermore, this takes into account the light scattering of cool and smoother products and the opposite object contrast on road surface. The pro/cons in using this type of coating on road surfaces is discussed in the conclusions in more detail, with consideration of the thermal impact and energy performances. Economic analyses have not been considered, but in case of a full feasibility study, the higher costs for pavements works, including periodic recoating, should be taken into account.

Table 9. Annual energy consumption indicator of the road lighting installation described in the text.

Sample	Energy Consumption	Energy Savings
	(kWh m^{-1})	(%)
R1_u	1.61	—
R1_g	0.52	67.5
R1_w	0.38	76.2

4. Discussion

The use of high reflective materials for road surfaces can reduce the UHI effect during the day and improve energy saving and traffic safety during night. Apart from the thermal aspects, high reflectance coating can reduce the energy consumption for roads or outdoor lighting installations. They can also improve visual perception of objects (obstacles) or pedestrians because of the opposite contrast against light background on the carriages and then increase road environment safety. Considering road lighting, it is possible:

- to reduce the installed luminous flux in addition to reaching the prescribed normative requirements about the road surface luminance;
- to improve the visual behavior due to the increased diffused part of the reflected light, with a consequent increase in the surrounding luminance with a reduction of glare (from lighting sources—luminaires) and improvement in the safety of pedestrians;
- to improve light pollution as a counter-effect, especially in extra-urban areas, where the diffused part of the reflected luminous flux would not be shielded by buildings as in the urban zone. This negative effect could be reduced or compensated optimization of the reflectance behavior in the specular directions. In this way, the reduction in the installed luminous flux could counterbalance the increment of the diffuse component of the light.

The introduction of high reflective and thus, brighter road surfaces would have strong influences in design and maintenance of road lighting installations and energy saving. However, for the large diffusion and application of this technology, at least three steps are necessary:

- the availability of reliable reference data of actual (or oncoming) high reflective coating materials, considering the influences of aging with time and maintenance;
- the strong control of the implementation techniques to guarantee repeatability of the optical properties of the coated road surface on the same site and between sites; and
- the development of luminaires with peculiar luminous intensity distributions to guarantee the normative requirements in the uniformity of the road surface luminance without changing the layout of road lighting installations, such as the inter-distance between consecutive columns.

The availability of reliable implementation techniques and reference data of current (or oncoming) high reflective coating materials will allow lighting designers to gather the energy savings and quality parameters forecasted in standards and meet the EU commitment to cut energy consumptions as committed by 2020 and 2030 Key Performances Target [46]. Thus, they will provide more efficient and safe road lighting for all users.

In a more general view, it can be observed that highly reflective coatings with reference to untreated road surfaces provide high potentials for thermal mitigation, with high energy savings for artificial lighting. However, they heavily affect the fraction of light scattered upwards, which is not useful for traffic safety. Grey and thus, moderately reflective coatings have a weaker impact on urban temperature than white coatings, but still provide relevant energy savings when road surfaces performances are considered with a viewing angle of 1° in road lighting calculations. Under this geometrical condition, the photometric performances of white and grey coatings are not significantly different, but grey has the advantage of being less disturbing to users. The trade-off between energy savings achievable by temperature mitigation and artificial lighting compared to the visual comfort for users is a crucial aspect that needs to be further explored, since few data are available yet. This topic will be addressed in the next phase of this research where different road surfaces painting techniques will also be considered.

Acknowledgments: Part of this work was funded by the Italian Ministry of Economic Development in the framework of the Program "RSE—Ricerca di Sistema Elettrico". Part of this work is part of project "16NRM02 Surface, Pavement surface characterization for smart and efficient road lighting" that has received funding from the EMPIR program. EMPIR program is co-financed by the Participating States and from the European Union's Horizon 2020 research and innovation program and funded the publishing fees.

Author Contributions: P.I., G.R. and M.Z. conceived and designed the experiments; P.I. performed the experiments; G.R. analyzed the data; M.Z. P.I. and G.R wrote the paper.

Conflicts of Interest: The authors declare no conflict of interest. The founding sponsors had no role in the design of the study; in the collection, analyses, or interpretation of data; in the writing of the manuscript, and in the decision to publish the results.

References

1. Oke, T.R. The energetic basis of the urban heat island. *Q. J. R. Meteorol. Soc.* **1982**, *108*, 1–24. [CrossRef]
2. Oke, T.R. The urban energy balance. *Prog. Phys. Geogr.* **1988**, *12*, 471–508. [CrossRef]
3. Taha, H. Urban climates and heat islands: Albedo, evapotranspiration and anthropogenic heat. *Energy Build.* **1997**, *25*, 99–103. [CrossRef]
4. Akbari, H.; Davis, S.; Dorsano, S.; Huang, J.; Winert, S. *Cooling Our Communities Guidebook on Tree Planting and White Coloured Surfacing*; US Environmental Protection Agency, Office of Policy Analysis, Climate Change Division: Washington, DC, USA, 1992.
5. Arnfield, A.J. Two decades of urban climate research: A review of turbulence, exchanges of energy and water, and the urban heat island. *Int. J. Climatol.* **2003**, *23*, 1–26. [CrossRef]
6. Santamouris, M. *Energy and Climate in the Urban Built Environment*; James and James Science Publishers: London, UK, 2001.
7. Akbari, H.; Kolokotsa, D. Three decades of urban heat islands and mitigation technologies research. *Energy Build.* **2016**, *133*, 834–842. [CrossRef]
8. Akbari, H.; Rose, L.S. *Characterizing the Fabric of the Urban Environment: A Case Study of Salt Lake City, Utah*; LBNL-47851; Lawrence Berkeley National Laboratory: Berkeley, CA, USA, February 2001.
9. Santamouris, M. Heat island research in Europe: The state of the art. *Adv. Build. Energy Res.* **2007**, *1*, 123–150. [CrossRef]
10. Hirano, Y.; Fujita, T. Evaluation of the impact of the urban heat island on residential and commercial energy consumption in Tokyo. *Energy* **2012**, *37*, 371–383. [CrossRef]
11. Santamouris, M.; Papanikolaou, N.; Livada, I.; Koronakis, I.; Georgakis, C.; Argiriou, A.; Assimakopoulos, D.N. On the impact of urban climate to the energy consumption of buildings. *Sol. Energy* **2001**, *70*, 3201–3216. [CrossRef]

12. Santamouris, M.; Paraponiaris, K.; Mihalakakou, G. Estimating the ecological footprint of the heat island effect over Athens, Greece. *Clim. Chang.* **2007**, *80*, 265–276. [CrossRef]
13. Sarrat, C.; Lemonsu, A.; Masson, V.; Guedalia, D. Impact of urban heat island on regional atmospheric pollution. *Atmos. Environ.* **2006**, *40*, 1743–1758. [CrossRef]
14. Santamouris, M.; Kolokotsa, D. On the impact of urban overheating and extreme climatic conditions on housing energy comfort and environmental quality of vulnerable population in Europe. *Energy Build.* **2015**, *98*, 125–133. [CrossRef]
15. Taleghani, M. Outdoor thermal comfort by different heat mitigation strategies—A review. *Renew. Sustain. Energy Rev.* **2018**, *81*, 2011–2018. [CrossRef]
16. Santamouris, M.; Ding, L.; Fiorito, F.; Oldfield, P.; Osmond, P.; Paolini, R.; Prasad, D.; Synnefa, A. Passive and active cooling for the outdoor built environment—Analysis and assessment of the cooling potential of mitigation technologies using performance data from 220 large scale projects. *Sol. Energy* **2017**, *154*, 14–33. [CrossRef]
17. Aflaki, A.; Mirnezhad, M.; Ghaffarianhoseini, A.; Ghaffarianhoseini, Z.; Omrany, H.; Akbari, H. Urban heat island mitigation strategies: A state-of-the-art review on Kuala Lumpur, Singapore and Hong Kong. *Cities* **2017**, *62*, 131–145. [CrossRef]
18. Gago, E.J.; Roldan, J.; Pacheco-Torres, R.; Ordóñez, J. The city and urban heat islands: A review of strategies to mitigate adverse effects. *Renew. Sustain. Energy Rev.* **2013**, *25*, 749–758. [CrossRef]
19. Zhai, J.; Previtali, J.M. Ancient vernacular architecture: Characteristics categorization and energy performance evaluation. *Energy Build.* **2010**, *42*, 357–365. [CrossRef]
20. Berdahl, P.; Bretz, S.E. Preliminary survey of the solar reflectance of cool roofing materials. *Energy Build.* **1997**, *25*, 149–158. [CrossRef]
21. Levinson, R.; Berdahl, P.; Akbari, H. Spectral solar optical properties of pigments Part II: Survey of common colorants. *Sol. Energy Mater. Sol. Cells* **2005**, *89*, 351–389. [CrossRef]
22. Synnefa, A.; Santamouri, M.; Apostolaki, K. On the development, optical properties and thermal performance of cool colored coatings for the urban environment. *Sol. Energy* **2007**, *81*, 488–497.
23. Synnefa, A.; Santamouris, M.; Livada, I. A study of the thermal performance and of reflective coatings for the urban environment. *Sol. Energy* **2006**, *80*, 968–981. [CrossRef]
24. Zinzi, M.; Agnoli, S. Cool and green roofs: An energy and comfort comparison between passive cooling and mitigation urban heat island techniques for residential buildings in the Mediterranean region. *Energy Build.* **2012**, *55*, 66–76. [CrossRef]
25. Santamouris, M. Using cool pavements as a mitigation strategy to fight urban heat island—A review of the actual developments. *Renew. Sustain. Energy Rev.* **2013**, *26*, 445–459. [CrossRef]
26. Qin, Y. A review on the development of cool pavements to mitigate urban heat island effect. *Renew. Sustain. Energy Rev.* **2015**, *52*, 445–459. [CrossRef]
27. Qin, Y. Urban canyon albedo and its implication on the use of reflective cool pavements. *Energy Build.* **2015**, *96*, 86–94. [CrossRef]
28. Mohajerani, A.; Bakaric, J.; Jeffrey-Bailey, T. The urban heat island effect, its causes, and mitigation, with reference to the thermal properties of asphalt concrete. *J. Environ. Manag.* **2017**, *197*, 522–538. [CrossRef] [PubMed]
29. Fintikakis, N.; Gaitani, N.; Santamouris, M.; Assimakopoulos, M.; Assimakopoulos, D.N.; Fintikaki, M.; Albanis, G.; Papadimitriou, K.; Chryssochoides, E.; Katopodi, K.; et al. Bioclimatic design of open public spaces in the historic centre of Tirana, Albania. *Sustain. Cities Soc.* **2011**, *1*, 54–62. [CrossRef]
30. Santamouris, M.; Xirafi, F.; Gaitani, N.; Spanou, A.; Saliari, M.; Vassilakopoulou, K. Improving the microclimate in a dense urban area using experimental and theoretical techniques. The case of Marousi, Athens. *Int. J. Ventilation* **2012**, *11*, 1–16. [CrossRef]
31. Shahidan, M.F.; Jones, P.J.; Gwilliam, J.; Salleh, E. An evaluation of outdoor and building environment cooling achieved through combination modification of trees with ground materials. *Build. Environ.* **2012**, *58*, 245–257. [CrossRef]
32. Zinzi, M.; Carnielo, E.; Rossi, G. Directional and angular response of construction materials solar properties: Characterisation and assessment. *Sol. Energy* **2015**, *115*, 52–67. [CrossRef]
33. CEN. *Road Lighting-Part 5: Energy Performance Indicators*; EN 13201-5:2015; CEN: Brussels, Belgium, 2015.
34. CEN. *Road Lighting-Part 3: Calculation of Performance*; EN 13201-3:2015; CEN: Brussels, Belgium, 2015.

35. Rossi, G.; Iacomussi, P.; Mancinelli, A.; DiLecce, P. Adaptive Systems in Road Lighting Installations. In Proceedings of the 28th Session of the CIE, Manchester, UK, 28 June–4 July 2015.
36. CEN. *Road Lighting-Part 2: Performance Requirements*; EN 13201-2:2015; CEN: Brussels, Belgium, 2015.
37. CEN. *Road Lighting-Part 4: Methods of Measuring Lighting Performance*; EN 13201-4:2015; CEN: Brussels, Belgium, 2015.
38. CIE. *Lighting of Roads for Motor and Pedestrian Traffic*, 2nd ed.; CIE 115:2010; CIE: Vienna, Austria, 2010.
39. Illuminating Engineering Society. *ANSI/IES RP-8-14:2014 Roadway Lighting*; Illuminating Engineering Society: New York, NY, USA, 2014.
40. UNI. *Illuminazione Stradale-Selezione Delle Categorie Illuminotecniche (Road Lighting–Selection of Lighting Classes)*; UNI 11248:2017; UNI: Milano, Italy, 2016.
41. BSI. *BS 5489-1:2013 Code of Practice for the Design of Road Lighting Part 1: Lighting of Roads and Public Amenity Areas*; BSI: London, UK, 2012.
42. CIE. *Recommended System for Mesopic Photometry Based on Visual Performance*; CIE 191:2010; CIE: Vienna, Austria, 2010.
43. CIE. *Road Surface and Road Marking Reflection Characteristic*; CIE 144:2001; CIE: Vienna, Austria, 2001.
44. CIE. *The Effect of Spectral Power Distribution on Lighting for Urban and Pedestrian Areas*; CIE 206:2014; CIE: Vienna, Austria, 2014.
45. ISO. *Glass in Building-Determination of Light Transmittance, Solar Direct Transmittance, Total Solar Energy Transmittance, Ultraviolet Transmittance and Related Glazing Factors*; ISO 9050, 2003; ISO: Geneva, Switzerland, 2003.
46. Avaliable online: http://ec.europa.eu/eurostat/statistics-explained/index.php/Europe_2020_indicators_-_background (accessed on 4 March 2018).

climate

MDPI

Article

A Multi-Criteria Approach to Achieve Constrained Cost-Optimal Energy Retrofits of Buildings by Mitigating Climate Change and Urban Overheating

Fabrizio Ascione [1], Nicola Bianco [1], Gerardo Maria Mauro [1,*], Davide Ferdinando Napolitano [1] and Giuseppe Peter Vanoli [2]

[1] Department of Industrial Engineering, Università degli studi di Napoli Federico II, Piazzale Tecchio 80, 80125 Naples, Italy; fabrizio.ascione@unina.it (F.A.); nicola.bianco@unina.it (N.B.); davide.f.napolitano@gmail.com (D.F.N.)
[2] Department of Medicine, Università degli studi del Molise, Via Cesare Gazzani 47, 86100 Campobasso, Italy; giuseppe.vanoli@unimol.it
*   Correspondence: gerardomaria.mauro@unina.it

Received: 23 March 2018; Accepted: 4 May 2018; Published: 8 May 2018

Abstract: About 40% of global energy consumption is due to buildings. For this reason, many countries have established strict limits with regard to building energy performance. In fact, the minimization of energy consumption and related polluting emissions is undertaken in the public perspective with the main aim of fighting climate change. On the other hand, it is crucial to achieve financial benefits and proper levels of thermal comfort, which are the principal aims of the private perspective. In this paper, a multi-objective multi-stage approach is proposed to optimize building energy design by addressing the aforementioned public and private aims. The first stage implements a genetic algorithm by coupling MATLAB® and EnergyPlus pursuing the minimization of energy demands for space conditioning and of discomfort hours. In the second stage, a smart exhaustive sampling is conducted under MATLAB® environment with the aim of finding constrained cost-optimal solutions that ensure a drastic reduction of global costs as well as of greenhouse gas (GHG) emissions. Furthermore, the impact of such solutions on heat emissions into the external environment is investigated because these emissions highly affect urban overheating, external human comfort and the livability of our cities. The main novelty of this approach is the possibility to properly conjugate the public perspective (minimization of GHG emissions) and the private one (minimization of global costs). The focus on the reduction of heat emissions, in addition to the assessment of energy demands and GHG emissions, is novel too for investigations concerning building energy efficiency. The approach is applied to optimize the retrofit of a reference building related to the Italian office stock of the 1970s.

Keywords: building energy performance; energy simulation; building retrofit; multi-objective optimization; genetic algorithm; urban overheating; cost-optimal analysis; lifecycle analysis; office buildings; sustainability

1. Introduction and State of the Art

Global energy consumption has strong implications on human socio-economic and political spheres. Improved data about the global energy consumption reveal systemic patterns and trends that can be useful for solving current energy issues. In this regard, looking at the worldwide scenario, the energy consumption increased by just 1% in 2016, by following a growth of 0.9% in 2015 and 1% in 2014, and the 10-year average is 1.8% per year [1]. Moreover, in the last decade, with reference to the European Union (EU) there was even a slow but continuous decrease of energy consumption (the 10-year average is −1.1% per year) [1]. This is due to the EU's rigorous policy regarding the reduction of energy consumption. The recent weak growth in energy demand implied that the global greenhouse gas (GHG) emissions from energy consumption were almost flat during 2016 for the third consecutive year. They increased by only 0.1% in 2016 and during 2014–2016 the average emission growth was the lowest over any three-year period since 1981–1983 [1]. This means that, during recent years, many governments are moving in the direction of sustainable development [2].

In this scenario, buildings bear a large responsibility as they account for about 40% of energy consumption and 36% of CO_2-eq emissions in the EU. For this reason, it is fundamental to act on them to strongly reduce energy consumption and polluting emissions. In fact, by improving the energy efficiency of buildings, it is possible to reduce total EU energy consumption by 5–6% as well as CO_2-eq emissions by about 5% [3]. Therefore, the mandatory improvement of energy performance of existing buildings, as well as the high energy quality of new constructions have been established by several European guidelines, Directives and regulations, starting from 2002, when the first version of the Energy Performance of Buildings Directive (EPBD), i.e., the Directive 2002/91/EU [4], was enacted. More recently, the EPBD recast 2010/31/EU [5] upgraded the previous version, introducing the concept of nearly zero-energy buildings and proposing the new methodology of cost-optimality, detailed in the Delegated Regulation 244/2012 [6]. Other mandatory prescriptions, mainly emphasizing the exemplary role of the public hand, were provided by the so-called energy-efficiency Directive, namely the 2012/27/EU [7].

The cited EU guidelines outline that the reduction of building energy consumption is a crucial issue of our generation and is fundamental to promote a sustainable development. In this regard, to optimize building energy performance, firstly it is important to act on building thermal envelope because this enables strong reductions of thermal energy demand (TED), since around 50% of a general-purpose building's energy needs depend on heat losses through the envelope [8]. However, a comprehensive intervention on the envelope composition is possible only for new buildings, while existing ones constitute the largest portion of buildings in service. This can be a problem, but it can be easily solved because there are many other solutions to reduce building energy consumption. In fact, while the envelope defines the TED, the primary energy consumption and the GHG emissions depend on the whole system "building + energy plants" and the designers operate on this whole system when investigating the retrofit of an existing building.

Generally, the minimization of energy consumption and related polluting emissions is the main objective of the public perspective, since it allows the fighting of crucial issues of contemporary society such as climate change and energy poverty. However, when the optimization of building energy performance is faced, it is fundamental to consider the cost-effectiveness of the design and the respect of a certain level of thermal comfort, which are the principal aims of the private perspective. For this reason, the optimization of building design is a complex multi-objective problem with a huge domain of design variables and several potential objective functions. In this regard, occupants' thermal comfort—which is defined as "the order at which occupants have no intention to modify their environment" [9]—represents a very critical aspect [10] because it has serious health-related consequences [9,11,12] and, as clear, a significant impact on energy consumption [13]. Aiming at understanding the extensive influence of occupants and thermal comfort on building energy performance, many studies have been performed in recent years. Unfortunately, there is a deep gap between the theoretical results of the researches and the practical aspects of their application to real-world buildings, mainly because of the financial implications. For this reason, the scientific literature provides many papers reporting attempts to design optimal financially-appealing retrofit strategies ([8,14,15] for instance). Finally, to develop a retrofit design that can be cost-effective and practically feasible—attracting investments from decision-makers—it is important to accurately estimate the energy savings and the financial benefits.

In addition, nowadays environmental problems, such as climate change and thus GHG emissions, have pushed designers to comprehensively assess the environmental impact of building designs according to current law prescriptions. At the same time, customer expectations on the design budget imposes higher pressure on designers to limit the project costs. Aiming at better understanding building environmental impact, different procedures and indicators have been implemented during the last decades, such as the life cycle assessment (LCA). "The LCA includes accumulating of all environmentally relevant streams inventory associated with production processes, transportation, and demolition of a product" [16]. According to this, a relevant number of studies has been conducted to identify the optimal designs with minimum life cycle cost (LCC) [17,18] and life cycle emission (LCE) [19]. The literature shows many investigations that implemented different strategies to support the professionals in determining the environmental impact of their designs (e.g., [17,18]). In the same vein, during the last few years, designs of high-energy performance using optimization techniques have had a significant diffusion. Many studies have been performed to find out which are the best strategies to minimize building energy needs and many other objective functions, such as global costs, environmental impact, occupants' discomfort. For instance, Asadi et al. [20] proposed an optimization technique that makes use of a genetic algorithm and artificial neural networks, with the aim of minimizing the retrofit cost, the energy consumption and the thermal discomfort hours. The two latter objectives were the main aims also of Delgarm et al. in [21], in which an artificial bee colony methodology was used.

More in general, Nguyen et al. [22] provided an accurate explanation concerning the optimization process in building design. To properly design a new building or to retrofit an existing one, dynamic energy simulation tools should be used by designers. The "parametric simulation method" approach is very common to improve building energy performance. According to this method, the designer must vary the input of each variable with the aim of highlighting the effect of the selected variable on the objective functions. This procedure can be iterated with all the variables. However, the limit is that this method often requires a huge computational time and it gives reliable results only in partial improvements because of the non-linear interactions among the different input variables. A different and more robust approach is the one known as "simulation-based optimization" or "numerical optimization", which performs sequences of progressively better approximations to a solution that satisfies an "optimality condition", previously-defined. This permits the attainment of the optimal solution to a problem (or a sub-optimal solution sufficiently close to the optimum [23]) with lower computational time and effort. The simulation-based optimization of building performance is usually conducted automatically by means of the coupling between a building simulation software and an optimization "engine", which implements one or several optimization algorithms that need to be properly set [24]. Generally, a simplification of the building model to be optimized should be done, but it is crucial to not over-simplify, to avoid the risk of inaccurate modeling of building phenomena. In addition, the convergence of the adopted optimization algorithm should be monitored. Convergence behaviors of different optimization algorithms are an extremely active research area [25,26]. Regarding errors, it is fundamental to say that they may occur because of infeasible combinations of variables (e.g., windows areas that extend the boundary of a surface), output reading errors (as in the coupling between MATLAB® [27] and EnergyPlus [28]), and so on. Furthermore, the entire optimization process may crash by a single simulation failure. To minimize such errors, some authors run parametric simulations to make sure that there are no failed simulation runs before running the optimization [22], or they make use of evolutionary algorithms because even the presence of a failed solution among the population does not interrupt the optimization process. Finally, it is important to verify if the found solutions are reliable and robust. There are no standard rules for this task, but the literature provides many strategies (for instance, the sensitivity-analysis [25,29], the brute-force search method [30], the comparison with different models [31]).

Since the building sector accounts for a large amount of global energy consumption and GHG emissions, the optimization of building energy design, in terms of minimization of energy demands and global costs, is strictly related to climatic conditions at large scale and can significantly support the mitigation of climate change and urban overheating. The proposed study aims to address this strong correlation between energy performance and climate to highlight that the optimization of building energy design is fundamental for solving the climatic issues of contemporary society.

Research Aim and Originality

Among the numerous optimization methodologies described in literature, the methodology proposed in this paper is structured in two consequent and interdependent stages, as in [32]. More precisely, during the first stage, there is the implementation of the GA and, by means of the continuous coupling between MATLAB® [27] and EnergyPlus [28], the thermal energy demands (TED) for heating and cooling, respectively, and the discomfort hours (DH) are minimized. Conversely, the second stage is entirely conducted under MATLAB® environment and enables the discovery of constrained cost-optimal solutions that ensure a drastic reduction of global cost (GC) as well as of CO_2-eq (i.e., GHG) emissions. Then, the effect of such solutions on building heat emissions into the external environment is assessed to evaluate the contribution to the mitigation of urban overheating, which highly affects the external human comfort and the livability of our cities. This is a crucial aspect, due to the constantly increasing urbanization, in fact more than half the global population (i.e., the 54%) lives in urban areas nowadays [33,34] and it is forecasted to be rising during the next few years [35,36], with obvious implications on environmental degradation, being the cities and their inhabitants the principal players in heat wasting and CO_2 emitting [35,37].

The main novelty of the proposed methodology consists of the possibility to satisfy both the perspectives, the public one (by reducing the GHG emissions) and the private one (by minimizing GC and reducing DH), thereby allowing to fight climate change and ensuring the design cost-effectiveness at the same time. The focus on the reduction of heat emissions is a further novel aspect for investigations concerning building energy efficiency. In this regard, since the second stage is conducted entirely in MATLAB® it is not time-consuming, thus many objective functions can be investigated and optimized without computational efforts. It should be noticed that similar optimization methodologies—based on the coupling of EnergyPlus and MATLAB® to implement a genetic algorithm—have been already proposed by the authors, such as in [32,38,39]. However, the frameworks and final purposes of these previous studies were different, such as to find cost-optimal retrofit solutions for single complex hospital [32] or educational buildings [38] or for a whole building category by using artificial neural networks [39]. The aforementioned studies applied a financial approach (detailed in the EU Commission Delegate Regulation [6]) in global cost assessment without considering the cost of GHG emissions and building heat emissions into the external environment. Therefore, they did not comprehensively address the issues of climate change and urban overheating related to building energy performance. Finally, the originality of this study is combining the optimization of building energy design in terms of global cost minimization to the reduction of building environment impact in terms of contribution to climate change and urban overheating. The global cost is assessed through a macro-economic approach by considering also the cost of CO_2eq emissions, and thus the achieved cost-optimal solution implies a drastic reduction of GHG emissions. Indeed, the results will show the solution that minimizes the global cost is very close to the one that minimizes CO_2-eq emissions, ensuring a very satisfying trade-off between the private and the public perspectives. Therefore, the application of the methodology at large scale can produce a significant reduction of building environmental impact since the detected solutions imply a drastic reduction of GHG emissions, thereby giving a strong support to the mitigation of climate change and urban overheating.

As a case study, the methodology is applied to a typical existing office building, representative of the Italian building stock since the 1970s.

2. Methodology

2.1. Framework

The proper choice of the energy efficiency measures to adopt for a new building or an existing one is a highly complicated issue, which affects two different perspectives:

- the private one, whose aim is to achieve financial benefits or minor indoor discomfort;
- the public one, whose aim is to reduce energy consumption, polluting emissions and to have an exemplary role for all citizens.

Regarding these differences, the methodology proposed in this paper ensures the best trade-off between these two perspectives, because it allows the addressing of more objectives at the same time, ensuring a good level of satisfaction for both the private and the public perspectives. Finding the cost-optimal solution usually requires high computational efforts, because of the huge amount of energy efficiency measures' combinations that must be simulated by means of building performance simulation (BPS) tools, which run time-costly dynamic simulations. For this computational issue, the cost-optimal analysis could not be applied to every building. Conversely, it should be limited to reference buildings (RBs) only, as established in the EPBD-recast. However, even when only RBs are examined, the robust assessment of cost-optimality is very time costly. For this reason, it is crucial to adopt proper building performance optimization (BPO) algorithms that can reduce the required computational efforts, by reducing, at the same time, the domain of the explored scenarios without affecting the detection of robust cost-optimal solutions. In this paper, a multi-objective and multi-stage optimization procedure is implemented to find a constrained cost-optimal solution that fulfils these three conditions:

- it ensures the Pareto optimization of TED (thermal energy demand) for heating, TED for cooling, discomfort hours (DH) if the retrofit involves the building envelope (in fact, it can be effective to act merely on the energy systems);
- it implies a drastic reduction of GHG emissions since a macroeconomic approach is applied for global cost (GC) assessment thereby considering the cost of such emissions [6];
- minimizes GC by respecting the first two conditions, which is why it is defined "constrained".

Once fixed the main boundary conditions, concerning geometry, occupancy profiles and climatic conditions, several energy efficiency measures are combined and examined. The considered energy efficiency measures concern all levers of energy efficiency in buildings, i.e.,

- the building envelope (e.g., new kind of low-emissive or selective glazing, addition of thermal insulation, particular plasters);
- the primary energy systems, considering also renewable energy sources (e.g., efficient air-source heat pumps, photovoltaic generators).

Specifically, EnergyPlus is used for dynamic energy simulations, because it ensures high accuracy and reliability, while MATLAB® is used to run the optimization algorithm and to perform the data-processing, because of its large opportunities of programming. Furthermore, MATLAB® is used to launch EnergyPlus simulations. Thus, the coupling of these two software allows the automatic running of a huge set of dynamic energy simulations that are managed by the optimization algorithm, developed directly in MATLAB® environment. More precisely, the methodology performs a multi-stage and multi-objective optimization by implementing a genetic algorithm (GA)—1st stage—and running a smart exhaustive sampling—2nd stage. The GA, born as a modification of NSGA-II, operates by iteratively improving the models of the building with the aim of identifying the non-dominated solutions (i.e., the Pareto front) for what concerns the building envelope design or retrofit, by minimizing TED for heating, TED for cooling and DH. Then, the smart exhaustive sampling stage allows the investigation of the Pareto front solutions obtained during the 1st stage and the baseline situation, aiming at reducing GC and GHG emissions thereby conducting a constrained cost-optimal analysis. Thus, decision making is performed by providing a recommended trade-off design/retrofit solution. A similar technique was used in [32,38], but this study addresses different objective functions to provide solutions that allow the fighting of climate change and ensure cost-effectiveness at the same time. This represents the main worthy and original contribution of the proposed approach that enables the conciliation of the private and public perspectives.

Since the 2nd stage is conducted entirely in MATLAB®, the required computational efforts are strongly reduced. The following subsections provide a description of the two methodology stages.

2.2. 1st Methodology Stage: Optimization Algorithm

In this stage, the baseline energy performance of the building ("as built") is assessed, in terms of TED for space cooling, TED for space heating and DH, respectively. The building is modeled in EnergyPlus by using the graphical interface DesignBuilder [40], that allows a careful definition of geometry and subdivision into thermal zones. It is quite important, for the EnergyPlus model, to set:

1. the thermo-physical characteristics of the building envelope;
2. the profiles of building use for each thermal zone, in terms of hourly schedules of occupancy, people activity, ventilation need, and so on;
3. the operation of HVAC (heating, ventilating and air conditioning) systems by setting the values of set-point temperatures;
4. the type of HVAC systems in terms of characteristics of the heating and cooling terminals as well as of the distribution network.

It should be noted that the heating/cooling primary systems are not modeled in this phase, because, during this stage, the aim is to calculate the thermal energy demand (i.e., the "net requirement") and not the primary energy consumption, which is assessed later by means of MATLAB®. After modeling the baseline building (BB), an EnergyPlus simulation is run by using a proper weather data file, usually available at the EnergyPlus online database. The annual values of TED for space heating (TED_{heat}), for space cooling (TED_{cool}) per unit of conditioned area, and DH are the simulation outputs. DH provides the annual percentage of discomfort hours. As done in [39], an occupied hour is considered a discomfort one if the average predicted mean vote (PMV) in the building thermal zones is out of the range $-0.85 \div 0.85$, implying a value of predicted percentage of dissatisfied (PPD) higher than 20%.

After the investigation of the energy behavior of the BB, a set of "n" energy efficiency measures for the reduction of TED_{heat}, TED_{cool} and DH is identified, based on the current energy performance, building peculiarities and best practices. A design variable is associated to each energy efficiency measure and it can be, potentially, discrete or continuous, even if in the case study presented in this paper all variables are considered as "discrete". Finally, "n" variables are introduced, and a range of variability is assigned to each of them, by defining the sample space that should be explored with the aim of examining the energy efficiency measures' combinations. At this point, the GA carries out a smart research within the entire solution domain by investigating only a limited number of solutions, properly selected by the optimization logic. As aforementioned, a large amount of computational time is saved if the method is compared to exhaustive researches. Since three objective functions are chosen—i.e., TED_{heat}, TED_{cool} and DH—the algorithm provides one three-dimensional (3-D) and three bi-dimensional (2D) Pareto fronts (one for each couple of objectives), by collecting the non-dominated solutions, which represent optimal packages of the investigated energy efficiency measures. Obviously, the goal is the minimization of all targets at the same time, but this is impossible because usually the objective functions are conflicting. Thus, the GA provides trade-off solutions collected in the aforementioned Pareto fronts (for this reason we call them "non-dominated"). The used GA has been already implemented by Ascione et al. [39] in MATLAB® environment according to the scheme reported in Figure 1, where the vector $\underline{F}$ collects the objective functions ($\underline{F}$ = [TED$_{heat}$, TED$_{cool}$, DH]) while the vector $\underline{x}$ is composed of bits that encode the design variables representing energy efficiency measures.

τ = 1 (index of generations)
Create the initial population $P^{(1)} \equiv \{\underline{x}_i^{(1)}\}_{i=1,\ldots,s}$
Calculate $\underline{F}(\underline{x}_i^{(1)})$ for i = 1, ..., s
Assess the rank value and the average crowding distance for each individual of $P^{(1)}$
DO UNTIL at least one stop criterion is fulfilled
$\tau = \tau + 1$
Select the parents from $P^{(\tau-1)}$
Generate $P^{(\tau)} \equiv \{\underline{x}_i^{(\tau)}\}_{i=1,\ldots,s}$ from crossover and mutation of the parents; elite parents survive
Calculate $\underline{F}(\underline{x}_i^{(\tau)})$ for i = 1, ..., s
Assess the rank value and the average crowding distance for each individual of $P^{(\tau)}$
END
Achieve the Pareto front

Figure 1. Scheme of the GA, adapted from [39].

Each design variable can assume a limited number of values, because this allows the reduction of the solution domain and it is much closer to reality and availability of the market. The possible values must be carefully chosen according to best practices and experiences. The GA performs, iteratively, an evolution of a population of "s" (population size) individuals, denoted as "chromosomes", each one characterized by a set of values of the vector $\underline{x}$, whose components are called "genes" and correspond to a combination of building energy efficiency measures. The process is performed through numerous iterations, the so-called "generations". It is required to improve the characteristics of the population by the selection of the best chromosomes as well as through the operations of mutation and crossover of their genes (e.g., the bits encoding the thicknesses of thermal insulation layer) in order to have new individuals that improve the energy and thermal performance of the building. The individuals that derive from crossover, called "children", are randomly generated by combining the design variables (i.e., bit strings) of two parents. The population fraction that originates from crossover is indicated by the crossover fraction "f_c". All other remaining individuals ("mutated children") are originated by the mutation of random parents, specifically by changing each bit with a mutation probability equal to "f_m". The best chromosomes are called "parents" and are chosen based on a rank assigned from the values of objective functions and from the average crowding distance among individuals. The best parents constitute the "elite" that survives to the generation. After the random creation of the initial population, the described "Darwinian evolution" occurs during each generation and ends when one of the following termination criteria is satisfied:

1. a threshold number of generations (g_{max}) is reached;
2. the Pareto front does not change significantly between two following generations. This means that the variation of the front spread is lower than a tolerance "tol".

In the case study here investigated, the used termination criterion is the first one and most GA parameters take the same values employed in [32,38], namely (n is the number of design variables):

- c_e = 2;
- f_c = 0.6:
- f_m = 0.1;
- s = 4·n;
- g_{max} = 20.

For what concerns the values of *s* and g_{max}, it is important to notice that these must be properly set depending on the complexity of the case study, because they crucially affect the reliability of the results and the required computational efforts. Ascione et al. [32] assessed that reliable "s" values are 2–6 times the number of design variables (in this study, it is set equal to 4), while reliable "g_{max}" values are included in the range 10–100 generations. In this paper, this is set equal to 20.

Specifically, for each energy efficiency measures' combination, which is encoded by certain values of the vector x, MATLAB® launches EnergyPlus to run a dynamic energy simulation. Then, the results of this simulation are post-processed for obtaining the values of the objective functions (i.e., TED_{heat}, TED_{cool}, DH) with reference to each examined combination. The coupling scheme between the two software is shown in Figure 2

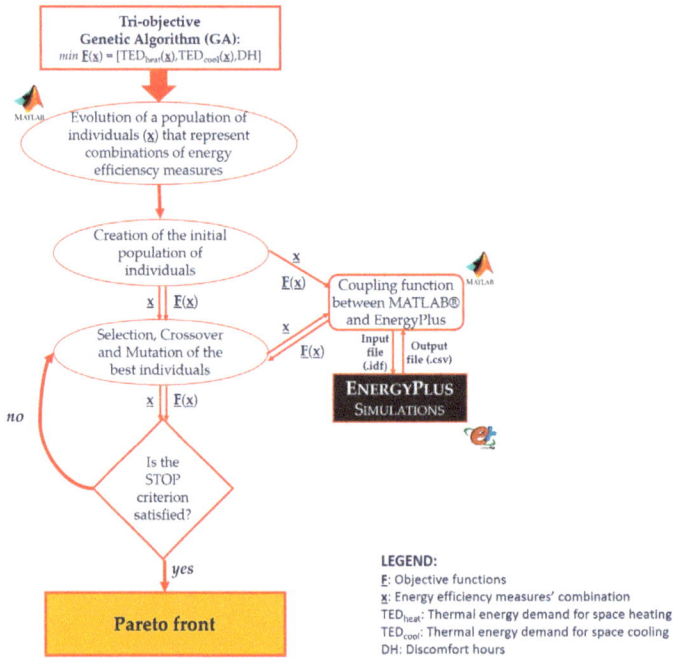

Figure 2. Scheme of the 1st optimization stage, adapted from [32].

The "coupling function" between EnergyPlus and MATLAB®" converts x into a new building model to be simulated (the ".idf" file) and consequently handles the output file of EnergyPlus (the ".csv" file) to calculate the values of the objectives contained in F. It is noticed that the energy efficiency measures are implemented and parametrized directly within the ".idf" EnergyPlus file. Moreover, also a constrain is defined, since all solutions that cause an increase of DH compared to the base building configuration are excluded. This constraint is set to ensure that the optimized energy retrofit does not cause a worsening of occupants' thermal comfort. Indeed, energy efficiency should not prejudice people well-being. Hence, the GA implementation must be followed by the decision-making process, which aims at selecting one recommended solution from the Pareto front. This process is performed during the second stage.

2.3. 2nd Methodology Stage: Decision-Making

In this phase, the decision-making process is performed, aiming at selecting one combination of energy efficiency measures among all non-dominated configurations. It is a crucial task and it can be carried out according to different criteria. Obviously, none of the solutions of the Pareto front can be chosen "a priori", because it cannot be defined better than another. For this reason, a selection criterion is essential. For what concerns the methodology described in this paper, the chosen criterion is the so-called "cost-optimality", which means that, at the end of the entire optimization process, the chosen solution (i.e., package of energy efficiency measures) is the one that minimizes the global cost (GC) over building predicted lifecycle, assessed according to a macro-economic approach [6]. The cost-optimal analysis is applied by means of a smart exhaustive sampling. This latter permits the investigation of further energy efficiency measures—addressed to primary energy systems—besides those examined in the 1st stage, which are addressed to the envelope and to the operation parameters of the HVAC systems. It is important to notice that GC considers the initial investment cost, the GHG emissions costs and the running costs, those latter evaluated for a certain number of years (depending on the category of the building) and actualized at the starting time. Such cost-optimal analysis is conducted according to EU guidelines, reported in the Energy Performance of Buildings Directive (EPBD) recast 2010/31/EU [5] and detailed in the Delegated Regulation 244/2012 [6]. In particular, in the global cost assessment, the macro-economic approach is used to consider the cost of CO_2-eq emissions. This allows the comprehensive consideration and minimization of the impact of building energy performance on climate change, which is one to of the main goals of the proposed study. Indeed, the achieved cost-optimal solution will imply a drastic reduction of GHG emissions, as shown in the Section 4, thereby ensuring a very satisfying trade-off between the private and the public perspectives.

Specifically, during this stage, a smart exhaustive sampling is carried out by investigating the energy performance of different solutions of primary energy systems, in presence of the non-dominated energy efficiency measures' combinations selected in the first stage, and in absence of energy efficiency measures (baseline configuration). For each combination, the GC and the GHG emissions are evaluated, and, finally, the cost-optimal solution is found. A sensitivity analysis is then performed in correspondence of different values of the discount rate, to investigate the robustness of the found cost-optimal solution. This stage is entirely implemented in MATLAB® environment, without launching further EnergyPlus simulations. For this reason, it needs a negligible computational time compared to the first stage (i.e., the order of magnitude is few seconds). The exhaustive sampling is "smart" [32] because:

1. it is performed in MATLAB® environment, without needing further EnergyPlus simulations;
2. it explores, besides the baseline building (BB), only the packages of energy efficiency measures that are properly selected through the GA implementation.

More precisely, the chosen energy efficiency package represents a "constrained" cost-optimal solution, since only suitable packages are selected for the cost-optimal analysis based on the results of the 1st methodology stage. Furthermore, the impact of such optimal retrofit solution on the annual heat emissions of building HVAC systems into the external environment is assessed. This analysis aims at investigating the contribution to the mitigation of urban overheating, which significantly affects the external thermal comfort of people, and thus the livability of our cities, as well as building energy needs.

3. Description of the Case Study

The case study is an existing office building, typical of the Italian building stock in reinforced concrete as structural material. It is theoretical reference building, provided by an accurate ENEA ("Italian National agency for new technologies, Energy and sustainable economic development") study [41], which examined the national building stock and proposed many reference buildings. The investigation of reference buildings can be particularly interesting because the achieved outcomes can be applied—with a good approximation and reliability—to several buildings (i.e., the ones represented by the investigated one).

3.1. Baseline Building (BB)

The building is supposed to be situated in Naples (South Italy) and it has five floors above the ground, each one having a net height of 3 m (see Figure 3). The building gross floor area is equal to 2400 m² (480 m² per level). It is possible to notice that the glazing area changes with the exposure. Specifically, for the west and the east façades, it is about the 55% of the whole area (i.e., about 128 m²), while, for the south exposure, it is about the 33%, and for the north side it is about the 30%. Shading systems are absent. For what concerns the air infiltration rate, according to common Italian values for existing buildings, it has been set at 0.5 air changes per hour (ACH). It is noticed that the building envelope, the schedules of building use and operation, the HVAC systems have been accurately modeled by considering the statistical analysis of ENEA—which developed the investigated reference building [41], the standard Italian constructive practice as well as the typical operation schedules for an office building taken from DesignBuilder [40].

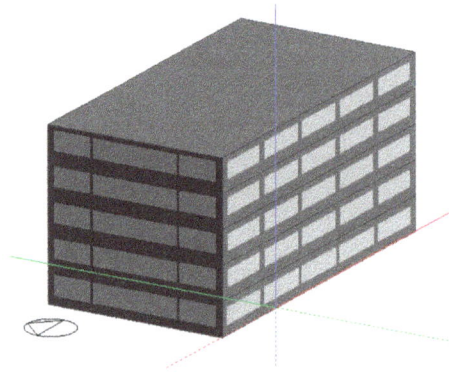

Figure 3. Overall building view.

As for the building use, 50 thermal zones can be individuated, and thus 10 for each floor. There are three different categories of thermal zones, as shown in Figure 4. On the other hand, the following Tables 1–3 show the composition of the opaque building envelope components. The attention is focused on ground floor, roof, and external walls by considering the necessity to rigorously respect the national law limits about the thermal transmittance (i.e., U-value). Finally, Table 4 reports the thermo-physical properties of the cited materials.

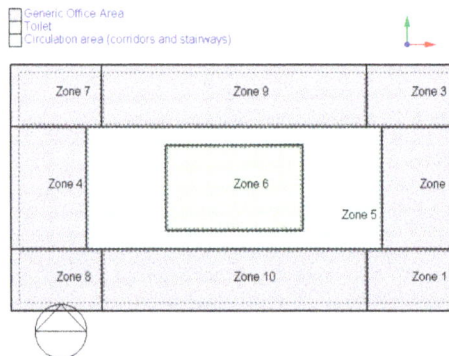

Figure 4. Floor subdivision in thermal zones.

Table 1. Baseline building: External walls composition, from the external to the internal layer.

Layer n°	Material	Thickness (m)
1	Plaster	0.025
2	Hollow bricks	0.12
3	Polystyrene	0.08
4	Air gap	0.12
5	Hollow bricks	0.08
6	Plaster	0.025

Table 2. Baseline building: Ground floor composition, from the external to the internal layer.

Layer n°	Material	Thickness (m)
1	Pebbles	0.18
2	Slab	0.30
3	Semi-rigid panels	0.05
4	Screed	0.03
5	Tiles	0.02

Table 3. Baseline building: Roof composition, from the external to the internal layer.

Layer n°	Material	Thickness (m)
1	Roof plaster	0.03
2	Roof slab	0.18
3	Semi-rigid panels	0.03
4	Screed	0.03
5	Cement	0.03

Table 4. Thermo-physical properties of the opaque building envelope materials.

Material	Density (kg/m³)	Specific Heat (J/kg K)	Conductivity (W/m K)
Plaster	2000	1000	1.40
Hollow bricks	2000	1000	0.90
Polystyrene	1100	1450	0.17
Pebbles	1500	1000	0.70
Semi-rigid panels	16	1660	0.046
Screed	1800	1000	0.90
Tiles	2300	840	1.00
Roof plaster	800	1000	0.70
Roof screed	400	1000	1.40
Cement	2000	1000	1.40

With regard to the transparent building envelope, the windows are double-glazed with clear float glasses, air-filling and aluminum frames. The window U-value is equal to 3.74 W/m^{2}K while the solar heat gain coefficient (SHGC) is equal to 0.76. Finally, the U-values of all envelope components are reported in Table 5, which provides an overview of the baseline configuration of the reference building, with regard to HVAC systems too. In this regard, there is a primary centralized system, which supplies hot and cold water to four-pipes fan coils. All building thermal zones are equipped with such terminals. The heating primary system is a traditional natural gas boiler, while the cooling one is an electric air-cooled chiller. The nominal efficiency (η) of the boiler at the LHV (lower heating value) is 0.85, the nominal coefficient of performance (COP) of the chiller is 2.3. The heating load of the entire building is about 220 kW, while the cooling load is about 235 kW.

Table 5. Characterization of the building.

Dimensions and Geometry			
Length (E-W direction)	30 m	Length (N-S direction)	16 m
Height	15 m (5 floors)	Total Area	2400 m^2
Surface to Volume Ratio	0.33 m^{-1}	Total Volume	7200 m^3
Main Boundary Conditions of Energy Simulations			
Climatic data	IWEC $\rightarrow$ EPW	Design occupancy	230 people
Number of thermal zones	50		
Winter setpoint temperature	20 °C (8 a.m.–1 p.m., 2 p.m.–7 p.m.)	Summer setpoint temperature	26 °C (8 a.m.–1 p.m., 2 p.m.–7 p.m.)
Artificial lighting, lighting levels and electric equipment are diversified depending on the thermal zone use			
Building Envelope			
U_{WALL}	0.97 W/m^{2}K	$U_{GROUNDFLOOR}$	0.51 W/m^{2}K
U_{ROOF}	0.85 W/m^{2}K	$U_{WINDOWS}$	3.74 W/m^{2}K
Shading systems	Absent	SHGC$_{WINDOWS}$	0.76
Infiltration rate	0.50 ACH		
HVAC System			
HVAC typology	Four pipe fan coils, hot and cold water loops, no heat recovery	Ventilation Air	2.5 m^3/s globally
Sensible load control	Yes	Latent load control	Not
Boiler nominal capacity	250 kW	Boiler type	Hot water, Gas fired η = 0.85
Chiller nominal capacity	260 kW	Chiller type	Electric air-cooled, COP = 2.3
Energy Prices, Conversion Factors and Emission Factors			
Electricity price	0.25 €/kWh	Gas price	0.90 €/Sm^3
Electricity selling price	0.07 €/kWh		
Electrical-to-primary energy conversion factor	1.95	Gas-to-primary energy conversion factor	1.05
Electricity LCA emission factor	0.708 t CO_2/MWh	Gas LCA emission factor	0.237 t CO_2 / MWh
Renewable electricity LCA emission factor	0.035 t CO_2 / MWh		
Baseline Performance Indicators			
TED$_{heat}$	10.7 kWh/m^{2}a	TED$_{cool}$	62.2 kWh/m^{2}a
DH	52.4%	CO_2-eq emissions	161.2 t/a
GC (r = 1%)	560.8 €/m^2	GC (r = 3%)	471.9 €/m^2
GC (r = 5%)	404.1 €/m^2		

With regard to the economic assumptions, the considered specific prices for electricity and natural gas are the following ones:

- 0.25 €/kWh$_{el}$ for the electricity;
- 0.90 €/Sm^3 for the gas with an LHV equal to 9.59 kWh/Sm^3.

In addition, as for the discount rate (denoted with r) applied in the assessment of global cost, three different values are considered (i.e., 1%, 3% and 5%). The assumed calculation period is 20 years, since the investigated building is an office [5,6].

Finally, Table 5 shows the explored performance indicators of the baseline building (BB), namely:

- thermal energy demand for space heating (TED$_{heat}$);
- thermal energy demand for space cooling (TED$_{cool}$);
- annual percentage of thermal discomfort hours (DH);
- global cost due to energy uses (GC);
- GHG emissions due to energy uses in terms of CO_2-eq emissions.

The GC is calculated for a long-time period τ of 20 years with the equation established by EU Guidelines [6] and reported below:

$$GC(\tau) = IC + \sum_j \left[\sum_i^{\tau} (RC(i) * R_d(i) + C_{c,i}(j)) - V_{f,\tau}(j) \right] \quad (1)$$

where:

- "IC" stands for the initial investment cost;
- "RC" is the running cost per year, and by means of R_d it is actualized for each year of the evaluating period;
- "R_d" is the actualization factor, which permits the actualization of the RC;
- "$C_{c,I}$" states for the cost of the GHG emissions. For $C_{c,i}(j)$ it is used a cost of 20 €/tCO$_2$-eq until the year 2025, 35 €/tCO$_2$-eq until 2030 and then 50 €/tCO$_2$-eq, as specified in [6];
- "$V_{f,\tau}$" is the residual value at the end of the evaluation period.

Equation (1) permits the adoption of a macro-economic approach, which is fundamental to choose proper energy efficiency measures aiming at reducing building environmental. In fact, mid-polluting measures' combinations—which could be the most efficient trade-offs between the two main objectives (minimization of GC and GHG emissions) in short-midterm evaluations—turn out to be inefficient for a long-term period, once considering also the cost of the GHG emissions.

3.2. Energy Retrofit Scenarios

With regard to building energy retrofit, 11 different design variables—representing retrofit measures for the reduction of thermal energy demands and/or discomfort—are considered to perform the 1st stage of the optimization process, namely:

1. setpoint temperature for space heating;
2. setpoint temperature for space cooling;
3. thermal emissivity of the most external layer of the vertical walls;
4. solar absorbance of the most external layer of the vertical walls;
5. thermal emissivity of the most external layer of the roof;
6. solar absorbance of the most external layer of the roof;
7. thickness of an additional external layer of thermal insulation for the vertical walls—polyurethane panels are considered (density = 25 kg/m^3, conductivity = 0.028 W/mK, specific heat = 1340 J/kgK);
8. thickness of an additional external layer of thermal insulation (polyurethane) for the roof;
9. type of windows;
10. type of shading systems;
11. position of the shading systems.

The values that the aforementioned variables can assume are all discrete as shown in the following Tables 6–8, where the acronym BB denotes the value of the baseline building configuration.

It should be noted that our main target was to propose a methodology. In future studies, the possible ranges and values that can be assumed by the variables can be better defined, according to the real availability of some solutions in the market of energy efficiency measures and building components. This may concern reflectance and emissivity of building external coatings that should comply with the fact that emissivity is high for almost all non-metal materials and that the soiling largely affects the solar absorptance. In this study, the ranges of variability of most design variables are set according to [38], which investigated the energy retrofit of an educational building for a similar climatic location.

Table 6. Characterization of the design variables of the 1st optimization stage.

Design Variables	Values
Setpoint temperature for space heating (°C)	19; 20 (BB); 21; 22
Set-point temperature for space cooling (°C)	24; 25; 26 (BB); 27
Emissivity of the vertical walls (-)	0.1; 0.25; 0.4; 0.5; 0.6; 0.7; 0.8; 0.9 (BB)
Absorbance of the vertical walls (-)	0.1; 0.25; 0.4; 0.5; 0.6 (BB); 0.7; 0.8; 0.9
Emissivity of the roof (-)	0.1; 0.25; 0.4; 0.5; 0.6; 0.7; 0.8; 0.9 (BB)
Absorbance of the roof (-)	0.1; 0.2; 0.3; 0.4; 0.5; 0.6 (BB); 0.75; 0.9
Additional insulation thickness of the vertical walls (m)	0 (BB); 0.03; 0.04; 0.05; 0.06; 0.08; 0.10; 0.12
Additional insulation thickness of the roof (m)	0 (BB); 0.03; 0.04; 0.05; 0.06; 0.08; 0.10; 0.12
Type of windows (-)	1 (BB); 2; 3; 4; 5; 6; 7; 8 (see Table 7)
Type of shading systems (-)	0 (BB); 1; 2; 3; 4; 5; 6; (see Table 8)
Position of the shading systems (-)	1 (internal); 2 (external)

Table 7. Investigated window types.

N°	TYPE	U (W/m²K)	SHGC (-)
1	Double-glazed with air-filling. Aluminum frame (BB)	3.74	0.76
2	Double-glazed with air-filling and low-e coating. PVC frame	2.12	0.69
3	Tinted double-glazed with air-filling and low-e coating. PVC frame	1.95	0.38
4	Selective double-glazed with air-filling and low-e coating. PVC frame	1.84	0.43
5	Double-glazed with argon-filling and low-e coating. PVC frame	1.90	0.69
6	Tinted double-glazed with argon-filling and low-e coating. PVC frame	1.72	0.37
7	Selective double-glazed with argon-filling and low-e coating. PVC frame	1.59	0.43
8	Triple-glazed with argon-filling and low-e coating. PVC frame	1.35	0.58

Table 8. Investigated shading systems.

N°	TYPE	Solar Transmittance	Solar Reflectance	Visible Transmittance	Visible Reflectance
0	Shading system is absent (BB)	/	/	/	/
1	Low reflect—Low trans shade	0.1	0.2	0.1	0.2
2	Low reflect—Medium trans shade	0.4	0.2	0.4	0.2
3	Low reflect—High trans shade	0.7	0.2	0.7	0.2
4	Medium reflect—Low trans shade	0.1	0.5	0.1	0.5
5	Medium reflect—Medium trans shade	0.4	0.5	0.4	0.5
6	High reflect—Low trans shade	0.1	0.8	0.1	0.8

Concerning the 2nd stage of the optimization process, 4 heating primary systems and 2 cooling primary systems are considered, as shown in Tables 9 and 10. When the air-source electric heat pump and the high-efficiency electric air-cooled chiller are implemented together, the installation of only one reversible heat pump is considered.

Table 9. Investigated heating primary systems.

Heating System	Efficiency
Traditional boiler (BB)	$\eta = 0.85$
High-efficiency natural gas boiler	$\eta = 0.95$
Condensing natural gas boiler	$\eta = 1.05$
Air-source electric heat pump	COP = 3.5

Table 10. Investigated cooling primary systems.

Cooling System	COP
Air-cooled electric chiller (BB)	2.3
High-efficiency electric air-cooled chiller	1.2

Furthermore, the 2nd optimization stage considers the installation of photovoltaic (PV) panels. Two different solutions are investigated—monocrystalline panels (more efficient) and polycrystalline ones—installed on the roof to satisfy the electricity needs of lighting, equipment and HVAC systems. In detail, 10 different roof coverage percentages are considered, from 10% to 100% by means of increments of 10%. The PV panels are installed with an inclination equal to 30°. In presence of PV panels, the price assumed for the electricity sold to the grid is 0.07 €/kWh$_{el}$ according to current Italian tariffs.

With regard to the cost-optimal analysis, the assumed values of investment costs for the energy retrofit measures are reported in Table 11. These values are taken partly from previous studies [42] and partly from quotations of suppliers. Finally, the conversion factors adopted to evaluate the GHG emissions due to the electricity and the gas needs are those reported in Table 5.

Table 11. Investment costs of energy retrofit measures.

ENVELOPE		
Energy Efficiency Measure	**Characterization**	**Investment Cost [€/m^2]**
Use of an additional insulation layer (roof, external walls) of thickness:	0.03 m	30
	0.04 m	35
	0.05 m	40
	0.06 m	45
	0.08 m	55
	0.10 m	65
	0.12 m	75
Replacement of the windows	Double-glazed with air-filling and low-e coating. PVC frame	250
	Tinted double-glazed with air-filling and low-e coating. PVC frame	260
	Selective double-glazed with air-filling and low-e coating. PVC frame	260
	Double-glazed with argon-filling and low-e coating. PVC frame	270
	Tinted double-glazed with argon-filling and low-e coating. PVC frame	280
	Selective double-glazed with argon-filling and low-e coating. PVC frame	280
	Triple-glazed with argon-filling and low-e coating. PVC frame	320
Installation of shading systems	Each type of considered shading system	50
HVAC SYSTEM + RES		
Energy Efficiency Measure	**Characterization**	**Investment Cost**
Replacement of the primary heating/cooling system *	High-efficiency natural gas boiler	12,390 €
	Condensing natural gas boiler	21,260 €
	Air-source electric heat pump	41,300 €
	High-efficiency electric air-cooled chiller	43,775 €
	Reversible air-source electric heat pump	65,662 €
Installation of PV panels	Polycrystalline PV panels	250 €/m^2
	Monocrystalline PV panels	430 €/m^2

* All the HVAC systems are oversized by considering an oversizing factor equal to 1.1.

Regarding the evaluation of GC, proper incentives are considered for each energy efficiency measure to be adopted, as established in the Italian economic balance law [43].

Aiming at investigating the urban overheating too, the heat emissions into the external environment due to HVAC systems are finally evaluated, with reference to both the heating and the cooling seasons. Specifically, only the direct thermal energy contributions are considered and thus merely the waste heat of the primary energy systems, namely:

- the thermal emissions of the gas boiler due to the smokes and the heat losses through the boiler metal box;
- the heat discharged into the ambient by the condenser of the cooling system.

For the BB, the heat emissions into the external environment are shown in Figure 5.

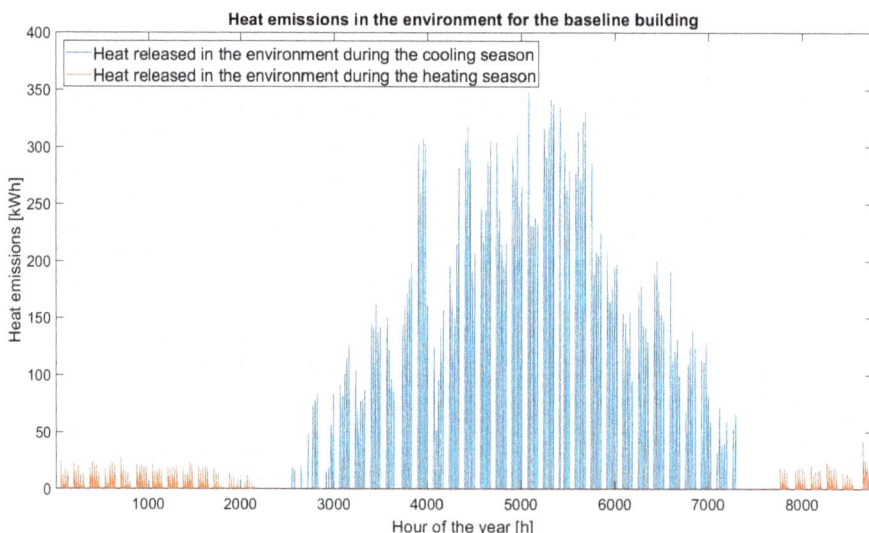

Figure 5. Heat emissions into the external environment for the baseline situation.

With reference to Figure 5, it should be noted that the heat loss emitted into the external environment due to the hot water gas fired boiler is much lower than the one due to the air-cooled chiller, i.e., for the heating season it is equal to 6.10 MWh, while for the cooling one it is around 189 MWh. In the intervals between 2000–3000 h and between 7000–8000 h, there are periods with no heat emissions, because during this mid-season climates both the heating and the cooling systems are supposed to be turned off.

4. Results and Discussion

During the 1st stage of the optimization process, the genetic algorithm (GA) is implemented to find optimal solutions for the building energy retrofit with regard to the minimization of thermal energy demands and discomfort hours. A starting population of 44 individuals is considered and 20 generations are set as termination criterion of the GA. Considering also the randomly generated starting population, more than 900 different dynamic energy simulations (through the automatic coupling between EnergyPlus and MATLAB®) are run to achieve the Pareto minimization of TED_{heat}, TED_{cool} and DH. The resulting Pareto fronts (one 3-D and three 2-D) for the multi-objective optimization are shown in Figures 6–9:

- Figure 6 outlines the non-dominated solutions that minimize all three objective functions (3D Pareto front);
- Figure 7 outlines the non-dominated solutions that minimize TED$_{heat}$ and TED$_{cool}$ (2D front);
- Figure 8 outlines the non-dominated solutions that minimize TED$_{cool}$ and DH (2D front);
- Figure 9 outlines the non-dominated solutions that minimize TED$_{heat}$ and DH (2D front).

Therefore, Figure 6 represents all non-dominated solutions, which for clarity reasons, are better represented in Figures 7–9. It is noticed (see Figure 9) that DH tends to increase with TED$_{heat}$. In this regard, DH refers to the whole year, also to the intermediate seasons (which have a high impact), and therefore it can increase or decrease with the heating demand (TED$_{heat}$). For instance, when the thickness of the thermal insulation layers decreases, clearly TED$_{heat}$ increases and it is very likely that also DH increases. Indeed, discomfort hours tend to increase in the heating and intermediate seasons (indoor air temperature and surface radiant temperatures tend to decrease when the envelope thermal resistance decreases) and can increase during the cooling season because of the indoor overheating effect. When the first effect is predominant, DH increases with the heating demand.

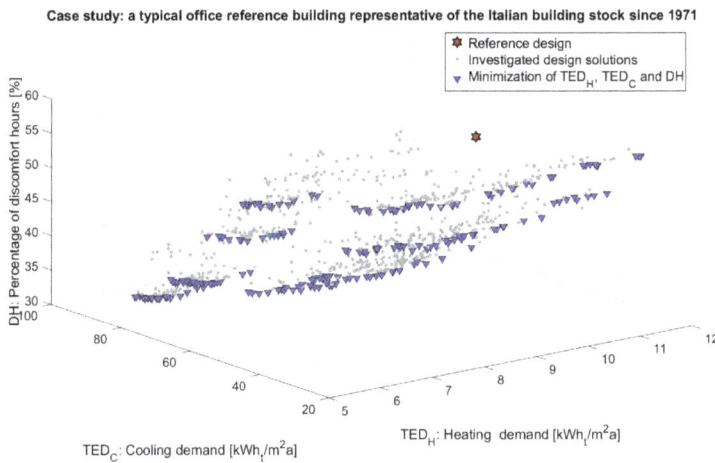

Figure 6. 3-D Pareto front.

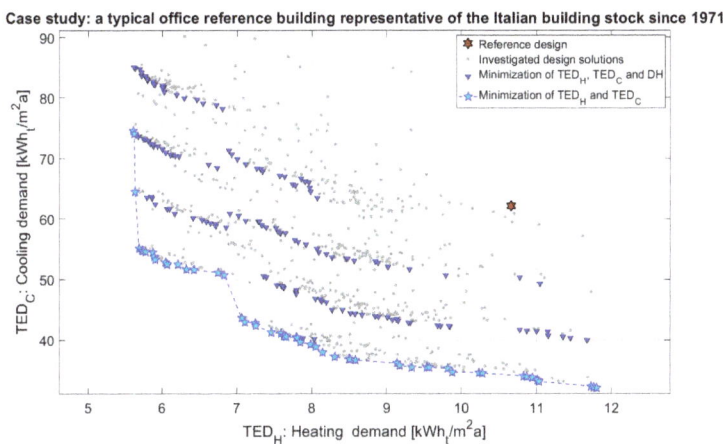

Figure 7. 2-D Pareto front TED$_{heat}$–TED$_{cool}$.

Case study: a typical office reference building representative of the Italian building stock since 1971

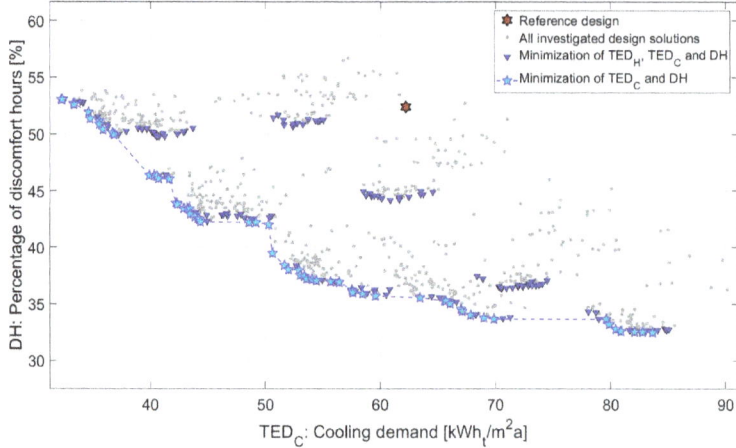

Figure 8. 2D Pareto front TED$_{cool}$–DH

Case study: a typical office reference building representative of the Italian building stock since 1971

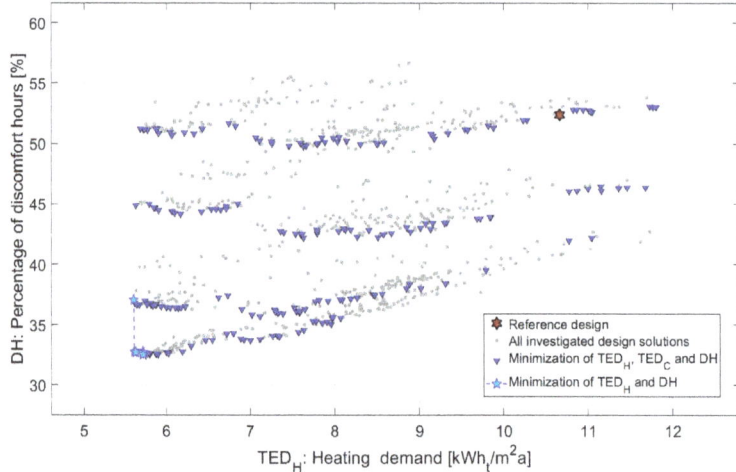

Figure 9. 2D Pareto front TED$_{heat}$–DH.

To have a general outline of the GA outcomes, the attention is focused on the solutions that produce the mono-objective minimization of TED$_{heat}$, TED$_{cool}$ and DH, respectively. These represent the "extreme" solutions of the Pareto fronts and allow understanding of how the Pareto non-dominated solutions vary depending on the weight given to each objective function.

The solution that minimizes TED$_{heat}$ provides the following energy efficiency measures:

- installation of a 0.12 m-thick external thermal insulation layer on the vertical walls;
- installation of a 0.10 m-thick external thermal insulation layer on the roof;
- installation of plasters with particular radiative properties, i.e., thermal emissivity (e) and solar absorptance (a). For the roof the provided values of e and a are 0.1 and 0.75, respectively, while for the vertical walls optimal values of e and a are 0.8 and 0.1, respectively;

- installation of triple-glazed windows with argon filling and low-e coatings in replacement of the existing ones;
- installation of external low reflection—medium transmittance shading system.

The set-point temperatures for space heating and cooling should be set equal to 19 °C and to 25 °C, respectively. This combination of retrofit measures produces the following outcomes:

- TED_{heat} passes from 10.7 kWh/m²a (BB, baseline building) to 5.6 kWh/m²a;
- TED_{cool} increases from 62.2 kWh/m²a (BB) to 74.5 kWh/m²a;
- DH passes from 52.4%(BB) to 37.1%.

The adoption of a high insulated envelope strongly reduces TED_{heat} and DH but causes an increase of TED_{cool} of around 20% because of the indoor overheating effect.

The solution that minimizes TED_{cool} provides the following energy efficiency measures:

- installation of a 0.05 m-thick external thermal insulation layer on the roof, while for the vertical walls no additional insulation is provided;
- installation of plasters with e equal to 0.8 and a equal to 0.1 for both roof and vertical walls; therefore, the use of cool plasters is recommended;
- installation of tinted double-glazed windows with argon filling and low-e coating coatings in replacement of the existing ones;
- installation of an external medium reflection—medium transmittance shading system.

In this case, the set-point temperatures for space heating and cooling should be higher than the previous ones, more precisely they should be set equal to 21 °C and to 27 °C, respectively. This combination of retrofit measures produces the following outcomes:

- TED_{heat} slightly increases from 10.7 kWh/m²a (BB) to 11.8 kWh/m²a;
- TED_{cool} decreases from 62.2 kWh/m²a (BB) to 32.1 kWh/m²a;
- DH passes from 52.4%(BB) to 53.0%.

Thus, the retrofit solution minimizing TED_{cool} does not exert significant effects on the other two objective functions, differently from the previous solution. However, this solution will be cut off during the 2nd optimization stage because it causes an increase of DH compared to the baseline.

Finally, the solution that minimizes DH provides the following energy efficiency measures:

- installation of a 0.12 m-thick external thermal insulation layer on both the vertical walls and the roof;
- installation of plasters with particular values of e and a. For the vertical walls the optimal values of e and a are 0.9 and 0.9, respectively, while for the roof the optimal values of e and a are 0.1 and 0.5, respectively;
- installation of triple-glazed windows with argon-filling and low-e coating in replacement of the existing ones;
- installation of an internal low reflection—high transmittance shading system;

In this final case, the set-point temperatures for space heating and cooling should be set equal to 19 °C and to 24 °C, respectively. This combination of retrofit measures produces the following outcomes:

- TED_{heat} strongly decreases from 10.7 kWh/m²a (BB) to 5.7 kWh/m²a;
- TED_{cool} increases from 62.2 kWh/m²a (BB) to 83.7 kWh/m²a;
- DH passes from 52.4% (BB) to 32.5%.

Finally, this solution is very similar to the one minimizing TED$_{heat}$. The installation of high-thick insulation layers and triple-glazed windows allow reduction of TED$_{heat}$ and, obviously, DH, while TED$_{cool}$ increases of around 35% mainly because of summer overheating.

The achieved Pareto non-dominated solutions are 224, most of which implies a significant improvement of occupants' thermal comfort compared to BB. Only twelve Pareto solutions cause an increase of DH compared to BB, and thus they are excluded in the 2nd methodology stage. This latter is implemented by conducting the smart exhaustive sampling. Thus, also the replacement of primary energy systems is considered, and globally 32,802 different energy retrofit scenarios are investigated by assessing GC and GHG emissions (see Figure 10). Specifically, the differences in global cost (dGC) and GHG emissions (dCO$_2$-eq) compared to the baseline are evaluated to obtain more representative results.

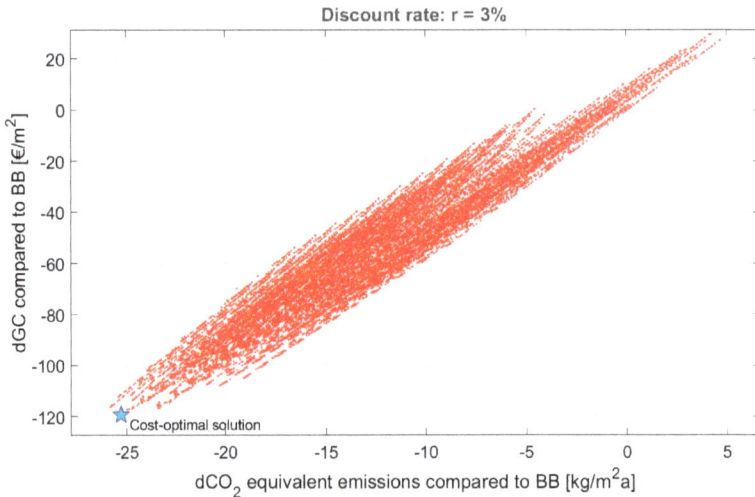

Figure 10. dGC vs dCO$_2$-eq emissions for all the investigated energy retrofit scenarios for r = 3%. The cost-optimal solution is highlighted by a bigger marker.

For a discount rate (r) equal to 3%, the resulting cost-optimal solution provides the following energy retrofit measures:

- installation of a 0.12 m-thick external thermal insulation layer on roof and vertical walls;
- installation of plasters with particular radiative properties, i.e., thermal emissivity (e) and solar absorptance (a). For the roof the optimal values of e and a are 0.7 and 0.1, respectively, while for the vertical walls optimal values of e and a are 0.8 and 0.1, respectively. Therefore, the optimization procedure recommends the use of cool plasters;
- installation of tinted double-glazed windows with argon-filling and low-e coating ones in replacement of the existing ones;
- installation of the reversible electric heat pump for both space heating and cooling;
- installation of PV monocrystalline panels covering the 100% of the usable roof area.

Finally, the set-point temperatures for heating and for cooling should be set equal to 20 °C and to 27 °C respectively. It should be noticed that the combination among envelope thermal insulation, cool plasters, tinted windows and high-efficiency primary systems is highly synergic and produces simultaneously reductions of:

- TED$_{heat}$ which passes from 10.7 kWh/m^{2}a (BB) to 8.5 kWh/m^{2}a;
- TED$_{cool}$ which passes from 62.2 kWh/m^{2}a (BB) to 36.8 kWh/m^{2}a;

- discomfort hours: DH passes from 52.4% (BB) to 49.9%;
- global cost: dGC = −119.3 €/m^2;
- GHG emissions: dCO$_2$-eq = −25.3 kg/m^{2}a;
- annual heat emissions into the external environment, which pass from 195 MWh (BB) to 105 MWh (in percentile terms, this means that the reduction is around 46%).

Regarding this last result, it is fundamental to highlight that the beneficial effect on heat emissions due to the installation of the air-source electric heat pump (which "removes" heat from the external environment) is not considered in order to avoid the overestimation of the goodness of the cost-optimal solution found in terms of urban overheating mitigation too. Hourly heat emissions into the external environment for the cost-optimal solution are reported in Figure 11.

It is highlighted that the achieved constrained cost-optimal solution produces a drastic reduction of GHG emissions. Indeed, as shown in Figure 10, the employed macro-economic approach for GC assessment implies that the solution minimizing GC is very close to the one that minimizes CO$_2$-eq emissions, thereby ensuring a very satisfying trade-off between the private and the public perspectives. Therefore, the application of the methodology at large scale can produce a significant reduction of building environmental impact, giving a strong support to the mitigation of climate change and urban overheating.

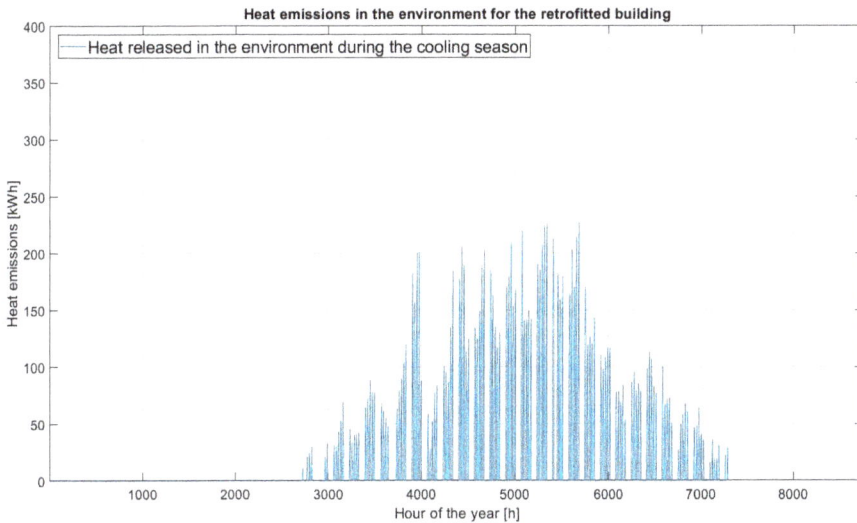

Figure 11. Heat emissions into the external environment in presence of the cost-optimal solution.

The robustness of the solution is examined by assessing the cost-optimal solution for other two values of the discount rate r (i.e., 1% and 5%). Specifically, when the discount rate is varied the cost-optimal solution remains the aforementioned one. Clearly, only the value of dGC changes and it is equal to −153.2 €/m^2 for r = 1% and to −93.2 €/m^2 for r = 5%.

Other different models of comfort and criteria for evaluating hygrothermal conditions in civil buildings, with reference to both cities and urban areas, can be used, as inferred by Paolini et al. [44] recently. Specifically, the authors performed a deep investigation of effects of local climate conditions, that, in cities, can be more significant in summertime (i.e., higher cooling loads) and less critical in wintertime (given the urban heat island effect) compared to rural zones. In this case, conditions of local climate (e.g., city, urban canyon, cooler backcountry or coastline) must be considered and the standard weather files cannot be considered as sufficient. Analogously, also improvement and variation of comfort models can be thought and considered. For instance, solar radiation entering into the building through fenestration can strongly affect human comfort, and this particular role was deeply studied by Arens et al. [45] who developed a new model for understanding the impacts of radiation on the energy balance of human body. According to the authors, the so-called "SolarCal", besides an improvement of comfort models, can help in choosing the most suitable kind of windows and fenestrations for a building, to improve its indoor thermal comfort conditions. However, in this study we used DH—which is based on Fanger theory [46]—because it is a well-accredited comfort index (used in several previous works, e.g., from Asadi et al. [20], Delgarm at al. [21], Ascione at al. [38,39] and Mauro et al. [42]) and provides an objective function to be minimized, which is fundamental for the implementation of the optimization procedure. Nevertheless, future optimization studies will be addressed to a more detailed characterization of thermal comfort. Another critical aspect concerning comfort is the risk of overheating during heatwaves in summertime. This risk can be significant for the considered case study, given the cooling-dominated climate as well the use destination (office buildings present high internal heat gains). Clearly, the proposed methodology cannot foresee extreme meteorological events because energy simulations are based on typical and average weather data files. However, the main aim of the multi-objective optimization is to design building energy refurbishments suitable for improving comfort, costs and energy demands, by reducing, at the same time, also the heat emission into the urban environment. Therefore, in the optimization algorithm, one objective function is the minimization of thermal discomfort, by considering all occupied hours of the year and, particularly, of the cooling season. Indeed, even if the building is fully equipped with air-conditioning systems (so that comfort conditions are allowed also during the cooling peaks, in the hottest hours of the summer period), the run period of the HVAC system is not 24 h/day, and thus it is recommended the refurbishment configuration that improves thermal conditions also during the off-periods of the cooling plant. In this way, by considering both energy demands and thermal comfort (for all the occupied hours), we can consider, in the same methodology and retrofit configuration, both passive and active energy conservation measures. This allows minimization of the risk of indoor overheating during heatwaves in summertime.

5. Conclusions

The paper proposes an optimization methodology for building energy design/retrofit based on two main objective functions, and thus the reduction of global costs and GHG emissions, to perfectly conjugate the two involved perspectives: the private one (minimization of financial expenditure) and the public one (minimization of pollution and environmental impacts of buildings).

The optimization process is structured in two consequent stages. The first one consists of the implementation of a genetic algorithm by means of the coupling of MATLAB® and EnergyPlus, while during the second stage a smart exhaustive sampling is conducted entirely in MATLAB®, thereby ensuring feasible computational times even when around 30,000 retrofit scenarios are investigated. As a case study, a typical office reference building representative of the Italian building stock since 1971 is investigated. The cost-optimal solution provided by the application of the proposed methodology to this case study permits a strong reduction in the GHG emissions, which change from 78.8 kg/m²a to 53.5 kg/m²a, as well as the global cost, which decreases by around 119 €/m² (assuming a discount rate equal to 3%).

The importance of the application of the proposed optimization methodology is that the reduction of the CO_2-eq emissions can enable the different countries to respect the limits imposed by the international agreements on polluting emissions for fighting climate change, while the minimization of the GC makes the adoption of proper energy efficiency measures more appealing to building owners, letting them play also an important role for the community.

Finally, the applied methodology enables the reaching of more than satisfying results not only in fighting climate change under a macroscopic approach, but also in contrasting the urban overheating by adopting a local-limited approach.

Author Contributions: All authors have contributed with the same weight and effort. In detail, G.M.M. and D.F.N. provided the literature investigations and the development of the numerical studies, together with F.A. that framed this study into the scientific vein in matter of impacts of buildings on the city energy balances and on the urban heating. N.B. and G.P.V. contributed, together with the other co-authors, in coordination, analysis of data and writing of the manuscript.

Conflicts of Interest: The authors declare no conflicts of interest.

Nomenclature

Symbols

dCO_2-eq	difference in CO_2-eq emissions compared to BB	kg/m^2a
dGC	difference in GC compared to BB	$€/m^2$
dGHG	difference in GHG emissions compared to BB	kg/m^2a
g_{max}	maximum number of generations	—
r	discount rate	—
s	population size	—
x	vector of design variables of the multi-objective optimization problem	—
DH	annual percentage of discomfort hours	%
F	vector of objective functions of the multi-objective optimization problem	—
GC	global cost	€
IC	initial investment cost	€
PEC	annual primary energy consumption per unit of net floor area	kWh/m^2a
TED	thermal energy demand	kWh/m^2a
U	thermal transmittance of building envelope components	W/m^2K

Subscripts

cool	referred to the energy needs for space cooling
heat	referred to the energy needs for space heating
roof	referred to the roof
tot	referred to the sum of the energy needs for space cooling and heating
wall	referred to the external vertical walls
windows	referred to the windows (frame + glasses)

Acronyms

BB	baseline building
GA	genetic algorithm
GHG	greenhouse gas
HVAC	heating, ventilating and air conditioning
NZEB	net zero-energy buildings
PV	Photovoltaic

References

1. BP Annual Report. 2016. Available online: https://www.bp.com/content/dam/bp/en/corporate/pdf/investors/bp-annual-report-and-form-20f-2016.pdf (accessed on 24 January 2018).
2. Asikainen, A.; Pärjälä, E.; Jantunen, M.; Tuomisto, J.T.; Sabel, C.E. Effects of Local Greenhouse Gas Abatement Strategies on Air Pollutant Emissions and on Health in Kuopio, Finland. *Climate* **2017**, *5*, 43. [CrossRef]
3. Europa. Available online: https://ec.europa.eu/energy/en/topics/energy-efficiency/buildings (accessed on 25 January 2018).
4. EU Commission and Parliament. Directive 2002/91/EC of the European Parliament and of the Council of 16 December 2002 on the Energy Performance of Buildings (EPBD). Available online: https://eur-lex.europa.eu/legal-content/EN/TXT/?uri=celex%3A32002L0091 (accessed on 8 May 2018).
5. EU Commission and Parliament. Directive 2010/31/EU of the European Parliament and of the Council of 19 May 2010 on the Energy Performance of Buildings (EPBD Recast). Available online: http://www.buildup.eu/en/practices/publications/directive-201031eu-energy-performance-buildings-recast-19-may-2010 (accessed on 8 May 2018).

6. Commission Delegated Regulation No 244/2012. Available online: http://www.buildup.eu/sites/default/files/content/l_08120120321en00180036.pdf (accessed on 8 May 2018).

7. EU Commission and Parliament. Directive 2012/27/EU of the European Parliament and of the Council of 25 October 2012 on Energy Efficiency, Amending Directives 2009/125/EC and 2010/30/EU and Repealing Directives 2004/8/EC and 2006/32/EC. Available online: https://eur-lex.europa.eu/legal-content/EN/TXT/?uri=celex:32012L0027 (accessed on 8 May 2018).

8. Fan, Y.; Xia, X. A multi-objective optimization model for energy-efficiency building envelope retrofitting plan with rooftop PV system installation and maintenance. *Appl. Energy* **2017**, *189*, 327–335. [CrossRef]

9. Jamei, E.; Rajagopalan, P. Thermal comfort of multiple user groups in indoor aquatic centres. *Energy Build.* **2015**, *105*, 129–138.

10. Asadi, S.; Mostavi, E.; Boussaa, D. Development of a new methodology to optimize building life cycle cost, environmental impacts, and occupant satisfaction. *Energy* **2017**, *121*, 606–615.

11. Artuso, P.; Santiangeli, A. Energy solutions for sports facilities. *Int. J. Hydrogen Energy* **2008**, *33*, 3182–3187. [CrossRef]

12. Nematchoua, M.K.; Orosa, J.A.; Tchinda, R. Thermal comfort and energy consumption in modern versus traditional buildings in Cameroon: A questionnaire-based statistical study. *Appl. Energy* **2014**, *114*, 687–699. [CrossRef]

13. Day, J.K.; Gunderson, D.E. Understanding high performance buildings: The link between occupant knowledge of passive design systems, corresponding behaviours, occupant comfort and environmental satisfaction. *Build. Environ.* **2014**, *84*, 114–124. [CrossRef]

14. Antunes, C.H.; Asadi, E.; da Silva, M.G.; Dias, L. Multi-objective optimization for building retrofit strategies: A model and an application. *Energy Build.* **2012**, *44*, 81–87.

15. Fan, Y.; Xia, X. A multi-objective optimization model for building envelope retrofit planning. *Energy Procedia* **2015**, *75*, 1299–1304. [CrossRef]

16. Hawkins, T.R.; Majeau-Bettez, G.; Singh, B.; Strømman, A.H. Comparative environmental life cycle assessment of conventional and electric vehicles. *J. Ind. Ecol.* **2013**, *17*, 53–64. [CrossRef]

17. Gálvez, D.M.; Kerdan, I.G.; Raslan, R.; Ruyssevelt, P. ExRET-Opt: An automated exergy/exergoeconomic simulation framework for building energy retrofit analysis and design optimisation. *Appl. Energy* **2017**, *192*, 33–58.

18. Attia, S.; Carlucci, S.; Hamdy, M.; O'Brien, W. Assessing gaps and needs for integrating building performance optimization tools in net zero energy buildings design. *Energy Build.* **2013**, *60*, 110–124. [CrossRef]

19. Amiri, S.S.; Asadi, S.; Mottahedi, M. On the development of multi-linear regression analysis to assess energy consumption in the early stages of building design. *Energy Build.* **2014**, *85*, 246–255.

20. Asadi, E.; da Silva, M.G.; Antunes, C.H.; Dias, L.; Glicksman, L. Multi-objective optimization for building retrofit: A model using genetic algorithm and artificial neural network and an application. *Energy Build.* **2014**, *81*, 444–456. [CrossRef]

21. Delgarm, N.; Sajadi, B.; Delgarm, S. Multi-objective optimization of building energy performance and indoor thermal comfort: A new method using artificial bee colony (ABC). *Energy Build.* **2016**, *131*, 42–53. [CrossRef]

22. Nguyen, A.T.; Reiter, S.; Rigo, P. A review on simulation-based optimization methods applied to building performance analysis. *Appl. Energy* **2014**, *113*, 1043–1058. [CrossRef]

23. Alcayde, A.; Baños, R.; Gil, C.; Gómez, J.; Manzano-Agugliaro, F.; Montoya, F.G. Optimization methods applied to renewable and sustainable energy: A review. *Renew. Sustain. Energy Rev.* **2011**, *15*, 1753–1766.

24. Wright, J.; Alajmi, A. Efficient Genetic Algorithm sets for optimizing constrained building design problem. *Int. J. Sustain. Built Environ.* **2016**, *5*, 123–131. [CrossRef]

25. Alajmi, A.; Wright, J.A. The robustness of genetic algorithms in solving unconstrained building optimization problems. In Proceedings of the 9th International Building Performance Simulation Association Conference, Montreal, QC, Canada, 15–18 August 2005; pp. 1361–1368.

26. Wetter, M.; Polak, E. A convergent optimization method using pattern search algorithms with adaptive precision simulation. *Build. Serv. Eng. Res. Technol.* **2004**, *25*, 327–338. [CrossRef]

27. MathWorks. *MATLAB-MATrixLABoratory (2015)-8.5.0*; User's Guide; MathWorks: Natick, MA, USA, 2015.

28. EnergyPlus 8.5.0. Available online: https://github.com/NREL/EnergyPlus/releases/tag/v8.5.0 (accessed on 8 December 2017).

29. Krarti, M.; Tuhus-Dubrow, D. Genetic-algorithm based approach to optimize building envelope design for residential buildings. *Build. Environ.* **2010**, *45*, 1574–1581.

30. Hasan, A.; Sirén, K.; Vuolle, M. Minimisation of life cycle cost of a detached house using combined simulation and optimisation. *Build. Environ.* **2008**, *43*, 2022–2034. [CrossRef]

31. Eisenhower, B.; Fonoberov, V.A.; Mezić, I.; Narayanan, S.; O'Neill, Z. A methodology for meta-model based optimization in building energy models. *Energy Build.* **2012**, *47*, 292–301. [CrossRef]

32. Ascione, F.; Bianco, N.; De Stasio, C.; Mauro, G.M.; Vanoli, G.P. Multi-stage and multi-objective optimization for energy retrofitting a developed hospital reference building: A new approach to assess cost-optimality. *Appl. Energy* **2016**, *174*, 37–68. [CrossRef]

33. Muscio, A. The Solar Reflectance Index as a Tool to Forecast the Heat Released to the Urban Environment: Potentiality and Assessment Issues. *Climate* **2018**, *6*, 12. [CrossRef]

34. United Nations. *World Urbanization Prospects: 2014 Revision*; United Nations: New York, NY, USA, 2014.

35. Vuckovic, M.; Maleki, A.; Mahdavi, A. Strategies for Development and Improvement of the Urban Fabric: A Vienna Case Study. *Climate* **2018**, *6*, 7. [CrossRef]

36. United Nations. *Department of Economic and Social Affairs, Population Division. World Population Prospects: The 2015 Revision, Key Findings and Advance Tables*; Working Paper No. ESA/P/WP.241; United Nations: New York, NY, USA, 2015.

37. Wilson, B.; Chakraborty, A. The Environmental Impacts of Sprawl. *Sustainability* **2013**, *5*, 3302–3327. [CrossRef]

38. Ascione, F.; Bianco, N.; De Masi, R.F.; Mauro, G.M.; Vanoli, G.P. Energy retrofit of educational buildings: Transient energy simulations, model calibration and multi-objective optimization towards nearly zero-energy performance. *Energy Build.* **2017**, *144*, 303–319. [CrossRef]

39. Ascione, F.; Bianco, N.; De Stasio, C.; Mauro, G.M.; Vanoli, G.P. CASA, cost-optimal analysis by multi-objective optimisation and artificial neural networks: A new framework for the robust assessment of cost-optimal energy retrofit, feasible for any building. *Energy Build.* **2017**, *146*, 200–219. [CrossRef]

40. *DesignBuilder Software-V. 5.0.3.7*; DesignBuilder Software Ltd.: Gloucestershire, UK, 2017; Available online: www.designbuilder.co.uk (accessed on 8 December 2017).

41. Citterio, M. Analisi Statistica sul Parco Edilizio non Residenziale e Sviluppo di Modelli di Calcolo Semplificati (Statistical Analysis on the Non-Residential Building Stock and Development of Simplified Calculation Tools). 2009. Available online: http://old.enea.it/attivita_ricerca/energia/sistema_elettrico/Condizionamento/RSE161.pdf (accessed on 8 May 2018). (In Italian)

42. Mauro, G.M.; Hamdy, M.; Vanoli, G.P.; Bianco, N.; Hensen, J.L.M. A new methodology for investigating the cost-optimality of energy retrofitting a building category. *Energy Build.* **2015**, *107*, 456–478. [CrossRef]

43. Italian Government Law. Legge 11 Dicembre 2016, n. 232. Bilancio di Previsione Dello Stato per l'nno Finanziario 2017 e Bilancio Pluriennale per il Triennio 2017–2019 (State Estimated Budget for the Financial Year 2017 and Multi-Year Budget for the Three-Year Period 2017–2019). Available online: http://www.agenziaentrate.gov.it/wps/file/Nsilib/Nsi/Schede/Agevolazioni/DetrRistrEdil36/NP/Articolo+1+commi+2+e+3+legge+232_2016/art+1+commi+2+e+3+legge+11+dicembre+2016.pdf (accessed on 8 May 2018). (In Italian)

44. Paolini, R.; Zani, A.; MeshkinKiya, M.; Castaldo, V.L.; Pisello, A.L.; Antretter, F.; Poli, T.; Cotana, F. The hygrothermal performance of residential buildings at urban and rural sites: Sensible and latent energy loads and indoor environmental conditions. *Energy Build.* **2017**, *152*, 792–803. [CrossRef]

45. Arens, E.; Hoyt, T.; Zhou, X.; Huang, L.; Zhang, H.; Schiavon, S. Modeling the comfort effects of short-wave solar radiation indoors. *Build. Environ.* **2015**, *88*, 3–9. [CrossRef]

46. Fanger, P.O. Thermal Comfort. Analysis and Applications in Environmental Engineering. 1970. Available online: http://journals.sagepub.com/doi/abs/10.1177/146642407209200337 (accessed on 8 May 2018).

climate

MDPI

Article

Multifractal Analysis of High-Frequency Temperature Time Series in the Urban Environment

Stavroula Karatasou [1,*] **and Mat Santamouris** [2]

[1] Group Building of Environmental Physics, University of Athens, Building 5, Physics University Campus, 15784 Athens, Greece
[2] High Performance Architecture, School of Built Environment, University of New South Wales, Sydney NSW 2052, Australia; m.santamouris@unsw.edu.au
* Correspondence: skarat@phys.uoa.gr; Tel.: +30-210-727-6995

Received: 2 May 2018; Accepted: 5 June 2018; Published: 8 June 2018

Abstract: Continuous monitoring systems have been regarded as a very useful tool to provide continuous high-frequency measurements of many parameters. Here, we analyze high-frequency time series of air temperature measurements, recorded every 10 min during 2003 in Athens (Greece) by an online monitoring system for the urban environment. We propose a set of time series analysis techniques, where missing data are well respected and information concerning the system's dynamics is preserved. A power spectral density analysis is performed over time scales spanning from 10 min to several days. A scale-invariant behavior of the form $E(f) \approx f^{-\beta}$ is revealed for scales below 9 h. Over this scaling range, we have performed structure functions analysis, and shown that air temperature data exhibit turbulent-like intermittent properties with multi-fractal statistics. The multifractal exponents obtained possess some similarities with passive scalar turbulence results. Although we illustrate the proposed approach using air temperature data, the method can be used as an efficient tool to analyse other environmental parameters monitored in urban environment.

Keywords: air temperature; spectral analysis; multifractal analysis; structure functions analysis

1. Introduction

Atmosphere is highly variable over a large range of time and space scales. To understand environmental variability, continuous sampling measurements taken over time and space scales, which vary by many orders of magnitude, are essential. Environmental data can now be collected at time scales on the order of minutes, giving access to microscale information related to physical processes important for the urban built environment.

Many studies statistically analyze ambient air temperatures and the urban heat island phenomenon, and propose some data-driven modelling techniques, based mainly on hourly records [1–3].

The high-frequency sampling is important to better understand small-scale couplings between different meteorological parameters (such as temperature, solar radiation, air velocity, and humidity) and buildings. In this framework, databases are being recorded so that multi-scale analyzing procedures can be applied, in order to be able to treat, exploit, and validate these datasets.

Even though the recording devices are often highly automated and give the means for online, real-time monitoring, they may possess a number of gaps, for many reasons. This being the case, many classical time series analysis methods, such as discrete-time Fourier transform, linear correlation analysis, multi-scale, or wavelet transform cannot be applied (see e.g., [4,5]).

Several authors have showed empirically that fluctuations of geophysical time series possess multifractal statistics.

Multifractals can be regarded as a rather considerable generalization of fractal geometry, developed for the description of geometrical patterns [6] and the scaling relationship between patterns and the scale of measurement; the scale of a fractal set varies with the scale at which it is examined and raised to a scaling exponent, given by the fractal dimension. The extended multifractal fields concept [7–13] can be considered as an infinite hierarchy of sets, each with its own fractal dimension. Thus, multifractals are described by scaling relationships that require a set of different exponents, rather than the single exponent of fractal patterns. Despite the apparent complexity of the multifractals, the distribution of a given scalar field can be wholly described by only three indices, using the universal multifractal formalism, as proposed by Schertzer and Lovejoy [14,15]. These indices resume the statistical behavior of turbulent fields from larger to smaller scales, as well as from extreme to mean behaviors.

The multifractal framework was largely developed for studying turbulent intermittency [11,16–21]; it has been shown to be well-adapted for providing a parsimonious description of the statistics at all orders and over a wide range of scales for air temperature, air velocity, rainfall [11,22,23], and for other intermittent geophysical fields [24–26].

The aim of this paper is to further investigate the scaling behavior of air temperature data and its evolution with respect to the urban built environment. We therefore propose an approach to analyze high-frequency meteorological data, by adopting and applying time series analysis techniques in which missing data are well-respected: a frequency analysis through power spectral density estimation and a multi-scale structure functions analysis. To apply and assess the proposed approach, we use air temperature data recorded every 10 min by an autonomous monitoring system.

2. Materials and Methods

Here, we study high-frequency ambient air temperature data continuously monitored in Athens, Greece, during the period 2003. We use two different data sets provided from two different locations. The first data set, time series A, comes from a location suited in Athens University Campus, a sub-urban area on the outskirts of Ymitos Mountain. The second data set, time series B, comes from a monitoring system placed at the Western Athens Municipality, Peristeri, which is an urban area characterized by a high-density built environment and population. Temperature sensors where housed in a Stevenson screen 1.5 m above ground level.

Both time series consisted of data sampled every 10 min from 24 June 2003 to 21 August 2003. This corresponds to 8202 data points.

We performed spectral analysis, using extended discrete Fourier transform (EDFT), and multifractal analysis, using structure functions. Here follows a short description of the methods.

Spectral analysis corresponds to an analysis of variance, in which the total variance of a process is portioned into contributions arising from processes with different periodicities. Spectral analysis separates and measures the amount of variability occurring in different wave numbers of frequency bands. It characterizes the dynamics of the investigated series by associated scales and emphasizes periodicities, as well as forcing and scaling ranges. When all or part of the spectrum follows a power law, such as

$$E(f) \approx f^{-\beta} \tag{1}$$

where f is the frequency, the data possess scaling statistics over this range. The velocity field for homogenous turbulence scales with a power spectrum slope of $\beta = 5/3$ [27,28]. The same slope is obtained for turbulent temperatures in homogeneous turbulence [29,30].

The fluctuation of time series that possess intermittent distributions can be studied using structure functions analysis, a simple yet powerful tool.

Assuming statistical time translational invariance, the structure function $S^{(q)}(\tau)$, i.e., the statistical moments of the increment of the original series $X(t)$, will depend only on the time lag τ; in addition, according to a power law, if the process is scaling, then

$$S^{(q)}(\tau) = \langle |X(t+\tau) - X(t)|^q \rangle = S^{(q)}(T) \left(\frac{\tau}{T} \right)^{\zeta(q)} \tag{2}$$

where T is the fixed, largest time scale of the system, $\langle . \rangle$ denotes the statistical average (for non-overlapping increments of length τ), q is the order of the moment (we take here $q \succ 0$), and $\zeta(q)$ is the scale-invariant structure function exponent. Structure function analysis corresponds, in fact, to studying "generalized" average volatilities at scale, since only moments of orders 1 or 2 are usually used to define the volatility. Furthermore, the present analysis consists in analyzing this generalized volatility for all time scales.

Intermittency can be characterized introducing scaling exponents $\zeta(q)$ from Equation (2). In this relation, since T is fixed, $S^{(q)}(T)/T^{\zeta(q)}$ is a constant, and the main information is the time increment (τ) dependence of the moments of fluctuations. The scaling exponent $\zeta(q)$ is estimated by the slope of the linear trends of $S^{(q)}(\tau)$ versus τ in a log–log plot.

The average of the fluctuations corresponding to $q = 1$, i.e., the first moment, gives the scaling exponent $H = \zeta(1)$, which is the so-called "Hurst" exponent characterizing the scaling non-conservation of the mean. The Hurst exponent provides a measure that characterizes the level of persistence in the time series. Values of $H < 0.5$ point to anti-persistency—i.e., a lower-than-average value tends to be followed by a higher-than-average value, and vice versa. A value of $H = 0.5$ signifies complete randomness, while values of $H > 0.5$ correspond to positive autocorrelation, and exhibit long memory that enables predictability.

The second moment is linked to the slope β of the Fourier power spectrum [16–28]:

$$\beta = 1 + \zeta(2) \tag{3}$$

The main property of a multifractal process is that it is characterized by a nonlinear $\zeta(q)$ function [16]. This function is convex, being proportional to the second Laplace characteristic function of the generator of cascade. Multifractals are the generic result of multiplicative cascades. A continuous-scale limit of such processes leads to the family of log-infinitely divisible distributions, among which are universal multifractals [31], which have a normal or Levy generator, and for which

$$\zeta(q) = qH - \frac{c_1}{\alpha - 1}(q^\alpha - q) \tag{4}$$

where c_1 is an intermittency parameter, and $0 < \alpha \leq 2$ is the basic parameter that characterizes the process. The index c_1 provides a measure that characterizes the mean inhomogeneity of the field: a value of $c_1 = 0$ corresponds to homogenous fields, while larger c_1 values correspond to more intermittent fields. The parameter α is the Levy index, and indicates the extent of multifractality. It takes a value between 0 and 2, and can be understood as an interpolation between two extremes and well-known cascade models of turbulence: the β-model ($\alpha = 0$) and the lognormal model ($\alpha = 2$). The case $1 \leq \alpha \leq 2$ corresponds to a log–Levy process with unbounded singularities.

The above formulas indicate that universal multifractals can be characterized by only three parameters, in which α and c_1 are the fundamental parameters, with α being the most basic.

For simple monofractal processes, $\zeta(q)$ is linear, e.g., $\zeta(q) = qH$ for fractional Brownian motion (fBm) or $\zeta(q) = q/2$ for Brownian motion (Bm).

The deviation of the experimentally obtained $\zeta(q)$ curve from the linear trend is an indication of intermittent multifractal fluctuations, whose statistics are characterized by $\zeta(q)$ estimated for non-integer moment orders, up to moments of order 4 or 5, depending on the data sets.

The structure functions can be estimated even when there are missing values in the original time series. This can be expressed as:

$$S^{(q)}(\tau) = \frac{1}{N_\tau} \sum_i |X(t_1) - X(t_2)|^q \tag{5}$$

where N_τ is the number of consecutive data points $(X(t_1), X(t_2))$ in the series, verifying $|t_1 - t_2| = \tau$.

3. Results

Data set A possesses a percentage of missing values, due to Internet connection problems.

Figure 1 shows the multi-scale fluctuations of temperature for both data sets, as well as the succession of missing values in the time series for data set A. For time series B, there are no missing data. The time increments between successive values, normalized by the sampling time scale of $t_0 = 10$ min, present an inhomogeneous distribution, as shown in Figure 2.

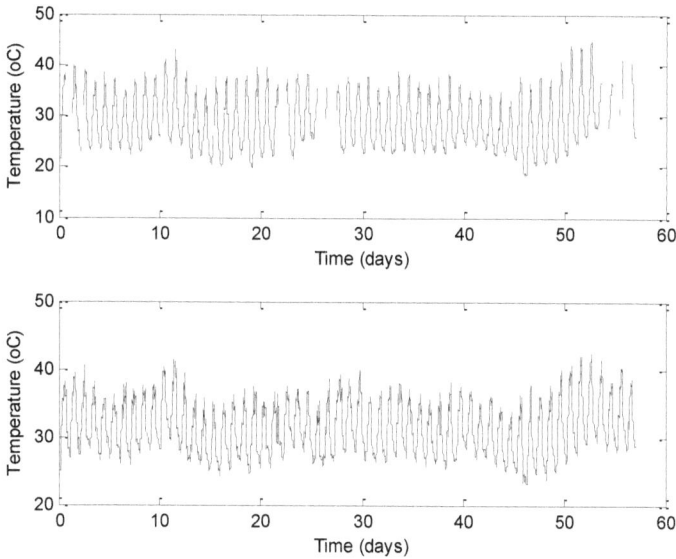

Figure 1. Air temperature values, measured by an online environmental monitoring system. (**Top** panel corresponds to time series A, **bottom** panel to time series B).

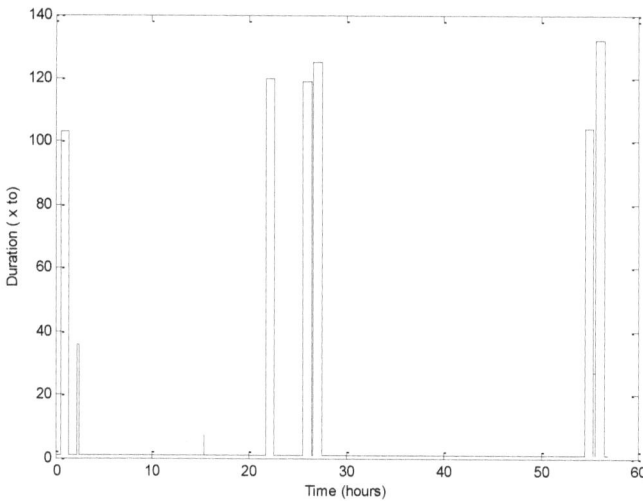

Figure 2. Duration of the interval that separates two consecutives values in an air temperature time series (Data Set A).

For time series A, the time interruption range was from $7t_o$ to $125t_o$, with a mean value of $80.1t_o$, which corresponds to 13.5 h mean interruption time.

3.1. Power Spectral Analysis of an Air Temperature Time Series

Power spectral analysis is often performed using a fast Fourier transform (FFT) algorithm. This algorithm cannot be directly applied to the data considered in the present work, because it requires continuously sampled data.

The problem of missing data is usually faced by averaging the data, or by filling in missing values with interpolated data. However, averaging corresponds to degraded resolution, while interpolation introduces additional artifacts. Another approach, adapted in the present work, is to directly analyze available data without losing scale information or introducing artificial correlations.

Thus, frequency spectra were obtained using extended discrete Fourier transform (EDFT), an algorithm for evaluating a Fourier transform even in the case of incomplete data and irregular sampling, as proposed by V. Liepins [32].

Estimated frequency spectra for the two different temperature time series data are shown in Figure 3, in a log–log plot. The frequency range is from 20 min to seven days.

In Figure 2, vertical dotted lines correspond to specific time scales associated to periodic events or solar forcing: the diurnal periodicity and its harmonics (shown in the graph is the 8 h periodicity). For scales smaller than the 8 h solar forcing, the temperature power spectrum exhibits scaling behavior. Small scale fluctuations do not permit accurate estimation of the slope β. Nevertheless, a linear fit through the data gives approximate values of $b = 1.93$ for time series A and $b = 1.87$ for time series B. The slopes (straight lines) are plotted according to the estimate (see below) of the structure function scaling exponent, e.g., $b = 2.7$ for time series A and $b = 2.5$ for time series B. The absence of characteristic time scales and the presence of a scaling range for scales smaller than 8 h lead us to perform multifractal analysis.

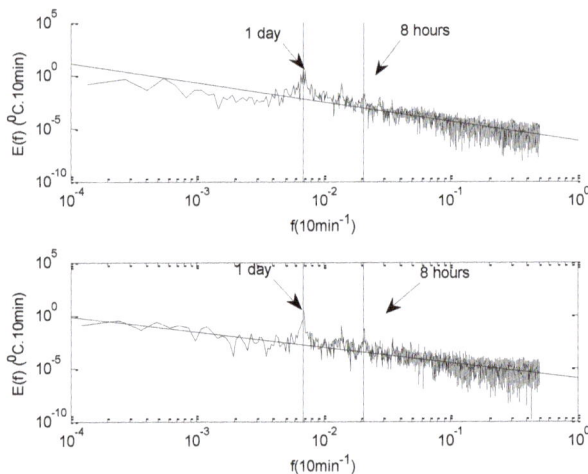

Figure 3. Power spectrum of the temperature time series, in log-log plot, estimated using extended discrete Fourier transform (EDFT). The **top** panel shows the spectrum for the time series A, the **lower** panel for the time series B.

3.2. Multifractal Data Analysis

Here, we present the structure functions analysis of the two considered time series. In creffig:climate-305557-f004,fig:climate-305557-f005, we show the structure functions in log–log plot for different orders of moments. The scaling range previously revealed by the spectral analysis is confirmed. The straight lines show that the scaling of Equation (2) is very well-respected; scaling behavior is visible for scales between 10 min and 9 h for time series A, and between 10 min and 7 h for time series B; we repeated this for moments up to 6.0, with a 0.2 increment. The scaling begins to break only for moments up to 6.0, because of the insufficient amount of data analyzed.

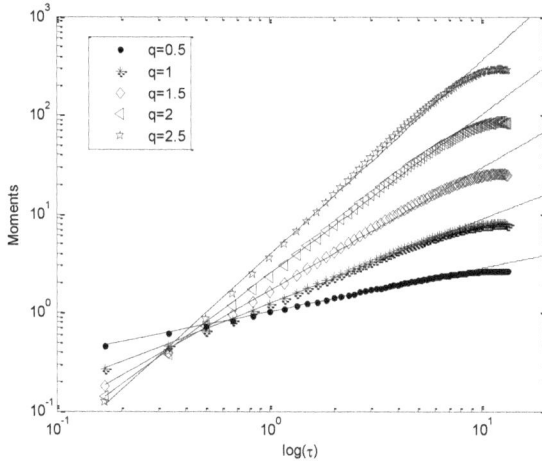

Figure 4. Scaling of the structure functions in log–log plot for moments of order 0.5, 1, 1.5, 2, and 2.5 for time series A.

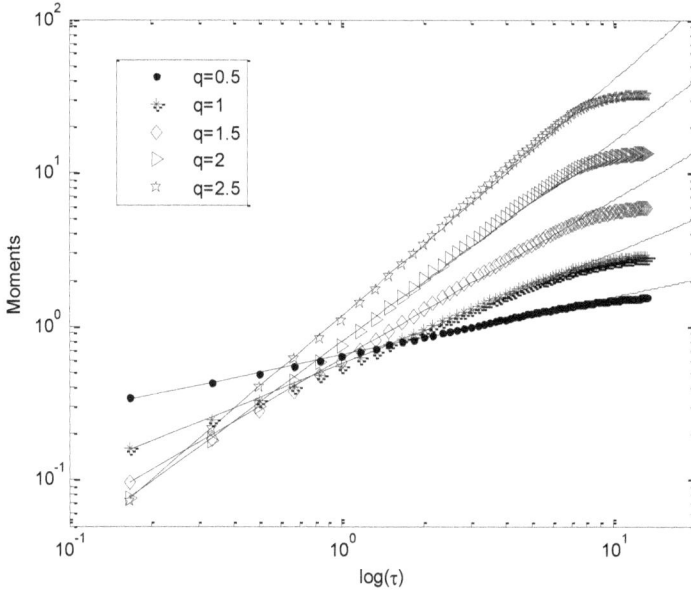

Figure 5. Scaling of the structure functions in log-log plot for moments of order 0.5, 1, 1.5, 2, and 2.5 for time series B.

The resulting $\zeta(q)$ functions are shown in Figures 6 and 7; nonlinearity is present. We also directly estimated the scaling exponent of the nonlinear term $\tau^{qH}/\langle(\Delta X_T)^q\rangle$, which is a convex function plotted on the same graph.

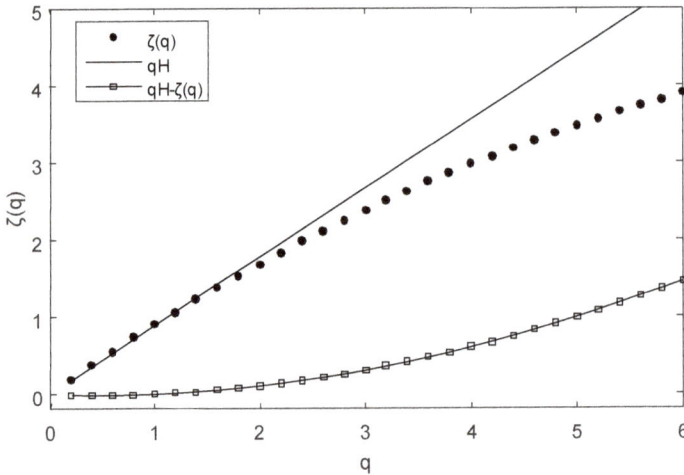

Figure 6. Empirical values of $\zeta(q)$ obtained here (dots) compared to qH (straight line). Also shown is the convex function qH-$\zeta(q)$ (Time series A, squares).

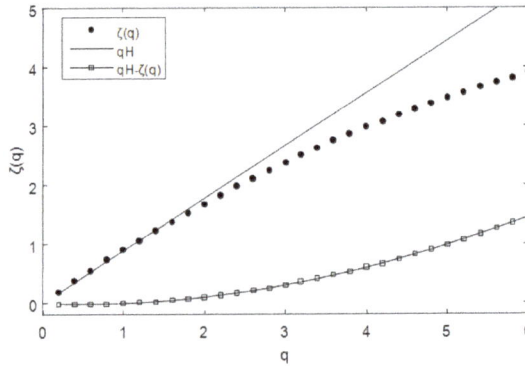

Figure 7. Empirical values of $\zeta(q)$ obtained here (dots) compared to qH (straight line). Also shown is the convex function qH-$\zeta(q)$ (Time series B, squares).

We obtained the following values for the two time series: $H = 0.89 \pm 0.01$ for time series A and $H = 0.74 \pm 0.01$ for time series B. Using specific analysis techniques, we obtained $c_1 = 0.0520 \pm 0.0004$ and $\alpha = 1.936 \pm 0.008$ for time series A, and $c_1 = 0.082 \pm 0.002$ and $\alpha = 1.79 \pm 0.02$ for time series B. We also obtained $\zeta(2) = 1.68$ and $\zeta(2) = 1.35$.

This indicates that in general, air temperature data are characterized by multifractal processes; the values of H are relative high, indicating also that the fluctuations are less variable than fluctuations found in homogenous turbulence.

In Figure 8, the normalized structure function scaling exponent $\zeta(q)/H$ is also shown for turbulent atmospheric temperature data ($H = 0.38$, $\alpha = 1.45$, $c_1 = 0.34$, [21]) and for laboratory passive scalar turbulence ($H = 0.37$, $\gamma+ = 0.31$, $c_1 = 0.84$, [33]). The strong correspondence observed in this figure indicates that the universal models with the above parameters are compatible with the data through the medium range of moments, and probably reflect a general property of temperature fluctuations, which could be valid for small-scale turbulence and for temperature fluctuations at larger scales.

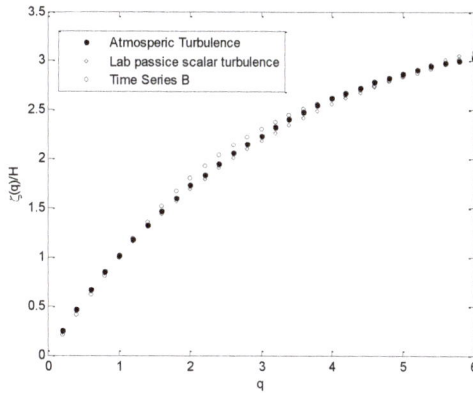

Figure 8. Our estimates of $\zeta(q)$ plotted versus q for moments up to 6.0, compared to other databases: atmospheric turbulent temperature data ($H = 0.38$, Schmitt et al. [21]) and laboratory passive scalar turbulence ($H = 0.37$, Ruiz et al. [33]).

Figure 9 shows the H spectra as a function of the temporal scale of ambient temperature for the two data sets. The scale range in the plots is from 1 h to around 10 h. For both data sets, a trend of decreasing H values after around 4 h is evident. The suburban location's air temperature plot show larger H values, compared to H values for urban location, for scales up to 4 h. For both data sets, this indicates increased persistence and good predictability up to these time scales. After this scale, H decreased smoothly and reached the $H = 0.5$ value, which implies random behavior and lack of predictability, at a time scale of about 9 h.

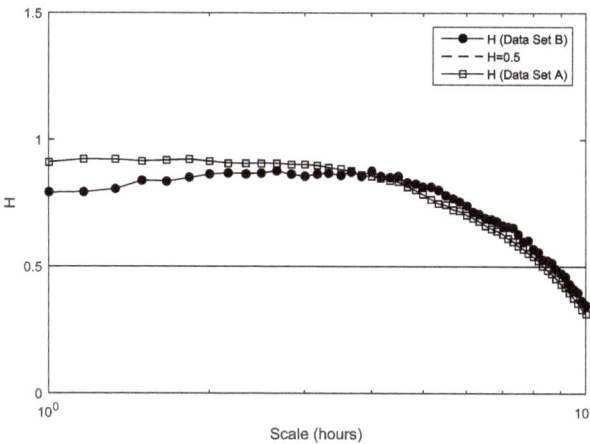

Figure 9. Spectrum of H values as a function of temporal scale for the two data sets. The dotted line at $H = 0.5$ denotes complete randomness and zero predictability.

4. Discussion

In the years of information and communication technology expansion, time series will be continuously longer, providing a challenge to better understand and predict urban climate complex processes. Finding optimal ways to analyze these complex and possible nonlinear environmental time series data sets is therefore very important.

To explore these methods, we have analysed here ambient air temperature time series, sampled every 10 min throughout 2003 by an online environmental monitoring system for suburban and highly-built urban environments, in Athens, Greece. Missing data are well-respected in the sense that we neither average the data nor fill the gaps with interpolations; thus, any information concerning system's dynamics is well-preserved.

Spectral analysis reveals a diffuse forcing around 12 h, while for smaller scales (from 10 min to 6–7 h), a turbulent-like behaviour has been observed. Multifractal structure functions analysis was applied, and provided the experimental estimate of scaling exponents $\zeta(q)$. We computed these scaling exponents for several values of q between 0 and 6; the clearly nonlinear bevariour of $\zeta(q)$ is direct evidence of the multifractal nature of air temperature.

We then studied and quantified this multifractality using the universal multifractal model. Using two different data sets, we obtained values of the three indices (H, c_1, α) which can completely characterize the statistics of air temperature time series data. The low c_1 values, 0.052 and 0.082 respectively for the air temperatures recorded at the suburban and the urban locations, indicate a homogenous field. Similarly, the high α values, 1.93 and 1.79, respectively, indicate that multifractality approaches its upper limit. Thus, both data sets present characteristics that are typical of log–Levy processes.

The differences between the calculated H, c_1 and α parameters, although small, are greater than the statistical error. This indicate that urban locations, compared to suburban, are characterized by an increased heterogeneity, and that there are more large deviations from the mean; the high values of the air temperature do not dominate as much as for suburban locations. However, more data are needed for further analysis.

We compared the normalized structure function scaling exponents with turbulent atmospheric temperature data and laboratory passive scalar turbulence (see Figure 8). The normalized values of the data are good, indicating that some of the general properties of temperature fluctuations valid for small-scale turbulence are preserved at larger scales.

Multifractal analysis can also be used to characterize the self-affinity of time series, through the detailed information about the Hurst parameter H. The Hurst parameter represents a temporal fractality measure, and thus can be used as a tool for assessing the time scale dependence of the predictability of air temperature. For the distinct range of time scales considered in the study, i.e., from 1 h up to around 9 h, air temperature behaves in a predictable way. Beyond this scale, H values become smaller than 0.5, indicating a decreasing predictability, probably due to the lack of local correlations. As far as it concerns quantitative predictability, our results indicate that statistical methods that are usually used to predict time series, like multiple linear regression (MLR), adaptive neuro-fuzzy inference systems (ANFIS), and neural networks (ANNs) can potentially be effective up to a prediction horizon of 10 h.

It appears that air temperature in urban and suburban environments is a complex time series, caused by the local correlations and interactions of various factors, such as topography, the built environment, and the atmosphere. More generally, the non-linearity of the empirical curve $\zeta(q)$ shows that the time series studied here can be considered as multifractals. Empirical curves are slightly different for the two time series, as the time series corresponding to the high-density built environment appear to be more convex. The differences between the calculated H, c_1 and α parameters may indicate increasing heterogeneity and decreasing multifractality from suburban to urban locations.

Online monitoring systems and Information and Communication Technology (ICT) provide the framework to obtain databases that can facilitate data availability, as well as investigation of new methods and techniques. The time series analysis methods presented here may be generally applied to other environmental parameters as well, thus providing a way to improve our knowledge about the dynamics of environmental processes at the urban scale. Our results pertain to a certain geographical location, though we estimate that they can be confirmed and generalized by further studies, using data from other sites and climate zones.

Author Contributions: The study was conceived by the authors together. S.K. conducted the study and wrote the paper. M.S. provided critical discussions.

Funding: This research received no external funding.

Conflicts of Interest: The authors declare no conflict of interest.

References

1. Livada, I.; Santamouris, M.; Assimakopoulos, M.N. On the variability of summer air temperature during the last 28 years in Athens. *J. Geophys. Res.* **2007**, *112*. [CrossRef]
2. Mihalakakou, G.; Santamouris, M.; Asimakopoulos, D. Modeling ambient air temperature time series using neural Networks. *J. Geophys. Res.* **1998**, *103*, 19509–19517. [CrossRef]
3. Santamouris, M.; Mihalakakou, G.; Papanikolaou, N.; Assimakopoulos, D.N. A Neural Network Approach for Modelling the Heat Island Phenomenon in Urban Areas during the Summer Period. *Geophys. Res. Lett.* **1999**, *26*, 337–340. [CrossRef]
4. Kantz, H.; Schreiber, T. *Nonlinear Time Series Analysis*, 2nd ed.; Cambridge University Press: Cambridge, UK, 2004.
5. Ingle, V.K.; Proakis, J.G. *Digital Signal Processing Using MATLAB*; Brooks/Cole Publishing Company: Pacific Grove, CA, USA, 2000.
6. Mandelbrot, B. *The Fractal Geometry of Nature*; Freeman: New York, NY, USA, 1983.
7. Grassberger, P. Generalized dimensions of strange attractors. *Phys. Lett.* **1983**, *97*, 227–230. [CrossRef]
8. Hentschel, A.W.; Procaccia, I. The infinite number of generalized dimensions of fractal and strange attractors. *Phys. D* **1983**, *8*, 435–444. [CrossRef]
9. Schertzer, D.; Lovejoy, S. The dimension and ittermittency of atmospheric dynamics. In *Turbulent Shear Flows 4*; Launder, B., Ed.; Spinger-Verlag: Karlsruhe, Germany, 1983; pp. 7–33.
10. Schertzer, D.; Lovejoy, S. Generalised scale invariance in turbulent phenomena. *Physicochem. Hydrodyn. J.* **1985**, *6*, 623–635.
11. Schertzer, D.; Lovejoy, S. Physical modeling and analysis of rain and clouds by anisotropic scaling of multiplicative processes. *J. Geophys. Res. D* **1987**, *92*, 9693–9714. [CrossRef]
12. Lovejoy, S.; Schertzer, D.; Allaire, V.C. The remarkable wide range spatial scaling of TRMM precipitation. *Atmos. Res.* **2008**, *90*, 10–32. [CrossRef]
13. Parisi, G.; Frisch, U. *Turbulence and Predictability of Geophysical Flows and Climatic Dynamics*; Ghil, N., Benzi, R., Parisi, G., Eds.; North Holland: Amsterdam, The Netherlands, 1985; pp. 84–87.
14. Schertzer, D.; Lovejoy, S. Physically based rain and cloud modelling by anisotropic scaling multiplicative processes. *J. Geophys. Res.* **1987**, *92*, 96–99. [CrossRef]
15. Schertzer, D.; Lovejoy, S. Nonlinear variability in geophysics: Multifractal analysis and simulation. In *Fractals, Physical Origin and Consequences*; Pietronero, L., Ed.; Plenum: New York, NY, USA, 1987; pp. 49–79.
16. Frisch, U.; Parisi, G. *Fully Developed Turbulence and Intermittency, Turbulence and Predictability in Geophysical Fluid Dynamics and Climate Dynamics*; Ghil, M., Benzi, R., Parisi, G., Eds.; North Holland: Amsterdam, The Netherlands, 1985; pp. 84–92.
17. Mandelbrot, B. Intermittent turbulence in self-similar cascades: Divergence of high moments and dimension of the carrier. *J. Fluid Mech.* **1974**, *62*, 331–358. [CrossRef]
18. Meneveau, C.; Sreenivasan, K.R. A simple multifractal cascade model for fully developed turbulence. *Phys. Rev. Lett.* **1987**, *59*, 1424–1427. [CrossRef] [PubMed]
19. Schertzer, D.; Lovejoy, S. Hard and soft multifractal processes. *Phys. A* **1992**, *185*, 187–194. [CrossRef]
20. Schmitt, F.; Lavallée, D.; Schertzer, D.; Lovejoy, S. Multifractal temperature and flux of temperature variance in fully developed turbulence. *Phys. Rev. Lett.* **1992**, *68*, 305–308. [CrossRef] [PubMed]
21. Schmitt, F.; Schertzer, D.; Lovejoy, S.; Brunet, G. Universal multifractal structure of atmospheric temperature and velocity fields. *Europhys. Lett.* **1996**, *34*, 195–200. [CrossRef]

22. Gupta, V.K.; Waymire, E.C. A statistical analysis of mesoscale rainfall as a random cascade. *J. Appl. Meteorol.* **1993**, *32*, 251–267. [CrossRef]

23. Marsan, D.; Schertzer, D.; Lovejoy, S. Causal space-time multifractal modelling of rain. *J. Geophys. Res. D* **1996**, *26*, 333–346.

24. Tessier, Y.; Lovejoy, S.; Schertzer, D. Universal multifractals: Theory and observations for rain and clouds. *J. Appl. Meteorol.* **1993**, *32*, 223–250. [CrossRef]

25. Dur, G.; Schmitt, F.G.; Souissi, S. Analysis of high frequency temperature time series in the Seine estuary from the Marel autonomous monitoring buoy. *Hydrobiologia* **2007**, *588*, 59–68. [CrossRef]

26. Seuront, L.; Schmitt, F.; Lagadeuc, Y.; Schertzer, D.; Lovejoy, S. Universal multifracatl analysis as a tool to characterise multiscale intermittent patterns; example of phytoplankton distribution in turbulent coastal waters. *J. Plankton Res.* **1999**, *21*, 877–922. [CrossRef]

27. Kolmogorov, A.N. Local structure of turbulence in incompressible fluid at very large Reynolds numbers. *Dokl. Acad. Sci. USSR* **1941**, *30*, 299. [CrossRef]

28. Frisch, U. *Turbulence, the Legacy of A. N. Kolmogorov*; Cambridge University Press: Cambridge, UK, 1995.

29. Obukhov, A.M. Structure of the temperature field in a turbulent flow. *Isv. Geogr. Geophys. Ser.* **1949**, *13*, 55–69.

30. Corrsin, S. On the spectrum of isotropic temperature fluctuations in an isotropic turbulence. *J. Appl. Phys.* **1951**, *22*, 469–473. [CrossRef]

31. Schertzer, D.; Lovejoy, S.; Schmitt, F.; Chigirinskaya, Y.; Marsan, D. Multifractal cascade dynamics and turbulent intermittency. *Fractals* **1997**, *5*, 427–471. [CrossRef]

32. Liepinsh, V. An algorithm for evaluation a discrete Fourier transform for incomplete data. *Autom. Control Comput. Sci.* **1996**, *30*, 27–40.

33. Ruiz Chavarria, G.; Baudet, C.; Ciliberto, S. Scaling laws and dissipation scale of a passive scalar in fully developed turbulence. *Phys. D* **1996**, *96*, 369–380. [CrossRef]

climate

MDPI

Article

Air-Temperature Response to Neighborhood-Scale Variations in Albedo and Canopy Cover in the Real World: Fine-Resolution Meteorological Modeling and Mobile Temperature Observations in the Los Angeles Climate Archipelago

Haider Taha [1,*], Ronnen Levinson [2], Arash Mohegh [3], Haley Gilbert [2], George Ban-Weiss [3] and Sharon Chen [2]

[1] Altostratus Inc., 940 Toulouse Way, Martinez, CA 94553, USA
[2] Lawrence Berkeley National Laboratory, 1 Cyclotron Road, Berkeley, CA 94720, USA; RMLevinson@LBL.gov (R.L.); HaleyGilbert@gmail.com (H.G.); SSChen@LBL.gov (S.C.)
[3] University of Southern California, Los Angeles, CA 90007, USA; mohegh@usc.edu (A.M.); banweiss@usc.edu (G.B.-W.)
* Correspondence: haider@altostratus.com; Tel.: +01-(925)-228-1573

Received: 11 May 2018; Accepted: 14 June 2018; Published: 17 June 2018

Abstract: To identify and characterize localized urban heat- and cool-island signals embedded within the temperature field of a large urban-climate archipelago, fine-resolution simulations with a modified urbanized version of the WRF meteorological model were carried out as basis for siting fixed weather monitors and designing mobile-observation transects. The goal was to characterize variations in urban heat during summer in Los Angeles, California. Air temperatures measured with a shielded sensor mounted atop an automobile in the summers of 2016 and 2017 were compared to model output and also correlated to surface physical properties focusing on neighborhood-scale albedo and vegetation canopy cover. The study modeled and measured the temperature response to variations in surface properties that already exist in the real world, i.e., realistic variations in albedo and canopy cover that are attainable through current building and urban design practices. The simulated along-transect temperature from a modified urbanized WRF model was compared to the along-transect observed temperature from 15 mobile traverses in one area near downtown Los Angeles and another in an inland basin (San Fernando Valley). The observed transect temperature was also correlated to surface physical properties characterizations that were developed for input to the model. Both comparisons were favorable, suggesting that (1) the model can reliably be used in siting fixed weather stations and designing mobile-transect routes to characterize urban heat and (2) that except for a few cases with opposite co-varying influences, the correlations between observed temperature and albedo and between observed temperature and canopy cover were each negative, ranging from −1.0 to −9.0 °C per 0.1 increase in albedo and from −0.1 to −2.2 °C per 0.1 increase in canopy cover. Observational data from the analysis domains pointed to a wind speed threshold of 3 m/s. Below this threshold the variations in air temperature could be explained by land use and surface properties within a 500-m radius of each observation point. Above the threshold, air temperature was influenced by the properties of the surface within a 1-km upwind fetch. Of relevance to policy recommendations, the study demonstrates the significant real-world cooling effects of increasing urban albedo and vegetation canopy cover. Based on correlations between the observed temperature (from mobile transects) and surface physical properties in the study domains, the analysis shows that neighborhood-scale (500-m) cooling of up to 2.8 °C during the daytime can be achieved by increasing albedo. A neighborhood can also be cooled by up to 2.3 °C during the day and up to 3.3 °C at night by increasing canopy cover. The analysis also demonstrates the suitability of using fine-resolution meteorological models to design mobile-transect routes or site-fixed weather monitors in order to quantify urban heat and the efficacy of albedo and canopy cover countermeasures. The results also show that the model is capable of accurately predicting the geographical locations and the magnitudes of localized urban heat and cool islands. Thus the model results can also be used to devise urban-heat mitigation measures.

Keywords: cool roofs; fine-resolution meteorological modeling; mobile temperature observations; urban climate archipelago; urban heat island; urban vegetation; urbanized WRF; Weather Research and Forecasting model

1. Introduction

Characterization of urban heat and its causes, such as land-surface properties, is an important first step towards designing countermeasures [1,2]. Understanding the correlation between urban heat and variations in land-cover and physical properties of the urban surface is also critical in understanding how future changes in land use can inadvertently impact urban heat, e.g., the heat island effect, and, hence, its mitigation.

Taha [3] shows that in California, the urban heat island (UHI) takes on different characteristics, viz: small, single cores, multiple cores, and climate archipelagos, and manifests itself differently with varying topography, urban morphology, coastal/inland situations, and land-cover properties. The Los Angeles area is one major urban-climate archipelago where it is difficult to define or even discern the UHI in conventional terms since there are no clear urban/non-urban demarcations in the region [3,4].

In this study, the goal was to identify air-temperature-based localized UHI and urban cool islands (UCI) at the intra-urban scale and to correlate their intensities with land-use/land-cover (LULC) and surface physical properties. Such characterizations have been undertaken elsewhere to facilitate planning for mitigation [5]. However, these studies were based on surface rather than air temperature, as done in this work.

Further, and unlike "standard" characterizations of the UHI effect as differences between some urban and non-urban temperatures [6,7], here we attempt to identify UHI and UCI signals that are embedded within the temperature field of an urban-climate archipelago [3]. A similar approach has recently been identified in other studies as well, such as by correlating UHI with local climate zones and land-use [8,9]. To cancel out the effects of larger-scale processes, such as onshore warming with distance from coastline, the archipelago effect, and time of day, the correlations we sought were examined at fine spatio-temporal scales—e.g., within a 500-m radii of influence and within time horizons briefer than 1 hour.

In this attempt at characterizing the roles of surface physical properties, such as albedo and vegetation-canopy cover, in localized UHI and UCI, we did not seek to idealize these parameters or use hypothetical values as many heat-island mitigation modeling studies typically do [7,10–12]. Instead, this study modeled and measured the temperature response to variations in surface properties that already exist in the real world, thus representing realistic modifications in surface albedo and canopy cover that are attainable with current building and urban design practices. The study relied on (1) fine-resolution meteorological modeling with an updated version of the urbanized WRF-ARW model and (2) 15 mobile-observation transects that were carried out on different dates (in summers of 2016 and 2017), times, and routes.

At the intra-urban scale of ~1 km or finer, three main factors dominate in terms of the effects on microclimate and heat. These are (1) albedo; (2) vegetation cover and related shading and evapotranspiration effects; and (3) urban morphology and related parameters such as sky-view factor and surface roughness [7,13]. At the local scales of analysis in this study—i.e., in the areas selected for mobile observations—the statistical and classification-and-regression-tree (CART) analyses showed that the changes in surface roughness were relatively small and that the main variables to consider as predictors to air temperature were albedo and canopy cover.

Several studies have relied on mobile platforms to carry out the measurement of urban microclimate parameters, especially air temperature. For example, Qiu et al. [14] carried out automobile observations to quantify the effects of green spaces on the UHI in Shenzhen, China. The study focused on evapotranspiration as the main cooling mechanism and found that green spaces were the coolest in the urban environment (by up to ~1.6 °C lower in diurnal-average air temperature). The study also found that green areas were cooler than open water bodies and also cooler at night than other urban land uses.

Tsin et al. [15] compared on-foot mobile transect air-temperature measurements in Vancouver with temperature readings from fixed monitors and land-surface temperature from Landsat. They found greater variability in temperature from the transect measurements relative to that from the fixed monitoring stations. The reason is because the mobile measurements were carried out in streets and urban canyons; thus temperatures were influenced by microclimate anomalies. Jonsson [16] used mobile measurements to show that the intra-urban variability in air temperature, as a result of changes in vegetation cover, was of the same magnitude as the urban-rural temperature difference. Furthermore, the study found that urban green spaces could be cooler than non-urban surroundings. In this case, a midday oasis of 2 °C was found to exist in the urban area relative to rural surroundings.

While many studies have characterized summer UHIs, because of interest in their negative impacts on cooling energy use and thermal comfort, some studies (e.g., Sun et al. [17]) have used mobile observations and fixed monitoring to characterize wintertime UHIs. They found that LULC was an important factor in their magnitudes. They also showed that vegetated parts were the coolest urban areas during both day and night times.

Ellis et al. [18] deployed 10 fixed weather stations at ~2 m above ground level in different land-use types across Knoxville, TN, for summertime observations of the UHI. They found that vegetation cover had a significant cooling effect, reducing maximum daytime temperature by up to ~1.2 °C, but with smaller effects at night. They also found that the distance from the city center was not a factor in UHI, but that the effects of LULC variations were dominant. This is similar to what Taha [3] found in the Los Angeles climate archipelago and other large urban regions in California. It is also what we found in this study and will be described in the following sections. For a review of other UHI investigations, the reader is referred to Taha [19].

While there are numerous studies, similar to the ones reviewed above, that used observations to evaluate real-world heat islands, two of the aspects that have not been sufficiently addressed are (1) the characterization of UHI and UCI within large urban-climate archipelagos and (2) characterizing UHI and UCI based on a combination of both atmospheric modeling and field observations, where the former guides the latter. These two aspects were the inspiration behind the work summarized in this paper.

2. Methodology

2.1. LULC Analysis and Bottom-Up Approach to Developing Input to WRF

Land-use and land-cover analysis was carried out following a bottom-up approach to develop surface input in the urbanized meteorological model. LULC and surface physical properties were used to (1) characterize study regions for targeted modeling; (2) develop urban-parameter inputs to the modified urban WRF model for the selected domains; and (3) use these computed parameters to correlate surface properties with observed air temperature from mobile transects.

In the bottom-up approach of Taha [3] and Taha and Freed [20], each model grid cell is characterized based on as much information as available from any and all sources. This is to directly characterize or scale each cell's properties in terms of parameters required by the meteorological and land-surface models and use them instead of the standard model's lookup parameters. These include, among others, surface albedo, urban morphology parameters, vegetation canopy cover, shade factor, view factor, roughness length or drag coefficient, and soil moisture. The following data sources were used:

- Fine-resolution (30 m) LULC classification of the Los Angeles region's six counties (134 classes total; 97 urban classes) generally following the Anderson Level-4 classification system [21]. The dataset was developed by the Southern California Association of Governments (SCAG) and the City of Los Angeles.
- Fine-resolution individual buildings footprint and height information throughout the Los Angeles County. This dataset was developed by SCAG and the County of Los Angeles.
- Building-specific roof albedo derived based on aerial imagery from the National Agriculture Imagery Program (NAIP). Albedo was developed for each roof within the boundaries of the City of Los Angeles. The dataset was generated by the Lawrence Berkeley National Laboratory (http://albedomap.lbl.gov) [22].
- Fine-resolution tree-canopy cover for urban areas in California characterizing each 1-m pixel as either canopy or non-canopy. The data were developed by EarthDefine and CAL FIRE/Fire Resource Assessment Program (http://frap.fire.ca.gov).
- National Land Cover Datasets (NLCD) and United States Geological Survey LULC [23] providing additional information in areas where other fine-resolution datasets are lacking or where such data are sparse.
- Fine-resolution, detailed Light Detection and Ranging (LiDAR)-derived urban morphology parameters for areas within the Los Angeles region (in Los Angeles and Orange counties). Data were based on the National Urban Data and Portal Tool (N/WUDAPT) effort of Ching et al. [24].
- Google Earth PRO 3-D building attributes information for site-specific building and urban canyon geometrical characterizations.

In the bottom-up approach, the above datasets were merged at the grid-cell level and used to derive model parameters as needed. This cell-specific approach does away with the typical look-up that the model uses and significantly improves the site-specificity of surface characterization. The parameters discussed above were computed at 500-m resolution including roof and non-roof albedo; vegetation cover; ground cover; shade factor; soil moisture; roughness length; view factor; building and vegetation-canopy plan-area, frontal-area, and top-area densities; anthropogenic heat flux profiles; street orientations; street widths; and building heights. The bottom-up approach followed in computing these parameters is discussed in Taha and Freed [20] and Taha [1,3,7].

Based on LULC analysis, two areas of interest in the Los Angeles region were identified for fine-scale meteorological modeling (Figure 1). One area is near downtown Los Angeles (quasi coastal) while the other is in an inland basin (San Fernando Valley). These two areas were selected (1) for contrast in vegetation cover (San Fernando Valley; see Figure 2) as well as variability in both albedo and vegetation cover (in the downtown area), and (2) because they represent contrasting local weather conditions: whereas the downtown area is subject to on-shore flow and sea breeze circulation, the San Fernando Valley is strongly influenced by locally-driven winds and up- and downslope flows.

The meteorological model was run with a nested-grid configuration of 27 km, 9 km, 3 km, and 500 m (only the 500 -m grids are shown in Figure 1). For the long-term, multi-seasonal simulations, the finest grids were run off-line after downscaling. For transect-specific simulations (discussed in Section 3.2), two-way feedback was applied. The WRF configuration used in this project is based on and discussed in a study by Taha [3] and will not be repeated here, except for a brief summary in Section 2.2. Here, we add two 500-m grids over the desired domains in San Fernando Valley and the west basin (as shown in Figure 1).

Figure 1. Urban WRF 500-m modeling domains for San Fernando Valley (left rectangle) and downtown area (right rectangle). Red diamonds locate the mesoscale network (mesonet) monitors closest to either domain. Downtown Los Angeles is near the northwest corner of the right rectangle.

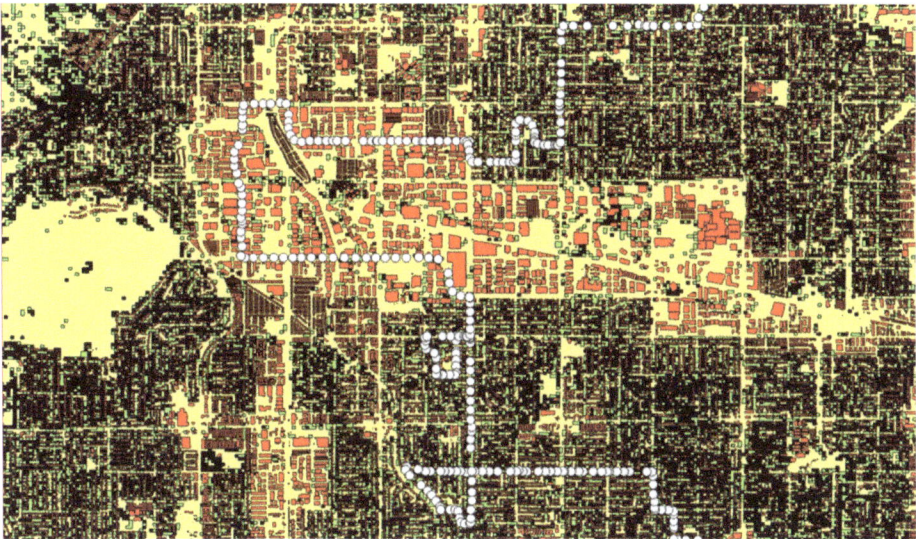

Figure 2. Detail from the San Fernando Valley domain in Figure 1, showing 30-m tree cover (yellow: <10%, light green: 10–20%, black: >20% cover), building-specific roof albedo (red: 0.05–0.25, orange: 0.25–0.50, light orange: 0.50–0.90), and a sample mobile transect segment (white dots).

2.2. Meteorological Modeling

The urbanized WRF-ARW model was modified at Altostratus Inc. and applied to study the domains identified above. The parameterizations, land-surface model, and input were modified and tailored specifically for this type of applications. The approach in this study overrides the typical LULC lookup in urban WRF and characterizes every grid cell independently using the bottom-up approach discussed above that is significantly more resolved than the standard approach in the WRF model.

Modifications in this study were carried out within the Noah land-surface model of Pleim et al. [25] and the urban canopy model of Kusaka et al. [26]. These modifications, discussed in Taha and Freed [20] and Taha [3,7], were carried out to (1) enable ingestion of new parameters computed in the bottom-up approach discussed above; (2) trigger the modified urban-canopy model for each grid cell based on a different set of physical criteria rather than simply as a function of land use, which is how the standard model operates; (3) account for wind direction in the calculations of roughness- and drag-related parameters; and (4) mesh the urban and non-urban parts of each grid cell using cell-specific surface properties rather than a single default value for the non-urban part as in the standard model. In addition, the urban-canopy model was also modified to directly use cell-specific building height, street width, orientation, building footprint, roof albedo, ground albedo, shade factor, vegetation cover, and anthropogenic heat flux, instead of the generic lookup values assigned to LULC types in the standard model.

The modified urban WRF was run to (1) characterize microclimate variations within the domains defined above to select the study areas; (2) help site the fixed weather monitors in the study region and design mobile-observation routes to characterize urban heat; and (3) provide a full 4-dimensional picture of the state of the urban atmosphere before, during, and after the specific time of each mobile-transect segment. For objectives 1 and 2, the model was run June through September, 2013–2015. For objective 3, transect-specific simulations were started one week prior to (leading up to) the time of the actual mobile transect and continued for two days past that time. Model validation and performance evaluation—i.e., comparison of model output with observations—are discussed in Section 3.2.

2.3. Mobile Observations

A mobile-observations apparatus (Figure 3) was designed in this project to measure air temperature at ~2 m above ground level by attaching a thermometer atop a vehicle, then logging (at 10-second intervals) temperature, position, elevation, speed, and time during a transect. To ensure that the thermometer accurately measures air temperature, its sensor is aspirated by the motion of the vehicle, shielded from the sun, and radiatively isolated from its shield. The sensor also responds quickly to air temperature changes to minimize spatial inaccuracies, or blurring, in the air temperature map induced by the motion of the vehicle.

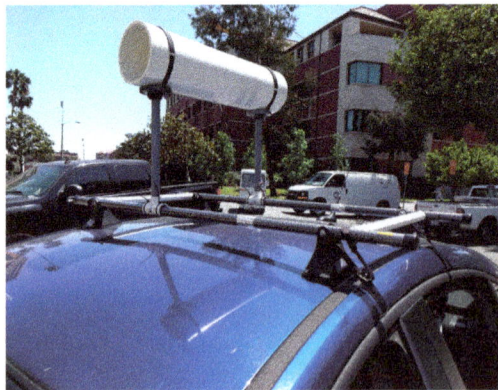

Figure 3. Mobile-transect apparatus.

The mobile apparatus contains five elements: (1) a shielded temperature sensor (2.5-mm diameter thermistor of nominal accuracy ±0.1 °C from 0 to 70 °C and still-air time constant 10 s, wrapped in aluminum foil to minimize exchange of longwave radiation with enclosure, and suspended in the center of a 17 cm diameter white PVC-pipe solar shield) that is aspirated by vehicle motion; (2) a quick-install mount to attach the shielded sensor to the roof of a vehicle; (3) a portable data logger to record the temperature time series; (4) a global positioning system (GPS) to record the position time series; and (5) a dash camera (dash-cam) to record a time-stamped video of the transect from the perspective of the driver.

Following testing and calibration of the apparatus, 15 transects were carried out for this analysis, some on duplicate routes (but different times and dates) and others on different routes as well as different dates. This is discussed in Section 3.1.

To ensure a uniform basis for analyzing the mobile observations, changes in elevation (within each transect segment) were constrained to under 15 m. To minimize time-of-day effects (solar radiation and background temperature) the analysis was done for segments of well under 1 hour at a time. To minimize the effects of anthropogenic heating changes on the observations and to ensure that measurements represent air temperature, observations readings at travel speeds of under 10 km/h (~3 m/s) were discarded. Finally, transects were carried out only during clear-sky conditions to avoid the effects of coastal stratus or other cloud cover.

Mobile temperature observations, subject to the above criteria, were compared to model temperature at the nearest grid points (based on specification of radii of influence, as will be discussed in Section 3.3). In addition to the mobile observations, microclimate readings from existing nearby fixed mesonet monitors (NOAA/MADIS) were used to determine background temperature, cloud cover, wind speed, and wind direction during each transect segment. The mesonet stations nearest to the modeling domains are located in Figure 1. As discussed in Section 3.1, the transects were designed to go through areas with varying temperature, albedo, and canopy cover. As an example, Figure 2 shows a segment from one of the transects in the San Fernando Valley study domain.

3. Results and Discussion

3.1. Predicted Urban Heat and Cool Islands

In this section, modeling results for the two 500-m resolution domains (Figure 1) are discussed. Four time intervals selected from 2006–2013 simulations are presented, where interval 4 is the California heat wave of 2006:

Interval 1: 2013-05-30_00:00 through 2013-06-16_00:00 UTC
Interval 2: 2013-06-29_00:00 through 2013-07-16_00:00 UTC
Interval 3: 2006-05-30_00:00 through 2006-06-16_00:00 UTC
Interval 4: 2006-07-14_00:00 through 2006-08-01_00:00 UTC

Across the different summer periods in May–August (MJJA) 2006–2013 (not shown), the model produced generally similar spatial patterns of air temperature in each region and relatively consistent geographical locations of UHI and UCI, as will be discussed next in this section. Since one goal of the modeling was to assist in siting fixed weather monitors and designing mobile-observations routes, the repeatability of these spatial patterns, i.e., areas with consistent hot or cool islands, could facilitate this task.

To provide a geographical context, Figures 4 and 5 depict the modeled 2-m air temperature field (shown as degree-hours, °C·h, to capture the cumulative rather than instantaneous signal) for interval 1 in the two 500-m domains. The purpose of these two figures is not to present the quantitative data per se but, rather, to give an idea where certain temperature patterns occur relative to the urban and geographical features in the area and also to show how the model captures the cooling and heating effects of certain land covers such as parks, large roof areas, and roadways. In subsequent graphs in

Figures 6 and 7 the corresponding domains are shown again but without the background for easier visualization and assessment of the temperature-field characteristics. The contours are color-coded ranging from low to high degree-hours (blue to red).

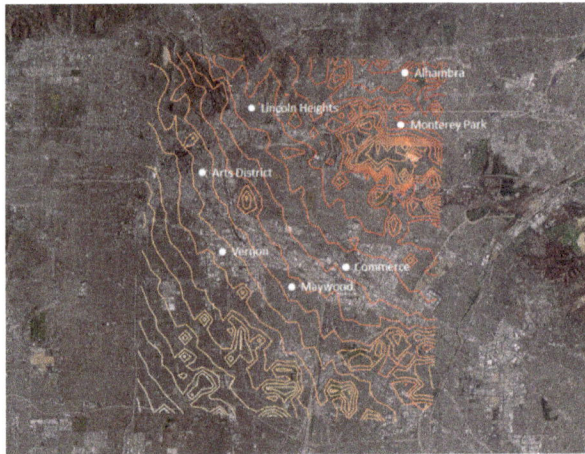

Figure 4. Total DH for interval 1 in the downtown area 500-m domain. The unlabeled contours are only meant to show the spatial pattern of the temperature field.

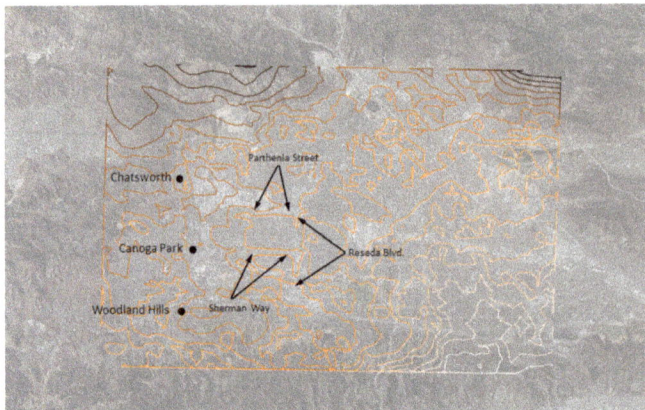

Figure 5. Total DH for interval 1 in the San Fernando Valley 500-m domain. The unlabeled contours are only meant to show the spatial pattern of the temperature field.

In the downtown area (Figure 6), the modeled air-temperature field captures the on-shore warming tendency (in the southwest-to-northeast direction) and also the consistently warmer commercial and industrial areas southeast of the Arts District, including the cities of Vernon, Maywood, and Commerce. In the northeastern parts of this domain, the model shows warmer urban areas in the region between the cities of Monterey Park and Alhambra and between Lincoln Heights and South Glendale. The temperature field also captures many areas of localized cool islands, mostly associated with open and/or green spaces and areas with higher urban albedo. Figure 6 also shows that the effects of on-shore warming are dominant and that the localized UHI/UCI effects are superimposed on this signal [3].

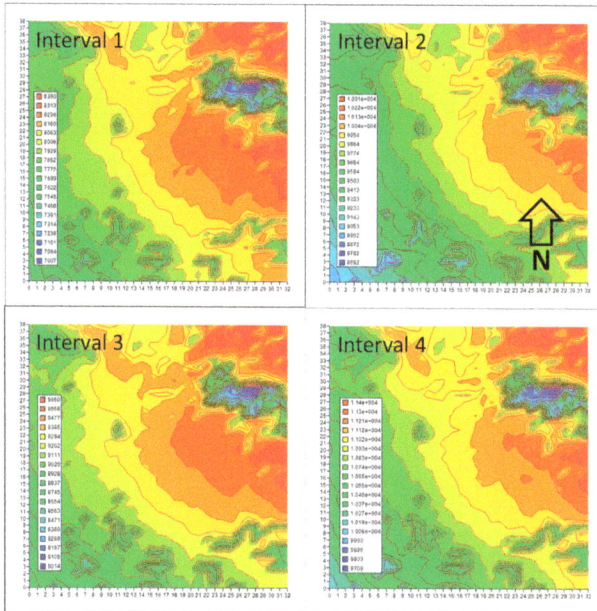

Figure 6. Modeled total degree-hours (DH) for intervals 1–4 in the downtown area. The derived interval-average temperature (computed as DH/hour), blue to red, is as follows: 17.2–20.6 °C for interval 1; 21.3–25.3 °C for interval 2; 19.6–23.7 °C for interval 3; and 22.5–27.9 °C for interval 4.

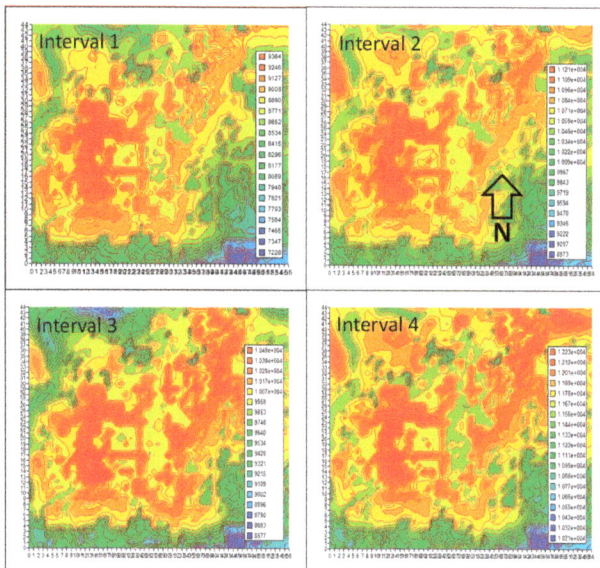

Figure 7. Modeled total degree-hours (DH) for intervals 1–4 in the San Fernando Valley domain. The derived interval-average temperature (computed as DH/hour), blue to red, is as follows: 17.7–22.9 °C for interval 1; 22.0–27.5 °C for interval 2; 21.0–25.7 °C for interval 3; and 23.6–30.0 °C for interval 4.

In the San Fernando Valley domain (Figure 7), sea-breeze effects are practically non-existent and temperature is influenced mainly by topography and variations in land use and surface properties. The graphs in Figure 7 show that the model predicts higher temperatures in the industrial and commercial areas from near Chatsworth in the north, to Canoga Park, and Woodland Hills in the south. Higher model temperatures are also seen along the major roadways, including, for example Sherman Way and Parthenia Street (running west to east) and Reseda Blvd. (running north to south). Cool islands in this domain are associated with areas of higher vegetation cover.

In Figures 6 and 7, the graphs show the 2-m model air-temperature field expressed as degree-hour totals (DH), in units of °C·h, over each of the four intervals identified at the beginning of this section. A temperature equivalent is also provided as DH/hour, that is, the interval-average temperature. Since one goal of the modeling was to assist in siting the fixed weather stations and in designing the mobile-observation routes, the model DH in these figures were plotted with the same number of levels (across the data range) for all periods. This serves to show that the model consistently defines certain areas as heat or cool islands (at fine scales) and that this information can be useful in selecting the monitoring sites and the transect routes. Plotting at a different number of levels will make it difficult to visually identify these consistent features.

It is of interest to evaluate how the gradients (spacing) in DH (or average temperature) vary across the four periods and the two domains. In the downtown area (Figure 6), the gradients are 77 DH in interval 1, 90 DH in interval 2, 92 DH in interval 3, and 94 DH in interval 4. In the San Fernando Valley area (Figure 7), the DH gradients are 119, 124, 106, and 110 DH in intervals 1, 2, 3 and 4, respectively. Converting these DH values into temperature (calculated as DH/hour), we obtain the following averaged temperature gradients (spacing) per contour level; in the downtown area: 0.189 °C, 0.220 °C, 0.225 °C, and 0.230 °C in intervals 1–4, respectively; in the San Fernando Valley: 0.291 °C, 0.303 °C, 0.259 °C, and 0.269 °C in intervals 1–4, respectively.

Based on model results, several transect routes were designed to pass through areas of varying temperature (UHI/UCI spots) and land use properties (albedo and/or canopy cover). Figure 8 is a composite of 15 transects showing the routes in the downtown area and the San Fernando Valley domains. Observations from these mobile transects and comparison with model results are discussed next in Sections 3.2 and 3.3.

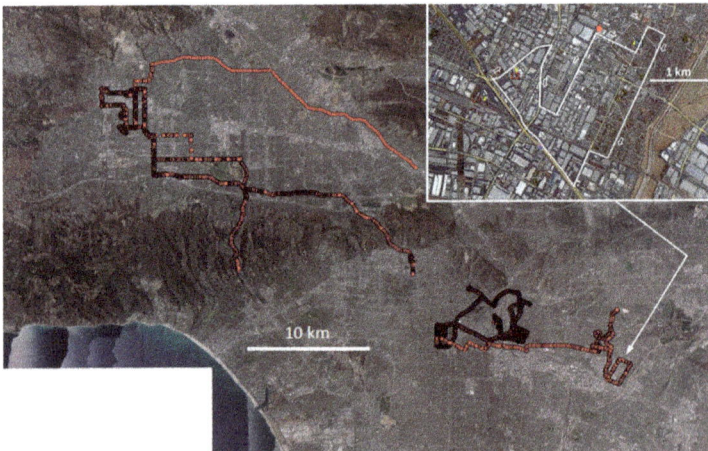

Figure 8. Superimposed routes of 15 transects. Inset: a randomly-selected transect detail.

3.2. Model Performance Evaluation and Validation against Mobile Observations

To demonstrate that the model correctly captures the features of urban heat (i.e., magnitudes and variations in air temperature), model performance evaluation for the simulations listed in Section 3.1 was carried out. This included evaluating the regional, seasonal runs for the coarse domains, as well as the fine-scale transect-specific simulations. For the seasonal (summer) coarse-scale WRF simulations (3-km resolution), performance indicators are shown in Table 1, averaged over the seasonal model runs.

Table 1. Model performance indicators for seasonal simulations: bias, error, RMSE (root mean square error), and IOA (index of agreement). Model results were compared to observations from fixed monitor networks.

Performance Indicator	Units	Computed Value
Wind speed bias	m/s	0.08
Wind speed RMSE	m/s	0.71
Wind speed IOA	-	0.80
Wind direction bias	°	0.92
Wind direction error	°	37.3
Temperature bias	°C	0.60
Temperature error	°C	2.10
Temperature IOA	-	0.95
Humidity bias	g/kg	−0.80
Humidity error	g/kg	1.41
Humidity IOA	-	0.71

The computed values for the metrics (in Table 1) compare favorably to the modeling community's recommended benchmarks [27]. For the transect-specific simulations (500-m resolution), modeled air temperature was compared to observations from mobile transects at the coincident times (sub-hourly intervals). The goal was to ascertain successful model capture of the micrometeorological variations in the urban areas and to validate the modified WRF-urban model against the mobile observations.

While the transect-specific model runs were initiated a week ahead of the actual transect time and were continued for 2 days past that, model performance evaluation for these runs was carried out only and specifically at the actual transect time. The statistics reported in Table 2 compare along-transect model temperature to along-transect observations and demonstrate a satisfactory performance for the approach adopted in this study. In Table 2, MAE is mean absolute error (°C) and RMSE is root mean square error (°C). Both MAE and RMSE are significantly better than the modeling-community-recommended performance benchmarks of MAE $\leq$ 2 °C and RMSE $\leq$ 2 °C [27].

In this paper, we describe a methodology that researchers could follow to model regions and urban areas of interest depending on local data availability. In this study, the Los Angeles region is a data-rich geographical area for which it is possible to develop sufficient fine-resolution, detailed characterizations of surface properties, urban morphology, canopy-layer properties, and microclimate input for use in urbanized WRF model.

For urban areas that lack detailed characterizations, it is possible to develop input to the WRF-urban model based on indirect approaches such as the local climate zones (LCZ) classification based on N/WUDAPT [24], or by developing crosswalks among various datasets—e.g., USGS L-II–L-IV [21], NLCD 2011 [23], and other LCZs [28].

In the rest of this section, we briefly assess the improvements in model performance when using the detailed urban surface characterizations and the updated/customized WRF-urban of Taha [3,7] relative to standard, non-urban WRF. Thus three "configurations" are compared in Table 3: (1) the standard non-urban WRF with default lookup values; (2) the standard WRF-urban with lookup values; and (3) the modified WRF-urban including parameterizations, triggers, and surface input (non-lookup) as modified and customized by Taha [3,7] and discussed in Section 2.2.

Table 2. Model performance metrics against observational mobile-transect temperature.

Transect	MAE (°C)	RMSE (°C)
2016_04_22 (west basin)	1.15	1.33
2017_06_14 Part 1 (west basin)	0.88	1.00
2017_06_14 Part 2 (west basin)	0.61	0.76
2017_06_14 Part 3 (west basin)	0.80	0.94
2017_06_14 Part 4 (west basin)	0.70	0.86
2017_06_21 (San Fernando)	1.73	2.00
2017_07_27 day Part 1 (San Fernando)	0.97	1.20
2017_07_27 day Part 2 (San Fernando)	0.92	1.10
2017_07_27 night Part 1 (San Fernando)	0.55	0.68
2017_07_27 night Part 2 (San Fernando)	0.85	1.00
2017_08_28 day Part 1 (west basin)	0.48	0.60
2017_08_28 day Part 2 (west basin)	0.71	0.94
2017_08_28 night Part 1 (west basin)	1.00	1.10
2017_08_28 night Part 2 (west basin)	0.82	0.92

Table 3. Comparison of model performance using standard and customized WRF-urban model.

	Standard WRF-Urban Relative to Standard Non-Urban WRF		Modified WRF-Urban (Taha [3,7]) Relative to Standard Non-Urban WRF	
Average over 2013, 2014, 2015	Change in mean bias (°C)	Change in mean error (°C)	Change in mean bias (°C)	Change in mean error (°C)
1–15 June	−40%	−4%	−98%	−28%
16–30 June	−25%	−6%	−72%	−31%
1–15 July	−68%	−4%	−126%	−15%
16–31 July	−46%	−7%	−97%	−26%
1–15 August	−40%	−7%	−87%	−26%
16–31 August	−33%	−6%	−84%	−25%
1–20 September	−36%	−5%	−82%	−25%

In Table 3, we provide metrics for 2-m air temperature, as an example to show how performance improves from configuration 1 to configuration 2 to configuration 3. In the table, reductions in mean bias and mean error of 2-m air-temperature are averaged for each 2-week time interval over 3 years (2013, 2014, and 2015) as identified in the first column. The results show that the approach of Taha [3,7] with improved fine-resolution data and parameterizations (last two columns) reduces mean bias by two- to three-fold and mean error by three- to seven-fold compared to the standard urban version of WRF (columns 2 and 3) that uses lookup values and has no specific model customizations.

3.3. Correlations of Observed Temperature with Albedo and Canopy Cover

Following model performance evaluation, correlations between observed temperature and surface physical properties were evaluated. In this case, two surface properties of interest were examined: neighborhood-scale albedo and vegetation canopy cover. The relationships between observed air temperature (dependent variable, or "predictand") and either albedo and/or canopy cover (independent variables, or "predictors") were examined in three manners: (1) simple linear regression, (2) multiple regression, and (3) CART analysis.

To develop these correlations, weighted albedo and canopy cover corresponding to each mobile-transect observation point were calculated based on Cressman-type analysis, where

$$W_{p,i} = \frac{R^2 - d_{p,i}^2}{R^2 + d_{p,i}^2} \tag{1}$$

for $d_{p,i} \leq R$, and $W_{p,i} = 0$ for $d_{p,i} > R$.

In Equation (1), $W_{p,i}$ is the weighting factor for the quantity of interest (e.g., albedo or canopy cover) at a model grid point, i, relative to a transect observation point, p; R is a pre-determined radius of influence; and $d_{p,i}$ is the distance from the transect observation point p to the grid point i. The weighted property, P_{wv} (i.e., weighted albedo or canopy cover) per each transect observation point is simply:

$$P_{wv,p} = \frac{\sum_i W_{p,i} V_i}{\sum_i W_{p,i}} \tag{2}$$

where V is the property in question (albedo or canopy cover) at the grid point i.

Thus, the albedo and canopy-cover values that we use to predict air temperature are not just those along the street, i.e., in the urban canyon. The values used are at neighborhood scale, i.e., computed at 500-m radius or larger if wind speed is higher than a certain threshold, as discussed later in Section 3.3.1. For each point of observation along a mobile transect, the area-wide albedo and/or canopy cover within a 500-m radius (or larger) of that point are correlated to the observed temperature from the transect. Thus, the temperature observed along a transect is influenced by albedo of all surfaces within the radius of influence, including albedo of roofs, pavements, vegetation, non-built surfaces, water, other ground cover, etc. The observed transect temperature was also correlated to area-wide vegetation-canopy cover within the radius of influence from each observation point.

Unless the air flow is perfectly aligned with the length of the road, the road's albedo and surface temperature will have little effect on the localized air temperature at 2 m because the residence time over the road will be very small (order seconds). Millstein and Levinson [29] found that air has to flow over a surface for about 1 km to get a noticeable change in 2 m air temperature. Furthermore, the transects in this study were conducted within uniform land uses in each segment with little variation in building geometry and heights. Thus, these effects tend to cancel out, leaving albedo and canopy cover as the main predictor variables.

Finally, in terms of urban block shape, our CART analysis (discussed in Section 3.3) shows that roughness length (a surrogate to urban geometry) has much smaller effects than area-wide albedo and/or vegetation cover, in part because the areas where we conducted the transects were relatively uniform with no significant contrasts in building geometry or heights. This cancels out the effects of changes in geometry on the flow (again, leaving albedo and canopy cover as the two main predictors to air temperature).

3.3.1. Simple Linear Regression

In Figure 9a–i, observed air temperature (°C) from mobile transects is plotted on the vertical axis against grid-level (neighborhood-scale) albedo (ALB) or canopy cover (VEG), where ALB and VEG are each computed via Equations (1) and (2). This analysis is for the downtown area defined by the rectangle on the right in Figure 1. For the San Fernando Valley (domain defined by white rectangle on the left in Figure 1), the observed temperature in Figure 10a–e is plotted against canopy cover only (as predictor) because albedo has a smaller variability in this domain.

The analysis in Figures 9 and 10 provides information for each transect segment including dates, slopes computed as temperature change per 0.1 increase in albedo or canopy cover (to normalize and facilitate inter-comparisons of effects across various transects), corresponding p-values (probability values), wind speed (WSP), and solar radiation (SOLRAD) at the time of the transect. The latter two were obtained from the NOAA/MADIS mesonet monitors closest to each mobile transect at the time it was conducted. In this analysis, a significance level of 0.05 was selected and, as such, a p-value < 0.05 represents a statistically significant correlation between observed temperature (from the transect) and surface properties (albedo and canopy cover).

The analysis in Figures 9 and 10 was based on the Cressman-weighting scheme discussed above and a radius of influence of 1 km. The fifteen transects are identified as TR01 through TR15 and if a transect is made up of parts carried out at different times, these will be indicated as P1, P2, etc. One transect (TR14) is not shown in this analysis because of missing data.

Table 4 summarizes the main takeaways from the analysis in Figures 9 and 10. It shows the response of observed air temperature (°C) to a 0.1 increase in neighborhood-scale albedo, symbolically written as $\Delta T/(0.1\Delta a)$, or in canopy cover, written as $\Delta T/(0.1\Delta \eta)$, in columns 2 and 6, along with the corresponding *p*-value for each transect (in columns 3 and 7, respectively). Again, the reason for selecting a denominator of 0.1 is simply to normalize the temperature sensitivity and facilitate inter-comparisons of the effects across various transects. Also, an increase of 0.1 in neighborhood-scale albedo and/or canopy cover is one assumption often made as a mitigation scenario in UHI studies. Hence, it is used here as an indicator to what the real-world impact might be on air temperature.

Except for two entries (contribution of canopy cover to air temperature in transect TR01 and contribution of albedo to air temperature in transect TR02, as seen in columns 7 and 3, respectively), all other entries are statistically significant. For these two transects, the *p*-values suggest that in TR01, albedo is the main driver of air temperature and in TR02, canopy cover is the main driver.

In addition, all correlations are negative (i.e., when albedo and/or canopy cover increase, temperature decreases) except for transects TR02 (for canopy cover, column 6) and TR03 (for albedo, column 2). As will be discussed next, these two transects were among those carried out during periods of higher wind speeds, which can weaken the correlations. In transects TR04 and TR05, the large temperature response (sensitivity) to albedo change is likely caused by the extensive freeway and roof cover in these areas (>95%).

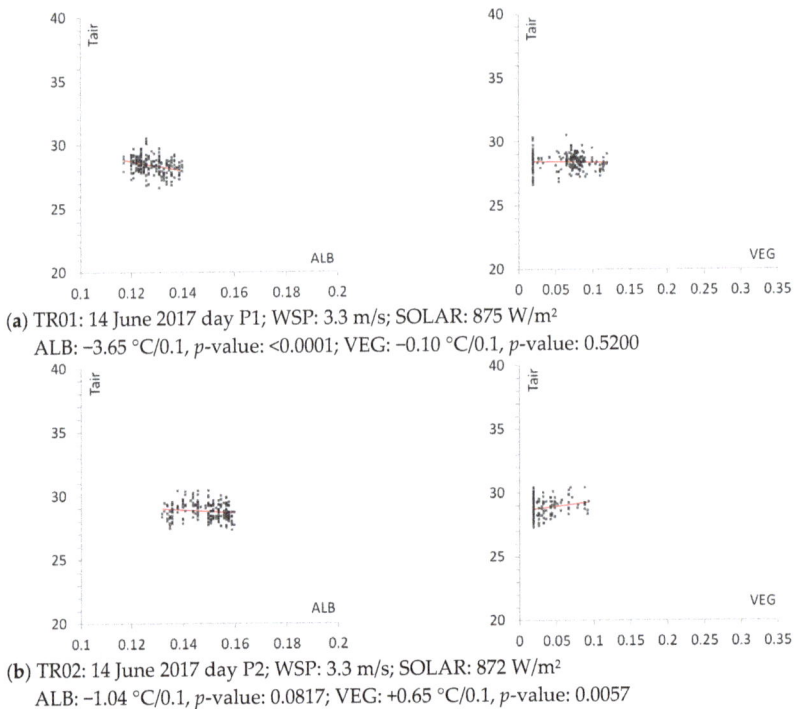

(a) TR01: 14 June 2017 day P1; WSP: 3.3 m/s; SOLAR: 875 W/m²
ALB: −3.65 °C/0.1, *p*-value: <0.0001; VEG: −0.10 °C/0.1, *p*-value: 0.5200

(b) TR02: 14 June 2017 day P2; WSP: 3.3 m/s; SOLAR: 872 W/m²
ALB: −1.04 °C/0.1, *p*-value: 0.0817; VEG: +0.65 °C/0.1, *p*-value: 0.0057

Figure 9. *Cont.*

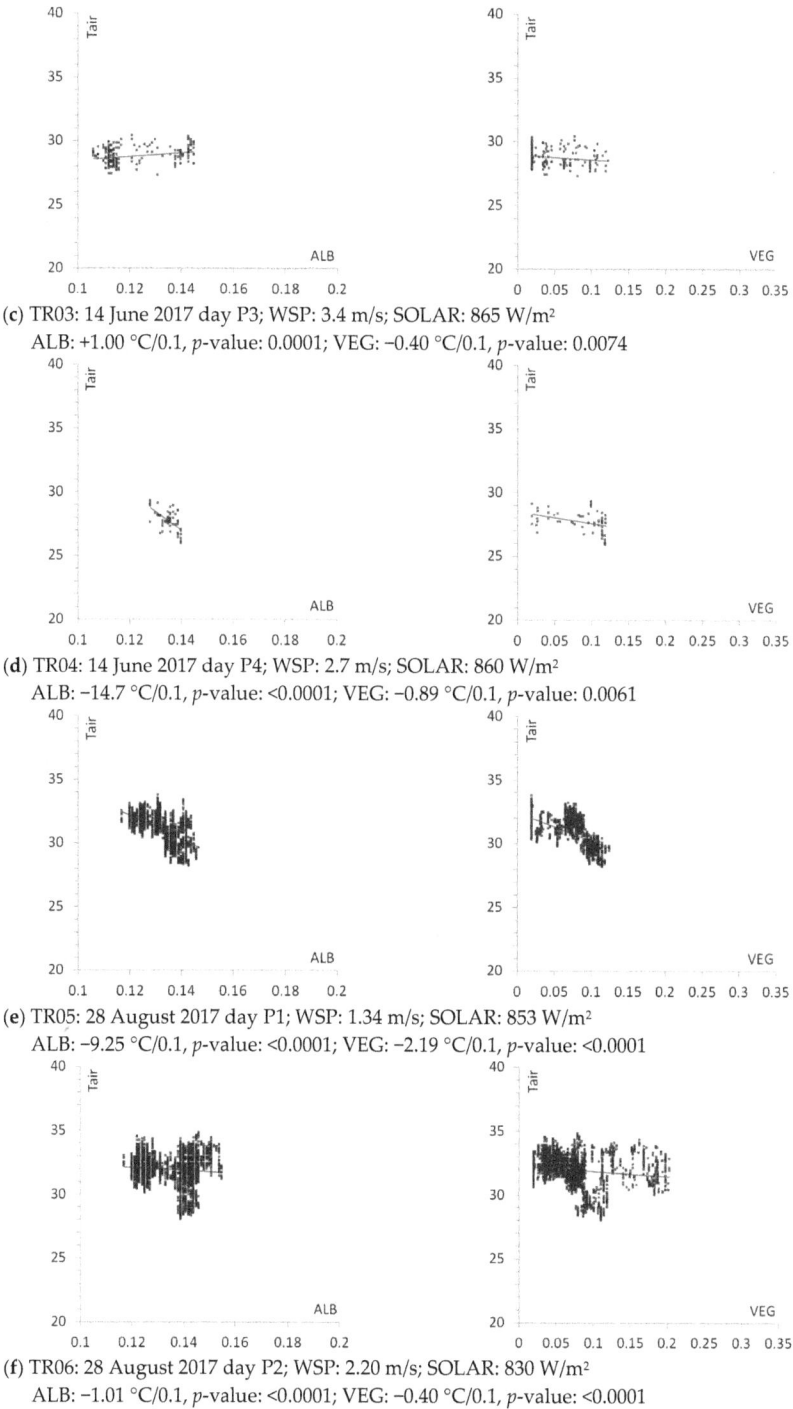

(**c**) TR03: 14 June 2017 day P3; WSP: 3.4 m/s; SOLAR: 865 W/m^2
ALB: +1.00 °C/0.1, *p*-value: 0.0001; VEG: −0.40 °C/0.1, *p*-value: 0.0074

(**d**) TR04: 14 June 2017 day P4; WSP: 2.7 m/s; SOLAR: 860 W/m^2
ALB: −14.7 °C/0.1, *p*-value: <0.0001; VEG: −0.89 °C/0.1, *p*-value: 0.0061

(**e**) TR05: 28 August 2017 day P1; WSP: 1.34 m/s; SOLAR: 853 W/m^2
ALB: −9.25 °C/0.1, *p*-value: <0.0001; VEG: −2.19 °C/0.1, *p*-value: <0.0001

(**f**) TR06: 28 August 2017 day P2; WSP: 2.20 m/s; SOLAR: 830 W/m^2
ALB: −1.01 °C/0.1, *p*-value: <0.0001; VEG: −0.40 °C/0.1, *p*-value: <0.0001

Figure 9. *Cont.*

(**g**) TR07: 28 August 2017 night P1; WSP: 1.10 m/s; SOLAR: 0 W/m²
ALB: −4.00 °C/0.1, *p*-value: <0.0001; VEG: −1.61 °C/0.1, *p*-value: <0.0001

(**h**) TR08: 28 August 2017 night P2; WSP: 1.00 m/s; SOLAR: 0 W/m²
ALB: −3.40 °C/0.1, *p*-value: <0.0001; VEG: −1.81 °C/0.1, *p*-value: <0.0001

(**i**) TR13: 22 April 2016 day; WSP: 2.70 m/s; SOLAR: 828 W/m²
ALB: −4.90 °C/0.1, *p*-value: <0.0001; VEG: −0.23 °C/0.1, *p*-value: <0.0001

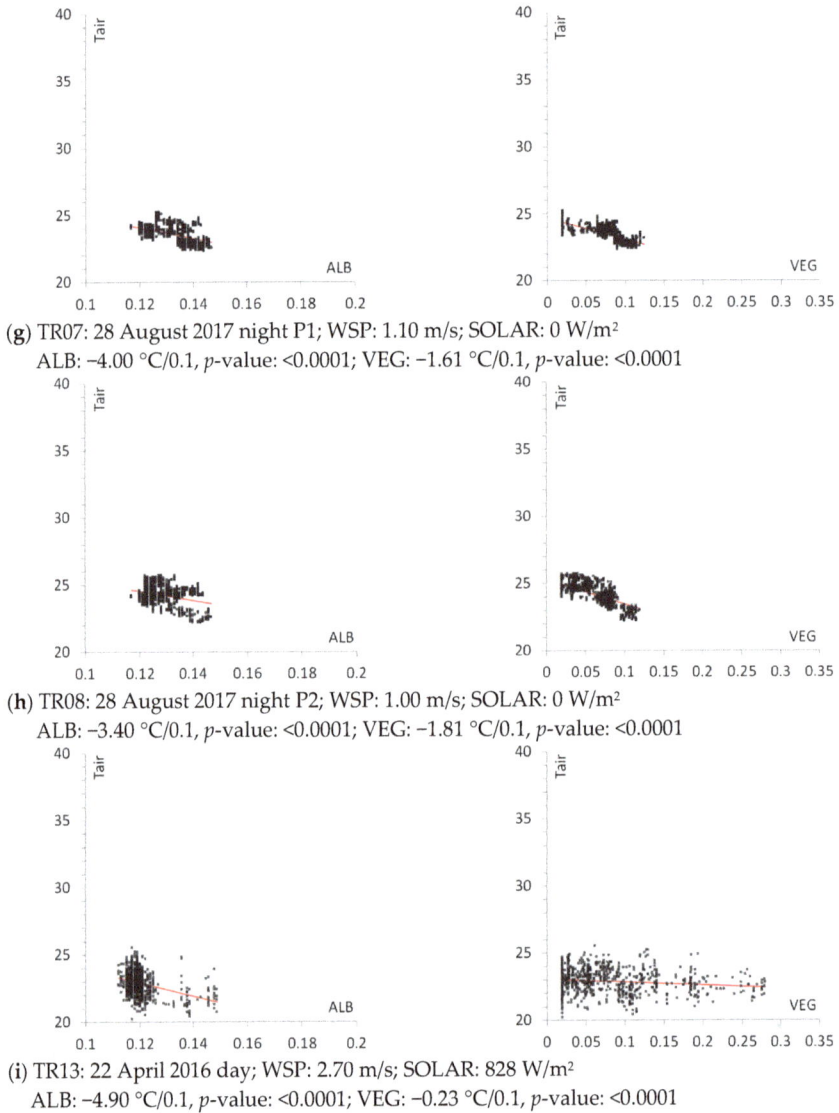

Figure 9. (**a–i**): Observed air temperature (from mobile transects) versus neighborhood-scale albedo and canopy cover in the downtown area. Vertical axis (T_{air}) is air temperature (°C); horizontal axis is albedo (ALB) or canopy cover (VEG), unitless.

(**a**) TR09: 27 July 2017 day P1; WSP: 1.70 m/s; SOLAR: 912 W/m²
VEG: −0.40 °C/0.1, *p*-value: <0.0001

(**b**) TR10: 27 July 2017 day P2 (NW 1.5 fetch); WSP: 4.00 m/s; SOLAR: 803 W/m²
VEG: −0.11 °C/0.1, *p*-value: 0.0105

(**c**) TR11: 27 July 2017 night P1; WSP: 1.30 m/s; SOLAR: 0 W/m²
VEG: −0.20 °C/0.1, *p*-value: <0.0001

(**d**) TR12: 27 July 2017 night P2; WSP: 0 m/s; SOLAR: 0 W/m²
VEG: −1.06 °C/0.1, *p*-value: <0.0001

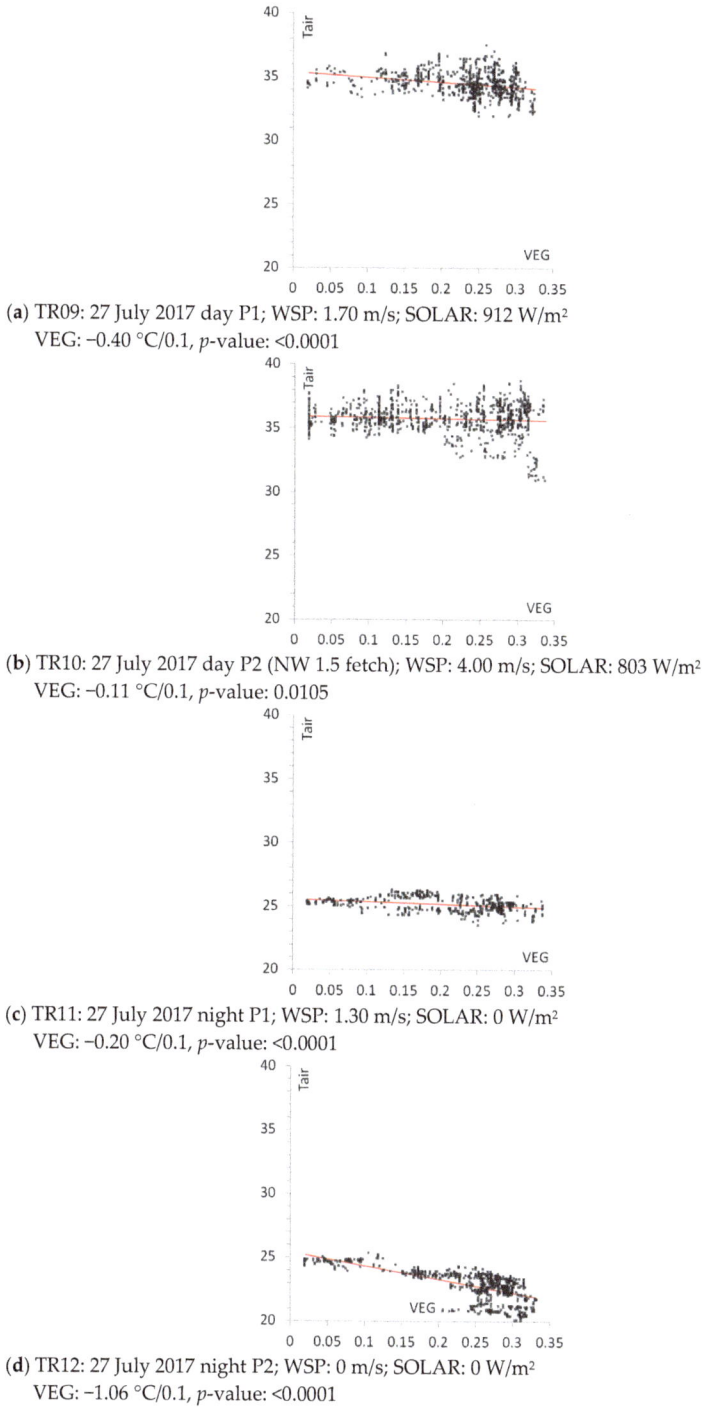

Figure 10. *Cont.*

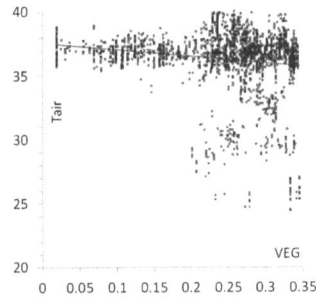

(e) TR15: 21 June 2017 day; WSP: 3.3 m/s; SOLAR: 721 W/m²
VEG: −0.53 °C/0.1, *p*-value: <0.0001

Figure 10. (a–e): Observed air temperature (from mobile transects) versus neighborhood-scale canopy cover in the San Fernando Valley domain. Vertical axis (T_{air}) is air temperature (°C); horizontal axis is canopy cover (VEG), unitless.

Table 4. Summary of observed transect temperature response to changes in neighborhood scale (500 m) albedo and/or canopy cover.

1	2	3	4	5	6	7	8	9
		Albedo Effects				Canopy-Cover Effects		
Transect	$\Delta T/(0.1\Delta a)$	*p*-value	Actual Range (*a*)	ΔT Bounded	$\Delta T/(0.1\Delta\eta)$	*p*-value	Actual Range (*η*)	ΔT Bounded
TR01	−3.65	<0.0001	0.023	−0.84	−0.10	0.5300	0.100	−0.10
TR02	−1.04	0.0817	0.021	−0.22	0.65	0.0057	0.054	0.35
TR03	1.00	0.0001	0.017	0.17	−0.40	0.0074	0.070	−0.28
TR04	−14.70	<0.0001	0.012	−1.76	−0.89	0.0061	0.100	−0.89
TR05	−9.24	<0.0001	0.030	−2.77	−2.20	<0.0001	0.105	−2.29
TR06	−1.10	<0.0001	0.038	−0.38	−0.40	<0.0001	0.183	−0.73
TR07	−4.00	<0.0001	0.030	−1.20	−1.61	<0.0001	0.105	−1.69
TR08	−3.40	<0.0001	0.030	−1.02	−1.81	<0.0001	0.100	−1.81
TR13	−4.89	<0.0001	0.037	−1.81	−0.22	<0.0001	0.260	−0.60
TR09					−0.40	<0.0001	0.330	−1.32
TR10					−0.11	0.0105	0.304	−0.33
TR11					−0.20	<0.0001	0.330	−0.66
TR12					−1.06	<0.0001	0.313	−3.31
TR15					−0.53	<0.0001	0.310	−1.76

Columns 4 and 8 show the actual range of albedo and canopy cover, respectively, associated with each specific transect instead of a hypothetical range of 0.1 as used in columns 2 and 6. In other words, the ranges of albedo and canopy cover in columns 4 and 8 are "bounded" by the values encountered in the real world at each of these transects.

Columns 5 and 9 are temperature changes (°C) computed by multiplying the corresponding actual range (from columns 4 or 8, respectively) by the slope given in columns 2 or 6 and dividing by 0.1. Thus the bounded temperature variations in columns 5 and 9 represent the maximum changes that can be expected in each specific transect (per its actual albedo or canopy-cover range) rather than the unbounded values in columns 2 and 6. By doing so, some of the unreasonably large slopes (unbounded) in column 2, e.g., for TR04 and TR05, become much more reasonable when bounded, as in column 5.

Thus, columns 5 and 9 represent the changes in temperature that one can expect as a result of increasing albedo and/or canopy cover by the amounts already encountered in the real world and that are achievable via current practices in building and planning. On the other hand, columns 2 and 6 represent the potential cooling effects that would result from implementations of high-albedo measures (cool roofs and cool pavements) and/or urban forestation, where each would be increased by 0.1.

Next, the correlations between mobile-observed temperature and albedo and/or canopy cover are re-examined but with a smaller radius of influence (<500 m). The goal is to evaluate whether a length-scale effect exists in these correlations. This analysis is done by comparing observations directly to surface properties at grid points, that is, without the Cressman analysis discussed above.

This analysis shows that the correlations between observed temperature and neighborhood-scale albedo and/or canopy cover are negative and statically significant at $R < 500$ m when wind speed is under 3 m/s locally—compare the p-values of rows 1, 3, 5, 7, and 9 in the right half of Table 5 with all the rows in the left half, suggesting weaker correlation between temperature and surface properties within 500 m at wind speeds exceeding 3 m/s. That is, when wind speed exceeds 3 m/s, the correlations either become weaker (for one of the predictors or the other) or statistically insignificant (see TR01, TR02, TR03, TR10, and TR15). For these cases, the analysis was repeated once more by (1) increasing the radius of influence from 500 m to 1 km and (2) restricting the correlations to upwind model points.

Table 5. Correlations between air temperature (°C) (observed from transects) and neighborhood-scale (500 m) albedo and/or canopy cover. Transects with wind direction represent effects from a 1-km upwind fetch.

Transect	$\Delta T/(0.1\Delta\alpha)$	p-value	$\Delta T/(0.1\Delta\eta)$	p-value	wind m/s	Transect	$\Delta T/(0.1\Delta\alpha)$	p-value	$\Delta T/(0.1\Delta\eta)$	p-value	wind m/s
		Wind speed < 3 m/s						Wind speed > 3 m/s			
TR04	−2.14	0.0120	−0.53	0.0017	2.7	TR01	−0.83	0.0500	0.12	0.0370	3.3
TR05	−5.03	<0.0001	−0.49	<0.0001	1.3	TR01-SE	−0.95	0.0003	−0.18	0.0087	
TR06	−1.08	<0.0001	−0.22	<0.0001	2.2	TR02	1.08	0.0007	−0.22	0.1000	3.3
TR07	−2.66	<0.0001	−0.41	<0.0001	1.1	TR02-NW	−0.44	0.0414	−0.11	0.6000	
TR08	−1.96	<0.0001	−0.50	<0.0001	1.0	TR03	1.41	0.0020	−0.19	0.0003	3.4
TR13	−2.35	<0.0001	−0.22	<0.0001	2.7	TR03-NE	1.15	0.0010	−0.30	0.0005	
TR09			−0.18	<0.0001	1.7	TR10			0.08	0.0030	4.0
TR11			−0.13	<0.0001	1.3	TR10-NW			−0.07	0.0800	
TR12			−0.39	<0.0001	0.0	TR15			0.04	0.0330	3.3
						TR15-NE			−0.18	<0.0001	

In this repeated analysis, the correlations improved further, i.e., some positive or weak correlations converted into negative or stronger ones as seen in Table 5 (compare TR01-SE to TR01, TR02-NW to TR02, TR03-NE to TR03, TR10-NW to TR10, and TR15-NE to TR10). In these cases, the p-values and slopes improved after increasing the upwind fetch (from 500 m to 1 km) and including only grid points within the wind approach direction. This also explains why these adjusted correlations are better than the corresponding ones in Table 4 (in some cases) for one or the other predictors, since the correlations in Table 4 were not limited to upwind points. For example, comparing TR01-SE from Table 5 to TR01 from Table 4, we can see that whereas the albedo-temperature correlation becomes a little weaker, the canopy-temperature correlation becomes much more significant (p-value changes from 0.53 to 0.0087). Comparing TR02-NW (Table 5) to TR02 (Table 4) shows a slight improvement in the significance of the albedo-temperature correlation and also a change in sign (from 0.65 to −0.11) for canopy-temperature correlation, albeit at a larger p-value. In this case, therefore, the temperature is driven by the albedo effect mainly. For TR03-NE there are mixed effects: whilst TR03-NE is improved relative to TR03 (in Table 5), it is not as good as the correlation for TR03 in Table 4 (this transect appears to be an exception). Finally, for TR10-NW and TR15-NW (in Table 5 relative to Table 4), the correlations do not change much.

Thus the results from this analysis suggest that when wind speed is less than 3 m/s, the observed temperature is influenced by the physical properties of the immediate surroundings (less than 500 m radius). At wind speeds exceeding 3 m/s, temperature is influenced by surface properties in a longer fetch (~1 km) in the upwind direction. Physically, this is because advective effects are smaller than the localized convective effects when wind speed is lower. At higher wind speed, in this case >3 m/s, temperature is influenced more by advective than localized effects. To provide context, the monthly average wind speeds in the study areas (from climatology) for January–December are 3.7, 3.5, 3.3, 3.1, 3.1, 2.6, 2.3, 2.2, 2.6, 2.6, 3.1, and 4.0 m/s, respectively. Thus the 3.3–4.0 m/s winds that were observed during some of the transects are relatively uncommon in this area and time of year, i.e., July and August, where 2.2–2.6 m/s is a more typical range. Of note, while there likely is a smooth transition around the 3-m/s threshold, we do not have observational values slightly under or over 3 m/s that can be used to evaluate this transition. Thus the suggested threshold of 3 m/s, while not a hard cutoff, is the most representative value that we determined based on the statistical analysis discussed above.

3.3.2. Multiple Regression

Multiple regression was carried out for albedo and canopy cover as predictors to observed air temperature from the transects. This analysis applies only to the downtown area since the San Fernando Valley analysis involved only variations in one predictor, canopy cover.

Here, the form of the correlation is given by Equation (3) where a is albedo and η is canopy cover. The coefficients ($C1$, $C2$, $C3$) and corresponding p-values ($p1$, $p2$, $p3$) are given in Tables 6 and 7. Coefficients $C1$, $C2$, and $C3$ are in °C, and the "0.1" denominator simply indicates that the changes in temperature ($C2$ and $C3$) correspond to a 0.1 increase in surface albedo or a 0.1 increase in canopy cover. This was done for reasons explained in Section 3.3.1.

$$T_{\text{air}} = C1 + \frac{C2}{0.1}\, a + \frac{C3}{0.1}\, \eta \tag{3}$$

Table 6. Equation (3) applied to a 1-km Cressman-type analysis of air temperature correlation to albedo and canopy cover. Coefficients *C2* and *C3* are, respectively, the responses of observed air temperature to a 0.1 increase in neighborhood-scale albedo and canopy cover.

Transect	C1 (°C)	p1	C2 (°C)	p2	C3 (°C)	p3
TR01	33.25	<0.0001	−3.70	<0.0001	−0.15	0.3350
TR02	31.65	<0.0001	−1.81	0.1204	−1.08	0.0154
TR03	29.42	<0.0001	−0.15	0.935	−0.75	0.0086
TR04	47.18	<0.0001	−13.95	<0.0001	−0.71	0.0052
TR05	43.17	<0.0001	−8.17	<0.0001	−1.97	<0.0001
TR06	33.25	<0.0001	−0.78	<0.0001	−0.36	<0.0001
TR07	27.83	<0.0001	−2.42	<0.0001	−1.47	<0.0001
TR08	30.04	<0.0001	−3.71	<0.0001	−1.85	<0.0001
TR13	28.52	<0.0001	−4.63	<0.0001	−0.10	0.0500

Table 7. Equation (3) applied to a 500-m analysis of air temperature correlation to albedo and canopy cover without weighting (without and with correction for wind speed and direction). Coefficients *C2* and *C3* are, respectively, the responses of observed air temperature to a 0.1 increase in neighborhood-scale albedo and canopy cover.

Transect	C1 (°C)	p1	C2 (°C)	p2	C3 (°C)	p3
TR01	29.28	<0.0001	−0.765	0.050	+0.119	0.0266
TR01-SE	29.59	<0.0001	−0.881	0.010	−0.030	0.7000
TR02	27.37	<0.0001	+1.087	0.008	−0.148	0.2660
TR02-NW	29.56	<0.0001	−0.459	0.036	−0.161	0.5000
TR03	29.94	<0.0001	−0.702	0.100	−0.207	0.0005
TR04	28.66	<0.0001	−0.450	0.055	−0.417	0.1600
TR05	37.41	<0.0001	−4.575	<0.0001	−0.238	<0.0001
TR06	33.98	<0.0001	−1.211	<0.0001	−0.353	<0.0001
TR07	25.93	<0.0001	−1.488	<0.0001	−0.334	<0.0001
TR08	27.21	<0.0001	−1.910	<0.0001	−0.491	<0.0001
TR13	25.49	<0.0001	−2.088	<0.0001	−0.120	<0.0001

Following the structure of the discussion above, results from the multiple-regression analysis are presented in two parts; Table 6 summarizes the analysis at a 1-km radius of influence using the Cressman-type weighting discussed earlier, while Table 7 summarizes the 500-m analysis without weighting. The results in Table 6 show that the correlations are overwhelmingly negative (as albedo and canopy cover increase, temperature decreases) and statistically significant except for three situations. These are in transect TR01, where the role of the canopy cover is insignificant, and in transects TR02 and TR03, where the role of albedo is insignificant. This was already discussed in Section 3.3.1, as were the large slopes in TR04 and TR05.

In Table 7, and as introduced in Section 3.3.1, the sign of some correlations change to negative or significance improves (smaller *p*-values) when the radius of influence is increased and wind from a specific direction is accounted for rather than an average from several directions (e.g., TR01-SE and TR02-NW). In other words, at wind speeds exceeding 3 m/s, extending the upwind fetch from 500 m to 1 km and accounting for LULC properties only in the wind-approach direction improves the *p*-values and/or the coefficients for albedo and/or canopy cover changes. This is similar to what was found in Section 3.3.1.

The values listed in Tables 6 and 7 are relative to changes of 0.1 in albedo or vegetation cover and, thus, are unbounded. These can be converted into transect-specific bounded values as was done in Table 4. Two examples are demonstrated here that will be referred to in the following discussion of example CART for transects TR04 and TR13 in Section 3.3.3. The first example is the coefficient *C3* (for canopy cover) in Table 7, corresponding to transect TR04 with a value of -0.417 °C. If this is multiplied by the actual canopy-cover range of 0.1 (from Table 4, TR04 under column 8) and divided by 0.1, we obtain a bounded value of -0.41 °C. The second example is the coefficient *C2* (for albedo) in Table 7, corresponding to transect TR13 (value of -2.088 °C). If this is multiplied by the actual albedo range of 0.037 (Table 4, transect TR13, under column 4), and divided by 0.1, we obtain a bounded value of -0.77 °C.

3.3.3. Classification and Regression Tree (CART)

A classification and regression tree (CART) analysis was also undertaken to assess the interactions among predictors of observed air temperature (from the mobile transects). The purpose of the CART analysis was to identify the main driver (albedo or canopy cover) of air temperature in each case, i.e., in various transect segments, because of variations in land use and surface physical properties. In addition, the roughness length parameter, computed following the approach of MacDonald et al. [30], was included in the following CART examples as an additional predictor to evaluate its role relative to that of albedo and canopy cover.

While the CART analysis was carried out for each of the 15 transects, two examples are presented here. Transect TR04 is a daytime transect at 14:00 Pacific Daylight saving Time (PDT) carried out on 14 June 2017 in the downtown area; TR13 also a daytime (13:00 PDT) transect in the downtown area, was carried out on 22 April 2016.

From the CART analysis of TR04 (Figure 11a), the following can be deduced:

- In this transect segment, canopy cover (top node) is the main splitting variable, i.e., the main driver of air temperature. The lowest temperatures are associated with the highest canopy cover (node 3).
- To estimate the influence of canopy cover on temperature in this sub-segment, we compare terminal nodes 3, 4, and 5. The difference between temperature at node 3 and the weighted temperatures at nodes 4 and 5 is -0.35 °C, which is the contribution of canopy-cover change to air temperature in this sub-segment. This is comparable to the value -0.41 °C computed above, in the last paragraph of Section 3.3.2 (based on the multiple regression in Table 7). The reduction in air temperature is significant, considering that this is a short segment (57 temperature observations).

- Roughness length has minor, secondary effects. Within the range of lower canopy cover (<0.055), roughness length (Zo) can play a role. In this case, the larger Zo (node 5) produces slightly higher air temperatures (0.23 °C warmer on average) during daytime.

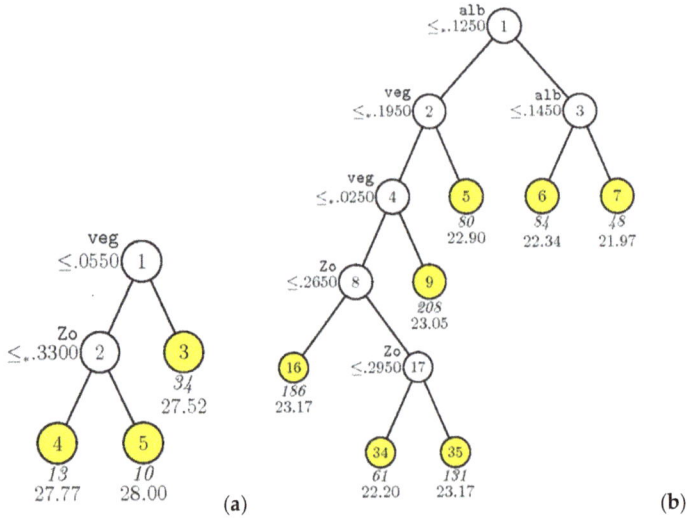

Figure 11. CART for transects (**a**) TR04 and (**b**) TR13.

Transect TR13 is a longer segment with 798 temperature observations. The following can be deduced from this CART analysis (Figure 11b):

- In this transect segment, the most influential variable on air temperature (top splitting node) is albedo.
- Observed temperature is lower where albedo is larger. For neighborhood-scale albedo greater than 0.125, the temperatures are lowest, e.g., compare terminal nodes 6 and 7 (temperatures of 22.34 and 21.97 °C) with nodes 5, 9, 16, 34, and 35 (with higher temperatures of 22.90, 23.05, 23.17, 22.20, and 23.17 °C, respectively).
- Calculating the weighted temperature differences for nodes with albedo higher then 0.125 (6 and 7) versus lower than 0.125 (nodes 5, 9, 16, 34, and 35) shows that the contribution of higher albedo in this transect is to lower air temperature by 0.81 °C, which is significant. This is comparable to the value of −0.77 °C computed in the last paragraph of Section 3.3.2, which was based on multiple regression.
- Furthermore, within each subtree, the effects of increased albedo are evident. For example, comparing terminal nodes 6 and 7 shows that observed temperatures are 0.37 °C lower where albedo is larger than 0.145 relative to where albedo is lower than that.
- In this transect, where albedo is lower than 0.125, canopy cover also has a significant effect. Comparing node 5 (where canopy cover is greater than 0.195) and nodes 9, 16, 34, and 35 (where cover is lower than 0.195), the contribution of canopy cover is to cool the air by 0.12 °C. This effect is smaller than the effects of albedo in this transect.
- The effects of roughness length are secondary and different from one subtree to another. For instance, whereas increased roughness length in node 35 relative to node 34 does show increased temperature, comparing the increased roughness in nodes 34 and 35 relative to node 16, shows the opposite effect.

4. Conclusions

Fine-resolution meteorological modeling with a modified urbanized version of WRF was carried out as basis for siting fixed weather monitors and designing temperature mobile-observation routes. In combination, the modeling and observations were carried out to study urban heat in the Los Angeles region. A total of 15 observational mobile transects were carried out in the summers of 2016 and 2017. The mobile-observation temperatures were compared to model fields and also correlated to surface physical properties focusing on neighborhood-scale albedo and vegetation canopy cover.

Evaluation of coarse-scale regional simulations and fine-scale, transect-specific simulations against observations shows a meteorological-model performance that meets or is better than community-recommended benchmarks. For example, the temperature mean absolute error for transect-specific simulations is 0.86 °C (compare to a recommended threshold of ≤ 2 °C). The performance evaluation demonstrates that the model can be used in siting fixed weather stations and designing mobile-observation routes to characterize urban heat and cool islands.

The model results and observational data analysis (measurements from mobile transects) show that as albedo increases, air temperature decreases during the day, but also at night in smaller extents likely because of lag effects or masking by other variables (co-variance). The increase in canopy cover lowers air temperature during both day and night via combined effects of canopy shading, soil moisture, roughness, and evapotranspiration. The unbounded correlation between temperature and albedo or canopy cover, are both negative, ranging from -1.0 to -9.0 °C per 0.1 increase in neighborhood-scale albedo (or larger) and from -0.1 to -2.2 °C per 0.1 increase in canopy cover, except for a few cases with co-varying influences. While both albedo and canopy cover have significant impacts on air temperature, CART analysis done in this study also shows that, depending on land-use, surface physical properties, and geographical location, that one or the other becomes the main driver of air temperature.

At the 500-m scale, the analysis of observed temperature from mobile transects indicates that the negative correlations between air temperature and surface albedo and/or canopy cover become weaker or statistically insignificant at wind speeds exceeding about 3 m/s. For observations when wind speeds exceeded this threshold, the analysis was repeated to increase the radius of influence (from 500 m to 1 km) and to consider land use and surface physical properties in the upwind fetch only (wind approach direction). This resulted in the correlations becoming stronger (lower p-values and statistically more significant) and, where they were positive, the correlations became negative.

Of relevance to policy recommendations, the results from model, simple and multiple regressions, and CART analyses demonstrate the significant real-world cooling potential of increasing neighborhood-scale albedo and canopy cover. Based on bounded correlations between observed temperature and transect-specific ranges of albedo and canopy cover, the analysis shows that cooling of up to 2.8 °C during the daytime can be achieved by increasing neighborhood-scale (500-m) albedo. Cooling of up to 2.3 °C during the day and up to 3.3 °C at night can be achieved by increasing canopy cover. The changes in albedo and canopy cover that resulted in these amounts of cooling are actual neighborhood-scale variations already encountered in the real world in the Los Angeles area. This ability to cool a neighborhood will have added significance during excessive heat events or future climates when the warmer weather exacerbates urban heat.

Author Contributions: H.T., R.L., A.M., H.G, and G.B-W. conceived and designed the study; H.T. performed the land-use analysis, atmospheric modeling, analysis of simulation results, and statistical analysis of observations; R.L. and S.C. designed and built the air temperature measurement apparatus; A.M. and G.B-W. conducted the mobile transects; H.G. managed the project and coordination with partners; H.T. wrote the paper.

Acknowledgments: This work was funded by the California Energy Commission under Contract No. EPC-14-073. The effort from Lawrence Berkeley National Laboratory was also supported by the Assistant Secretary for Energy Efficiency and Renewable Energy, Office of Building Technology, State, and Community Programs, of the U.S. Department of Energy under Contract No. DE-AC02-05CH11231. We would like to acknowledge the support of the City of Los Angeles' Department of Water and Power Bureau of Street Services; City of Los Angeles' Mayor's Office of Sustainability and Office of Resilience, County of Los Angeles, Los Angeles Unified School

District, and California's Department of Forestry and Fire Protection for providing data and informing selection of study areas. We also acknowledge the Southern California Association of Governments, the City of Los Angeles, and EarthDefine/CAL FIRE for providing high-resolution land-use and land-cover datasets. Tianbo Tang and Joseph Ko (University of Southern California) and Angie Rodriguez (National Autonomous University of Mexico) are acknowledged for assisting in carrying out mobile-transect observations and in the installation of fixed weather monitors.

Conflicts of Interest: The authors declare no conflicts of interest.

References

1. Taha, H. Meteorological, emissions, and air-quality modeling of heat-island mitigation: Recent findings for California, USA. *Int. J. Low Carbon Technol.* **2013**, *10*, 3–14. [CrossRef]

2. Sailor, D.; Shepherd, M.; Sheridan, S.; Tone, B.; Kalkstein, L.; Russell, A.; Vargo, J.; Andersen, T. Improving heat-related health outcomes in an urban environment with science-based policy. *Sustainability* **2016**, *8*, 1015. [CrossRef]

3. Taha, H. Characterization of Urban Heat and Exacerbation: Development of a Heat Island Index for California. *Climate* **2017**, *5*, 59. [CrossRef]

4. Vahmani, P.; Ban-Weiss, G.A. Impact of remotely sensed albedo and vegetation fraction on simulation of urban climate in WRF-urban canopy model: A case study of the urban heat island in Los Angeles. *J. Geophys. Res.* **2016**, *121*, 1511–1531. [CrossRef]

5. Mavrakou, T.; Polydoros, A.; Cartalis, C.; Santamouris, M. Recognition of thermal hot and cold spots in urban areas in support of mitigation plans to counteract overheating: Application for Athens. *Climate* **2018**, *6*, 16. [CrossRef]

6. Vahmani, P.; Sun, F.; Hall, A.; Ban-Weiss, G.A. Investigating the climate impacts of urbanization and the potential for cool roofs to counter future climate change in Los Angeles. *Environ. Res. Lett.* **2016**, *11*. [CrossRef]

7. Taha, H. Meso-urban meteorological and photochemical modeling of heat island mitigation. *Atmos. Environ.* **2008**, *42*, 8795–8809. [CrossRef]

8. Alexander, P.G.; Mills, G. Local climate classification and Dublin's urban heat island. *Atmosphere* **2014**, *5*, 755–774. [CrossRef]

9. Georgakis, C.; Santamouris, M. Determination of the surface and canopy urban heat island in Athens central zone using advanced monitoring. *Climate* **2017**, *5*, 97. [CrossRef]

10. Mohegh, A.; Rosado, P.; Jin, L.; Millstein, D.; Levinson, R.; Ban-Weiss, G.A. Modeling the climate impacts of deploying solar reflective cool pavements in California cities. *J. Geophys. Res.* **2017**. [CrossRef]

11. Zhang, J.; Zhang, K.; Liu, J.; Ban-Weiss, G.A. Revisiting the climate impacts of cool roofs around the globe using an earth system model. *Environ. Res. Lett.* **2016**, *11*. [CrossRef]

12. Taleghani, M.; Sailor, D.; Ban-Weiss, G.A. Micrometeorological simulations to predict the impacts of heat mitigation strategies on pedestrian thermal comfort in a Los Angeles neighborhood. *Environ. Res. Lett.* **2016**, *11*, 1–12. [CrossRef]

13. Jin, H.; Cui, P.; Wong, N.H.; Ignatius, M. Assessing the effects of urban morphology parameters on microclimate in Singapore to control the urban heat island effect. *Sustainability* **2018**, *10*, 206. [CrossRef]

14. Qiu, G.Y.; Zou, Z.; Li, X.; Li, H.; Guo, Q.; Yan, C.; Tan, S. Experimental studies on the effects of green space and evapotranspiration on urban heat island in a subtropical megacity in China. *Habitat Int.* **2017**, *68*, 30–42. [CrossRef]

15. Tsin, P.K.; Knudby, A.; Krayenhoff, E.S.; Ho, H.C.; Brauer, M.; Henderson, S.B. Microscale mobile monitoring of urban air temperature. *Urban Clim.* **2016**, *18*, 58–72. [CrossRef]

16. Jonsson, P. Vegetation as an urban climate control in the subtropical city of Gaborone, Botswana. *Int. J. Climatol.* **2004**, *24*, 1307–1322. [CrossRef]

17. Sun, C.-Y.; Brazel, A.J.; Chow, W.T.L.; Hedquist, B.C.; Prashad, L. Desert heat sialnd study by mobile transect and remote sensing techniques. *Theor. Appl. Climatol.* **2009**, *98*, 323–335. [CrossRef]

18. Ellis, K.N.; Hathaway, L.; Mason, R.; Howe, D.A.; Epps, T.H.; Brown, V.M. Summer temperature variability across four urban neighborhoods in Knoxville, Tennessee, USA. *Theor. Appl. Climatol.* **2017**. [CrossRef]

19. Taha, H. Cool cities: Counteracting potential climate change and its health impacts. In *Current Climate Change Report*; Invited Paper; Springer International Publishing: Cham Switzerland, 2015; Volume 1, pp. 163–175.

20. Taha, H.; Freed, T. Creating and Mapping an Urban Heat Island Index for California. Report Prepared by Altostratus Inc. for the California Environmental Protection Agency (Cal/EPA); 2015. Available online: https://calepa.ca.gov/wp-content/uploads/sites/62/2016/10/UrbanHeat-Report-Report.pdf (accessed on 5 January 2018).

21. Anderson, J.R.; Hardy, E.E.; Roach, J.T.; Witmer, R.E. *A Land Use and Land Cover Classification System for Use with Remote Sensor Data*; USGS Professional Paper 964; U.S. Government Printing Office: Washington, DC, USA, 2001.

22. Ban-Weiss, G.A.; Woods, J.; Levinson, R. Using remote sensing to quantify albedo of roofs in seven California cities, Part 1: Methods. *Sol. Energy* **2015**, *115*, 777–790. [CrossRef]

23. Multi-Resolution Land-Characteristics Consortium (MRLC), 2011. National Land Cover Databases. Available online: www.mrlc.gov (accessed on 15 June 2017).

24. Ching, J.; Brown, M.; Burian, S.; Chen, F.; Cionco, R.; Hanna, A.; Hultgren, T.; McPherson, T.; Sailor, D.; Taha, H.; et al. National urban database and access portal tool, NUDAPT. *Bull. Am. Meteorol. Soc.* **2009**. [CrossRef]

25. Pleim, J.; Xiu, A.; Finkelstein, P.; Otte, T. A coupled land-surface and dry deposition model and comparison to field measurements of surface heat, moisture, and ozone fluxes. *Water Air Soil Pollut.* **2001**, *1*, 243–252. [CrossRef]

26. Kusaka, H.; Kondo, H.; Kikegawa, Y.; Kimura, F. A simple single-layer urban canopy model for atmospheric models: Comparison with multi-layer and slab models. *Bound.-Lay. Meteorol.* **2001**, *101*, 329–358. [CrossRef]

27. Tesche, T.W.; McNally, D.E.; Emery, C.A.; Tai, E. *Evaluation of the MM5 Model over the Midwestern U.S. for Three 8-Hour Oxidant Episodes*; Prepared for the Kansas City Ozone Technical Workgroup; Alpine Geophysics LLC: Arvada, CO, USA; Environ Corp.: San Rafael, CA, USA, 2001.

28. Stewart, I.D.; Oke, T.R. Local climate zones for urban temperature studies. *Bull. Am. Meteorol. Soc.* **2012**. [CrossRef]

29. Millstein, D.; Levinson, R. Preparatory meteorological modeling and theoretical analysis for a neighborhood-scale cool roof demonstration. *Urban Clim.* **2018**, *24*, 616–632. [CrossRef]

30. Macdonald, R.W.; Griffiths, R.F.; Hall, D.J. An improved method for estimation of surface roughness of obstacle arrays. *Atmos. Environ.* **1998**, *32*, 1857–1864. [CrossRef]

![climate logo] *climate*

MDPI

Article

Evaluation and Modeling of Urban Heat Island Intensity in Basel, Switzerland

Andreas Wicki *, Eberhard Parlow and Christian Feigenwinter

Department of Environmental Science, University Basel, CH-4056 Basel, Switzerland; eberhard.parlow@unibas.ch (E.P.); christian.feigenwinter@unibas.ch (C.F.)
* Correspondence: a.wicki@unibas.ch; Tel.: +41-61-207-0752

Received: 1 June 2018; Accepted: 18 June 2018; Published: 21 June 2018

Abstract: An increasing number of people living in urban environments and the expected increase in long lasting heat waves makes the study of temperature distribution one of the major tasks in urban climatology, especially considering human health and heat stress. This excess heat is often underestimated because stations from national meteorological services are limited in numbers and are not representing the entire urban area with typically higher nocturnal temperatures, especially in densely built-up environments. For a majority of the population, heat stress is consequently monitored insufficiently. In this study, the factors influencing the nocturnal urban heat island have been evaluated in detail and have been tested using different spatial resolutions. A multiple linear regression model has been developed with predictors resulting from different data sources to model the urban air temperature distribution continuously. Results show that various datasets can be used for the prediction of the heat island distribution with comparable results, ideally run on a 200 m grid. Validation using random sampling indicated a RMSE clearly below the standard deviation of the measurements with an average around ~0.15 °C. The regression coefficients are varying within the nocturnal runs with best results around 22:00 CET ($R^2 > 0.9$).

Keywords: multiple linear regression; urban heat island; urban climatology; urban energy balance; air temperature; land cover fraction; urban morphology; land surface temperature; heat stress

---

1. Introduction

Though only a very limited proportion of the global land surface is covered by urban areas, the majority of the world's population is spending most of its time in such environments. Currently, over 75% of the people in Central Europe are city dwellers, with proportionally increasing tendencies [1,2].

Cities are known to have a special microclimate with higher nocturnal air temperatures (T_a) compared to their rural surroundings, which has been frequently discussed in literature and described with the well-established term "urban heat island" (UHI) [3–12]. This effect may be pleasant for urban dwellers during transition periods or in winter, because it reduces cold stress and space heating is less required due to the additional warmth [5]. During certain conditions, i.e., synoptically stable and long lasting anticyclonic weather situations in summer, this extra heat load can be threatening for elderly and vulnerable people [13]. The maximum temperatures are thereby not necessarily the main problem, but rather the lack of relief when the minimum temperatures are not declining below a certain level. Clarke and Bach (1971) focused their early studies on the influence of urban climate on human health. Based on a small data record, they found an increase in heat related excess mortality due to high nocturnal temperatures, either as a primary effect or as a contributing factor in heart disease, strokes or pulmonary disorders [14]. Scherer (2014) stated that extreme heat was responsible for 5% of all deaths from 2001 to 2010 in Berlin, affecting mainly people exceeding the age of 65 [15]. During extreme heat waves, the linear relationship between minimum temperature and mortality becomes an exponential function, indicating that every tenth of a degree increase has a dramatic influence on the total death rates [15,16]. Additional heat related deaths—ranging from 22,000 [17] to 70,000 [18] according to different studies—and an excess mortality of up to 30% [19] were found during the extreme heat waves in Central Europe in 2003 and 2015, respectively. Due to climate change, such events increased in the 20th century and will be an usual phenomenon in the 21st century [20]. These findings are also confirmed by statistical analysis considering the duration of large-scale weather patterns in Europe, which indicate that long-lasting weather situations (i.e., >10 d) are currently doubled in their frequency compared to the first half of the last century [21]. Studies show that this trend is ongoing in the future with heat waves that are more intense, longer lasting, and/or more frequent [22,23]. The combination of those three observations—i.e., the increasing urban population, the higher nocturnal T_a in cities and the increasing frequency, duration and intensity of heat waves—makes the study of urban climate, the resulting impacts on the urban population and especially the study of possible mitigation strategies indispensable.

To understand the physical processes behind the development of the UHI and to investigate the use of remote sensing and GIS data to study urban climate dynamics, it is fundamental to understand the concept of urban energy balance. Due to urbanization and the modification of natural vegetated surfaces to heterogeneous three-dimensional and largely sealed surfaces, vertical fluxes of heat, mass and momentum are altered dramatically compared to natural conditions [24]. The urban morphology, artificial surface materials and human behavior are the main factors influencing these exchanges [25]. Additionally, the nearly doubled urban surface, compared to a plane surface, increases the amount of energy stored in the urban fabric and within the urban canopy layer (UCL) [26].

The urban surface as a total can be described as a box with variable height (depending on the processes of interest) including streets, walls, roofs and all natural and artificial elements. It can be treated as an integrated system [24,27]. At about roof level, the energy balance is dominated by site-specific microscale effects [6]. On top of this UCL, the urban energy balance can be written as follows:

$$Q^* (+ Q_F) = Q_H + Q_E + \delta Q_S (+ \delta Q_A),\qquad(1)$$

with:

Q^* = Net radiation,
Q_F = Anthropogenic heat release by combustion, traffic and human metabolism,
Q_H = Turbulent sensible heat flux density,
Q_E = Turbulent latent heat flux density,
δQ_S = Net heat storage,
δQ_A = Net heat advection.

The net radiation can be measured with net radiometers and/or computed with remote sensing data:

$$Q^* = SW \downarrow - SW \uparrow + LW \downarrow - LW \uparrow,\qquad(2)$$

with:

$SW \downarrow$ = Solar irradiance,
$SW \uparrow$ = Shortwave reflection,
$LW \downarrow$ = Atmospheric counter radiation,
$LW \uparrow$ = Terrestrial emission.

The input terms $SW \downarrow$ and $LW \downarrow$ can be measured or modelled using topographic features combined with the geographical position and the actual solar altitude or boundary layer air temperature [28].

The longwave radiative loss, i.e., $LW \uparrow$, is a function of the land surface temperature (LST) and emissivity:

$$LW \uparrow = \sigma * \varepsilon * T^4,\qquad(3)$$

with σ as the Stefan-Boltzmann constant ($5.67 \cdot 10^{-8}$ W m^{-2} K^{-4}), ε as the emissivity and T as the LST in K. Since EO-derived LST is strongly dependent on adequate atmospheric correction, this term is a possible source of errors. Even small variations have significant effects on the energy balance [29,30].

A crucial part in Equation (2) is $SW \uparrow$. It is calculated by the multiplication of $SW \downarrow$ with the broadband surface albedo (α), computed using the VIS and NIR channels from satellite data or measured by a pyranometer (Section 2.3.1, Equation (8)) [31,32]:

$$SW \uparrow = \alpha * SW \downarrow . \tag{4}$$

The term Q_F is currently under investigation and so far is not adequately described by remote sensing data [33]. It can be assumed, that in moderate climate cities like Basel, the summer Q_F input is relatively small, because cooling systems producing a lot of anthropogenic heat in summer are uncommon in Switzerland. The largest amount of Q_F in Basel is most likely generated by heating, which is limited to winter situations. Christen and Vogt (2004) found an annual average amount of Q_F in the range of 5 to 20 W m^{-2} in suburban and dense urban environments, respectively [26].

Urban fabrics comprise of special thermal properties, which makes the urban surface an ideal heat saver. The ratio $\delta Q_s / Q^*$, indicating the amount of available energy used for storage, may reach ratios up to 0.58, 0.48 and 0.31 for measurement sites in Mexico City, Vancouver and Los Angeles, respectively [34]. During the BUBBLE measurement campaign in the city of Basel, values of 0.3 to 0.4 resulted at different urban sites [35]. This means, that only 60 to 70% of the net radiation is available for partitioning into the turbulent fluxes of sensible and latent heat. In case of extremely hot urban surfaces, this value might be even lower and if a minimum evapotranspiration is assumed on rural sites, which is likely to be the case during extreme heat waves and/or dry conditions, the sensible heat flux in an urban environment is even lower compared to its rural counterpart.

The turbulent fluxes Q_H and Q_E are related through the Bowen ratio ($\beta = Q_H / Q_E$). It is defined for specific surfaces and derived from flux tower measurements. Oke (1988) reported values for β of 0.6 to 1.0 in suburbs and 1.8 in the city core [24]. During the REKLIP project, Parlow (1996) found values of 1.5 in high-density urban to 0.8 in low-density urban areas and a maximum β of 1.8 in industrial areas in and around Basel [36]. On forests and natural surfaces, the values ranged between 0.4 above deciduous forests to 0.7 on arable land [24,36]. During one month of intensive measurements in Basel, Christen (2004) found daily variations of β, most pronounced at the rural site (0.5 to 0.2) and less pronounced at the urban site (~2) [26]. Taking these values ($\beta_U = 1.5$, $\beta_R = 0.6$) and comparing them to the differences in available energy due to lower Q_S on rural sites ($Q_{S\_U} = 40\%$, $Q_{S\_R} = 10\%$), Q_H is in the same range on grassland compared to dense urban areas.

Net advection (δQ_A) is neglected here, because heat wave events are normally coupled to synoptically calm conditions with very weak winds, especially during the night (ignoring micro-scale advection [24]).

In this study, the theoretical knowledge about the physical processes happening at the transition layer between the urban surface and the overlying air is used to investigate the development of the frequently evolving UHI in Basel, Switzerland. Thereby, various data sources are tested using a multiple linear regression (MLR) approach within different resolutions and at different time steps during the development of the nocturnal UHI. The investigation period was during a minor heat wave at the end of August 2016. The investigation area, the meteorological measurement stations and the complete setup of the method are described in Section 2. Results—including statistics, validation and exemplary maps—are presented in Section 3, followed by a discussion and concluding remarks in Sections 4 and 5.

2. Materials and Methods

2.1. Site Description

Basel is located at the southern end of the Upper Rhine Valley between the bordering hills of the Vosges (France), the Black Forest (Germany), and the Jura Mountains (Switzerland) at about 250 m a.s.l. (Figure 1).

The climate is considered oceanic, with mild winters and warm and sunny summers (Köppen Cfb climate) [37]. The annual mean temperature is 10.5 °C with an annual precipitation rate of 842 mm for the reference period 1981 to 2010 [38]. The warmest month is July with average maximum temperatures above 25 °C, though record-high temperatures during heat waves may also occur in June or August. Cold stress situations are less pronounced in the city, occasionally such phenomena may occur due to the outbreak of cold air masses from continental Russia.

Long-term measurements and several measurement campaigns in the region describe a distinct UHI, which establishes typically during the night with maximum temperature differences of up to 8 °C between rural and urban reference stations [12,39,40].

The investigation area contains the metropolitan area of Basel with approximately 730,000 inhabitants in total, including the bordering suburbs located in the Canton of Basel-Land (BL), the German cities Weil-am-Rhein and Lörrach, the French city Saint-Louis, and the city of Basel [41,42]. From north to south and west to east the area covers 21 by 20 km^2, respectively, spanning from the upper left 47°38'35" N 7°29'17" E to the lower right 47°27'25" N 7°45'32" E.

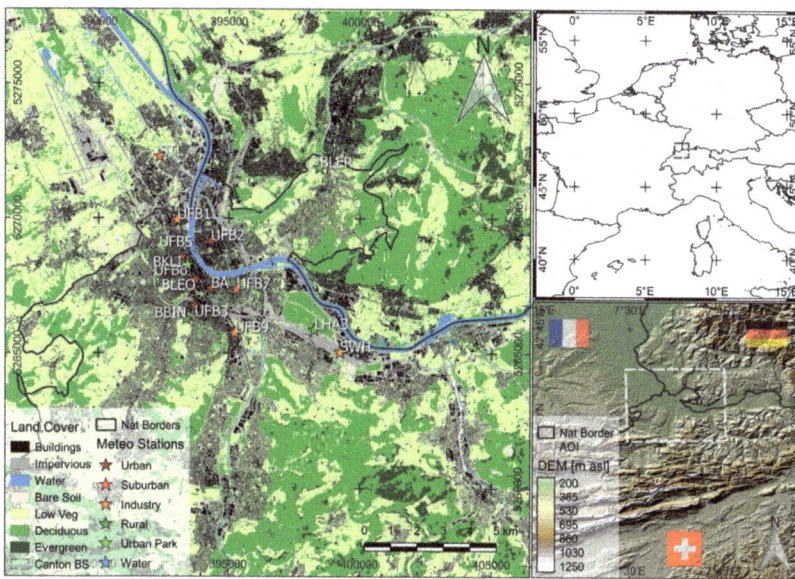

Figure 1. Land cover of the investigation area including classified measurement stations (left, grid reference UTM-32N) and the area of interest (AOI) as a grey rectangle on the topographic map (ASTER GDEM) of the tri-national region France, Germany and Switzerland (lower right), indicated as a grey rectangle in the upper right overview map. Station name and detailed classification can be found in Table 1.

2.2. Air Temperature Measurements

Below a blending height z*, the urban turbulence field must be considered three-dimensional and small-scale fluctuations may occur [43]. In this roughness sublayer (RSL), the measured temperature signal is directly influenced by the underlying surface, with a variable footprint due to different wind directions. Ideally, comparative measurements should be carried out at the same relative position in the RSL, depending on the anticipated model resolution. Only stations located at the transition level between the UCL and the RSL at about roof level are considered in this study. At this height, the sensors are not located directly in the wakes of the individual surface elements, but a local mixture of influences is still captured [6]. Note that the lower boundary of the RSL used here is defined after Oke (1988) as the top of the UCL, i.e., roughly the mean building height (z_H). Christen and Vogt (2004) indicate that the measurement heights (z) of $0.8 < z/z_H < 1.4$ are representative for the local urban climate [26]. Stations located in large open spaces are used even if they are only located at 2 m measurement height, because the UCL may be entirely absent in such environments [4]. The main criterion for exclusion for those stations is cold air accumulation (Table 1).

All data are adjusted to the lowest station height adding a lapse rate of 0.6 °C per 100 m. To test the approach and to minimize the influence of temperature fluctuations, a mean nocturnal course of 6 consecutive nights during a heat wave starting on 23 August 2016 is used. During that period, the daytime maximum temperatures were mostly above 30 °C and several tropical nights (i.e., $T_{min} > 20$ °C) were measured at the urban reference station (i.e., BKLI, Section 4.1).

Table 1. List of all stations used for this study with the specific code as an abbreviation. The geographic location of the stations can be found in Figure 1. The local climate zones (LCZ) definition is based on previous work and expert knowledge after the guidance of Stewart and Oke (2012) [44,45].

Station	Code	Height [a.s.l.]	Height [a.g.l.]	Sensor	LCZ
Basel-Binningen	BBIN	316	2	ASPTC	D_9
Feldbergstrasse	UFB2	256	15	Decagon VP3	2
A2 Hard	LHA3	275	6	Thygan	A_E
Klingelbergstrasse	BKLI	286	28	ASPTC	2
Lange Erlen	BLER	273	10	HMP45C	D
Johanniter-Brücke	UFB5	256	19(4) *	WXT520	G_A
Meret-Oppenheim	UFB3	280	20	WXT520	2_{10}
Hirzbodenweg	UFB7	270	25	WXT520	6
Novartis	UFB1	256	75	WXT520	10_2
Dreispitz	UFB9	284	18	Decagon VP3	10_2
Petersplatz	UFB6	268	24	Decagon VP3	2_A
Aeschenplatz	BA	270	40	HMP45C	2
Leonhard	BLEO	273	41	HMP45C	2_3
Schweizerhalle	SWH	280	25	HMP45C	E_{10}
St. Louis	STL	247	4	HMP45C	6

* Located on a bridge with a height of about 15 m.

2.3. Urban Surface Properties

Different types of surface properties are calculated, collected and tested separately to find reasonable input parameters for a MLR model. Three groups of surface descriptors emerged, which are described in the following sections.

2.3.1. Landsat Data

LST, NDVI and albedo is derived from satellite data acquired by the Landsat 8 operational land imager (OLI) and thermal infrared sensor (TIRS) at the beginning of the measurement period on 23 August 2016 10:16 UTC.

Atmospheric effects on the TIR signal are removed using a correction parameter accessed through the web-based atmospheric correction tool provided by NASA, which is based on NCEP reanalysis data and MODTRAN code. The conversion of top-of-atmosphere radiance (L_{TOA}) to surface-leaving radiance (L_λ) is done solving Equation (5):

$$L_{TOA} = \tau \varepsilon L_\lambda + L_u + \tau(1 - \varepsilon)L_d,$$ (5)

with L_u as the upwelling or atmospheric path radiance and L_d as the downwelling or sky radiance, which is reflected on the surface. Attenuating effects of the atmosphere are incorporated by the atmospheric transmission (τ) [46]. The emissivity map (ε) is based on a modified NDVI approach by Sobrino et al. (2008) [47]. To convert L_λ to LST, the Landsat-specific estimate of the Planck curve including the calibration constants k_1 and k_2, derived from Landsat metadata, is used:

$$\text{LST} = \frac{k_2}{\ln\left(\frac{k_1}{L_\lambda} + 1\right)}. \tag{6}$$

The complete procedure is described and discussed in detail in Wicki and Parlow (2017) and reveals effectively corrected LST based on in situ measurements and models of atmospheric conditions [48]. The LST represents the surface UHI (SUHI), which has different mechanisms compared to the canopy UHI [12,49]. LST is used as a predictor, because the principles of energy fluxes and the expected UHI effect are based on different surface properties and their interaction with the incoming solar radiation. Dark impervious surfaces normally have a large heat capacity and high thermal conductivity rates, which enables the surface to absorb the incoming radiant energy more effectively [49]. The LST is the best estimate for this interaction available from remote sensing data. The energy stored in the urban fabric during the day and released during the night is one of the main causes of the nocturnal UHI, especially during heat wave conditions and large solar irradiance on the surface. Even though LST is not a perfect representation for the energy stored in the urban fabric, because thermal properties are only partly represented by LST, it is certainly connected to the soil energy flux, especially if used in combination with other surface properties [50].

The NDVI represents the different reflection properties of vegetated surfaces in the visible (RED) and near infrared (NIR) range. It can be written as:

$$\text{NDVI} = \frac{\text{NIR} - \text{RED}}{\text{NIR} + \text{RED}}. \tag{7}$$

A water mask based on a land cover classification is used to exclude water bodies from the analysis.

Surface broadband albedo (α_{short}) is calculated using calibrated reflectance of five channels from the VIS to the SWIR range (ρ_{2-7}) as a weighted average after Liang (2001) and adapted to Landsat 8 data [31]:

$$\alpha_{short} = 0.356 * \rho_2 + 0.130 * \rho_4 + 0.373 * \rho_5 + 0.085 * \rho_6 + 0.072 * \rho_7 - 0.0018. \tag{8}$$

The calibration applied on the Landsat data for α estimation includes the sun elevation, but no correction for distracting effects due to aerosols, atmospheric gases or topography was applied.

For this analysis, Landsat 8 data from a single acquisition was used. Previous studies have shown that during comparable weather situations, the LST distribution follows a certain pattern and does not essentially change [48]. The NDVI and α are even more stable, beside variation due to different solar angle, in a near-term period and do not change fundamentally within 6 days.

2.3.2. Urban Morphology

The sky view factor (ψ_{svf}), i.e., the proportion of the observed sky divided by the total upper hemisphere, is calculated using the Urban Multi-scale Environmental Predictor (UMEP) and a digital surface model (DSM) with 3 m resolution [51]. During the resampling (averaging) to coarser resolutions, roof pixels are omitted to represent for dense urban areas with narrow streets, and therefore low ψ_{svf}.

A digital object model (DOM), the product of subtracting a digital elevation model from the DSM, was used for the computation of z_H. The datasets are representative for the years 2008 to 2016. The averaging is done omitting the ground pixels, but including trees.

The plan area fraction (λ_P) is derived using the raster DOM or high-resolution remote sensing data. This fraction is both a morphological variable (density) and part of the land cover fraction (surface cover), therefore, it is computed in two different ways and added to both groups (i.e., Morphology and Land Cover), separately. Since the land cover fraction is derived from remote sensing data, the resulting coefficients differ. The abbreviation BSF (building to surface fraction) is used for λ_P in the Land Cover group.

The frontal area index (λ_F) or roughness density [52] combines the mean height, breadth and density of the roughness elements in per unit ground area and describes how much of the area reached by a fictional air parcel is covered by walls (or vegetation) for a specific wind direction [53]. In this case, the isotropic λ_F—which is not depending on the wind direction—is calculated using UMEP [51].

The two parameters λ_P and λ_F are used to determine the roughness length (z_0) and zero displacement height (z_d). These roughness parameters are calculated using a method described by Kanda (2013) [54].

The tree fraction (λ_T), i.e., the number of trees per hectare, is estimated using high-resolution (10^{-2} m) aerial LIDAR data from 2010 [55]. The processed vector dataset consists of point data including position, height and crown diameter of every single tree of a major part within the investigation area (i.e., the Canton Basel-Stadt). Areas that are not covered by this dataset are derived from the DOM with a discrimination between vegetation and buildings based on a high-resolution land cover classification, described in Section 2.3.3.

2.3.3. Land Cover

A land cover classification is used to compute the BSF, the impervious surface fraction (ISF) and pervious surface fraction (PSF) by summing up the respective surfaces. The land cover classification is part of the Copernicus contributing datasets allocated with the Horizon2020 project URBANFLUXES [56]. The classification technique is based on a neural network approach. It is applied on SPOT data from 26 May 2012 and supplemented by WorldView imagery and seasonally variable Landsat data (2013–2016). The map with an original ground resolution of 2.5 m can be found in Figure 1 (left).

2.4. Method

2.4.1. State of the Art

Urban measurement networks are usually not dense enough to cover intra-urban differences effectively and denser networks are very cost intensive and impossible to maintain. Mobile measurements with devices mounted on cars [4,57–59], bicycles, different public transport vehicles [60,61] and/or by hand/foot are alternatives to cover small-scale differences despite some limitations (spatial, temporal and synoptic representativeness [57], duration and costs).

The link between the surface and the above T_a based on remote sensing or GIS data for comprehensive T_a estimations has been intensively investigated since the launch of the first satellites and Rao's (1972) analysis using ITOS [62]. Karl (1988) used urban population as a surrogate for urbanization and tried to remove the "urban bias" in T_a measurements using 1200 urban and rural stations around the USA. Gallo (1993 and 1999) investigated the linear correlations between T_a and LST, but also between T_a and NDVI, which provided better results [63,64]. The same objective was pursued by Epperson (1995), who added nighttime brightness data from the Defense Meteorological Satellite Program to the NDVI analysis, which led to an MLR analysis [65].

Better correlation is usually found between minimum T_a (T_{min}) and nighttime LST compared to maximum T_a (T_{max}) and daytime LST [8,59,66–70]. The different behavior of surface and air temperature in urban environments, i.e., the shift in maximum heat island of the surface (day) to the maximum heat island of the air temperature (night), is crucial and mentioned explicitly in different investigations [8,57,67,71].

Regarding the application of such methods in highly heterogeneous urban areas, the coarse resolution and increased viewing angle of Meteosat SEVIRI, NOAA AVHRR or Terra/Aqua MODIS conceal important details, especially in medium-sized and small cities. Ma (2016) demonstrated the possibility of downscaling MODIS LST to 250 m resolution using a TsHARP algorithm and found better results compared to the original 1-km resolution data [72]. The possibility of using helicopter or aerial imagery is very cost intensive, difficult to conduct due to flight restrictions and often unrepeatable and thus unique. ASTER and Landsat data are most suitable for applications in urban areas regarding the spatial resolution [59,69,73,74], but ASTER has no constant revisit rate and Landsat is normally only acquiring daytime images with a revisit rate of maximum twice within 16 days (due to overlapping paths, which is the case for Basel). Nevertheless, multi-temporal LST analysis has shown that the general intra-urban LST patterns in an urban area are not varying substantially during comparable weather conditions [48].

However, LST or the built environment should not be used solely to explain the UHI effect [74]. Cresswell (1999) found an improvement of 50% less errors and higher correlation coefficients from predicted to actual temperature (0.80 to 0.84) if an additional proxy to LST versus T_a analysis was added [75]. Cristobal (2008) added remote sensing variables such as α, LST and NDVI to a multiple regression model based on altitude, latitude, continentality, solar radiation and cloudiness [76]. The use of multiple predictors was previously mentioned in the outlook section of several publications, e.g., Bechtel (2017) and Vogt (1997), and was applied before in other studies [76–79].

2.4.2. Method Description

The GIS dataset in this study is based on factors proposed by Stewart and Oke (2012) for the classification of local climate zones (LCZ) and accounts for the causes of the nocturnal UHI as discussed in many previous studies [5,6,26,27,35,44,80]. Cities are a system of different surface units—or morphological forms—that are repeated and arranged differently throughout the urban area [27]. These units define and control the urban climate as an integration of several influences, which can be disaggregated using multiple regression approaches. Urban meteorological measurement stations represent different types of urban composition that control the nocturnal temperature. The MLR approach decomposes, or unmixes, the specific portion of different land use types to the temperature signal. Thereby, the controlling units—hereafter referred to as the predictors—are treated as separate GIS layers. The number of dependent variables, i.e., measured T_a, must exceed the number of predictors (Equation (10)). Ideally, the measurement stations should represent different types of urban controlling units and a certain amount of variance is required.

The model finally enables the production of two-dimensional UHI maps based on the resulting coefficients applied on GIS and remote sensing data. Several runs with different combinations of surface properties and different spatial resolutions, which was done by resampling based on the average of every grid cell, were tested and evaluated in order to find the ideal configuration.

MLR is preferred instead of other methods (e.g., machine learning), because the individual behavior of the predictors is of major interest and not only calculations resulting from a black box.

2.4.3. Suitability Tests and Predictor Selection

A major problem of MLR is multi-collinearity, i.e., the explanation of one predictor by another, which can increase the estimates variance. The result is an adversely better statistic. If the data is to a large amount explained by one of the already existing variables, this excess data does not help to better explain the independent variable [81].

To build the final MLR model, two suitability tests have been applied: A multi-collinearity analysis for each group including a variance inflation factor (VIF) test and a multi-temporal simple linear correlation (SLC) analysis between the potential predictors and the measured T_a.

The SLC is used to reduce the number of predictors in advance. Potential predictors providing no explanation for the variance in the multi-temporal analysis are excluded before the collinearity analysis. In the final decision process, the behavior of the predictors in the SLC analysis was a criterion for exclusion if multi-collinearity was detected. The VIF is used to estimate how much of the variance is inflated due to the presence of correlation within the predictors of the model. This indicates how much of the estimated variance of the specific regression coefficient is increased, compared to the case that R^2 would equal zero. For every predictor i in the model, the VIF formula can be written as:

$$VIF = \frac{1}{1 - R_i^2}.$$ (9)

For the interpretation of the VIF values, several rules of thumb and the respective thresholds are described in literature. Most commonly the rule of >10 is applied to detect harmful multi-collinearity within the dataset. Since this must be considered dataset specific, models with a VIF > 10 could still include useful information and the thresholds need to be applied with caution. Nevertheless, we decided that a VIF of 10 can be used as a threshold, since this value was frequently applied [81,82].

A further method to reduce useless predictors would be a stepwise MLR (SMLR). This was also applied for testing, but not for the final model, because the model setup is intended to be stable over time and using SMLR changes the predictor composition with every run. However, the results are used to verify the decisions about inclusion or exclusion of predicting parameters.

2.4.4. Multiple Linear Regression

For every regression model, the LST is added as an additional predictor to account for the heat released from storage [12,24,27,71]. The variables that passed the suitability test are used for a MLR analysis by solving a set of linear equations (Equation (10)):

$$\begin{aligned}
X_{11}T_1 + X_{12}T_2 + \ldots + X_{1m}T_m &= \beta_1 \\
X_{21}T_1 + X_{22}T_2 + \ldots + X_{2m}T_m &= \beta_2 \\
X_{n1}T_1 + X_{n2}T_2 + \ldots + X_{nm}T_m &= \beta_n.
\end{aligned}$$ (10)

The dependent variables (T_{1-m}) and the input predictors (X_{11-nm}) are used to determine the coefficients (β_{1-n}) for the best fit. The set can be solved, if more equations (i.e., T_a measurements, m) than unknown variables (i.e., input predictors, n) exist (m > n).

$$T_{pre} = \beta_0 + \beta_1 X_1 + \beta_2 X_2 + \ldots + \beta_n X_n.$$ (11)

The MLR returns a vector of coefficient estimates (β_{1-n}) for each predictor (X_{1-n}) to solve a multiple linear regression equation (Equation (11)) and to find the responses (T_{pre}) for every grid point in a two-dimensional raster environment. The intercept value (β_0) is the expected value if all predictors are set to zero and close to the average of the measurements [48].

The model is tested iteratively with varying conditions considering time, resolution and predictor combination to find the best results for the three groups defined as Land Cover, Landsat Data and Morphology.

2.4.5. Validation

The model output is validated using in situ measurements, dividing the T_a measurements in two clusters (i.e., 2/3 input and 1/3 validation). To ensure a certain variation within the model, the stations are classified into four groups (similar to Figure 1, but combining Park with Suburban) and one station per group is used for the validation cluster. The group Urban supplied two stations for the validation process, since it is the most populated class. The model is run five times with different input predictors, chosen by randomly generated numbers (MATLAB function *randperm*), and the resulting predictions are compared to the independently measured T_a.

3. Results

3.1. Suitability Test and Predictor Selection

The suitability test reveals important information about the significance of individual predictors and allows assembling of the different model groups. Figure 2 describes the results for the SLC as a surface plot including all correlation coefficients for six resolutions (subplots) at different steps in time (rows) for every predictor (columns). Above the columns and on the right side of the rows, mean values indicate the average behavior for every predictor and time step, respectively (displacement due to space limitation). The total average of all predictors for all time steps for every model resolution is printed as a bold number at the upper right of every subplot.

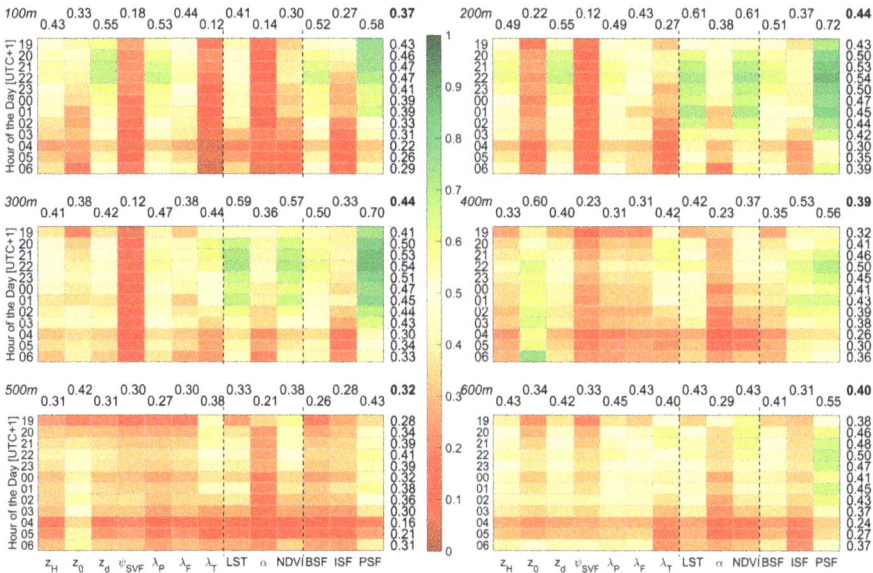

Figure 2. Correlation coefficients of all linear regressions for the 100 to 600 m cell size runs as a suitability test. The columns represent the individual predictors and the rows represent the different time of the day, colored after the specific correlation coefficient (red = 0 to green = 1). The three different groups are separated by vertical dotted lines and the abbreviations are mentioned in the text.

For the first group (Morphology), z_d, λ_P and z_H reveal the best results in the runs with a resolution below 400 m. With increasing cell size, z_0 is gaining importance. The lowest correlations for almost every run are found for ψ_{svf}. Especially low correlations are found at resolutions below 400 m. Weak correlations between ψ_{svf} and urban T_a are similarly described in Nichol (2005). They found the same cooling rates on parking lots and between high-rise buildings at 21:40 local time in Hong Kong, which was explained by the early time of comparison and the intense heating of the surface [69]. The second group (Landsat Data) showed a high abundance of nocturnal T_a to LST and NDVI. Albedo revealed very low correlation in the 100 m run and performed better at 200 and 300 m resolution. For the Land Cover group, the pattern changes with different resolutions. In most cases, the PSF and BSF are explaining the variance better than the ISF. Only for a resolution of 400 m, the ISF outperforms the BSF. The PSF revealed very high correlation coefficients and, especially for a resolution of 200 and 300 m, explains already a lot of the variance within the nocturnal T_a distribution. As a total, the multi-temporal analysis showed variations of regression results during the night. The best results are usually found between 20 UTC+1 and midnight. The best overall results are found at 200 and 300 m resolutions. The resulting coefficients are in good accordance to comparable SLC analyses [74].

The evaluation of multi-collinearity showed high collinearity for z_H, z_d, λ_F and λ_P. The effects on the microclimate described by these parameters are therefore already explained to a large amount by one of these parameters. High VIF values up to 72 resulted from the multi-collinearity analysis. Reducing these parameters to only one, where the SLC regression factors are used as a criterion, helped to lower the VIF value to a maximum of 3.9 (at 200 m cell size). The resulting model includes the parameter z_d, z_0 and λ_T for the group Morphology (Figure A1).

No problems with collinearity are found for the Landsat Data group. Therefore, all three predictors (LST, NDVI and α) are included (Figure A2).

The third group Land Cover reached values close to the exclusion criterion, which was still a considered valuable. Increasing the cell size also increases the VIF, which needs to be taken into account during the interpretation of the results. Therefore, better regression results at higher resolutions might be the effect of collinearity for this group in special (Figure A3).

3.2. Multiple Linear Regression

Three groups, six different resolutions (100 to 600 m grid cells), and 12 different points in time (19:00 to 06:00 CET) are tested. Figure 3 shows the complete statistics of this analysis including correlation, significance and error variance—i.e., total deviation from prediction to in situ measured temperatures—of the model runs. The two significance levels (0.05 and 0.01, as a red and black line, respectively) indicate the performance considering the specific criteria. A run falling below the 0.05 threshold is considered significant and valuable, a run falling below the 0.01 criteria is considered excellent. Runs above the 0.05 significance line should be treated with caution.

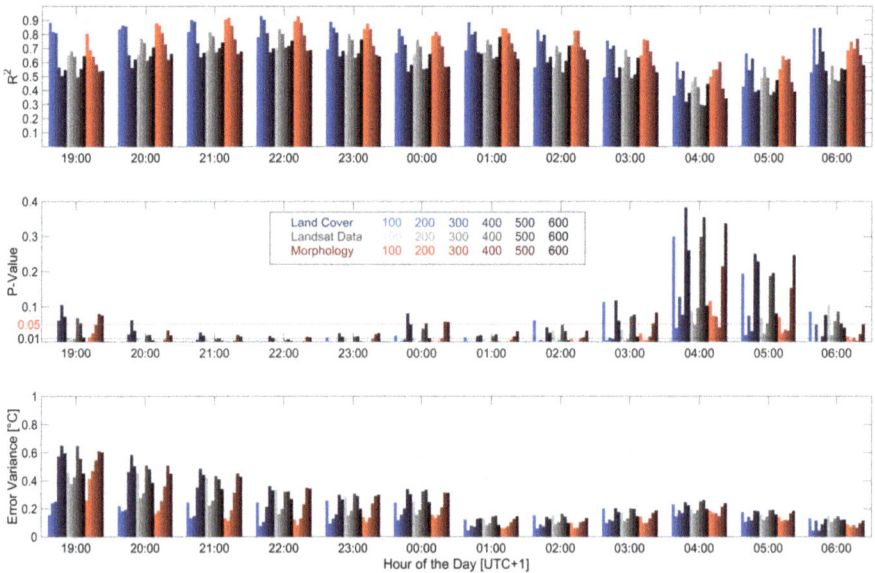

Figure 3. Statistics for all multiple regression runs shown as bar plots with R^2 in the upper plot, *p*-value in the middle and error variance in the lower plot. The colors are corresponding to different input datasets and the brightness of the specific colors decreases with decreasing resolution. In the center plot, the 0.05 and 0.01 significance level is indicated by red and black dotted lines, respectively.

According to the statistics, best results are found for the Land Cover and Morphology (slightly lower) model with a R^2 of 0.93 (significant on the 0.01-level) at 22:00 CET with 200 m resolution. The 200 m resolution showed the highest correlation factors also for the Landsat Data group (0.84) with decreasing tendencies at increasing resolutions. The significance is also reduced with coarser resolutions (except 100 m), with the same tendencies for the error variance. Highest correlations are found around 22:00 with a minimum at 04:00 CET. The same drop for significance is found at around 03:00 to 06:00 CET. The error variance typically decreases during the night since the temperature decreases, but the lower model success at around 04:00 CET is captured as well. This might be due to the loss of 'fuel', i.e., heat stored in the urban fabric, and the resulting decrease of the urban heat island and/or katabatic wind systems generated after substantial cooling on the surrounding hillsides. In Section 4.1, meteorological measurements during the investigation period are described and they demonstrate the higher UHI intensity at the same time the correlation reaches the highest numbers, or vice versa. As the UHI intensity decreases, the variance within the response is reduced, which depreciates the model statistics. The analysis of the wind systems indicates slightly stronger winds after midnight, but not with a significant trend. However, the wind direction—an indicator for cooling slope wind systems—does not change dramatically throughout the night.

Average statistics for the entire model run (i.e., 19:00 to 06:00 CET) are shown in Table 2. The R^2-values and mean error variance are temporally averaged and the two significance levels (0.05 and 0.01) indicate how much percent of the overall runs have fulfilled the specific criteria for the different model resolutions. Again, the 200 m runs show highest correlation with almost 100% significance for all runs. More than three out of four runs show very high significance. The scales of 200 and 300 m seem to capture the temperature influence properly, because the error variance throughout all runs is the lowest for this resolution with errors ranging from 0.11 to 0.20 °C. Lowest correlations, accompanied by low significance and highest error variances, are found using 500 m grid cells.

Figure 4 shows the evolution of T_a throughout the night with a resolution of 200 m. This predicted T_a results from the regression formula (Equation (11)) applied with the resulting coefficients and the specific predictor maps. Significant height differences are taken into account by a lapse rate of 0.6 °C per 100 m—the same cooling rate used to normalize the measured T_a. Therefore, a temperature bias can be addressed to elevated areas in the south and northeast of the city. However, the intra-urban differences are not resulting from height differences, because the urban area is relatively flat (Figure 1, Section 2.1). In this area, land cover is the main driver for regulating the T_a distribution and thus the UHI evolution. The old town core in the center of the image is clearly defined and remains always one of the warmest regions until the distribution is getting flatter around 03:00 CET. This is a typical feature of the nocturnal UHI, which often shows largest urban to rural differences in the first half of the night. Urban parks can clearly be detected as cool spots close to the city center. Figure 5 includes the difference of every raster cell to the modelled temperature at BBIN (i.e., an official weather station of MeteoSwiss). It indicates that the official temperature measurement for the city underestimates the heat load for most of the urban dwellers by 2–3 degrees. The temperature of this station would only be valid for the urban park sites, which have an influence of less than the park extent [83]. Large residential areas like Bachletten or St. Alban reveal a lower temperature difference to BBIN, and, therefore, less heat stress for the population. Highly densified areas like the old town (i.e., Altstadt) are clearly on top of the heat scale. The different railway stations located in the Rosental, St. Johann and Gundeldingen neighborhoods also show high positive deviations. Additionally, the industrial areas in St. Johann (at station UFB1) and at the eastern city limits are about 2 degrees warmer than BBIN in this model run.

Table 2. Statistics for six different model resolutions derived from hourly model runs between 19:00 and 06:00 CET and three different input datasets. The R^2 statistics and error variances are averaged. The significance-analysis shows the precentral number of runs fulfilling the 0.05/0.01-significance level during the modelling period of 12 h.

Statistics	Method	100 m	200 m	300 m	400 m	500 m	600 m
	Land Cover	0.64	0.82	0.75	0.72	0.56	0.58
Mean R^2	Landsat Data	0.60	0.71	0.66	0.54	0.57	0.66
	Morphology	0.75	0.79	0.76	0.70	0.60	0.57
0.05-level	Land Cover	58	100	83	83	50	58
significance fulfilled [%]	Landsat Data	75	100	75	58	50	83
	Morphology	83	92	92	100	58	50
0.01-level	Land Cover	33	83	67	67	0	0
significance fulfilled [%]	Landsat Data	25	75	75	17	17	50
	Morphology	67	75	67	67	0	0
Mean Error	Land Cover	0.20	0.11	0.14	0.21	0.31	0.28
Variance [°C]	Landsat Data	0.26	0.17	0.20	0.30	0.29	0.23
	Morphology	0.14	0.13	0.16	0.21	0.28	0.29

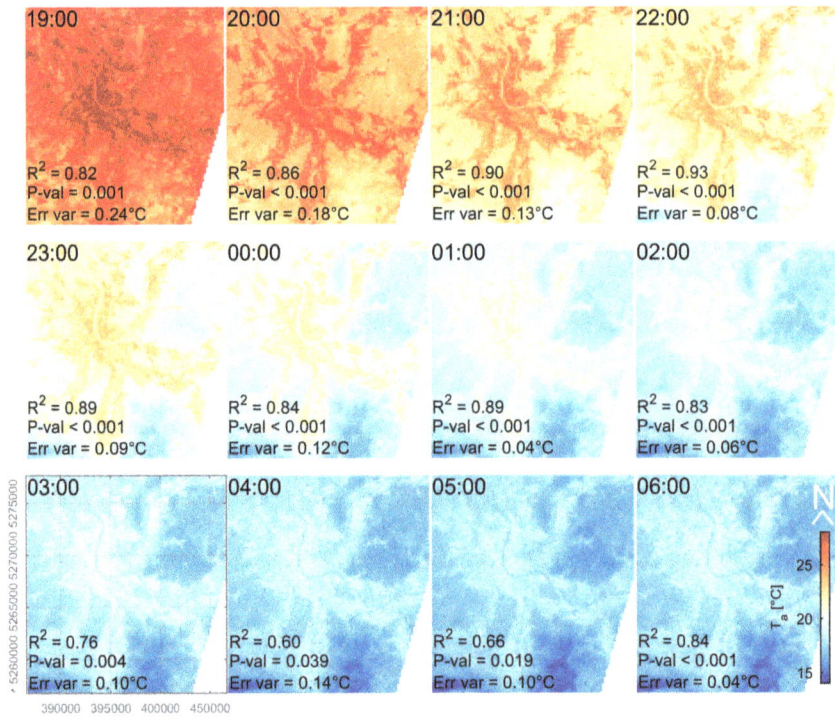

Figure 4. Outputs for the Land Cover model representing the about-roof level temperature for the specific heat wave in August 2016 as an average with 200 m resolution. The coordinate system indicated in the lower left map is UTM-32N and the time is CET.

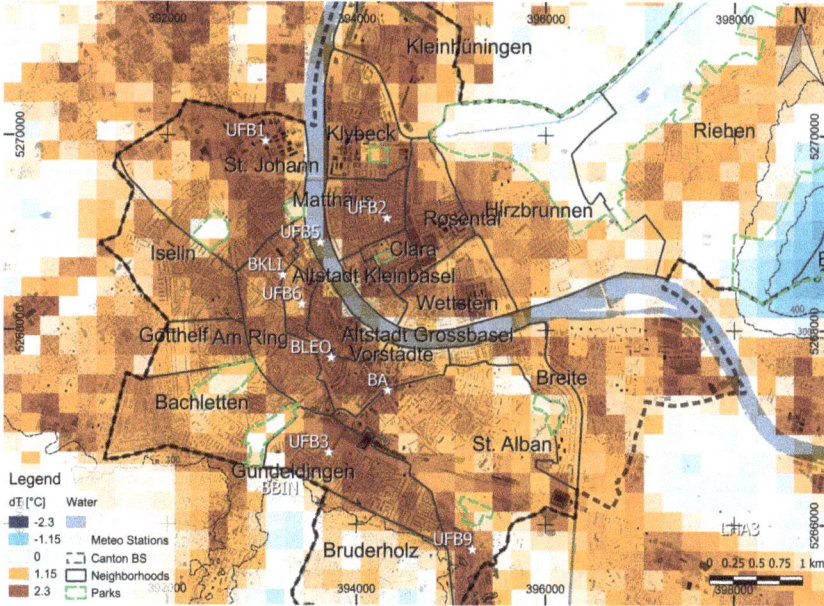

Figure 5. Zoom to the city center showing the difference between each grid point and the MeteoSwiss station Basel/Binningen (BBIN, lower left third of the map) based on the model output with 200 m grid size and the Land Cover predictors at 22:00 CET. The border of the Canton Basel-Stadt (BS) is shown as a dashed line and the buildings are plotted as a dimmed overlay. Large parks within BS are manually included as green polygons and the different neighborhoods are indicated as labelled dark grey polygons. The white stars are measurement stations used for the analysis. The river Rhine is shown as a transparent blue overlay and the coordinate system is UTM-32N.

3.3. Validation

The RMSEs of the validation runs are clearly below the standard deviation of the measurements at the validation sites for all input predictor groups with the highest errors for the Morphology group and the best results using the Land Cover group, on average. The latter performed with RMSE values ranging from 0.07 to 0.27 °C in the exemplary run. The validation shows that some stations are specifically important for the model, especially if they are located in a unique environment. For a successful model run, a certain variation of the input predictor is needed. RMSEs are varying from one run to the other, because the different predictor setup creates slightly different coefficients for the regression formula. Nevertheless, the tendencies seem to remain constant and the results are still reasonable, even though a third of the input predictors are removed. Figure 6 shows deviation plots for five random sample sets on the 200 m grid, which is the resolution that performed best with the lowest RMSE. The largest errors are typically found at the beginning of the night, when the UHI is not yet fully developed. The error often decreases with decreasing temperature and increasing UHI intensity.

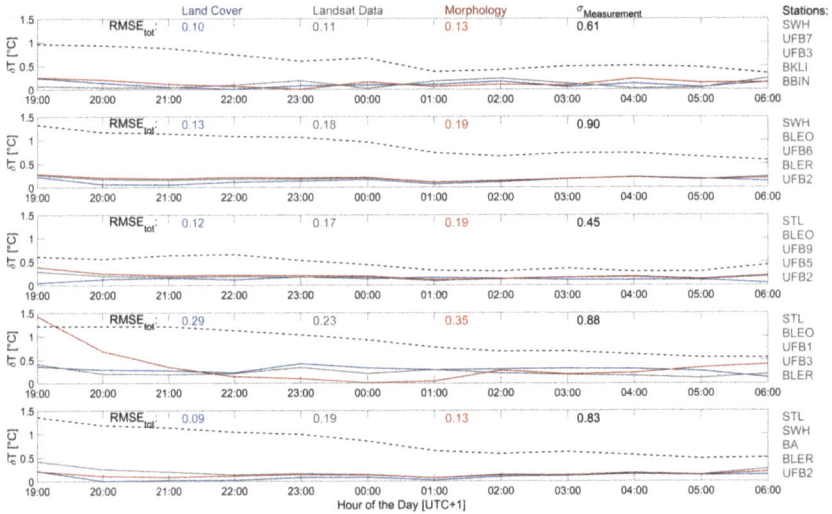

Figure 6. Validation of the model output using five different random samples on a 200 m grid for three groups. For every step in time (*x*-axis), the RMSE resulting from the difference between predicted T_a to measured T_a is indicated for each group. The resulting average is shown as numbers referring to the specific groups according to the color (blue = Land Cover, grey = Landsat Data and red = Morphology) and the stations used for validation are listed at the right side of the plots. The black dotted line depicts the standard deviation of the measured T_a used for validation.

4. Discussion

4.1. Synoptic Conditions and Meteorological Measurements

The synoptic conditions during the investigation period were very calm with a persistent high-pressure system centering around 1025 hPa around northeast Germany. Due to the orientation of the high-pressure system, winds were expected to be very weak and coming from the eastern direction. The end of the period resulted from an eastward movement of the high-pressure system and the afterwards approaching cold front on 29 August 2016. During this small heat wave, a strong upper-atmospheric ridge developed with a stable omega-like pattern providing Central Europe with warm and dry air. The meteorological conditions measured at the urban reference station above roof-level delivered the same results as expected from the weather maps. The nocturnal wind was usually blowing from the eastern and east-southeastern sector with a mean velocity of 1.7 m s^{-1}, almost never exceeding 4 m s^{-1}. A wind system providing the city with fresh air regularly established in the afternoon, which is also one reason for the almost absent heat island during the day and the homogeneous temperature distribution. The development of the observed heat island in Basel started shortly after T_{max} at around 18:00 CET (Figure 7) and generally peaked at around 22:00 CET. A maximum heat island intensity of 5.3 °C was found in this period, with a total average nocturnal heat island intensity of about 2 °C. During 15% of the time, the temperature difference between the urban and the rural station was larger than 3 °C between 19:00 and 06:50 CET. Most of the nights, the UHI intensity decreased steadily until it almost turned into a cooling island after sunrise in the morning hours. With increasing turbulence in the afternoon and the above-described wind system establishing at about noon, no significant difference between the urban and the rural measurement stations are apparent. On average, a slight cooling island with −0.2 °C was measured between 07:00 and 18:50 CET.

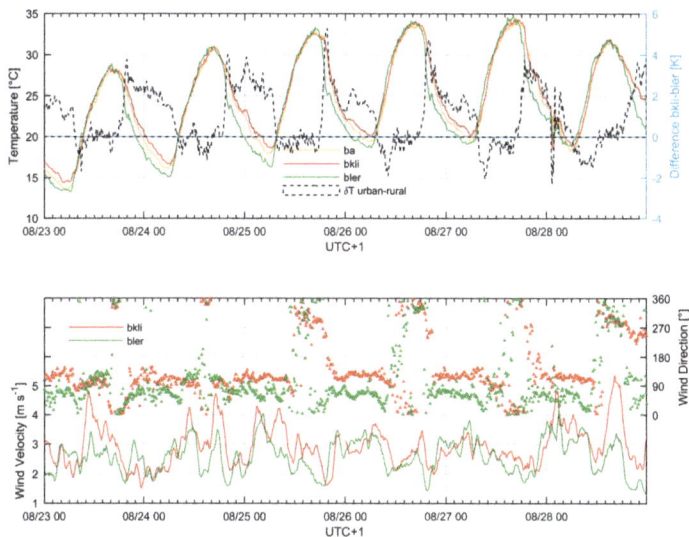

Figure 7. Time series of air temperature (upper plot) and wind (lower plot) measurements at urban (red and orange) and rural (green) reference stations for the entire investigation period. The light blue area in the upper plot represents the difference between the rural and the urban reference station. The x-axis label indicates the time in the following format: *mm/dd/hh*.

4.2. Model Output

The model output depicts a typical UHI development during the night with some colder areas in the suburbs and hot spots in the center and at industrial sites, including railway stations. T_a distribution seems to be rather a "heat archipelago of a city", as mentioned by Kuttler (2012) [84]. Other studies using coarser datasets, such as MODIS or AVHRR, found one single UHI [85]. Within this urban archipelago, urban parks are well distinguished as cool ponds. This shows the strength of the model, because measurement sites in urban parks are very limited and the cooling effect of urban parks (i.e., park cooling island) is well described in previous urban climate studies [43,83,86]. Due to the small size of the parks in Basel, i.e., below 10 ha, these features only become apparent with increasing resolution. Since also the R^2/p-value decrease/increase with coarser resolution, a minimum cell size of 200 m for medium-sized European cities, i.e., cities with a population <500,000 inhabitants, is proclaimed. Depending on the predictor used and the measurement height of T_a, a narrow resolution of 100 m—representing only the specific environment (i.e., the specific street canyon)—might be a good solution. For comprehensive maps with T_a measurements at about roof level, 100 m cell size is not ideal. Brazel and Stoll (1992) found a resolution of around 250 m reasonable, which was identified for North American cities with larger city structures [67]. Therefore, a 200 m resolution in this model seems to be realistic.

Eliasson and Upmanis (2000) found the greatest difference between urban parks and the surrounding urban areas between 2 and 6 h after sunset, which is exactly the time where the MLR model performed the best, considering the statistical analysis [86]. Park breezes, as mentioned by Eliasson & Upmanis (1999) and Oke (1989), and cold air drainage flows in general, are difficult to include in the model since they are allochthonous and have, besides barrier and channeling effects, no relation to the surface properties [43,86]. This might be a future improvement in the model.

4.3. Statistical Measures

It is difficult to contextualize the resulting statistical measures since most studies focused on different methods, periods, resolutions and/or datasets. If larger datasets (in number and spatial) are used, a higher error variance is expected, which cannot be explained with a better or worse model performance. On the other hand, if seasonal averages are used, less fluctuation and higher regression results are expected. This is also responsible for the usually lower RSME in our investigation (~0.1 to ~0.3 °C) compared to other studies, even though the regression coefficient is mediocre (between 0.71 and 0.82 with 200-m grid size on average), compared to other 'successful' studies.

Nevertheless, as important as high regression coefficients are the significance is expressed through low *p*-values. Thereby, the null hypothesis states that there is no relationship between the measurements and the predictors. A low *p*-value indicates that this null hypothesis is wrong and the model is significant. In this analysis, 0.05 was defined as the upper benchmark. If this benchmark is fulfilled, the specific predictors are useful for predicting T_a. If this threshold is not met, the model output should not be considered, even if the error variance is low or the R^2 is very high. The analysis of multi-collinearity showed high VIF for the class Land Cover at higher resolution. The improving statistics at a resolution of 600 m might be explained by mapping unit problems and not be related to a better description of the model in this case.

Other studies also included different types of MLR approaches. Stoll and Brazel (1992) found correlation coefficients >0.9 with a stepwise multiple regression using helicopter and in situ TIR data [67]. Epperson (1995) has found R^2 up to 0.92 using NDVI and nighttime brightness with an error of 1.84 °C, which can be explained through the large variance using 1000 stations [65]. Blennow (1998) achieved comparable results (R^2 = 0.87, RMSE = 1.08 °C) in a rural surrounding and Cresswell (1999) found R^2 between 0.80 and 0.84, but with a substantial error of 3.3 to 4.2 °C, which can also be explained with the large variance (modelled T_a for entire Africa) [75,87]. Ninyerola (2000), and later Christobal (2008) with a comparable approach, reached correlation coefficients up to 0.97 and 0.86, respectively [76,79]. They applied monthly and annual minimum, mean and maximum T_a and used different climatological factors such as altitude, latitude, continentality and solar radiation. Cristobal (2008) added remote sensing variables and called his model a "remote sensing and GIS hybrid", which is comparable to our idea [76]. Many studies are concentrated on monthly or daily mean, minimum or maximum values and often found best correlation predicting daily minimum (i.e., nocturnal) T_a (e.g., [64,65,68,76,88–91] etc.). Modelling diurnal (or nocturnal) courses are seldom done. Bechtel (2014 and 2017) has applied MLR based on multi-temporal Meteosat SEVIRI data to predict short term T_a with very high correlation (R^2 between 0.97 to 0.98) [77,92]. Nichol (2005 and 2009) used ASTER nighttime data in Hong Kong and compared the remote sensing data to T_a—measured during vehicle traverses and at measurement stations with R^2 between 0.80 and 0.94 [59,69]. Ma (2015) noted that the results of single predictor studies could be improved using multiple predictors and found better correlations using higher resolution data [72]. This is confirmed by our study, but with a lower limit in spatial resolution of 200 m.

Even though a comparison of our regression approach with other studies is difficult, the statistical parameters are very promising and allow rejecting a model output in case of low significance and/or low correlation. The good performance is particularly confirmed by the random sampling validation, which indicates that for the tested grid cells T_a is predicted with high accuracy.

4.4. Potentials

The analysis of different model setups to evaluate the best method for extrapolating measured temperature into a two-dimensional environment has shown that the type of input predictor is not the most critical decision, but that it is essential to have multiple predictors. The selection of the responses are very critical, because a certain variation in the surrounding land cover of the measurement stations is crucial for the success of the model. Linear correlations between LST and T_a must be handled carefully, because the amount of thermal infrared radiation does not include the thermal properties of the specific surface (e.g., thermal inertia) and the interaction with the overlying air. For example, surfaces with a high thermal admittance readily accept energy during the day, transmit it to the substrate and release it at night with similar capability [24]. If a surface has a low thermal admittance, it reacts very fast to energy input and is relatively warm during the day, but it loses this energy similarly very quickly and is therefore unable to maintain positive energy fluxes to the surface during the night [6]. The ability of storing this energy might be explained by other parameters, such as the building density (λ_P).

An important application of the resulting dataset is the assessment of heat stress due to high nocturnal T_{min}. Small increase in urban T_{min} during strong heat waves increases the mortality and the morbidity substantially [14,16,19]. During the night, most people are located inside their building and therefore exposed to the respective indoor climate, which does not necessarily need to be the same as the outdoor climate [15]. Nevertheless, the outdoor temperature is the best measure of nocturnal heat stress that is widely available since not every household can be monitored. The resulting maps offer a high-resolution view on the heat distribution throughout a large region and they can be used for risk maps or intervention guidelines. For example, during heat waves, nocturnal heat stress in every grid cell can be summarized after every night, and areas that transcend a certain threshold can be considered for special care or even evacuation of vulnerable people.

The findings are also suitable for surface flux estimation, where the difference between T_a and LST is an important component required to compute the sensible heat flux [93].

The model is evaluated using average temperatures of six consecutive days to exclude singularity effects and compare different setups for testing. The model is also tested as a heat stress predictor with individual hourly means during the evaluation period. During the night with the strongest heat island intensity (25 to 26 August 2016), the correlation coefficients reach values of 0.97 with very high significance. This proves that the model is also applicable on single day events with hourly temperature measurements as the response. As a measure for difference in heat stress, maps with the number of tropical nights during the specific heat wave are developed. Despite the higher elevated areas, in almost every grid cell at least one tropical night happened. In some areas, the number of tropical nights reached three during this period. The official measurement station (i.e., BBIN) is located in an area where only one tropical night was measured and this might lead to underestimation of heat stress in other areas (Figure 8).

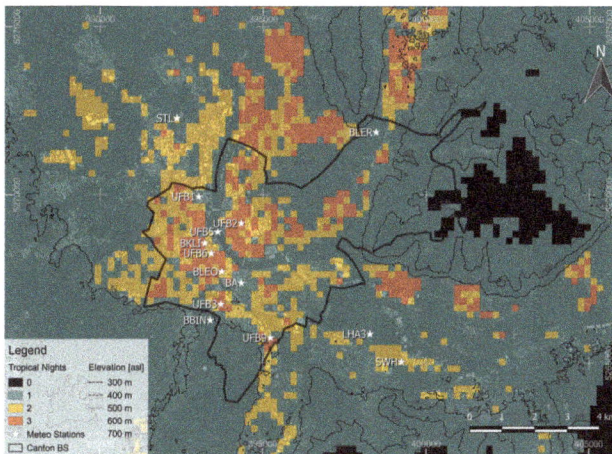

Figure 8. Number of tropical nights ($T_{min} > 20\,^\circ C$) between the 23 and the 29 August 2016 based on the model prediction in a 200-m grid. A DOM behind the data layer depicts the city structure and contour lines describe the topography. The coordinate system is UTM-32N.

5. Conclusions

The evaluation of multiple predictors influencing the development of the urban heat island (UHI) is shown and described in detail. Based on an empirical approach using in situ measurements, different groups of input predictors are tested within different grid resolutions to find the best parameters, resolution and time for a multiple linear regression model (MLR) to predict the UHI distribution and intensity.

Average correlation coefficients are ranging from 0.71 to 0.82 for the nocturnal model run and for the model resolution with a 200 m grid size, which emerged as the optimum spatial model resolution. The highest correlation coefficients are found at 22:00 CET, indicating that the nocturnal UHI phenomenon is most pronounced during that time under typical heat wave conditions, with 0.93 for the best model run during the entire measurement period (average diurnal course) and 0.97 for single nights. All model runs show high significance if applied on the 200 m resolution during the entire night.

Other studies based on multiple regression often predict monthly mean or climatological data with less fluctuations, larger measurement differences and less outliers, which provides higher correlations, but also larger error bias. In this study, the aim of the evaluation is a model applicable on near real-time air temperatures with a resolution representing different urban structures. The resulting two-dimensional maps are an innovation considering temporal and spatial resolution and they are useful for medical staff or, if used for statistical analysis of heat waves, as guidelines for urban planners and politics. Based on the above presented results, the implementation of a spatially coherent UHI surveillance can be further investigated.

An important finding of the study is that different datasets can be used with comparable results, which means the model is not tied to specific data and the user can decide which model setup provides the best results for his location. In our case, the use of land cover fraction including the distribution of sealed and built-up surfaces plus the LST offered best results.

The recently intensified discussion about the classification of different thermal zones throughout a city, i.e., the local climate zone scheme [44], improves the scientific knowledge about the UHI distribution and supported the discussion beyond urban to rural differences. The findings within this study are adding important insights to this discussion and they offer a quantification of theoretical considerations. Since an increasing number of people are living in the urban area compared to its rural counterpart, it is crucial to account for inner-urban differentiations.

Future improvement will be the testing of the method with less air temperature input data and the application in other cities.

Based on this study, tests and evaluations during the next summer are planned to further improve the model and make it a valuable tool for urban planners, the local administration and medical care.

Author Contributions: A.W. conceived and designed the experiments, conducted the analysis and is responsible for writing; E.P. helped developing the idea, supervised the analyses and the writing process and he is responsible for parts of the text; C.F. is responsible for parts of the measurement network, helped developing ideas and is responsible for the review of the text.

Funding: The project leading to this application has received funding from the European Union's Horizon 2020 research and innovation program under grant agreement No. 637519 (URBANFLUXES).

Acknowledgments: Many thanks go to Dieter Scherer for the initial idea and Fred Meier for a lot of discussion at the beginning of the project, and Stavros Stagakis providing ideas to improve the study. The whole process always includes many persons responsible for the database, maintenance and fruitful talks during lunch breaks. Therefore, I would like to thank all members of the MCR research group at the University Basel.

Conflicts of Interest: The authors declare no conflict of interest.

Appendix A

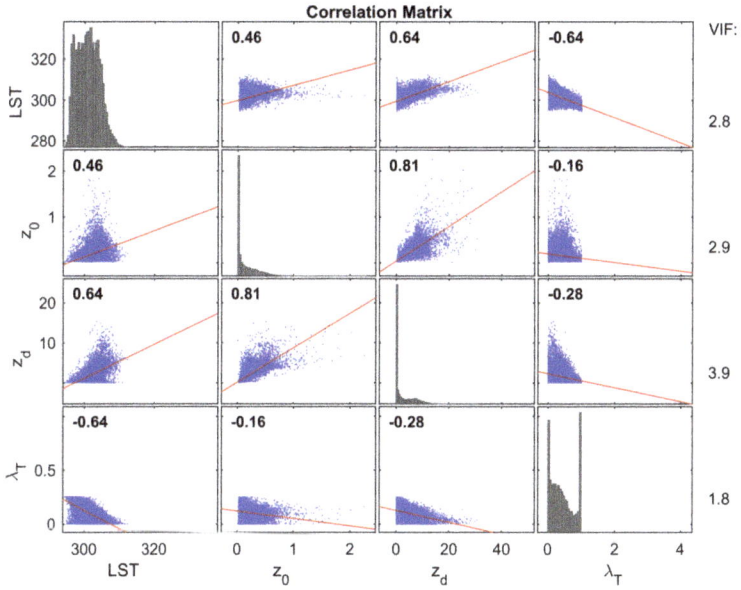

Figure A1. Correlation matrix and VIF analysis for the group Morphology at a resolution of 200 m.

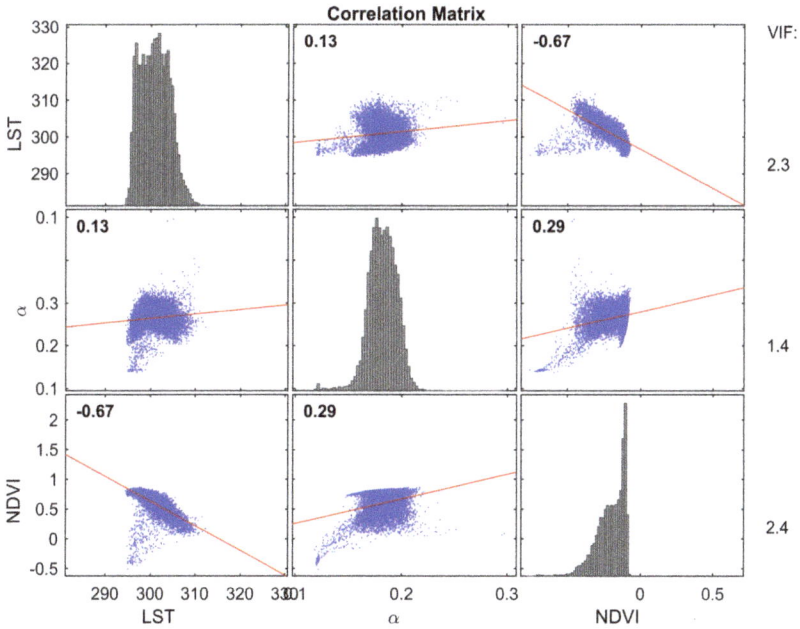

Figure A2. Correlation matrix and VIF analysis for the group Landsat Data at a resolution of 200 m.

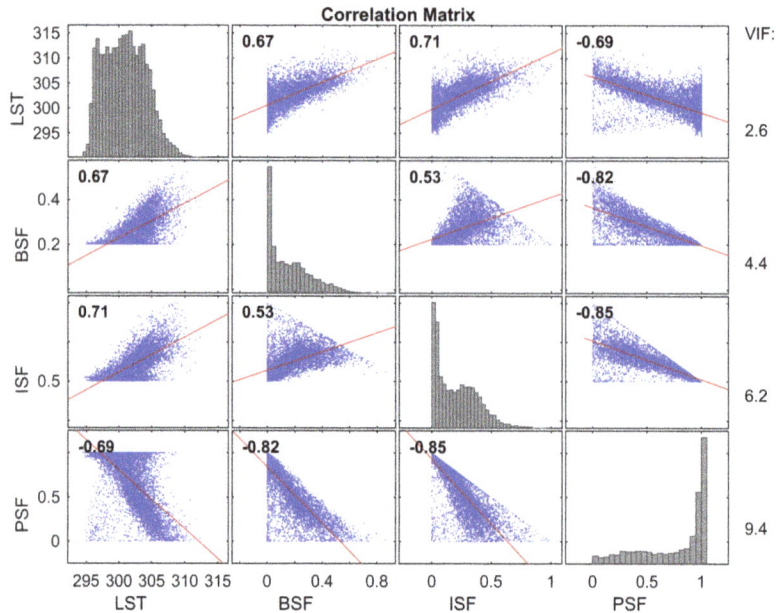

Figure A3. Correlation matrix and VIF analysis for the group Land Cover at a resolution of 200 m.

References

1. WHO. Global Health Observatory Data Repository. Available online: http://apps.who.int/gho/data/node. main.nURBPOP?lang=en (accessed on 13 December 2016).
2. López-Moreno, E.; Oyeyinka, O.; Mboup, G. *State of the World's Cities 2010/2011. Bridging the Urban Divide*; Earthscan: London, UK, 2008; p. 220. ISBN 978-1-84971-176-0.
3. Oke, T.R.; Fuggle, R.F. Comparison of urban/rural counter and net radiation at night. *Bound.-Layer Meteorol.* **1972**, *2*, 290–308. [CrossRef]
4. Oke, T.R. The distinction between canopy and boundary-layer urban heat islands. *Atmosphere* **1976**, *14*, 268–277. [CrossRef]
5. Oke, T.R. Canyon geometry and the nocturnal urban heat island: Comparison of scale model and field observations. *J. Climatol.* **1981**, *1*, 237–254. [CrossRef]
6. Oke, T.R. Street design and urban canopy layer climate. *Energy Build.* **1988**, *11*, 103–113. [CrossRef]
7. Oke, T.R.; Johnson, G.T.; Steyn, D.G.; Watson, I.D. Simulation of surface urban heat islands under 'ideal' conditions at night Part 2: Diagnosis of causation. *Bound.-Layer Meteorol.* **1991**, *56*, 339–358. [CrossRef]
8. Roth, M.; Oke, T.R.; Emery, W.J. Satellite-derived urban heat islands from three coastal cities and the utilization of such data in urban climatology. *Int. J. Remote Sens.* **1989**, *10*, 1699–1720. [CrossRef]
9. Karl, T.R.; Diaz, H.F.; Kukla, G. Urbanization: Its detection and effect in the united states climate record. *J. Clim.* **1988**, *1*, 1099–1123. [CrossRef]
10. Arnfield, A.J. Two decades of urban climate research: A review of turbulence, exchanges of energy and water and the urban heat island. *Int. J. Climatol.* **2003**, *23*, 1–26. [CrossRef]
11. Ren, C.; Ng, E.; Katzschner, L. Urban climatic map studies: A review. *Int. J. Climatol.* **2011**, *31*, 2213–2233. [CrossRef]
12. Parlow, E.; Vogt, R.; Feigenwinter, C. The urban heat island of Basel—Seen from different perspectives. *DIE ERDE J. Geogr. Soc. Berl.* **2014**, *145*, 96–110. [CrossRef]
13. Mayer, H.; Holst, J.; Dostal, P.; Imbery, F.; Schindler, D. Human thermal comfort in summer within an urban street canyon in Central Europe. *Meteorol. Z.* **2008**, *17*, 241–250. [CrossRef]

14. Clarke, J.F.; Bach, W. Comparison of the comfort conditions in different urban and suburban microenvironments. *Int. J. Biometeorol.* **1971**, *15*, 41–54. [CrossRef] [PubMed]

15. Scherer, D.; Fehrenbach, U.; Lakes, T.; Lauf, S.; Meier, F.; Schuster, C. Quantification of heat-stress related mortality hazard, vulnerability and risk in Berlin, Germany. *DIE ERDE J. Geogr. Soc. Berl.* **2014**, *144*, 238–259. [CrossRef]

16. Ragettli, M.S.; Vicedo-Cabrera, A.M.; Schindler, C.; Roosli, M. Exploring the association between heat and mortality in Switzerland between 1995 and 2013. *Environ. Res.* **2017**, *158*, 703–709. [CrossRef] [PubMed]

17. Schär, C.; Jendritzky, G. Climate change: Hot news from summer 2003. *Nature* **2004**, *432*, 559–560. [CrossRef] [PubMed]

18. Robine, J.-M.; Cheung, S.L.; Roy, S.L.; Oyen, H.V.; Griffiths, C.; Michel, J.-P.; Herrmann, F.R. Death toll exceeded 70,000 in Europe during the summer of 2003. *C. R. Biol.* **2008**, *331*, 171–178. [CrossRef] [PubMed]

19. Vicedo-Cabrera, A.M.; Ragettli, M.S.; Schindler, C.; Roosli, M. Excess mortality during the warm summer of 2015 in Switzerland. *Swiss Med. Wkly* **2016**, *146*, 12. [CrossRef] [PubMed]

20. Stocker, T.F.; Qin, D.; Plattner, G.-K.; Tignor, M.; Allen, S.K.; Boschung, J.; Nauels, A.; Xia, Y.; Bex, V.; Midgley, P.M. *Climate Change 2013: The Physical Science Basis. Contribution of Working Group I to the Fifth Assessment Report of the Intergovernmental Panel on Climate Change*; Cambridge University Press: Cambridge, UK; New York, NY, USA, 2013; p. 1535.

21. Gerstengarbe, F.-W.; Werner, P.C. *Katalog der Großwetterlagen Europas nach Paul Hess und Helmuth Brezowsky (1881–2004)*; Potsdam Institude for Climate Impact Research (PIK): Potsdam, Germany, 2005.

22. Karl, T.R.; Trenberth, K.E. Modern global climate change. *Science* **2003**, *302*, 1719–1723. [CrossRef] [PubMed]

23. Meehl, G.A.; Tebaldi, C. More intense, more frequent and longer lasting heat waves in the 21st century. *Science* **2004**, *305*, 994–997. [CrossRef] [PubMed]

24. Oke, T.R. The urban energy balance. *Prog. Phys. Geogr.* **1988**, *12*, 471–508. [CrossRef]

25. Grimmond, C.S.B.; Best, M.; Barlow, J.; Arnfield, A.J.; Baik, J.-J.; Baklanov, A.; Belcher, S.; Bruse, M.; Calmet, I.; Chen, F.; et al. Urban surface energy balance models: Model characteristics and methodology for a comparison study. In *Meteorological and Air Quality Models for Urban Areas*; Baklanov, A., Grimmond, C.S.B., Mahura, A., Athanassiadou, M., Eds.; Springer: Berlin, Germany, 2009; pp. 97–124.

26. Christen, A.; Vogt, R. Energy and radiation balance of a central European city. *Int. J. Climatol.* **2004**, *24*, 1395–1421. [CrossRef]

27. Nunez, M.; Oke, T.R. The energy balance of an urban canyon. *J. Appl. Meteorol.* **1977**, *16*, 11–19. [CrossRef]

28. Parlow, E. Correction of terrain controlled illumination effects in satellite data. In *Progress in Environmental Remote Sensing Research and Applications*; Parlow, E., Ed.; Balkema: Rotterdam, The Netherlands, 1996; pp. 139–145.

29. Jiménez-Muñoz, J.C.; Sobrino, J.A. Error sources on the land surface temperature retrieved from thermal infrared single channel remote sensing data. *Int. J. Remote Sens.* **2006**, *27*, 999–1014. [CrossRef]

30. Li, Z.-L.; Tang, B.-H.; Wu, H.; Ren, H.; Yan, G.; Wan, Z.; Trigo, I.F.; Sobrino, J.A. Satellite-derived land surface temperature: Current status and perspectives. *Remote Sens. Environ.* **2013**, *131*, 14–37. [CrossRef]

31. Liang, S. Narrowband to broadband conversions of land surface albedo I. *Remote Sens. Environ.* **2001**, *76*, 213–238. [CrossRef]

32. Chrysoulakis, N. Estimation of the all-wave urban surface radiation balance by use of ASTER multispectral imagery and in situ spatial data. *J. Geophys. Res.* **2003**, *108*. [CrossRef]

33. Chrysoulakis, N.; Marconcini, M.; Gastellu-Etchegorry, J.-P.; Grimmong, C.S.B.; Feigenwinter, C.; Lindberg, F.; Del Frate, F.; Klostermann, J.; Mi, Z.; Esch, T.; et al. Anthropogenic heat flux estimation from space. In Proceedings of the 2017 Joint Urban Remote Sensing Event (JURSE), Dubai, UAE, 6–8 March 2017; pp. 1–4. [CrossRef]

34. Grimmond, C.S.B.; Oke, T.R. Heat storage in urban areas: Local-scale observations and evaluation of a simple model. *J. Appl. Meteorol.* **1999**, *38*, 922–940. [CrossRef]

35. Christen, A.; Vogt, R.; Rotach, M.; Parlow, E. First results from BUBBLE II: Partioning of turbulent heat fluxes over urban surfaces. In Proceedings of the Fourth Symposium on the Urban Environment, Norfolk, VA, USA, 19 May 2002; American Meteorological Society: Norfolk, VA, USA, 2002; pp. 137–138.

36. Parlow, E. Net radiation in the REKLIP-area—A spatial approach using satellite data. In *Progress in Environmental Remote Sensing Research and Applications*; Parlow, E., Ed.; Balkema: Rotterdam, The Netherlands, 1996.

37. Kottek, M.; Grieser, J.; Beck, C.; Rudolf, B.; Rubel, F. World map of the Köppen-Geiger climate classification updated. *Meteorol. Z.* **2006**, *15*, 259–263. [CrossRef]

38. MeteoSwiss. *Climate Normals Basel/Binningen. Reference Period 1981–2010*; Federal Office of Meteorology and Climatology MeteoSwiss: Zurich, Switzerland, 2016.

39. Scherer, D.; Fehrenbach, U.; Beha, H.D.; Parlow, E. Improved concepts and methods in analysis and evaluation of the urban climate for optimizing urban planning processes. *Atmos. Environ.* **1999**, *33*, 4185–4193. [CrossRef]

40. Rotach, M.W.; Vogt, R.; Bernhofer, C.; Batchvarova, E.; Christen, A.; Clappier, A.; Feddersen, B.; Gryning, S.E.; Martucci, G.; Mayer, H.; et al. BUBBLE—An urban boundary layer meteorology project. *Theor. Appl. Climatol.* **2005**, *81*, 231–261. [CrossRef]

41. Chrysoulakis, N.; Feigenwinter, C.; Triantakonstantis, D.; Penyevskiy, I.; Tal, A.; Parlow, E.; Fleishman, G.; Düzgün, S.; Esch, T.; Marconcini, M. A conceptual list of indicators for urban planning and management based on Earth observation. *ISPRS Int. J. Geo-Inf.* **2014**, *3*, 980–1002. [CrossRef]

42. Basel-Stadt. *Basel-Stadt in Zahlen 2015*; Statistisches Amt des Kantons Basel-Stadt: Basel, Switzerland, 2015.

43. Oke, T.R.; Cleugh, H.A.; Grimmond, S.; Schmid, H.P.; Roth, M. Evaluation of spatially-averaged fluxes of heat, mass and momentum in the urban boundary layer. *Weather Clim.* **1989**, *9*, 14–21.

44. Stewart, I.; Oke, T. Local climate zones for urban temperature studies. *Bull. Am. Meteorol. Soc.* **2012**, *93*, 1879–1900. [CrossRef]

45. Wicki, A.; Parlow, E. Attribution of local climate zones using a multitemporal land use/land cover classification scheme. *J. Appl. Remote Sens.* **2017**, *11*, 026001. [CrossRef]

46. Barsi, J.; Barker, J.; Schott, J. An atmospheric correction parameter calculator for a single thermal band earth-sensing instrument. In Proceedings of the Geoscience and Remote Sensing Symposium, Toulouse, France, 21–25 July 2003; IEEE: Toulouse, France, 2003; Volume 5, pp. 3014–3016.

47. Sobrino, J.A.; Jiménez-Muñoz, J.C.; Sòria, G.; Romaguera, M.; Guanter, L.; Moreno, J.; Plaza, A.; Martìnez, P. Land surface emissivity retrieval from different VNIR and TIR sensors. *IEEE Trans. Geosci. Remote Sens.* **2008**, *46*, 316–327. [CrossRef]

48. Wicki, A.; Parlow, E. Multiple regression analysis for unmixing of surface temperature data in an urban environment. *Remote Sens.* **2017**, *9*, 684. [CrossRef]

49. Wang, C.; Myint, S.; Wang, Z.; Song, J. Spatio-Temporal Modeling of the Urban Heat Island in the Phoenix Metropolitan Area: Land Use Change Implications. *Remote Sens.* **2016**, *8*, 185. [CrossRef]

50. Bastiaanssen, W.G.M.; Pelgrum, H.; Wang, J.; Ma, Y.; Moreno, J.F.; Roerink, G.J.; van der Wal, T. A remote sensing surface energy balance algorithm for land (SEBAL). *J. Hydrol.* **1998**, *212*, 213–229. [CrossRef]

51. Lindberg, F.; Grimmond, C.S.B.; Gabey, A.; Huang, B.; Kent, C.W.; Sun, T.; Theeuwes, N.E.; Järvi, L.; Ward, H.C.; Capel-Timms, I.; et al. Urban Multi-scale Environmental Predictor (UMEP): An integrated tool for city-based climate services. *Environ. Model. Softw.* **2018**, *99*, 70–87. [CrossRef]

52. Raupach, M.R. Simplified expressions for vegetation roughness length and zero-plane displacement as functions of canopy height and area index. *Bound.-Layer Meteorol.* **1994**, *71*, 211–216. [CrossRef]

53. Grimmond, C.S.B.; Oke, T.R. Aerodynamic properties of urban areas derived from analysis of surface form. *J. Appl. Meteorol.* **1999**, *38*, 1262–1292. [CrossRef]

54. Kanda, M.; Inagaki, A.; Miyamoto, T.; Gryschka, M.; Raasch, S. A New Aerodynamic Parametrization for Real Urban Surfaces. *Bound.-Layer Meteorol.* **2013**, *148*, 357–377. [CrossRef]

55. Basel-Stadt. *Geoportal Kanton Basel-Stadt*; Grundbuch- und Vermessungsamt des Kantons Basel-Stadt: Basel, Switzerland, 2016.

56. Marconcini, M.; Heldens, W.; Del Frate, F.; Latini, D.; Mitraka, Z.; Lindberg, F. EO-based products in support of urban heat fluxes estimation. In Proceedings of the Joint Urban Remote Sensing Event (JURSE), Dubai, UAE, 6–8 March 2017; IEEE: Dubai, UAE, 2017; pp. 1–4.

57. Caselles, V.; Lopez Garcia, M.J.; Melia, J.; Perez Cueva, A.J. Analysis of the heat-island effect of the city of Valencia, Spain, through air temperature transects and NOAA satellite data. *Theor. Appl. Climatol.* **1991**, *43*, 195–203. [CrossRef]

58. Ben-Dor, E.; Saaroni, H. Airborne video thermal radiometry as a tool for monitoring microscale structures of the urban heat island. *Int. J. Remote Sens.* **1997**, *18*, 3039–3053. [CrossRef]

59. Nichol, J.E.; Fung, W.Y.; Lam, K.-S.; Wong, M.S. Urban heat island diagnosis using ASTER satellite images and 'in situ' air temperature. *Atmos. Res.* **2009**, *94*, 276–284. [CrossRef]

60. Seidel, J.; Ketzler, G.; Bechtel, B.; Thies, B.; Philipp, A.; Böhner, J.; Egli, S.; Eisele, M.; Herma, F.; Langkamp, T.; et al. Mobile measurement techniques for local and micro-scale studies in urban and topo-climatology. *DIE ERDE J. Geogr. Soc. Berl.* **2016**, *147*, 15–39. [CrossRef]

61. Maras, I.; Buttstädt, M.; Hahmann, J.; Hofmeister, H.; Schneider, C. Investigating public places and impacts of heat stress in the city of Aachen, Germany. *DIE ERDE J. Geogr. Soc. Berl.* **2013**, *144*, 290–303. [CrossRef]

62. Gallo, K.P.; Tarpley, J.D.; McNab, A.L.; Karl, T.R. Assessment of urban heat islands: A satellite perspective. *Atmos. Res.* **1995**, *37*, 37–43. [CrossRef]

63. Gallo, K.P.; McNab, A.L.; Karl, T.R.; Brown, J.F.; Hood, J.J.; Tarpley, J.D. The use of a vegetation index for assessment of the urban heat island effect. *Int. J. Remote Sens.* **1993**, *14*, 2223–2230. [CrossRef]

64. Gallo, K.P.; Owen, T.W. Satellite-based adjustments for the urban heat island temperature bias. *J. Appl. Meteorol.* **1999**, *38*, 806–813. [CrossRef]

65. Epperson, D.L.; Davis, J.M.; Bloomfield, P.; Karl, T.R.; McNab, A.L.; Gallo, K.P. Estimating the urban bias of surface shelter temperatures using upper-air and satellite data. Part II: Estimation of the urban bias. *J. Appl. Meteorol.* **1995**, *34*, 340–357. [CrossRef]

66. Dousset, B. AVHRR-derived cloudiness and surface temperature patterns over the Los Angeles area and their relationships to land use. In Proceedings of the 12th Canadian Symposium on Remote Sensing Geoscience and Remote Sensing Symposium, Vancouver, BC, Canada, 10–14 July 1989; Volume 4, pp. 2132–2137. [CrossRef]

67. Stoll, M.J.; Brazel, A.J. Surface-air temperature relationships in the urban environment of Phoenix, Arizona. *Phys. Geogr.* **1992**, *13*, 160–179. [CrossRef]

68. Vancutsem, C.; Ceccato, P.; Dinku, T.; Connor, S.J. Evaluation of MODIS land surface temperature data to estimate air temperature in different ecosystems over Africa. *Remote Sens. Environ.* **2010**, *114*, 449–465. [CrossRef]

69. Nichol, J. Remote sensing of urban heat islands by day and night. *Photogramm. Eng. Remote Sens.* **2005**, *71*, 613–621. [CrossRef]

70. Fu, G.; Shen, Z.; Zhang, X.; Shi, P.; Zhang, Y.; Wu, J. Estimating air temperature of an alpine meadow on the Northern Tibetan Plateau using MODIS land surface temperature. *Acta Ecol. Sin.* **2011**, *31*, 8–13. [CrossRef]

71. Voogt, J.A.; Oke, T.R. Thermal remote sensing of urban climates. *Remote Sens. Environ.* **2003**, *86*, 370–384. [CrossRef]

72. Ma, W.; Zhou, L.; Zhang, H.; Zhang, Y.; Dai, X. Air temperature field distribution estimations over a Chinese mega-city using MODIS land surface temperature data: The case of Shanghai. *Front. Earth Sci.* **2015**, *10*, 38–48. [CrossRef]

73. Mao, K.B.; Tang, H.J.; Wang, X.F.; Zhou, Q.B.; Wang, D.L. Near-surface air temperature estimation from ASTER data based on neural network algorithm. *Int. J. Remote Sens.* **2008**, *29*, 6021–6028. [CrossRef]

74. Myint, S.W.; Wentz, E.A.; Brazel, A.J.; Quattrochi, D.A. The impact of distinct anthropogenic and vegetation features on urban warming. *Landsc. Ecol.* **2013**, *28*, 959–978. [CrossRef]

75. Cresswell, M.P.; Morse, A.P.; Thomson, M.C.; Connor, S.J. Estimating surface air temperatures, from Meteosat land surface temperatures, using an empirical solar zenith angle model. *Int. J. Remote Sens.* **1999**, *20*, 1125–1132. [CrossRef]

76. Cristóbal, J.; Ninyerola, M.; Pons, X. Modeling air temperature through a combination of remote sensing and GIS data. *J. Geophys. Res.* **2008**, *113*. [CrossRef]

77. Bechtel, B.; Zakšek, K.; Oßenbrügge, J.; Kaveckis, G.; Böhner, J. Towards a satellite based monitoring of urban air temperatures. *Sustain. Cities Soc.* **2017**, *34*, 22–31. [CrossRef]

78. Vogt, J.V.; Viau, A.A.; Paquet, F. Mapping regional air temperature fields using satellite-derived surface skin temperatures. *Int. J. Climatol.* **1997**, *17*, 1559–1579. [CrossRef]

79. Ninyerola, M.; Pons, X.; Roure, J.M. A methodological approach of climatological modelling of air temperature and precipitation through gis techniques. *Int. J. Climatol.* **2000**, *20*, 1823–1841. [CrossRef]

80. Wang, C.; Middel, A.; Myint, S.W.; Kaplan, S.; Brazel, A.J.; Lukasczyk, J. Assessing local climate zones in arid cities: The case of Phoenix, Arizona and Las Vegas, Nevada. *ISPRS J. Photogramm. Remote Sens.* **2018**, *141*, 59–71. [CrossRef]

81. O'brien, R.M. A caution regarding rules of thumb for Variance Inflation Factors. *Qual. Quant.* **2007**, *41*, 673–690. [CrossRef]

82. Craney, T.A.; Surles, J.G. Model-dependent Variance Inflation Factor cutoff values. *Qual. Eng.* **2002**, *14*, 391–403. [CrossRef]

83. Spronken-Smith, R.A.; Oke, T.R. The thermal regime of urban parks in two cities with different summer climates. *Int. J. Remote Sens.* **1998**, *19*, 2085–2104. [CrossRef]

84. Kuttler, W. Climate change on the urban scale—Effects and counter-measures in Central Europe. In *Human and Social Dimensions of Climate Change*; InTech: Garching bei München, Germany, 2012. [CrossRef]

85. Pichierri, M.; Bonafoni, S.; Biondi, R. Satellite air temperature estimation for monitoring the canopy layer heat island of Milan. *Remote Sens. Environ.* **2012**, *127*, 130–138. [CrossRef]

86. Eliasson, I.; Upmanis, H. Nocturnal airflow from urban parks-implications for city ventilation. *Theor. Appl. Climatol.* **2000**, *66*, 95–107. [CrossRef]

87. Blennow, K. Modelling minimum air temperature in partially and clear felled forests. *Agric. For. Meteorol.* **1998**, *91*, 223–235. [CrossRef]

88. Yao, Y.; Zhang, B. MODIS-based estimation of air temperature of the Tibetan Plateau. *J. Geogr. Sci.* **2013**, *23*, 627–640. [CrossRef]

89. Mostovoy, G.V.; Filippova, M.G.; Kakani, V.G.; Reddy, K.R.; King, R.L. Statistical estimation of daily maximum and minimum air temperatures from MODIS LST data over the State of Mississippi. *GIScience Remote Sens.* **2006**, *43*, 78–110. [CrossRef]

90. Emamifar, S.; Rahimikhoob, A.; Noroozi, A.A. Daily mean air temperature estimation from MODIS land surface temperature products based on M5 model tree. *Int. J. Climatol.* **2013**, *33*, 3174–3181. [CrossRef]

91. Benali, A.; Carvalho, A.C.; Nunes, J.P.; Carvalhais, N.; Santos, A. Estimating air surface temperature in Portugal using MODIS LST data. *Remote Sens. Environ.* **2012**, *124*, 108–121. [CrossRef]

92. Bechtel, B.; Wiesner, S.; Zaksek, K. Estimation of dense time series of urban air temperatures from multitemporal geostationary satellite data. *IEEE J. Sel. Top. Appl. Earth Obs. Remote Sens.* **2014**, *7*, 4129–4137. [CrossRef]

93. Feigenwinter, C.; Parlow, E.; Vogt, R.; Schmutz, M.; Chrysoulakis, N.; Lindberg, F.; Marconcini, M.; del Frate, F. Spatial distribution of sensible and latent heat flux in the URBANFLUXES case study city Basel (Switzerland). In Proceedings of the 2017 Joint Urban Remote Sensing Event (JURSE), Dubai, UAE, 6–8 March 2017; pp. 1–4. [CrossRef]

climate

MDPI

Article

The Effect of Building Facades on Outdoor Microclimate—Reflectance Recovery from Terrestrial Multispectral Images Using a Robust Empirical Line Method

Jonathan Fox [1,2,*], Paul Osmond [1] and Alan Peters [3]

[1] Faculty of Built Environment, The University of New South Wales (UNSW), Sydney NSW 2052, Australia; p.osmond@unsw.edu.au
[2] CRC for Low Carbon Living Ltd, Tyree Energy Technology Building—UNSW, Sydney NSW 2052, Australia
[3] School of Architecture and Built Environment, The University of Adelaide, Adelaide SA 5005, Australia; alan.peters@adelaide.edu.au
* Correspondence: jonathan.fox@unsw.edu.au; Tel.: +61-(02)-9385-5247

Received: 6 June 2018; Accepted: 21 June 2018; Published: 25 June 2018

Abstract: Climate change and the urban heat island effect pose significant health, energy and economic risks. Urban heat mitigation research promotes the use of reflective surfaces to counteract the negative effects of extreme heat. Surface reflectance is a key parameter for understanding, modeling and modifying the urban surface energy balance to cool cities and improve outdoor thermal comfort. The majority of urban surface studies address the impacts of horizontal surface properties at the material and precinct scales. However, there is a gap in research focusing on individual building facades. This paper analyses the results of a novel application of the empirical line method to calibrate a terrestrial low-cost multispectral sensor to recover spectral reflectance from urban vertical surfaces. The high correlation between measured and predicted mean reflectance values per waveband (0.940 (Red) $< r_s >$ 0.967 (NIR)) confirmed a near-perfect positive agreement between pairs of samples of ranked scores. The measured and predicted distributions exhibited no statistically significant difference at the 95% confidence level. Accuracy measures indicate absolute errors within previously reported limits and support the utility of a single-target spectral reflectance recovery method for urban heat mitigation studies focusing on individual building facades.

Keywords: urban heat mitigation; albedo; cool facades; spectral reflectance; urban remote sensing; empirical line method; building scale

---

1. Introduction

Anthropogenic alterations to the optical, thermal, moisture and aerodynamic properties of city surfaces generate distinct urban climates, typically characterized by the urban heat island (UHI) effect [1]. The UHI effect refers to hotter air (and surface) temperatures observed in cities compared to non-urban surroundings [2]. UHI spatial and temporal characteristics are influenced by synoptic weather conditions [3,4] but UHI formation is attributed to differences in urban surface structure (3-D geometry), cover (land use and permeability), fabric (optical and thermal properties of materials) and metabolism (human activity) compared to non-urban surroundings [5,6]. Geometric and surface characteristics regulate the partitioning of the surface energy balance (SEB) [2] and at any given location surface temperature—and near-surface air temperature under calm conditions [7,8]—is controlled by the surface's SEB [9].

UHIs develop in most cities regardless of climate [10] and have been observed globally in over 400 urban areas [4]. The magnitude of the screen height air temperature difference between urban and non-urban locations, or between different Local Climate Zones [10], is quantified by the "UHI intensity" which is most pronounced during calm, clear summer nights [2]. The average maximum UHI intensity for 87 European cities has been reported to be almost 6.2 K within a range of 2.8 K to 12 K [11]. Similar magnitude UHI intensities were reported for 101 Asian and Australian cities [12].

The superimposition of inadvertent urban heating (i.e., UHIs) and global warming, including more frequent, longer and intense heat waves [13–15] elevates the risk of heat-related mortality and illness in cities [16–18] and extreme urban heat has detrimental outdoor comfort, economic and building energy impacts [19–21]. The intentional modification of urban surface geometry, cover and fabric to reduce urban heat and decouple heat waves from amplified UHIs [22,23] is now a policy priority for many cities [24,25].

1.1. Urban Heat Mitigation—Current Status and Cooling Magnitude

In summer, solar absorption by urban surfaces is the dominant cause of the UHI effect [5]. Recent efforts to mitigate the formation of urban heat at different spatial scales have focused on changes to urban surface geometry and fabric [26–28] with the primary aim of controlling the absorption of solar radiation and increasing moisture availability [22,29]. However, due to the significant urban surface and air temperature reduction potential of reflective technologies [19,30] heat mitigation solely focusing on "cool" materials—those with high solar reflectance and high infrared emittance [31–33]—applied to building envelopes (roofs and walls) and urban structures (roads, squares and footpaths) dominates current scientific research and the global implementation of UHI mitigation technologies [34].

An analysis of 75 simulation studies using reflective materials on roofs and pavements reported average peak and absolute maximum screen-height air temperature reductions of 1.43 K and 3.4 K respectively for their combinations [35]. The same study reported average reductions in air temperature of 0.23 K and 0.27 K per 10% increase in albedo for cool roofs and cool pavement technologies respectively. However, despite widespread acceptance of the cooling benefits of reflective technologies and the implementation of numerous large-scale projects using reflective roof and pavement materials [36], more rigorous experimental monitoring, consistent metadata reporting and detailed information on their reduced performance over time and potential outdoor comfort impacts are still required [37–39].

1.2. Urban Heat Mitigation—Principles of Reflective Technology

The primary summer daytime energy input into the urban canopy layer (UCL) is solar radiation, which is reflected or absorbed by solar-exposed terrestrial surfaces [1]. Depending on surface characteristics, solar radiation is partitioned into radiative, sensible, latent or storage heat fluxes [2,40,41]. Radiative and sensible heat fluxes dominate the SEB in the absence of moisture [42,43]. Sensible heat—or the perceptible rise in air temperature—is amplified when the difference between surface and ambient temperatures is large [44,45]. A reduction in the surface's surface temperature, which is achieved by shade or increasing the surface's reflectance [46,47] constrains the convective transfer intensity to the surrounding air [44], thus limiting the transfer of heat to the adjacent air volume [42,43,48] although the relationship between surface and near-surface air temperature is complex [49].

Conversely, surfaces with lower solar reflectance absorb more solar energy, thereby heating the surface, and through strengthened convective transfer [45], warm the adjacent air volume [46,50]. Additionally, via long-wave radiative transfer, hotter surfaces emit infrared radiation to cooler objects within view [51]. In summary, increasing surface reflectance—which in the solar spectrum is referred to as the surface's albedo—potentially reduces the surface's surface temperature, the convective transfer intensity to the air and the emitted infrared radiation to surfaces in view.

However, higher solar reflectance within cities may also increase reflected solar radiation to near-surface facets and people [37,51–59] although the magnitude of reflected solar radiation and its impact on human outdoor thermal comfort is highly context-dependent [37,53,60] and few studies have been experimentally determined [52,54].

1.3. Methods for Measuring Solar Reflectance—Nomenclature

The albedo of a surface is defined as its hemispherical and wavelength-integrated reflectance [50] and broadband albedo is the ratio of reflected to incident (direct and diffuse) solar radiation (250–3000 nm), or the fraction of incident sunlight reflected by the surface quantified from 0 to 1 [61]. The albedo of a terrestrial surface may vary with the wavelength (λ) of incident radiation (spectral dependence), the angle of incidence (θ) of radiation (angular dependence of direct and diffuse components) and the surface's surface structure and roughness [61,62].

In remote sensing and field measurements reflectance quantities are acquired under sky conditions and surface albedo is influenced by atmospheric turbidity, solar position, surface orientation and the geometry and optical properties of the surrounding urban form [62,63]. Solar irradiance consists of both direct beam and diffuse components and therefore the solar reflectance of a surface is variable in time and place (as atmospheric conditions and solar position change) and constant albedo values assume spectral and angular independence [62].

The use of laboratory, field or remote sensing methods for the measurement of surface reflectance is determined by the experimental design, material characteristics (e.g., sample size, etc.), spatial scale and wavelength and directional parameters of interest [64–66]. Reflectance nomenclature describes measured reflectance values first by the incident, and secondly by the reflected, angular distribution of radiation [63]. Levinson et al. [66] provide a comprehensive review of standard reflectance measurement methods and only remotely-sensed reflectance methods are briefly discussed further here.

1.4. Methods for Measuring Solar Reflectance—Remote Sensing of Urban Surfaces Using Narrow Field of View (FOV) Sensors

The standards-based laboratory [67] and field methods [68,69] for albedo measurement provide single reflectance quantities of small (0.1–5 cm^2), flat, homogenous surfaces or the hemispherically-integrated value of larger (>1 m^2) horizontal and low-sloped homogeneous diffusely-reflecting surfaces or the aggregate quantity of non-uniform horizontal surfaces [70] with relatively high accuracy (error <2%) [66]. However, the aforementioned methods have several limitations when applied to real urban surfaces [64,66,71,72].

Since the microclimate impacts of urban vertical surfaces are spatially dependent [54,73–76] and may be significantly influenced by microstructure heterogeneity [77] the ability to compute the spatial and geometric distribution of reflectance at sub-facet scale (<10 m) is desirable [78–82]. Satellite, aerial and ground-based remote sensing technologies permit increasing spatial, spectral, radiometric and temporal resolution with greater spatial coverage [48,83,84]. Ground-based imaging sensors are lightweight and mobile enabling in-canyon observations of surfaces with relative operational simplicity [85].

Remote sensors record wavelength-dependent energy emanating from a surface within the sensor's FOV. At short path lengths near the ground (<100–200 m) atmospheric absorption may be considered to be negligible [84,86,87]. Image data from image sensors are composed of discrete picture elements (pixels) each with a potentially unique brightness (or digital number, DN) value and ground resolution or instantaneous field of view (IFOV) determined by the sensor optics [84]. Depending on target distance, a ground-based image of a building wall or horizontal urban surface is "automatically" resolved into sub-facet (IFOV) scale surface brightness values that, once calibrated to spectral radiance [63] and geolocated, can be used to derive spatially-registered per-pixel spectral reflectance quantities based on the experimentally determined correlation between surface reflectance and at-sensor radiance [48,88,89].

Obtaining reflectance data in both the visible and NIR wavelengths improves surface albedo estimates, since many urban materials strongly reflect in the NIR region [31,48,90]. Multispectral (MS) sensors are spectrally selective and simultaneously record reflected radiation in several discrete narrow bandwidths, for example in Green (520–600 nm), Red (630–690 nm) and NIR (760–920 nm) [91,92]. Reflectance quantities derived from ground-based sensors with narrow FOV (and with knowledge of the surfaces' spectral and angular dependence [48]) are "hemispherical-canonical" reflectance values (per [63]) where direct and diffuse sky radiation is reflected into a narrow sensor viewing geometry [63]. A maximum absolute error of 14% has been reported for remotely measured albedo values of horizontal and low-sloped urban roofs from radiometrically calibrated high-resolution (1 m) aerial imagery with a smaller error (<2%) for low albedo (<0.2) surfaces [48].

1.5. Methods for Measuring Solar Reflectance—Overview of the Emprical Line (EL) Method

Remotely sensed image data of reflected energy received by a passive sensor is typically represented by a matrix of pixels containing DNs that are in value proportional to the intensity of energy reflected by the surface within view [92]. However, the proportionality relationship between image DNs and physical units such as surface reflectance is influenced by camera characteristics (i.e., the spectral response curve and analogue-to-digital signal conversion of the sensor [92]), sun-surface-sensor geometry (i.e., anisotropy of reflected radiation and illumination conditions [93]) and, for larger path distances, atmospheric transmittance [48]. Hence, DNs alone contain little meaningful quantitative information about surface reflectance unless the sensor is radiometrically calibrated [88].

Empirical (or "vicarious") radiometric calibration using in-situ reference targets of known spectral reflectance correlated to at-sensor radiance for each sensor waveband robustly accounts for atmospheric and illumination effects and produces acceptable radiance-to-reflectance conversion results [48,88]. Despite its widespread use the method is error prone if implemented without logistical and methodological considerations [88,94]. For many inexpensive, commercially available MS cameras the relationship between at-sensor spectral radiance and image DNs is not readily available [95,96], is onerous to obtain [48,97] and may be non-linear [98] and the unique camera response function—the gain and offset coefficients used to convert the radiance-as-electrical signal to output digital numbers [92]—requires determination before DNs can be converted to reflectance units [99,100]. In this case, the reflectance-based calibration method can be used to predict at-sensor radiance (expressed as image DNs) by measuring the reflectance of a calibration target [94,100].

Typically, an ordinary least squares (OLS) regression equation of spectral reflectance (y-axis) against DNs (x-axis) is computed for each sensor waveband from the mean spectral reflectance values of at least two spectrally distinct ground calibration targets and the mean DNs from the corresponding region of interest (ROI) in the image [88]. The EL equation is then validated in the field using spectral reflectance values measured by a field spectrometer or supplementary targets of known reflectance [86,99]. Mean DNs retrieved from remotely sensed MS images are then used as input data to the per-waveband predictive equations to generate per-pixel spectral reflectance values [86,96,101,102]. Several authors have developed simplified protocols using vicarious calibration methods to reduce the number of high-cost, in-situ and in-image targets [86,98].

While studies using EL methods for reflectance recovery are increasingly common to derive vegetation characteristics in open fields ([86] and references therein) the use of the EL method for reflectance recovery from urban surfaces is less common due to logistical and methodological challenges posed by urban areas [5] but interest in applications to man-made and urban surfaces is growing [94,103,104].

1.6. Purpose and Significance of the Work

Surface properties and impacts of cool roof and paving technologies have been extensively studied and applied [19,26,32,36,46,105–107]. However, research into the microclimate impacts of urban vertical surfaces and the effect of building facade geometry and fabric remain relatively underdeveloped [44]. While research into the optical properties, thermal and energy performance and outdoor thermal comfort impacts of cool materials for use in building envelopes is more advanced [36,37,54,72,73,108–113], the conceptualisation of building facades as more than merely an ensemble of material "facets" (or discrete, homogenous, surfaces [5]) and the development of climate sensitive architectural design applications that progress beyond a singular focus on cool and smart material specification [114–116] are still emerging (e.g., [117–120]).

While microclimate modelling and simulation tools exist and are fundamentally useful [121–123] comparatively few have been validated in the field [124] and none explicitly provide architects with meaningful "thermo-semantics"—the synthesis of building envelope thermal and optical performance computation and architectural design—at scales and interests commensurate with architects' decision-making [125].

This paper is part of ongoing research into the development of a thermo-semantic tool intended to assist architects to evaluate the outdoor thermal impacts of building facade design based on a vertical surface thermal typology supported by a predictive statistical model. The reflectance recovery method described here was applied to multispectral image panoramas of sampled building facades to create reflectance datasets for input to a probabilistic model.

The development of low-cost, replicable methods for reflectance recovery from real urban vertical surfaces within the UCL addresses the need for improved observation, understanding and transdisciplinary communication of urban atmospheric process at multiple scales and particularly at the street-level human scale [126] and contributes to the development of a predictive science of microclimatology [125]. Close-range ground-based reflectance recovery complements but also has some advantages over emerging unmanned aerial vehicle (UAV) technologies, particularly in relation to the statutory height and air-space restrictions applicable to UAVs in dense urban areas [86,127].

2. Materials and Methods

2.1. Overview of Method

To recover per-pixel spectral reflectance from MS digital images the relationship between known reflectance values per camera waveband and image DN distributions was statistically modelled as EL equations [88]. Two sets of alternative camera calibration targets were used to derive the theoretical constant "camera response function" and a second calibration target of higher reflectance was used to derive the variable "slope coefficient" [98]. The reflectance recovery method was evaluated by comparing known reflectance values of common building materials used in building facades with those derived from the final empirical line equations applied in ArcMap (ArcGIS by ESRI) to colour-processed MS images of the same materials. Predictive model precision and accuracy measures [128,129] were used to determine the optimum equations for later application to recover per-pixel spectral reflectance from MS panoramas of sixty mid-rise building facades (not discussed further here).

2.2. Camera Description

Tetracam Inc.'s Agricultural Digital Camera (ADC) (Tables 1 and 2) was used to capture images of the calibration targets. Images were stored using an uncompressed 10-bit per-pixel RAW (RAW) file format in automatic exposure mode (set to average).

Table 1. Tetracam ADC multispectral camera specifications [1].

Description	Specification
Image resolution	2048 × 1536 pixels (3.2 Megapixels)
Spectral range	0.52–0.92 µm
Lens focal length (f)	8 mm
Instantaneous field of view (IFOV)	0.3975 mrad
Image storage format	RAW 10 (10 bits)
Processed image format	8 bit JPEG (256 units)
Horizontal field of view (HFOV)	44.75°

[1] Source: Tetracam data sheet for ADC camera S/N 221215.

Table 2. Tetracam ADC camera waveband, bandwidth and colour channel specifications [1].

Waveband	ADC Bandwith (Range)	Colour Channel	Landsat TM
Green	520–600 nm (80 nm)	B	TM2
Red	630–690 nm (60 nm)	G	TM3
NIR	760–920 nm (160 nm)	R	TM4

[1] Source: Tetracam data sheet for ADC camera S/N 221215.

2.3. Camera Calibration Target Sets and Spectrophotometer Measurements

Two sets of camera calibration targets were used to develop the optimal EL model of the camera response function that describes the theoretical minimum reflectance per waveband detectable by the sensor [98]. Set 1 consisted of twelve A4-size stiff cardboard cards factory-painted with selected Dulux acrylic colours. The cards had a wide range (approx. 13% to 100%) of mean near-normal beam hemispherical spectral reflectance values (Refer Table 3, Figure 1) measured using a Perkins Elmer Lambda 950 UV/VIS/NIR spectrophotometer over a 250–1000 nm range at 5 nm intervals with a 150 mm integrating sphere calibrated with a 99% Spectralon certified reflectance standard in compliance with [67].

Table 3. Dulux colour card IDs and measured mean spectral reflectance values [1].

ID	Colour Name	Dulux Code	%R-G$_\lambda$	%R-R$_\lambda$	%R-NIR$_\lambda$
1	Black	PN2A9	13.852	14.147	15.338
2	Mt Eden	PN2A7	18.984	19.140	19.150
3	Klute	PN2A5	30.356	29.294	28.015
4	Stepney	PN2A3	37.794	37.083	35.243
5	Warm Granite	PN2C6	46.216	47.241	45.783
6	Soft Beige	PN2C5	56.919	57.773	57.367
7	Bleaches	PN2D5	65.578	65.421	65.816
8	Terrace White	PN2H2	78.332	76.312	78.940
9	Manorburn	PN2H1	80.151	79.341	80.364
10	Snowy Mountains	PN2B2	87.562	87.693	89.320
11	Lexicon Quarter	SW1E1	99.232	98.952	100.00

[1] Mean spectral reflectance values over ADC camera bandwidths.

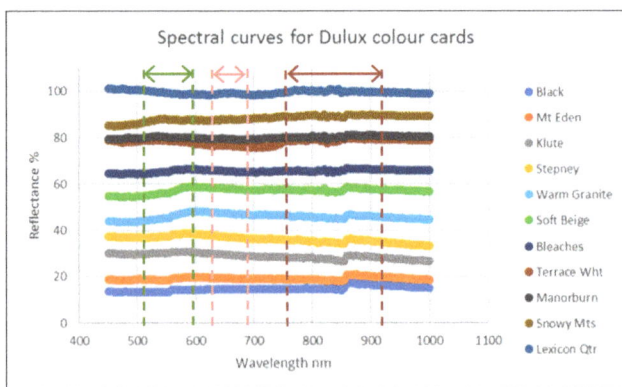

Figure 1. Dulux colour cards reflectance curves 450–1000 nm at 5 nm intervals.

The Dulux cards were selected for conformity to the recommended calibration target qualities of uniform, matte surface (assumed here to be diffusely reflecting or Lambertian), near-uniform reflectance over the bandwidths of interest and with high spectral contrast [88]. Measuring spectral reflectance at 5 nm intervals adequately records the spectral characteristics of most surfaces [66].

Camera calibration target Set 2 consisted of one 127 × 127 mm Spectralon multi-step diffuse calibration target (Labsphere Inc., North Sutton, NH, USA) with four side-by-side vertical panels of 12%, 25%, 50% and 99% nominal spectral reflectance values measured at 1 nm intervals from 250 to 2500 nm using a Perkins Elmer Lambda 900B UV/VIS/NIR dual beam spectrophotometer with 150 mm integrating sphere that collects 8-degree beam hemispherical reflectance (spectral and diffuse) calibrated with a Spectralon reflectance standard [130] (Refer Table 4, Figure 2).

Table 4. Spectralon multi-step target ID and mean spectral reflectance values [1].

ID	Panel Name	%R-G$_\lambda$	%R-R$_\lambda$	%R-NIR$_\lambda$
1	SRT-12	11.860	12.305	13.198
2	SRT-25	27.029	28.303	30.545
3	SRT-50	50.320	51.820	54.204
4	SRT-99	99.073	99.006	98.942

[1] Mean spectral reflectance values over ADC camera bandwidths.

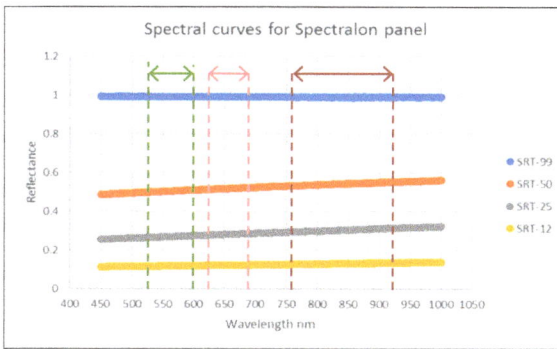

Figure 2. Spectralon multi-step reflectance curves 450–1000 nm at 1 nm intervals.

2.4. Camera Calibration Image Acquisition

The Dulux cards and the Spectralon multi-step target were placed vertically on a horizontal ledge adjacent a sun-exposed building wall (Figure 3).

(a)   (b)

Figure 3. Dulux colour cards (**a**) and Spectralon multi-step calibration target (striped) (**b**).

Images were taken of the calibration targets outdoors under clear-sky conditions on 15 October 2016 between 12:00 and 12:15 p.m. The solar altitude and azimuth at 12:05 p.m. were 63°23′06″ and 20°09′15″ respectively at 33°55′00″ south latitude and 151°13′00″ east longitude [131]. Multiple images were taken normal to the Dulux and Spectralon targets at distances of 3.2 m and 1 m respectively. The target sizes satisfied the recommendations in the literature to be a minimum 3 to 10 times the sensor IFOV [87–98,132] to ensure that as many pixels as possible associated with each calibration target were selected [88] and to avoid atmospheric effects [87]. Each Dulux panel and the Spectralon target measured 293 (h) × 210 mm (w) and 127 × 127 mm respectively (with a single multi-step column measuring 30 mm (w) × 120 mm (h)). Table 5 lists the camera calibration target distances, IFOV and minimum recommended target sizes.

Table 5. Camera calibration target distances and IFOV to satisfy target size requirements.

Target	Target Size	Distance	Sensor IFOV [1]	Min. 3 × IFOV	Min. 10 × IFOV
Spectralon	30 mm	1 m	0.4 mm	1.2 mm	4 mm (<30 mm)
Dulux	210 mm	3.2 m	1.27 mm	3.81 mm	12.7 mm (<210 mm)

[1] IFOV calculated from FOV optical calculator in Tetracam PW2 software.

2.5. Multispectral Image Pre-Processing

The ADC image data were stored in RAW format on a single SanDisk compact flash card and downloaded via a card reader for processing on a PC. The ADC is supplied with a camera-specific colour process file (CPF) that evaluates each RAW pixel value using a colour processing algorithm to extract NIR (760–920 nm), Red (630–690 nm) and Green (520–600 nm) waveband values for each pixel [133]. Colour-processed images were analysed in 10-bit RAW format and later saved in 8-bit JPEG format for import into ArcMap.

To minimize the effects of saturation on the colour processed images the "Auto Adjust Scaling" (AAS) box was checked on the Matrix page of the proprietary image editing software PixelWrenchII (PW2) [133]. As previously reported, saturation of the Tetracam ADC images may occur for high reflectance surfaces when the DNs exceed 256 resulting in the actual reflectance values exceeding the dynamic range of the sensor [98,134].

2.6. Derivation of Camera Calibration Equations

The mean DN per waveband was calculated from a uniform polygon region of interest (ROI) within the borders of the colour-processed image of each calibration target using the Histogram Tool in PW2. For each calibration sample the spectrophotometer-measured mean spectral reflectance per-waveband (on the y-axis) was regressed against the corresponding mean DN (on the x-axis) to derive the per-waveband "camera response function" or empirical line 1 (EL1) using the curve estimation function in the statistical software IBM SPSS. The OLS regression equations and scatter plots are shown in Tables 6 and 7 and Figures 4 and 5 below.

Table 6. OLS equations (EL1) of camera response function per waveband for Dulux samples.

Wavelength (ADC Channel)	Dulux-Generated EL1	R	R^2; AdjR^2
Green λ 520–600 nm (Blue)	$\%RG_\lambda = 7.7353 + 0.3607DN_{aveB}$	0.989	0.978; 0.976
Red λ 630–690 nm (Green)	$\%RR_\lambda = 5.7211 + 0.4878DN_{aveG}$	0.986	0.971; 0.968
NIR λ 760–920 nm (Red)	$\%R_{NIR\lambda} = 7.1711 + 0.5143DN_{aveR}$	0.988	0.977; 0.974

Table 7. OLS equations (EL1) of camera response function per waveband for Spectralon samples.

Wavelength (ADC Channel)	Dulux-Generated EL1	R	R^2; AdjR^2
Green λ 520–600 nm (Blue)	$\%RG_\lambda = 6.7622exp^{(0.0132 \cdot DNaveB)}$	0.982	0.964; 0.947
Red λ 630–690 nm (Green)	$\%RR_\lambda = 0.4761DN_{aveG} - 8.4403$	0.999	0.998; 0.997
NIR λ 760–920 nm (Red)	$\%R_{NIR\lambda} = 0.4803DN_{aveR} - 5.1695$	0.999	0.998; 0.998

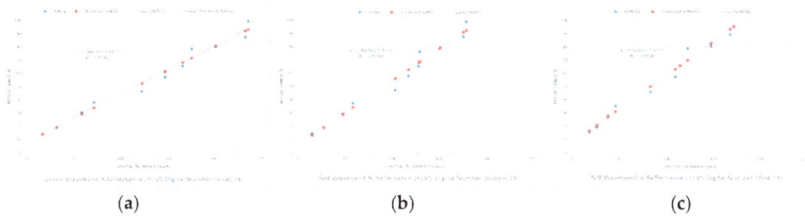

(a)   (b)   (c)

Figure 4. Plots of: Green (**a**); Red (**b**); and NIR (**c**) camera response functions for the Dulux targets.

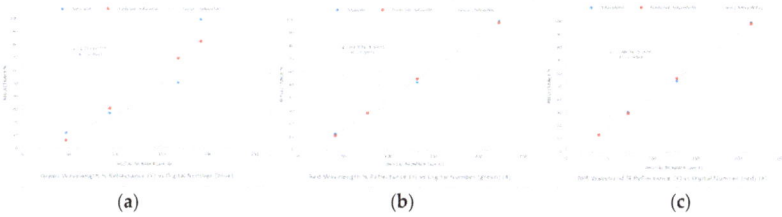

(a)   (b)   (c)

Figure 5. Plots of: Green (**a**); Red (**b**); and NIR (**c**) camera response functions for the Spectralon targets.

The plots of spectral reflectance against sensor DN for the Dulux targets depict a strong positive linear relationship for all three wavebands as measured by Pearson's correlation coefficient ($0.986_{red} < R_{Dulux} > 0.989_{green}$). For all sensor wavebands the coefficient of determination exceeds 0.971. The plots of spectral reflectance against sensor DN for the Spectralon calibration target depict a strong linear correlation for Red and NIR wavebands (R = 0.999) and an exponential relationship for the Green waveband. R^2 for Red and NIR were both 0.998 and lowest at 0.964 for Green.

Assessment of Camera Calibration Equations

When the x-intercept (DN) equals 0 the corresponding y-intercept values are theoretically the minimum reflectance values detectable by the unique image sensor per waveband [98]. Comparative results of the Dulux and Spectralon-derived camera response equations (Tables 6 and 7) indicate that in this instance, the y-intercept is not a true "constant calibration parameter" [98], since under identical illumination conditions the minimum possible detectable mean spectral reflectance varies depending on the properties of the calibration target [92,135].

In consideration that the spectral curves of both calibration target sets (Figures 1 and 2) indicate near-uniform reflectance over the bandwidths of interest and the Spectralon target is a diffuse reflector, the difference between the camera response to the calibration target sets may be accounted for, in part, by the non-Lambertian reflectance and material properties of the Dulux cards [86,88,136] although this reflectance anisotropy has not been experimentally verified.

2.7. Derivation of "Slope Coefficient" from In-Situ Calibration Target—Theory and Method

When two or more calibration targets with large spectral contrast are used the EL method permits the computation of the sensor-specific relationship between image DNs and surface spectral reflectance [88]. In many instances, however, it is impractical or expensive to acquire every image containing its own calibration targets [86]. Wang and Myint [98] developed a simplified method to convert image DNs to spectral reflectance by constructing an EL equation per image waveband that consists of the theoretically minimum possible detectable reflectance or "constant calibration parameter" and one additional coordinate—the "slope coefficient"—derived from a single field calibration target of higher reflectance. The slope coefficient accounts for the variable illumination effects on sensor DNs when the camera settings are invariant [98].

The MS images of the target building facades included a mobile meteorological station crowned by an aluminium "calibration bracket" (CB) measuring 97 (w) × 95 mm (h) with a white powder-coated low-sheen finish. The CB dimensions satisfied the target distance recommendations in the literature for this validation study and final reflectance recovery from images of building facades (Table 8).

Table 8. Facade target distance and IFOV to satisfy minimum target size requirement.

Target	Target Size	Distance	Sensor IFOV [1]	Min. 3 × IFOV	Min. 10 × IFOV
CB	95 mm	18.5 m (mean)	7.40 mm	22.20 mm	74 mm (<95 mm)

[1] IFOV calculated from FOV optical calculator in Tetracam PW2 software.

The mean near-normal beam hemispherical spectral reflectance values per waveband of the CB were measured using a Perkins Elmer Lambda 1050 UV/VIS/NIR spectrophotometer over a 250–2500 nm range at 1 nm intervals with a 150 mm integrating sphere calibrated with a 99% Spectralon certified reflectance standard compliant with [67] (Table 9 and Figure 6).

Table 9. Calibration bracket (CB) mean spectral reflectance values [1] per camera waveband.

Item	Coordinate	%R-G$_\lambda$	%R-R$_\lambda$	%R-NIR$_\lambda$
%R$_{CB}$	"By"	89.061	86.868	84.113

[1] Mean spectral reflectance values over ADC camera bandwidths.

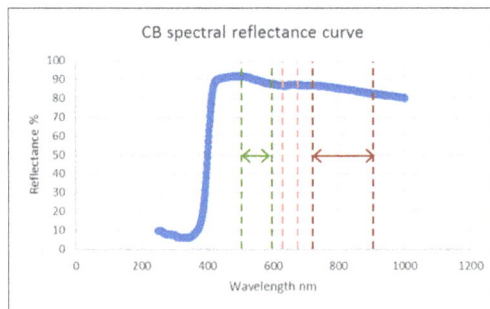

Figure 6. Calibration bracket reflectance curve 250–1000 nm at 1 nm intervals.

The following equations derived from [98] constitute the final EL equations per waveband:

$$\%R_\lambda = m \times DN_{ave(\lambda)} + Ay \tag{1}$$

$$m = slope = \Delta Y / \Delta X = (By - Ay)/Bx \tag{2}$$

where *Ay* is the y-intercept of EL1, when *Ax* = 0. *By* is the measured mean spectral reflectance of the calibration bracket and *Bx* is the mean DN corresponding to *By*.

Multispectral images of the CB were acquired at the same time as those of the material validation samples. The mean DN per waveband for the CB (Table 10) was calculated from a uniform polygon ROI within the borders of the colour-processed image of the CB using the Histogram Tool in PW2 software.

Table 10. Calibration bracket mean digital number (DN) values [1] per camera waveband.

Item	Coordinate	DN$_{G\lambda}$	DN$_{R\lambda}$	DN$_{NIR\lambda}$
DN	"Bx"	254	211	199

[1] Mean DNs were calculated from the average of five ROIs.

The final EL equations per waveband ("EL2") derived from the intercept obtained from EL1 equations and the slope coefficient derived from the slope values calculated from Equation (2) above are summarized in Tables 11 and 12 below. These equations were then applied in ArcGIS to the colour-processed MS images of the validation samples to recover per-pixel per-waveband spectral reflectance and then compared to the values obtained using ASTM E903-12 [67].

Table 11. EL2 equations from Dulux targets and calibration bracket per camera waveband.

Item	%R-Green$_\lambda$	%R-Red$_\lambda$	%R-NIR$_\lambda$
Intercept	7.7353	5.7211	7.1711
Slope	0.3202	0.3846	0.3866
EL2$_{Dulux\lambda}$	%R$_{G\lambda}$ = 7.7353 + 0.3202(DN$_B$)	%R$_{R\lambda}$ = 5.7211 + 0.3846(DN$_G$)	%R$_{NIR\lambda}$ = 7.1711 + 0.3866(DN$_R$)

Table 12. EL2 equations from Spectralon targets and calibration bracket per camera waveband.

Item	%R-Green$_\lambda$	%R-Red$_\lambda$	%R-NIR$_\lambda$
Intercept	6.7622	−8.4403	−5.1695
Slope	0.0102	0.4517	0.4487
EL2$_{Spectralon\lambda}$	ln(%R$_{G\lambda}$) = ln(6.7622) + 0.0102(DN$_B$)	%R$_{R\lambda}$ = 0.4517(DN$_G$) − 8.4403	%R$_{NIR\lambda}$ = 0.4487(DN$_R$) − 5.1695

2.8. Validation Using Material Samples of Known Reflectance Values

2.8.1. Part 1: Validation Sample Image Acquisition and Reflectance Measurement

To validate the EL equations, 13 samples of common building materials were used to compare spectrophotometer-measured [67] spectral reflectance values with reflectance quantities recovered from the EL2 equations.

Plywood-mounted material validation samples, several full-size samples of the same materials and two calibration brackets were placed vertically on a horizontal ledge adjacent a sun-exposed building wall (Figure 7).

Figure 7. Plywood-mounted and full size validation samples and CBs.

Multispectral images were taken of the validation samples outdoors under clear-sky conditions on 9 February 2017 between 12:00 and 12:15 p.m. The solar altitude and solar azimuth at 12:10 p.m. were 66°32′25″ and 38°27′47″ respectively at 33°55′00″ south latitude and 151°13′00″ east longitude [131]. Images were taken normal to the calibration samples at a distance of 2.9 m. The target sizes satisfied the recommendations in the literature [88] (Table 13).

Table 13. Validation sample target distance and IFOV to satisfy minimum target size requirements.

Target	Target Size	Distance	Sensor IFOV [1]	Min. 3 × IFOV	Min. 10 × IFOV
Various	45 mm (min)	2.9 m	1.16 mm	3.48 mm	11.6 mm (<45 mm)

[1] IFOV calculated from FOV optical calculator in Tetracam PW2 software.

The validation samples had a wide range (approx. 7% to 78%) of mean spectral reflectance values measured using a Perkins Elmer Lambda 1050 UV/VIS/NIR spectrophotometer over a 250–2500 nm range at 2.5 nm intervals with a 150 mm integrating sphere calibrated with a 99% Spectralon certified reflectance standard compliant with Reference [67] (Table 14 and Figure 8).

Table 14. Material validation sample ID and measured mean spectral reflectance values [1].

ID	Name	Description	%R-G$_\lambda$	%R-R$_\lambda$	%R-NIR$_\lambda$
V1	Colorbond "Windspray"	Factory painted steel sheet	19.2386	22.5158	40.7007
V2	Colorbond "Woodland Grey"	Factory painted steel sheet	12.1960	13.1979	28.8771
V3	Natural anodized aluminium	45 × 300 mm door frame profile	71.3176	70.5578	67.4945
V4	Powder-coated aluminium ("Cream")	45 × 300 mm door frame profile	75.2613	77.9694	75.3959
V5	Sydney sandstone (beige)	140 mm (diam.) core sample	55.6602	60.2846	65.5658
V6	Granite stone (white & grey colour)	140 mm (diam.) core sample	39.1646	38.8916	39.8336
V7	PGH dry-pressed brick "McGarvie Red"	Clay brick biscuit 76 × 110 mm	13.5092	25.1807	32.5668
V8	PGH dry-pressed brick "Copper Glow"	Clay brick biscuit 76 × 110 mm	21.4442	33.9654	39.2103
V9	PGH dry-pressed brick "Mowbray Blue"	Clay brick biscuit 76 × 110 mm	12.1330	14.0271	17.6589
V10	6.38 mm (th) clear laminated glass	Clear glass on black backing	8.8608	8.2229	7.1488
V11	5.0 mm (th) clear glass	Clear glass on black backing	8.7144	8.2300	7.4313
V12	Cement aggregate (white)	45 mm (diam.) core sample	49.3766	49.5922	47.8815
V13	Granite stone (red)	50 mm (diam.) core sample	18.2971	21.6636	20.0889

[1] Mean spectral reflectance values over ADC camera bandwidths.

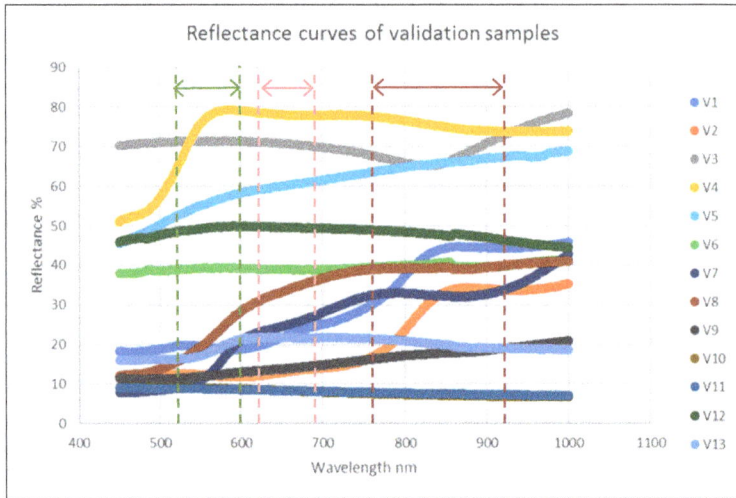

Figure 8. Validation samples reflectance curves 450–1000 nm at 2.5 nm intervals.

2.8.2. Part 2: Reflectance Recovery from MS Images in ArcMap

The multispectral images of the validation samples were pre-processed in accordance with the method previously outlined. The colour-processed images were imported in JPEG format into ArcMap for further processing. Discrete ROI raster files were created for each material sample (from a composite image) using the "Clip" function in ArcMap. These files were used as inputs into a raster algebra expression of the empirical line equations (EL2 from Tables 11 and 12) to create spectral reflectance raster files per waveband.

3. Results

The measured (M) and predicted (P) mean spectral reflectance values of the 13 validation samples for all camera wavebands using the Dulux-derived EL2 are shown in Table 15 below. Common statistical measures of correlation (Pearson's r and Spearman's rho), predictive precision (coefficient of determination, R^2), distribution difference (Mann-Whitney's U test) and prediction accuracy (Willmott's index of agreement d, root mean square (RMSE) and mean absolute error (MAE)) are shown in Table 16 for evaluation of model performance [128,129,137] based on the OLS regression of measured against EL2-predicted reflectance values [138].

Table 15. Measured (M) and EL2-predicted (P) mean spectral reflectance (%) of 13 validation samples using EL1$_{Dulux}$ intercept.

Waveband	Green			Red			NIR		
Sample ID	M	P	AR[1]	M	P	AR	M	P	AR
V1	19.237	37.778	18.541	22.516	28.898	6.382	40.701	52.101	11.400
V2	12.196	27.145	14.949	13.198	17.670	4.472	28.877	39.703	10.826
V3	71.318	70.411	0.907	70.558	62.952	7.606	67.495	59.313	8.182
V4	75.261	69.702	5.560	77.969	91.490	13.521	75.396	90.843	15.447
V5	55.660	62.596	6.936	60.285	62.741	2.456	65.566	61.489	4.077
V6	39.165	44.223	5.058	38.892	42.304	3.412	39.834	41.017	1.184
V7	13.509	26.531	13.022	25.181	34.055	8.874	32.567	36.513	3.946
V8	21.444	32.155	10.710	33.965	40.947	6.981	39.210	41.689	2.479
V9	12.133	20.282	8.149	14.027	19.200	5.173	17.659	22.848	5.189
V10	8.861	19.832	10.971	8.223	13.064	4.841	7.149	29.292	22.143
V11	8.714	20.409	11.694	8.230	11.781	3.551	7.431	33.470	26.039
V12	49.377	64.151	14.775	49.592	62.009	12.416	47.882	52.978	5.097
V13	18.297	39.395	21.098	21.664	42.434	20.771	20.089	35.495	15.406
Mean	31.167	41.124	10.952	34.177	40.734	7.728	37.681	45.904	10.108
Sum AR			142.37			100.46			131.41
RMSE [1]			12.228			9.177			12.561
d [1]			0.920			0.960			0.892

[1] AR: absolute residual. RMSE and d from [129].

Table 16. Rank correlation and OLS regression statistics M-P from Table 15 data.

Statistic	Green		Red		NIR	
Rank Correlation						
Mann-Whitney U [1]	$U_c = 45$	$U = 54$	$U_c = 45$	$U = 69$	$U_c = 45$	$U = 64$
Test result		$U > U_c$		$U > U_c$		$U > U_c$
Z-score	1.53846		0.76923		1.02564	
p-value	0.12356		0.4413		0.30302	
Spearman's rho (r_s)	0.945		0.940		0.967	
OLS regression M-P						
RMSEs [1]	5.993		6.346		9.261	
MAE	4.893		4.188		8.092	
Pearson's r	0.966		0.960		0.900	
R^2	0.933		0.922		0.810	
d	0.982		0.979		0.945	
Intercept (a)	-18.372		-4.81		-13.878	
Slope (b)	1.2046		0.9224		1.1232	

[1] RMSEs from [129]. U test: no statistical difference in distributions at 95% confidence level.

Plots of the measured and Dulux EL2 equation-predicted mean spectral reflectance values of the 13 validation samples are shown in Figures 9 and 10 below.

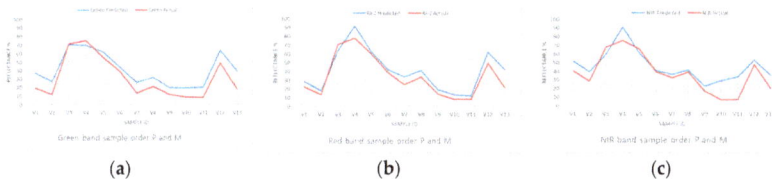

(a)	(b)	(c)

Figure 9. M (red line) and P reflectance plots for G (a); R (b) and NIR (c) bands from EL2$_{Dulux}$.

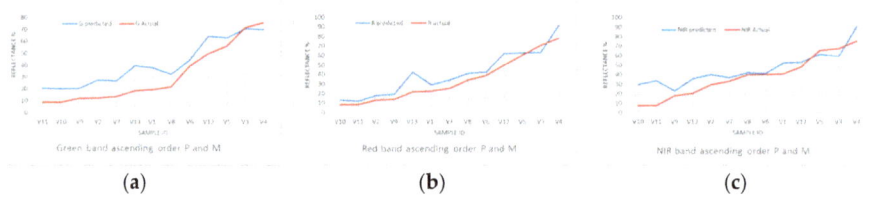

Figure 10. Ascending order M (red line) and P reflectance plots for G (**a**); R (**b**) and NIR (**c**) bands from EL2$_{\text{Dulux}}$.

The OLS plots of measured (M) regressed against predicted (P) reflectance values per camera waveband are shown in Figure 11 below.

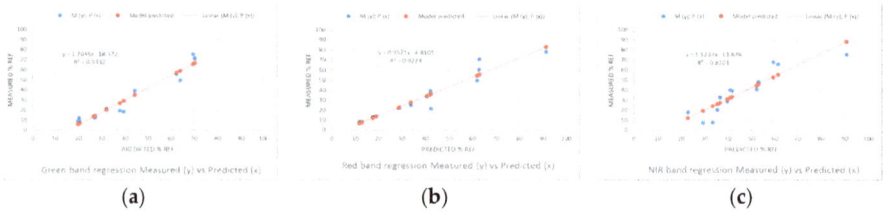

Figure 11. OLS regression of M against P reflectance per sample for G (**a**); Red (**b**) and NIR (**c**) from EL2$_{\text{Dulux}}$.

The measured and predicted mean spectral reflectance values of the 13 validation samples for the three camera wavebands using the intercept value from the Spectralon-derived EL2 equation and prediction model evaluation statistics are shown in Tables 17 and 18 below.

Table 17. Measured (M) and EL2-predicted (P) mean spectral reflectance (%) of 13 validation samples using EL1$_{\text{Spectralon}}$ intercept.

Waveband	Green			Red			NIR		
Sample ID	M	P	AR	M	P	AR	M	P	AR
V1	19.237	17.527	1.712	22.516	12.275	10.241	40.701	46.276	5.576
V2	12.196	12.512	0.315	13.198	−3.218	16.416	28.877	29.583	0.705
V3	71.318	49.314	22.003	70.558	59.251	11.307	67.495	55.987	11.507
V4	75.261	48.218	27.043	77.969	98.619	20.649	75.396	98.442	23.046
V5	55.660	38.494	17.166	60.285	58.960	1.325	65.566	58.917	6.649
V6	39.165	21.499	17.665	38.892	30.768	8.124	39.834	31.352	8.481
V7	13.509	12.270	1.239	25.181	19.389	5.792	32.567	25.287	7.280
V8	21.444	14.665	6.779	33.965	28.896	5.070	39.210	32.257	6.953
V9	12.133	10.065	2.068	14.027	−1.103	15.130	17.659	6.887	10.772
V10	8.861	9.922	1.062	8.223	−9.568	17.791	7.149	15.564	8.415
V11	8.714	10.106	1.391	8.230	−11.338	19.568	7.431	21.190	13.759
V12	49.377	40.439	8.938	49.592	57.950	8.358	47.882	47.458	0.424
V13	18.297	18.448	0.152	21.664	30.948	9.284	20.089	23.916	3.827
Mean	31.167	23.345	8.272	34.177	28.602	11.465	37.681	37.932	8.262
Sum AR			142.37			100.46			131.41
RMSE			12.228			9.177			12.561
d			0.920			0.960			0.892

Table 18. Rank correlation and OLS regression statistics M-P from Table 17 data.

Statistic	Green		Red		NIR	
Rank correlation						
Mann-Whitney U [1]	$U_c = 45$	$U = 73$	$U_c = 45$	$U = 70$	$U_c = 45$	$U = 80$
Test result		$U > U_c$		$U > U_c$		$U > U_c$
Z-score	0.5641		0.71795		0.20513	
p-value	0.57548		0.47152		0.83366	
Spearman's rho (r_s)	0.945		0.940		0.967	
OLS regression M-P						
RMSEs	4.648		6.346		9.261	
MAE	3.442		4.188		8.092	
Pearson's r	0.979		0.960		0.900	
R^2	0.960		0.922		0.810	
d	0.989		0.979		0.945	
Intercept (a)	−5.475		14.332		6.039	
Slope (b)	1.5696		0.6938		0.8342	

[1] U test: no statistical difference in distributions at 95% confidence level.

Plots of the measured and Spectralon EL2 equation-predicted mean spectral reflectance values of the 13 validation samples are shown in Figures 12 and 13 below.

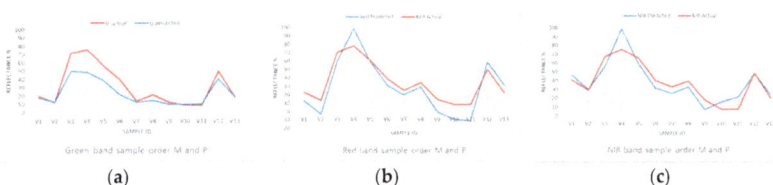

(a)   (b)   (c)

Figure 12. M (red line) and P reflectance plots for G (**a**); R (**b**) and NIR (**c**) bands from EL2$_{Spect}$.

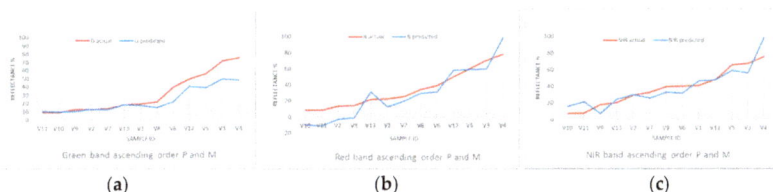

(a)   (b)   (c)

Figure 13. Ascending order M (red line) and P plots for G (**a**); R (**b**) and NIR (**c**) bands from EL2$_{Spect}$.

The OLS plots of measured (M) regressed against predicted (P) reflectance values per camera waveband are shown in Figure 14 below.

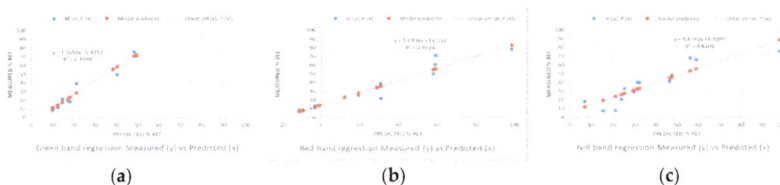

(a)   (b)   (c)

Figure 14. OLS regression of M against P reflectance per sample for G (**a**); Red (**b**) and NIR (**c**) from EL2$_{Spect}$.

In addition to linear covariance tests (r, R^2, intercept a and slope b), OLS model prediction error was evaluated using supplementary measures recommended by [128,129]. These include the systematic root mean square error (RMSEs) and the unsystematic root mean square error (RMSEu). For "good" model performance the RMSE should be low, RMSEs should tend to zero, RMSEu should approach RMSE and Willmott's index of agreement (d), a measure of prediction accuracy, should approach unity [128]. MAE provides an estimate of mean differences between measured and predicted values and is an "intuitive", reliable measure of average model prediction error [139]. While RMSE is a widely reported measure of average model error, it is less reliable than MAE since it exaggerates differences due to higher sensitivity of RMSE to outliers [129,140] and due to the inconsistent variance of RMSE with average error [139]. Table 19 below summarizes the performance of both models per waveband to evaluate the most accurate and precise predictive EL equations.

Table 19. Summary statistics of EL2-equations and OLS model performance.

Equations	EL2-Equation Measures				OLS Regression Parameters				OLS Difference Measures			
	[1] M_m	[1] P_m	RMSE	MAE	r	R^2	a	b	RMSE$_s$	RMSE$_u$	MAE	d
Green$_{Dulux}$	31.167	41.124	12.228	10.952	0.966	0.933	−18.372	1.205	5.993	10.659	4.893	0.982
Green$_{Spectralon}$	31.167	23.345	12.279	8.272	0.979	0.960	−5.475	1.570	4.649	11.364	3.442	0.989
Red$_{Dulux}$	34.177	40.734	9.177	7.728	0.960	0.922	−4.81	0.922	6.346	6.630	4.188	0.979
Red$_{Spectralon}$	34.177	28.602	12.830	11.465	0.960	0.922	14.332	0.6938	6.346	11.151	4.188	0.979
NIR$_{Dulux}$	37.681	45.904	12.561	10.108	0.900	0.810	−13.878	1.123	9.261	8.486	8.092	0.945
NIR$_{Spectralon}$	37.681	37.932	10.014	8.262	0.900	0.810	6.038	0.834	9.261	3.810	8.092	0.945

[1] M_m = Measured mean, P_m = Predicted mean.

Comparing the model results for the Green waveband, the lower Spectralon-derived OLS equation RMSEs and MAE, higher RMSEu with higher d in agreement with higher regression R^2 indicate its unambiguous predictive superiority (bold in Table 19). Comparing the model results for the Red waveband, considering equivalence of linear regression measures RMSEs, MAE and d, the lower Dulux-derived EL2-equation MAE (lowest of all wavebands) and regression parameters (intercept a (−4.81) and slope b (0.922)) closer to the ideal 1:1 correspondence line (when a = 0 and, more importantly, b = 1) for goodness of fit [141] indicate its superior predictive precision and accuracy (bold in Table 19). The above interpretation was supported when comparing the additive and proportional components [128] of the MSEs for each model.

Comparing the results for the NIR waveband, despite a lower MAE for the EL2-Spectralon equation estimates and considering the equivalence of OLS correlation and difference measures, the significantly lower RMSEu for Spectralon-derived OLS suggests that the Dulux-derived EL2 equation contains proportionally less systematic error overall and hence improved accuracy (bold in Table 19). This conclusion is augmented by computation of the ratio MSEs/MSE [129] that indicates that the Dulux-derived NIR predictive model contains a significantly lower proportion (54.4% vs. 85.5%) of systematic error and hence greater comparative model accuracy.

In summary, the optimal predictive models per waveband indicate a strong monotonic agreement (0.940 < Spearman's Rho > 0.967) and linear association (0.900 < Pearson's r > 0.966) between the measured and EL2-predicted reflectance values that confirm a near-perfect positive agreement between pairs of samples of ranked scores and a strong linear correlation [142]. The regression measures imply distribution near-equivalence and high confidence in trend predictions [143]. Regardless of the residual magnitudes, the distributions (Figures 9, 10, 12 and 13) show a strong positive agreement that supports reflectance predictions on a relative (high/low) scale. The non-parametric Mann-Whitney U-test for difference between two independent groups exhibits no statistically significant difference in distributions at the 95% confidence level, reinforcing the above interpretation [144]. The coefficients of determination (R^2) of the OLS models for all wavebands exceed 81% (NIR). Green and Red band R^2 exceed 92.2%. This suggests stronger agreement in the visible bands between measured and predicted values and higher confidence in the precision of the linear covariance between measured and predicted values [138].

Comparative Assessment of Measured and EL2 Predicted Reflectance Distributions

The mean bias error (MBE) [139] between measured and Dulux-derived EL2 equation predicted reflectance values for all wavebands indicates a systematic overestimation of measured values (Figures 9 and 10), a trend corroborated by the regression model (Figure 11) and regression slope and intercept parameters (Table 16). The overestimation bias derives from the camera calibration model (Table 6). However, higher reflectance samples (V3 (nominal reflectance 70%); V4 (75%) and V5 (60%) in Table 15) underestimate measured values. Underestimation of high-albedo surfaces by the EL method has been reported previously [94]. Sample V13, red Granite stone (nom. reflectance 20%) exhibits the largest absolute residual for the Green and Red wavebands. The high error may be theoretically accounted for by the reflectance anisotropy common to natural materials [135] in combination with artifacts of the prediction model. However, the low absolute residual magnitude of the Spectralon-derived estimate for the same sample (Table 17) implicates the latter. The largest absolute residual for the NIR waveband (Dulux-derived) occurs for sample V11, 5 mm Clear Glass (nom. reflectance 8%). The specular behaviour of glass at oblique incident angles (as occurred in the validation experiment, where the solar altitude exceeded 65°) may account for the larger difference in the NIR region [48]. Further, the low measured reflectance of glass falls outside the reflectance data range used to build the EL equations which may increase prediction error for this sample [98].

The MBE between measured and Spectralon-derived EL2 equation predicted reflectance values for the Green and Red wavebands indicates a systematic underestimation of measured values (Figures 12 and 13), a trend corroborated by the regression model (Figure 14) and regression slope and intercept parameters (Table 18). The underestimation bias derives from the camera calibration model (Table 7). MBE for the NIR waveband indicates marginal average overestimation of measured reflectance values, with some exceptions. Sample V13, red granite stone, is overestimated in all three wavebands and largest for the Red waveband. The highest reflectance sample, Sample V4 (powder coated aluminium, cream colour, nominal reflectance 75%), exhibits the largest absolute residual for all camera wavebands and is overestimated in the Red and NIR bands. The Spectralon model prediction of maximum error per sample is inconsistent with the Dulux-derived response to maximum error but consistent with prior studies that found predictions of high albedo surfaces have higher errors than low albedo surfaces [48].

The range of measured mean spectral reflectance values (approx. 7% to 78%) of the validation samples across all wavebands was comfortably within the upper reflectance limit of the calibration targets used in the development of the predictive equations (approx. 99% for EL1 and 89% for EL2). However, the two glass samples had measured reflectance values below the lower limit of the model data range (approx. 12% for EL1). For all samples (excluding glass) the predicted values were interpreted within the limits of the calibration equations [88,98].

4. Discussion

In this study a terrestrial application of the EL method [88] was developed to radiometrically calibrate and validate close-range remotely sensed images of vertical surface materials obtained from a narrow FOV MS camera using a single in-situ calibration target [98]. The y-intercept of the camera response function, although variable with camera calibration target properties, was shown to be the minimum reflectance detectable by the camera sensor per waveband and was used as the lower data point. The second coordinate of the EL equation was derived from the per-waveband spectral reflectance values and DNs of the CB. While the CB was assumed to be Lambertian, the spectrophotometer measurements of the CB indicated that mean spectral reflectance was unequal across sensor wavebands (Table 9) with a declining reflectance trend observed from shorter (visible) to longer (NIR) wavelengths (Figure 6) necessitating a separate EL equation for each waveband [98]. The optimal EL2 equations per waveband (Table 19) strongly linearly correlated ($0.900 < r > 0.979$) per-pixel spectral reflectance to image DNs with absolute prediction accuracies ($7.728\% < MAE > 10.108\%$) within the range reported in the literature [88,94].

Theoretically, improvements in accuracy are achievable if the calibration target is accurately characterized [136] and more than one in-situ calibration target is used [132] and future experimental design could, without much logistical effort, incorporate additional portable near-Lambertian targets, selected for spectral uniformity within sensor wavebands [135]. Instrument characteristics have been identified as a common source of experimental uncertainty [145]. Accuracy improvements may result from the optimisation of camera settings (i.e., manual exposure) within the limitations of the sensor radiometric resolution [92,98,134] and from corrections for sensor background noise and vignetting [87]. However, the anticipated absolute accuracy improvements are not crucial for the "relative magnitude" of elements of the scene approach adopted here where distribution equivalence is more highly prized.

The variable per-waveband mean spectral reflectance of the CB, characterized by greater reflectance in the visible band (Table 9), its smooth, low-sheen surface finish (potentially with a non-negligible specular component at large incident angles [48,146]) and the non-uniform Tetracam ADC CMOS sensor response function to different sensor bandwidths (with lowest sensitivity in the Green waveband) may account for the relative distribution of errors per waveband [86,87]. Importantly, due to near-horizontal viewing geometry, it is speculated that the accuracy of the reflectance recovery computations is strongly influenced by angular and adjacency effects (in particular from ground reflections) [75,85] and the impacts of in-situ spectral, diffuse and direct effects that are not explicitly accounted for in the laboratory measurements [62,146,147].

The majority of prior ground- or UAV-based studies utilising the vicarious EL method with narrow FOV MS sensors for reflectance recovery have been concerned with horizontal vegetated surfaces [86,87,95,96,102] and fewer with horizontal urban surfaces [48,104]. Furthermore, the use of the EL method applied to terrestrial close-range MS sensors with near-horizontal viewing geometry of urban building facades has not been previously reported. However, there is growing interest in the use of ground-based, close range narrow FOV sensors for retrieval of the spatio-temporal optical and thermal characteristics of vertical surfaces at the facet and sub-facet scales for building damage assessment [148], geological surveys [85] and urban climatology [78].

As there is a gap in the literature addressing, and no standard method exists for, facet and sub-facet scale terrestrial reflectance recovery from real building facades, the value of this paper is its contribution to the development of a logistically simple, replicable, validated reflectance recovery method, with explicit error magnitude and source estimation, suitable for climatology studies of building facades using a single mobile calibration target and high-resolution close-range images obtained from a relatively low-cost MS camera. However, since each camera has a unique sensor response-function [92] and irradiance may vary under different environmental conditions, the determination of the relationship between reflectance and DNs necessitates the computation of new equations for each new sensor and every unique environment [98].

5. Conclusions

This paper described a novel application of the EL method for radiometric calibration of a relatively low-cost MS sensor applied to close-range images of vertical urban surface materials. The performance of the single-target EL calibration equations was evaluated and quantified using validation samples of common building materials. Confidence in the precision and accuracy of the EL equations per-waveband was assessed using covariance statistics and error measures [128,129,137] and absolute prediction accuracies (7.728% < MAE > 10.108%) were within the range reported in the literature [88,94,149] and above prediction accuracies (15–20%) reported for single-target calibration methods [88,103].

Based on the optimum equations (Table 19) the results are encouraging and indicate that the prediction equations (Tables 11 and 12) satisfactorily characterize the per waveband reflectance differences of commonly used materials found on Australian building facades. For example, dry-pressed clay bricks (samples V7, V8, and V9, Table 14) are frequently used in low to medium-rise construction of residential buildings in Sydney and are satisfactorily predicted in the green waveband within a range of 1.24% to 6.68% absolute error (AE) (Table 17) and OLS model absolute residuals (OMAR) of 0.27%, 3.9% and 1.8% respectively. Importantly, the NIR waveband reflectance values are satisfactorily predicted within a range of 2.48% to 5.19% AE (Table 15) and OMAR in the range of 5.4% to 6.2%. This relative predictive strength is significant since dry-pressed clay bricks are both pervasive and more reflective in the NIR region (Table 14 and Figure 8).

Dark-grey pre-painted steel sheeting (sample V2) is another prevalent material used for building facade cladding that is more reflective in the NIR region (Table 14 and Figure 8). Green and NIR waveband reflectance values are satisfactorily predicted with a 0.32% and 0.71% AE (Table 17) and OMAR of 1.96% and 1.84% respectively. Glass (samples V10 and V11) is reasonably well predicted in the visible spectrum with a 1.06% and 4.84% AR in the Green (Table 17) and Red (Table 15) bands with an OMAR of 1.24% and 0.53% respectively for sample V10.

While the results confirm the findings of prior studies supporting the utility of a single in-situ target for radiometric calibration [86,98] it may be concluded from method and model evaluation that improvements in measurement protocols and calibration target specification would reduce prediction model error, although the anticipated absolute accuracy improvements are not warranted here since the "relative magnitude" of elements of the scene and distribution equivalence are desired for the development of a larger statistical model. The regression statistics (Tables 16, 18 and 19) imply distribution near-equivalence between measured and predicted reflectance values and high confidence in trend predictions. This demonstrates that a single-target EL method can be applied to recover spectral reflectance from terrestrial close-range MS images of vertical surfaces with satisfactory results.

Author Contributions: Conceptualization; Methodology; Validation; Formal Analysis; Investigation; Writing-Original Draft Preparation; Writing-Review & Editing and Funding Acquisition, J.F.; Resources; Funding Acquisition; Writing-Review & Editing; Supervision, P.O., Resources; Supervision and Funding Acquisition, A.P.

Funding: This research was funded by the Cooperative Research Centre (CRC) for Low Carbon Living, project number RP2005, and the Australian Government Research Training Program (RTP) Scholarship.

Acknowledgments: The authors wish to acknowledge Bill Joe, Research Support Engineer at the School of Material Science and Engineering, UNSW, and Alan Yee, Professional Officer at the School of Photovoltaics and Renewable Energy Engineering (SPREE), UNSW, for their technical support during the laboratory-based spectrophotometer measurements.

References

1. Oke, T.R. *Boundary Layer Climates*, 2nd ed.; Methuen: London, UK, 1987; ISBN 0-416-04422-0.
2. Oke, T.R. The energetic basis of the urban heat island. *Q. J. R. Meteorol. Soc.* **1982**, *108*, 1–24. [CrossRef]
3. He, B.-J. Potentials of meteorological characteristics and synoptic conditions to mitigate urban heat island effects. *Urban Clim.* **2018**, *24*, 26–33. [CrossRef]
4. Santamouris, M.; Haddad, S.; Fiorito, F.; Wang, R. Urban Heat Island and Overheating Characteristics in Sydney, Australia. An Analysis of Multiyear Measurements. *Sustainability* **2017**, *9*, 712. [CrossRef]
5. Oke, T.R.; Mills, G.; Christen, A.; Voogt, J.A. *Urban Climates*; Cambridge University Press: Cambridge, UK, 2017; ISBN 978-0-521-84950-0.
6. Oke, T.R. Towards a better scientific communication in urban climate. *Theor. Appl. Climatol.* **2006**, *84*, 179–190. [CrossRef]
7. Karimi, M.; Vant-Hull, B.; Nazari, R.; Mittenzwei, M.; Khanbilvardi, R. Predicting surface temperature variation in urban settings using real-time weather forecasts. *Urban Clim.* **2017**, *20*, 192–201. [CrossRef]
8. Doulos, L.; Santamouris, M.; Livada, I. Passive cooling of outdoor urban spaces. The role of materials. *Sol. Energy* **2004**, *77*, 231–249. [CrossRef]
9. Roth, M.; Oke, T.R.; Emery, W.J. Satellite-derived urban heat islands from three coastal cities and the utilization of such data in urban climatology. *Int. J. Remote Sens.* **1989**, *10*, 1699–1720. [CrossRef]
10. Stewart, I.D.; Oke, T.R. Local Climate Zones for Urban Temperature Studies. *BAMS* **2012**, *93*, 1879–1900. [CrossRef]

11. Santamouris, M. Innovating to zero the building sector in Europe: Minimising the energy consumption, eradication of the energy poverty and mitigating the local climate change. *Sol. Energy* **2016**, *128*, 61–94. [CrossRef]

12. Santamouris, M. Analyzing the heat island magnitude and characteristics in one hundred Asian and Australian cities and regions. *Sci. Total. Environ.* **2015**, *512–513*, 582–598. [CrossRef] [PubMed]

13. Mora, C.; Dousset, B.; Caldwell, I.R.; Powell, F.E.; Geronimo, R.C.; Bielecki, C.R.; Counsell, C.W.; Dietrich, B.S.; Johnston, E.T.; Louis, L.V.; et al. Global risk of deadly heat. *Nat. Clim. Chang.* **2017**, *7*, 501–506. [CrossRef]

14. Intergovernmental Panel on Climate Change (IPCC). Summary for policymakers. In *Climate Change 2014: Impacts, Adaptation, and Vulnerabilit*; Contribution of Working Group II to the Fifth Assessment Report of the Intergovernmental Panel on Climate Change; Cambridge University Press: Cambridge, UK, 2014.

15. Steffen, W.; Hughes, L.; Perkins, S. *Heatwaves: Hotter, Longer, More Often*; Climate Council of Australia: Sydney, Australia, 2014; ISBN 978-0-9924142-2-1.

16. Paravantis, J.; Santamouris, M.; Cartalis, C.; Efthymiou, C.; Kontoulis, N. Mortality associated with high ambient temperatures, heatwaves, and the urban heat island in Athens, Greece. *Sustainability* **2017**, *9*, 606. [CrossRef]

17. Heaviside, C.; Macintyre, H.; Vardoulakis, S. The urban heat island: Implications for health in a changing environment. *Curr. Environ. Health Report* **2017**, *4*, 296–305. [CrossRef] [PubMed]

18. Tan, J.; Zheng, Y.; Tang, X.; Guo, C.; Li, L.; Song, G.; Zhen, X.; Yuan, D.; Kalkstein, A.J.; Li, F. The urban heat island and its impact on heat waves and human health in Shanghai. *Int. J. Biometeorol.* **2010**, *54*, 75–84. [CrossRef] [PubMed]

19. Akbari, H.; Cartalis, C.; Kolokotsa, D.; Muscio, A.; Pisello, A.L.; Rossi, F.; Santamouris, M.; Synnefa, A.; Wong, N.H.; Zinzi, M. Local climate change and urban heat island mitigation techniques—The state of the art. *J. Civ. Eng. Manag.* **2016**, *22*, 1–16. [CrossRef]

20. Santamouris, M.; Kolokotsa, D. (Eds.) *Urban Climate Mitigation Techniques*; Routledge: New York, NY, USA, 2016; ISBN 978-0-415-71213-2.

21. Santamouris, M. Regulating the damaged thermostat of the cities—Status, impacts and mitigation strategies. *Energy Build.* **2015**, *91*, 43–56. [CrossRef]

22. Sun, T.; Kotthaus, S.; Li, D.; Ward, H.C.; Gao, Z.; Ni, G.H.; Grimmond, C.S.B. Attribution and mitigation of heat wave-induced urban heat storage change. *Environ. Res. Lett.* **2017**, *12*, 114007. [CrossRef]

23. Founda, D.; Santamouris, M. Synergies between urban heat island and heat waves in Athens (Greece), during and extremely hot summer. *Sci. Rep.* **2017**, *7*, 10973. [CrossRef] [PubMed]

24. Gunawardena, K.R.; Wells, M.J.; Kershaw, T. Utilising green and bluespace to mitigate urban heat island intensity. *Sci. Total Environ.* **2017**, *584–585*, 1040–1055. [CrossRef] [PubMed]

25. Georgescu, M.; Morefield, P.E.; Bierwagen, B.G.; Weaver, C.P. Twenty-first century megapolitan expansion. *Proc. Natl. Am. Sci. USA* **2014**, *111*, 2909–2914. [CrossRef] [PubMed]

26. Kyriakodis, G.-E.; Santamouris, M. Using reflective pavements to mitigate urban heat island in warm climates—Results from a large-scale urban mitigation project. *Urban Clim.* **2017**, *24*, 326–339. [CrossRef]

27. Chatzidimitriou, A.; Yannas, S. Microclimate design for open spaces: Ranking urban design effects on pedestrian thermal comfort in summer. *Sustain. Cities Soc.* **2016**, *26*, 27–47. [CrossRef]

28. Takebayashi, H. High-Reflectance Technology on Building Façades: Installation Guidelines for Pedestrian Comfort. *Sustainability* **2016**, *8*, 785. [CrossRef]

29. Santamouris, M. Urban warming and mitigation: Actual status, impacts and challenges. In *Urban Climate Mitigation Techniques*; Santamouris, M., Kolokotsa, D., Eds.; Routledge: New York, NU, USA, 2016; pp. 1–25, ISBN 978-0-415-71213-2.

30. Morini, E.; Touchaei, A.-G.; Rossi, F.; Cotana, F.; Akbari, H. Evaluation of albedo enhancement to mitigate impacts of urban heat island in Rome (Italy) using WRF meteorological model. *Urban Clim.* **2017**, *24*, 551–566. [CrossRef]

31. Muscio, A. The solar reflectance index as a tool to forecast the heat released to the urban environment: Potentiality and assessment issues. *Climate* **2018**, *6*, 12. [CrossRef]

32. Santamouris, M.; Synnefa, A.; Karlessi, T. Using advanced cool materials in the urban built environment to mitigate heat islands and improve thermal comfort conditions. *Sol. Energy* **2011**, *85*, 3085–3102. [CrossRef]

33. Bretz, S.E.; Akbari, H.; Rosenfeld, A. Practical issues for using solar-reflective materials to mitigate urban heat islands. *Atmos. Environ.* **1998**, *32*, 95–101. [CrossRef]

34. Aleksandrowicz, O.; Vuckovic, M.; Kiesel, K.; Mahdavi, A. Current trends in urban heat island mitigation research: Observations based on a comprehensive research repository. *Urban Clim.* **2017**, *21*, 1–26. [CrossRef]

35. Santamouris, M.; Ding, L.; Fiorito, F.; Synnefa, A. Passive and active cooling for the outdoor environment—An analysis and assessment of the cooling potential of mitigation technologies using performance data from 220 large-scale projects. *Sol. Energy* **2017**, *154*, 14–33. [CrossRef]

36. Pisello, A.L. State of the art on the development of cool coatings for buildings and cities. *Sol. Energy* **2017**, *144*, 660–680. [CrossRef]

37. Lee, H.; Mayer, H. Thermal comfort of pedestrians in an urban street canyon is affected by increasing albedo of building wall. *Int. J. Biometeorol.* **2018**. [CrossRef] [PubMed]

38. Lontorfos, V.; Efthymiou, C.; Santamouris, M. On the time varying mitigation performance of reflective geoengineering technologies in cities. *Renew. Energy* **2018**, *115*, 926–930. [CrossRef]

39. Yang, J.; Wang, Z.H.; Kaloush, K.E. Environmental impacts of reflective materials: Is high albedo a "silver bullet" for mitigating urban heat island? *Renew. Sustain. Energy Rev.* **2015**, *47*, 830–843. [CrossRef]

40. Nunez, M.; Oke, T.R. The energy balance of an urban canyon. *J. Appl. Meteorol.* **1977**, *16*, 11–19. [CrossRef]

41. Anandakumar, K. A study on the partition of net radiation into heat fluxes on a dry asphalt surface. *Atmos. Environ.* **1999**, *33*, 3911–3918. [CrossRef]

42. Brownlee, J.; Ray, P.; Tewari, M.; Tan, H. Relative role of turbulent and radiative flux on the near-surface temperature in a single-layer urban canopy model over Houston. *J. Appl. Meteorol. Climatol.* **2017**, *56*, 2173–2187. [CrossRef]

43. Qin, Y.; Hiller, J.E. Understanding pavement-surface energy balance and its implications on cool pavement development. *Energy Build.* **2014**, *85*, 389–399. [CrossRef]

44. Synnefa, A.; Santamouris, M. Mitigating the urban heat with cool materials for the building's fabric. In *Urban Climate Mitigation Techniques*; Santamouris, M., Kolokotsa, D., Eds.; Routledge: New York, NY, USA, 2016; pp. 67–91, ISBN 978-0-415-71213-2.

45. Costanzo, V.; Evola, G.; Marletta, L.; Gagliano, A. Proper evaluation of the external convective heat transfer for the thermal analysis of cool roofs. *Energy Build.* **2014**, *77*, 467–477. [CrossRef]

46. Akbari, H.; Kolokotsa, D. Three decades of urban heat islands and mitigation technologies research. *Energy Build.* **2016**, *133*, 834–842. [CrossRef]

47. Takebayashi, H.; Moriyama, M. Relationships between the properties of an urban street canyon and its radiant environment: Introduction of appropriate urban heat island mitigation technologies. *Sol. Energy* **2012**, *86*, 2255–2262. [CrossRef]

48. Ban-Weiss, G.A.; Woods, J.; Levinson, R. Using remote sensing to quantify albedo of roofs in seven California cities, Part 1: Methods. *Sol. Energy* **2015**, *115*, 777–790. [CrossRef]

49. Crum, S.M.; Jenerette, G.D. Microclimate Variation among Urban Land Covers: The Importance of Vertical and Horizontal Structure in Air and Land Surface Temperature Relationships. *J. Appl. Meteorol. Climatol.* **2017**, *56*, 2531–2543. [CrossRef]

50. Taha, H. Urban climates and heat islands: Albedo, evapotranspiration, and anthropogenic heat. *Energy Build.* **1997**, *25*, 99–103. [CrossRef]

51. Levinson, R.M. Near-Ground Cooling Efficacies of Trees and High-Albedo Surfaces. Ph.D. Thesis, University of California, Berkeley, CA, USA, 1997.

52. Lee, I.; Voogt, J.A.; Gillespie, T.J. Analysis and comparison of shading strategies to increase human thermal comfort in urban areas. *Atmosphere* **2018**, *9*, 91. [CrossRef]

53. Taleghani, M. The impact of increasing urban surface albedo on outdoor summer thermal comfort within a university campus. *Urban Clim.* **2018**, *24*, 175–184. [CrossRef]

54. Rosso, F.; Golasi, I.; Castaldo, V.L.; Piselli, C.; Pisello, A.L.; Salata, F.; Ferrero, M.; Cotana, F.; de Lieto Vollaro, A. On the impact of innovative materials on outdoor thermal comfort of pedestrians in historical urban canyons. *Renew. Energy* **2018**, *118*, 825–839. [CrossRef]

55. Taleghani, M.; Berardi, U. The effect of pavement characteristics on pedestrians' thermal comfort in Toronto. *Urban Clim.* **2017**, *24*, 449–459. [CrossRef]

56. Hardin, A.W.; Vanos, J.K. The influence of surface type on the absorbed radiation by a human under hot, dry conditions. *Int. J. Biometeorol.* **2018**, *62*, 43–56. [CrossRef] [PubMed]

57. Yoshida, S.; Yumino, S.; Uchida, T.; Mochida, A. Numerical Analysis of the Effects of Windows with Heat Ray Retro-reflective Film on the Outdoor Thermal Environment within a Two-dimensional Square Cavity-type Street Canyon. *Procedia Eng.* **2016**, *169*, 384–391. [CrossRef]

58. Salata, F.; Golasi, I.; Vollaro, E.D.L.; Bisegna, F.; Nardecchia, F.; Coppi, M.; Gugliermetti, F.; Vollaro, A.D.L. Evaluation of different urban microclimate mitigation strategies through a PMV analysis. *Sustainability* **2015**, *7*, 9012–9030. [CrossRef]

59. Erell, E.; Pearlmutter, D.; Boneh, D.; Kutiel, P.B. Effect of high-albedo materials on pedestrian heat stress in urban street canyons. *Urban Clim.* **2014**, *10*, 367–386. [CrossRef]

60. Laureti, F.; Martinelli, L.; Battisti, A. Assessment and mitigation strategies to counteract overheating in urban historical areas in Rome. *Climate* **2018**, *6*, 18. [CrossRef]

61. Coakley, J.A. Reflectance and albedo, surface. In *Encyclopedia of Atmospheric Sciences*; Holton, J.R., Ed.; Academic Press: Oxford, UK, 2003; pp. 1914–1923, ISBN 9780122270901.

62. Levinson, R.M.; Akbari, H.; Berdahl, P. Measuring solar reflectance—Part 1: Defining a metric that accurately predicts solar heat gain. *Sol. Energy* **2010**, *84*, 1717–1744. [CrossRef]

63. Schaepman-Strub, G.; Schaepman, M.E.; Painter, T.H.; Dangel, S.; Martonchik, J.V. Reflectance quantities in optical remote sensing—definitions and case studies. *Remote Sens. Environ.* **2006**, *103*, 27–42. [CrossRef]

64. Sailor, D.J.; Resh, K.; Segura, D. Field measurement of albedo for limited extent test surfaces. *Sol. Energy* **2006**, *80*, 589–599. [CrossRef]

65. Qin, Y.; He, H. A new simplified method for measuring the albedo of limited extent targets. *Sol. Energy* **2017**, *157*, 1047–1055. [CrossRef]

66. Levinson, R.; Akbari, H.; Berdahl, P. Measuring solar reflectance—Part 2: Review of practical methods. *Sol. Energy* **2010**, *84*, 1745–1759. [CrossRef]

67. ASTM International. *ASTM E903-12, Standard Test Method for Solar Absorptance, Reflectance and Transmittance of Materials Using Integrating Spheres*; ASTM International: West Conshohocken, PA, USA, 2012.

68. ASTM International. *ASTM C1549-16, Standard Test Method for Determination of Solar Reflectance Near Ambient Temperature Using a Portable Solar Reflectometer*; ASTM International: West Conshohocken, PA, USA, 2016.

69. ASTM International. *ASTM E1918-16, Standard Test Method for Measuring Solar Reflectance of Horizontal and Low-Sloped Surfaces in the Field*; ASTM International: West Conshohocken, PA, USA, 2016.

70. Akbari, H.; Levinson, R.; Berdahl, P. Cool materials rating instrumentation and testing. In *Advances in the Development of Cool Materials for the Built Environment*; Kolokotsa, D., Santamouris, M., Akbari, H., Eds.; Bentham Science Publishers: Oak Park, IL, USA, 2013; pp. 47–76, ISBN 978-1-60805-597-5.

71. Mei, G.; Wu, B.; Ma, S.; Qin, Y. A simplified method for the solar reflectance of a finite surface in field. *Measurement* **2017**, *110*, 211–216. [CrossRef]

72. Qin, Y.; Liang, J.; Tan, K.; Li, F. A side-by-side comparison of the cooling effect of building blocks with retro-reflective and diffuse-reflective walls. *Sol. Energy* **2016**, *133*, 172–179. [CrossRef]

73. Rossi, F.; Castellani, B.; Presciutti, A.; Morini, E.; Filipponi, M.; Nicolini, A.; Santamouris, M. Retroreflective façades for urban heat island mitigation: Experimental investigation and energy evaluations. *Appl. Energy* **2015**, *145*, 8–20. [CrossRef]

74. Pisello, A.L.; Goretti, M.; Cotana, F. A method for assessing buildings' energy efficiency by dynamic simulation and experimental activity. *Appl. Energy* **2012**, *97*, 419–429. [CrossRef]

75. Voogt, J.A.; Oke, T.R. Radiometric temperatures of urban canyon walls obtained from vehicle traverses. *Theor. Appl. Climatol.* **1998**, *60*, 199–217. [CrossRef]

76. Johnson, G.T.; Watson, I.D. The Determination of View-Factors in Urban Canyons. *J. Clim. Appl. Meteorol.* **1984**, *23*, 329–335. [CrossRef]

77. Krayenhoff, E.S.; Voogt, J.A. Daytime thermal anisotropy of urban neighbourhoods: Morphological causation. *Remote Sens.* **2016**, *8*, 108. [CrossRef]

78. Adderley, C.A.; Christen, A.; Voogt, J.A. The effect of radiometer placement and view on inferred directional and hemispheric radiometric temperatures of an urban canopy. *Atmos. Meas. Tech. Discuss.* **2015**, *8*, 2699–2714. [CrossRef]

79. Hénon, A.; Mestayer, P.G.; Lagouarde, J.P.; Voogt, J.A. An urban neighborhood temperature and energy study from the CAPITOUL experiment with the Solene model. Part 2: Influence of building surface heterogeneities. *Theor. Appl. Climatol.* **2012**, *110*, 197–208. [CrossRef]

80. Christen, A.; Meier, F.; Scherer, D. High-frequency fluctuations of surface temperatures in an urban environment. *Theor. Appl. Climatol.* **2012**, *108*, 301–324. [CrossRef]
81. Lagouarde, J.P.; Hénon, A.; Kurz, B.; Moreau, P.; Irvine, M.; Voogt, J.; Mestayer, P. Modelling daytime thermal infrared directional anisotropy over Toulouse city centre. *Remote Sens. Environ.* **2010**, *114*, 87–105. [CrossRef]
82. Voogt, J.A. Assessment of an urban sensor view model for thermal anisotropy. *Remote Sens. Environ.* **2008**, *112*, 482–495. [CrossRef]
83. Allen, M.A.; Voogt, J.A.; Christen, A. Time-Continuous hemispherical urban surface temperatures. *Remote Sens.* **2018**, *10*, 3. [CrossRef]
84. Richards, J.A. *Remote Sensing Digital Image Analysis, an Introduction*, 5th ed.; Springer: Heidelberg, Germany, 2013; ISBN 978-3-642-30062-2.
85. Kurz, T.H.; Buckley, S.J.; Howell, J.A. Close-range hyperspectral imaging for geological field studies: Workflow and methods. *Int. J. Remote Sens.* **2013**, *34*, 1798–1822. [CrossRef]
86. Iqbal, F.; Lucieer, A.; Barry, K. Simplified radiometric calibration for UAS-mounted multispectral sensor. *Eur. J. Remote Sens.* **2018**, *51*, 301–313. [CrossRef]
87. Del Pozo, S.; Rodríguez-Gonzálvez, P.; Hernández-López, D.; Felipe-García, B. Vicarious radiometric calibration of a multispectral camera on board an unmanned aerial system. *Remote Sens.* **2014**, *6*, 1918–1937. [CrossRef]
88. Smith, G.M.; Milton, E.J. The use of the empirical line method to calibrate remotely sensed data to reflectance. *Int. J. Remote. Sens.* **1999**, *20*, 2653–2662. [CrossRef]
89. Brest, C.L.; Goward, S.N. Deriving surface albedo measurements from narrow band satellite data. *Int. J. Remote Sens.* **2007**, *8*, 351–367. [CrossRef]
90. Synnefa, A.; Santamouris, M.; Apostolakis, K. On the development, optical properties and thermal performance of cool colored coatings for the urban environment. *Sol. Energy* **2007**, *81*, 488–497. [CrossRef]
91. Tetracam Inc. Agricultural Digital Camera (ADC) Specifications. 2016. Available online: www.tetracam.com (accessed on 25 November 2015).
92. Del Pozo, S.; Sánchez-Aparicio, L.J.; Rodríguez-Gonzálvez, P.; Herrero-Pascual, J.; Muñoz-Nieto, A.; González-Aguilera, D.; Hernández-López, D. Multispectral imaging: Fundamentals, principles and methods of damage assessment in constructions. In *Non-Destructive Techniques for the Evaluation of Structures and Infrastructure*; Riveiro, B., Solla, M., Eds.; CRC Press: London, UK, 2016; pp. 139–166, ISBN 9781315685151.
93. Corripio, J.G. Snow surface albedo estimation using terrestrial photography. *Int. J. Remote Sens.* **2004**, *25*, 5705–5729. [CrossRef]
94. Pompilio, L.; Marinangeli, L.; Amitrano, L.; Pacci, G.; D'andrea, S.; Iacullo, S.; Monaco, E. Application of the empirical line method (ELM) to calibrate the airborne Daedalus-CZCS scanner. *Eur. J. Remote Sens.* **2018**, *51*, 33–46. [CrossRef]
95. Berra, E.F.; Gaulton, R.; Barr, S. Commercial off-the-shelf digital cameras on unmanned aerial vehicles for multi-temporal monitoring of vegetation reflectance and NDVI. *IEEE Trans. Geosci. Remote Sens.* **2017**, *55*, 4878–4886. [CrossRef]
96. Mathews, A.J. A practical UAV remote sensing methodology to generate multispectral orthophotos for vineyards: Estimation of spectral reflectance using compact digital cameras. *Int. J. Appl. Geospace Res.* **2015**, *6*, 65–87. [CrossRef]
97. Stow, D.; Hope, A.; Nguyen, A.T.; Phinn, S.; Benkelman, C.A. Monitoring detailed land surface changes using an airborne multispectral digital camera system. *IEEE Trans. Geosci. Remote Sens.* **1996**, *34*, 1191–1203. [CrossRef]
98. Wang, C.; Myint, S.W. A Simplified empirical line method of radiometric calibration for small unmanned aircraft systems-based remote sensing. *IEEE J. Sel. Top. Appl. Earth Obs. Remote Sens.* **2015**, *8*, 1876–1885. [CrossRef]
99. Herrero-Huerta, M.; Hernández-López, D.; Rodriguez-Gonzalvez, P.; González-Aguilera, D.; González-Piqueras, J. Vicarious radiometric calibration of a multispectral sensor from an aerial trike applied to precision agriculture. *Comput. Electron. Agric.* **2014**, *108*, 28–38. [CrossRef]
100. Honkavaara, E.; Arbiol, R.; Markelin, L.; Martinez, L.; Cramer, M.; Bovet, S.; Chandelier, L.; Ilves, R.; Klonus, S.; Marshal, P.; et al. Digital airborne photogrammetry—A new tool for quantitative remote sensing?—A state-of-the-art review on radiometric aspects of digital photogrammetric images. *Remote Sens.* **2009**, *1*, 577–605. [CrossRef]

101. Zaman, B.; Austin, J.; Clemens, S.; McKee, M. Retrieval of spectral reflectance of high-resolution multispectral imagery acquired with an autonomous unmanned aerial vehicle: Aggieair™. *Photogramm. Eng. Remote Sens.* **2014**, *80*, 1139–1150. [CrossRef]

102. Laliberte, A.S.; Goforth, M.A.; Steele, C.M.; Rango, A. Multispectral remote sensing from unmanned aircraft: Image processing workflows and applications for rangeland environments. *Remote Sens.* **2011**, *3*, 2529–2551. [CrossRef]

103. Mei, A.; Bassani, C.; Fortinovo, G.; Salvatori, R.; Allegrini, A. The use of suitable pseudo-invariant targets for MIVIS data calibration by the empirical line method. *ISPRS J. Photogramm. Remote Sens.* **2016**, *114*, 102–114. [CrossRef]

104. Gaitani, N.; Burud, I.; Thiis, T.; Santamouris, M. High-resolution spectral mapping of urban thermal properties with Unmanned Aerial Vehicles. *Build. Environ.* **2017**, *121*, 215–224. [CrossRef]

105. Qin, Y.; Liang, J.; Tan, K.; Li, F. The amplitude and maximum of daily pavement surface temperature increase linearly with solar absorption. *Road Mater. Pavement Des.* **2017**, *18*, 440–452. [CrossRef]

106. Kolokotsa, D.D.; Giannariakis, G.; Gobakis, K.; Giannarakis, G.; Synnefa, A.; Santamouris, M. Cool roofs and cool pavements application in Acharnes, Greece. *Sustain. Cities Soc.* **2018**, *37*, 466–474. [CrossRef]

107. Qin, Y. A review on the development of cool pavements to mitigate urban heat island effect. *Renew. Sustain. Energy Rev.* **2015**, *52*, 445–459. [CrossRef]

108. Pisello, A.L.; Castaldo, V.L.; Piselli, C.; Fabiani, C.; Cotana, F. Thermal performance of coupled cool roof and cool façade: Experimental monitoring and analytical optimization procedure. *Energy Build.* **2017**, *157*, 35–52. [CrossRef]

109. Zinzi, M. Exploring the potentialities of cool facades to improve the thermal response of Mediterranean residential buildings. *Sol. Energy* **2016**, *135*, 386–397. [CrossRef]

110. Zinzi, M. Characterisation and assessment of near infrared reflective paintings for building facade applications. *Energy Build.* **2016**, *114*, 206–213. [CrossRef]

111. Rossi, F.; Pisello, A.L.; Nicolini, A.; Filipponi, M.; Palombo, M. Analysis of retro-reflective surfaces for urban heat island mitigation: A new analytical model. *Appl. Energy* **2014**, *114*, 621–631. [CrossRef]

112. Hernández-Pérez, I.; Álvarez, G.; Xamán, J.; Zavala-Guillén, I.; Arce, J.; Simá, E. Thermal performance of reflective materials applied to exterior building components—A review. *Energy Build.* **2014**, *80*, 81–105. [CrossRef]

113. Georgakis, C.; Zoras, S.; Santamouris, M. Studying the effect of "cool" coatings in street urban canyons and its potential as a heat island mitigation technique. *Sustain. Cities Soc.* **2014**, *13*, 20–31. [CrossRef]

114. Fiorito, F.; Santamouris, M. High performance technologies and the future of architectural design. *TECHNE J. Technol. Archit. Environ.* **2017**, *80*, 72–76. [CrossRef]

115. Sung, D. A New Look at Building Facades as Infrastructure. *Engineering* **2016**, *2*, 63–68. [CrossRef]

116. Libbra, A.; Muscio, A.; Siligardi, C. Energy performance of opaque building elements in summer: Analysis of a simplified calculation method in force in Italy. *Energy Build.* **2013**, *64*, 384–394. [CrossRef]

117. Halawa, E.; Ghaffarianhoseini, A.; Ghaffarianhoseini, A.; Trombley, J.; Hassan, N.; Baig, M.; Yusoff, S.Y.; Ismail, M.A. A review on energy conscious designs of building façades in hot and humid climates: Lessons for (and from) Kuala Lumpur and Darwin. *Renew. Sustain. Energy Rev.* **2018**, *82*, 2147–2161. [CrossRef]

118. Castaldo, V.L.; Pisello, A.L.; Piselli, C.; Fabiani, C.; Cotana, F.; Santamouris, M. How outdoor microclimate mitigation affects building thermal-energy performance: A new design-stage method for energy saving in residential near-zero energy settlements in Italy. *Renew. Energy* **2018**, *127*, 920–935. [CrossRef]

119. Lamarca, C.; Quense, J.; Henríquez, C. Thermal comfort and urban canyons morphology in coastal temperature climate, Conception, Chile. *Urban Clim.* **2018**, *23*, 159–172. [CrossRef]

120. Song, D.; Han, S. The Analysis of Reactive Factors between Architectural Envelop Condition and Urban Microclimate. *Procedia Eng.* **2016**, *169*, 125–132. [CrossRef]

121. Garuma, G.M. Review of urban surface parameterizations for numerical climate models. *Urban Clim.* **2017**, *24*, 830–851. [CrossRef]

122. Bozonnet, E.; Musy, M.; Calmet, I.; Rodriguez, F. Modeling methods to assess urban fluxes and heat island mitigation measures from street to city scale. *Int. J. Low-Carbon Technol.* **2015**, *10*, 62–77. [CrossRef]

123. Best, M.J.; Grimmond, C.S.B. Key conclusions of the first international urban land surface model comparison project. *BAMS* **2015**, *96*, 805–819. [CrossRef]

124. Velasco, E. Go to field, look around, measure and then run models. *Urban Clim.* **2018**, *24*, 231–236. [CrossRef]

125. Page, J.K. The fundamental problems of building climatology considered from the point of view of decision-making by the architect or urban designer. In *Building Climatology, Proceedings of the Symposium on Urban Climates and Building Climatology*; WMO Technical Note 10; WHO: Switzerland, Geneva, 1968; Volume 2, pp. 9–21.

126. Barlow, J.; Best, M.; Bohnenstengel, S.I.; Clark, P.; Grimmond, S.; Lean, H.; Christen, A.; Emeis, S.; Haeffelin, M.; Harman, I.N. Developing a research strategy to better understand, observe and simulate urban atmospheric processes at kilometre to sub-kilometre scales. *BAMS* **2017**, *98*, ES261–ES264. [CrossRef]

127. Australian Government Civil Aviation Safety Authority. CASR Part 101—Unmanned Aircraft and Rockets. Available online: https://www.casa.gov.au/standard-page/casr-part-101-unmanned-aircraft-and-rocket-operations (accessed on 2 April 2018).

128. Willmott, C.J. On the validation of models. *Phys. Geogr.* **1981**, *2*, 184–194. [CrossRef]

129. Willmott, C.J. Some comments on the evaluation of model performance. *BAMS* **1982**, *63*, 1309–1313. [CrossRef]

130. Labsphere Inc. *8°/Hemispherical Reflectance Multi-Step Target Certificate*; Serial number MS050-0916-6020; Labsphere Inc.: North Sutton, NH, USA, 2016.

131. Geoscience Australia, Commonwealth of Australia. Available online: http://www.ga.gov.au/geodesy/astro/smpos.jsp (accessed on 3 June 2017).

132. Baugh, W.M.; Groeneveld, D.P. Empirical proof of the empirical line. *Int. J. Remote Sens.* **2008**, *29*, 665–672. [CrossRef]

133. *PixelWrench II, Proprietary Image Editing Software*, version 1.2.1.8; Tetracam Inc.: Chatsworth, CA, USA, 2015.

134. Huang, Y.; Thomson, S.J.; Lan, Y.; Maas, S.J. Multispectral imaging systems for airborne remote sensing to support agricultural production management. *Int. J. Agric. Biol. Eng.* **2010**, *3*, 50–62. [CrossRef]

135. Honkavaara, E.; Hakala, T.; Peltoniemi, J.; Suomalainen, J.; Ahokas, E.; Markelin, L. Analysis of properties of reflectance reference targets for permanent radiometric test sites of high resolution airborne imaging systems. *Remote Sens.* **2010**, *2*, 1892–1917. [CrossRef]

136. Moran, M.S.; Bryant, R.; Thome, K.; Ni, W.; Nouvellon, Y.; Gonzalez-Dugo, M.P.; Qi, J.; Clarke, T.R. A refined empirical line approach for reflectance factor retrieval from Landsat-5 TM and Landsat-7 ETM+. *Remote Sens. Environ.* **2001**, *78*, 71–82. [CrossRef]

137. Willmott, C.J.; Ackleson, S.G.; Davis, R.E.; Feddema, J.J.; Klink, K.M.; Legates, D.R.; O'donnell, J.; Rowe, C.M. Statistics for the evaluation and comparison of models. *J. Geophys. Res.* **1985**, *90*, 8995–9005. [CrossRef]

138. Piñeiro, G.; Perelman, S.; Guerschman, J.P.; Paruelo, J.M. How to evaluate models: Observed vs. predicted or predicted vs. observed? *Ecol. Model.* **2008**, *216*, 316–322. [CrossRef]

139. Willmott, C.J.; Matsuura, K. Advantages of the mean absolute error (MAE) over the root mean square error (RMSE) in assessing average model performance. *Clim. Res.* **2005**, *30*, 79–82. [CrossRef]

140. Hyslop, N.P.; White, W.H. Estimating Precision Using Duplicate Measurements. *J. Air Waste Manag. Assoc.* **2009**, *59*, 1032–1039. [CrossRef] [PubMed]

141. Smith, E.P.; Rose, K.A. Model goodness-of-fit analysis using regression and related techniques. *Ecol. Model.* **1995**, *77*, 49–64. [CrossRef]

142. Chen, P.Y.; Popovich, P.M. *Correlation: Parametric and Non-Parametric Measures*; Sage University Paper Series on Quantitative Applications in the Social Sciences, Series No. 07-139; Sage: Irvine, CA, USA, 2002; ISBN 0761922288.

143. Lewis-Beck, M.S. *Data Analysis, an Introduction*; Sage University Paper Series on Quantitative Applications in the Social Sciences, Series No. 07-103; Sage: Irvine, CA, USA, 1995; ISBN 0803957726.

144. Gibbons, J.D. *Nonparametric Statistics: An Introduction*; Sage University Paper Series on Quantitative Applications in the Social Sciences, Series No. 07-090; Sage: Irvine, CA, USA, 1993; ISBN 0803946643.

145. Hueni, A.; Damm, A.; Kneubuehler, M.; Schlapfer, D.; Schaepman, M.E. Field and airborne spectroscopy cross validation—Some considerations. *IEEE J. Sel. Top. Appl. Earth Obs. Remote Sens.* **2017**, *10*, 1117–1135. [CrossRef]

146. Zinzi, M.; Carnielo, E.; Rossi, G. Directional and angular response of construction materials solar properties: Characterisation and assessment. *Sol. Energy* **2015**, *115*, 52–67. [CrossRef]

147. Dangel, S.; Verstraete, M.M.; Schopfer, J.; Kneubuhler, M.; Schaepman, M.; Itten, K.I. Toward a direct comparison of field and laboratory goniometer measurements. *IEEE Trans. Geosci. Remote Sens.* **2005**, *43*, 2666–2675. [CrossRef]

148. Del Pozo, S.; Herrero-Pascual, J.; Felipe-García, B.; Hernández-López, D.; Rodríguez-Gonzálvez, P.; González-Aguilera, D. Multispectral radiometric analysis of façades to detect pathologies from active and passive remote sensing. *Remote Sens.* **2016**, *8*, 80. [CrossRef]

149. Honkavaara, E. Calibrating Digital Photogrammetric Airborne Imaging Systems Using a Test Field. Ph.D. Thesis, University of Technology Espoo, Espoo, Finland, 2008.

![climate logo] *climate*

MDPI

Article

Sky View Factor Calculation in Urban Context: Computational Performance and Accuracy Analysis of Two Open and Free GIS Tools

Jérémy Bernard [1,†], Erwan Bocher [1,*,†], Gwendall Petit [2,†] and Sylvain Palominos [2,†]

[1] CNRS, UMR 6285 Vannes, France; jeremy.bernard@univ-ubs.fr
[2] Université de Bretagne Sud, UMR 6285 Vannes, France; gwendall.petit@univ-ubs.fr (G.P.); sylvain.palominos@univ-ubs.fr (S.P.)
* Correspondence: erwan.bocher@univ-ubs.fr
† These authors contributed equally to this work.

Received: 31 May 2018; Accepted: 2 July 2018; Published: 4 July 2018

Abstract: The sky view factor (SVF) has an important role in the analysis of the urban micro-climate. A new vector-based SVF calculation tool was implemented in a free and open source Geographic Information System named OrbisGIS. Its accuracy and computational performance are compared to the ones of an existing raster based algorithm used in SAGA-GIS. The study is performed on 72 urban blocks selected within the Paris commune territory. This sample has been chosen to represent the heterogeneity of nine of the ten Local Climate Zone built types. The effect of the algorithms' input parameters (ray length, number of directions and grid resolution) is investigated. The combination minimizing the computation time and the SVF error is identified for SAGA-GIS and OrbisGIS algorithms. In both cases, the standard deviation of the block mean SVF estimate is about 0.03. A simple linear relationship having a high determination coefficient is also established between block mean SVF and the facade density fraction, confirming the results of previous research. This formula and the optimized combinations for the OrbisGIS and the SAGA-GIS algorithms are finally used to calculate the SVF of every urban block of the Paris commune.

Keywords: local climate zone; urban climate; sky view factor; morphological indicator; open science; GIS

1. Introduction

Urbanization is often characterized by a low vegetation density combined with a high fraction of impervious surfaces, a high rugosity (due to buildings' size and height), and a denser energy consumption. These particular features affect the climate of the urban areas. Compared to the rural one, the urban microclimate shows a lower evaporation heat flux, a higher trapping of short and long-wave radiation, a lower ventilation and a larger amount of heat released in the atmosphere [1]. As a result, the temperature is often higher in urban areas than in its surrounding rural areas. This phenomenon is called Urban Heat Island (UHI) and may impact outdoor comfort [2], human health especially during heat waves [3], plant phenology [4], precipitation [5], energy consumption [6] and indirectly air pollution [7].

One of the climate change effects is the increase in the frequency and intensity of the heat waves, thus there is an urgent need to decrease the UHI phenomenon [8]. Several techniques have been investigated and tested all over the world: roof, facade and ground greening, roof and wall painting to alter the radiation absorption in both short and long-wave spectrum, water based solution such as fountains or sprinklers, building density and orientation optimization for shading purpose, etc. [9]. The performance of each technique may vary a lot both depending on the regional climate and the built environment where they are implemented. For example, Kikegawa et al. [10] showed that pavement watering was more adapted to residential areas than to narrow street canyons; Bernard et al. [11] have highlighted that the cooling induced by a given park would better propagate within streets having a large aspect ratio (building height to street width ratio—H/W); Ng et al. [12] have demonstrated that above a one-aspect ratio, the cooling benefit of a green roof at grade is low. To identify the best mitigation solutions adapted to a particular building environment, the efficiency of each solution may be tested running micro-climate models. Although this method is probably the most accurate, its applicability to each neighborhood of each city could be very expensive. Another solution is conceivable: based on a given set of climate-related parameters such as aspect ratio, albedo or vegetation density [1], each neighborhood may be classified in a specific built environment. Then, the efficiency of each mitigation solution may be tested running micro-climate models on a limited number of built environments, thus decreasing the number of simulations to perform for each city. In order to achieve such work, Stewart [13] proposed the concept of Local Climate Zone (LCZ): he defined ten built types and seven land-cover types based on a set of geometric and land-cover properties. These types and properties are detailed in Stewart and Oke [14]. One of the geometric properties, called Sky View Factor (SVF), defines the ratio of sky hemisphere visible from the ground (not obstructed by buildings, terrain or trees). This parameter is of major importance for urban climate applications since the long-wave radiation term is directly impacted by its value: the higher the SVF, the lower the long-wave radiation flux emitted by built surfaces to the sky during night-time. Oke [15] showed that the net long-wave radiation received on a given point located on the floor of a canyon is almost linearly related to the SVF of the point. At block scale (elementary area resulting from the division of the territory by the road network), Bernabé et al. [16] also showed a linear relationship (R^2 = 0.939) between the block mean SVF (the building roof and facades as well the ground surfaces were meshed in 10 m² triangles) and the net solar radiation. The SVF is then a very interesting parameter regarding urban climate applications. In order to obtain an accurate SVF value to each block of a city, it should be calculated for a high number of points. However, its calculation being computationally costly, it has been only performed on a limited sample of points for several years. This problem was partly solved for street canyons using an analytical solution [15]: the SVF of the mid-width of the floor of a canyon with symmetric cross-section and infinite length can be calculated using the Equation (1):

$$SVF_{mid-canyon} = \cos(\mathrm{atan}(2 \cdot H/W)), \tag{1}$$

where H/W is the canyon aspect ratio defined as the building height (H) divided by the street width (W).

Another solution to quickly calculate an approximate value of the block mean SVF is to use an empirical relationship with explaining variables being simple indicators calculated at block level. Bernabé et al. [16] found a linear relationship (R^2 = 0.8) between the block mean SVF and the block facade density fraction (defined Equation (2)):

$$F_F = \frac{S_F}{S_F + S_{REF}}, \tag{2}$$

where S_F the area of free external facade (m²), and S_{REF} is the area of the urban block (m²).

However, these methods have not been largely used for LCZ classification yet since they have major limitations:

- the analytical solution is only adapted for urban canyons having an infinite length, which is rarely the case in real urban tissues. Moreover, when averaged at block scale, the $SVF_{mid-canyon}$ is underestimated since it is only calculated at the center of the street.
- Concerning the empirical method, the relationship proposed by Bernabé et al. [16] is not valid for the SVF as defined in the LCZ properties. Bernabé et al. [16] calculated the SVF averaged to all

urban surfaces (roofs, facades and ground), whereas, in the LCZ classification, it should only be averaged to ground surfaces.

- Last but not least, these methods are not adapted to all complexes urban tissues: the estimated SVF for some of them can be far from the real SVF value.

Several algorithms have been developed to calculate the SVF at any space position. Some of them use images (fish-eye, street view, etc.) as input data [17–19], and the others use either raster or vector data to identify the location of the sky obstacles [17,18,20–23]. The accuracy of the SVF values depends on the quality of the input data, on the method chosen and the value of its associated parameters (for example, the number of meshes of the sky hemisphere in the case of a ray throwing method). In these studies, the performance of the existing algorithms has been evaluated at point scale but never when averaged at block scale, while this information is of major importance for LCZ classification. Hämmerle et al. [17] selected several LCZ within the city of Szeged (Hungary), but they did not present their results through this variable. The objective of this work is then to evaluate the ability of two algorithms to accurately (and in a minimum computational time) estimate the SVF of a block independently of its LCZ type. In addition, the influence of the input data (raster or vector) used by the algorithm on the resulting SVF is also investigated.

The results of this work are addressed to any researcher or urban planner willing to apply a LCZ classification. Thus, the choice was made to use two SVF algorithms developed within two different free and open source GIS. The first is embedded in the software SAGA-GIS [24] and is also available as a plugin in QGIS. SAGA-GIS has the advantage of being known again and is often used by the urban climate community—for instance, in the framework of the World Urban DataBase and Access Portal Tools—WUDAPT (http://www.wudapt.org/—accessed in April 2018). Its algorithm is based on raster input data (information on building and terrain vertical elevation is stored in every pixel of the raster) and calculations are performed using two input parameters: the number of directions and the distance of search. The larger their value and the more accurate the SVF of each pixel, the longer it takes. The resolution of the raster is also a critical parameter both regarding accuracy and computational duration. It is a major limitation of the algorithm: since a building's footprint is vector-based data, it has to be rasterized in order to be used in SAGA-GIS, thus inducing a loss of information proportional to the raster resolution. The second algorithm is based on vector data and has been developed within the OrbisGIS platform [25], which is also a free and open source GIS.

First, the effect of the algorithm input parameters (number of directions, search distance and grid resolution) on the SVF accuracy will be evaluated at point scale. Then, the block SVF accuracy and the processing time spent using each combination will be calculated. The results will be sorted by LCZ type in order to identify whether or not an algorithm is biased for any specific LCZ type. Finally, because the fastest method to calculate the SVF at block scale remains the use of morphological indicators, linear regression models will be used to test the intensity of their relationship with the block SVF.

2. Material

The dataset used to calculate the SVF in urban areas should contain the information concerning the location and the height of the following sky obstacles: trees, buildings and terrain. Except for some specific cities located in mountainous areas, the slope of the terrain is most of the time negligible compared to the angle created by a building seen from the street. On the contrary, trees may largely influence the SVF even in cities, but their influence fluctuates along the years because they grow, along the seasons because they lose their leaves (deciduous trees) and along each day because of the wind. For this reason and because trees also have very complex three-dimensional shapes, it is complicated to correctly represent them in geographical databases. This study is dedicated to the comparison of the accuracy and the time duration of SVF calculation method rather than to calculate accurately the SVF value of a given area. Thus, for the sake of simplicity, only buildings will be considered as sky obstacles. Depending on the buildings' arrangement (distance between buildings and variability of this distance within a given area) and the buildings' characteristics (building width, height, orientation and variability of these parameters within a given area), each algorithm could lead to more or less large errors when estimating the SVF. In order to identify the specific fail cases of each algorithm, they are tested on all LCZ built types (Figure 1).

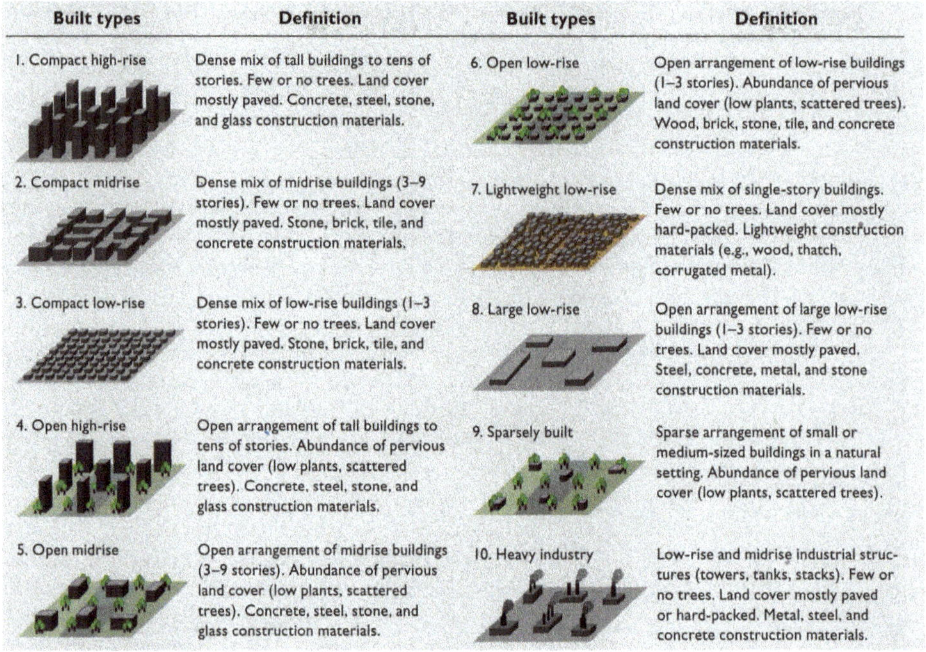

Built types	Definition	Built types	Definition
1. Compact high-rise	Dense mix of tall buildings to tens of stories. Few or no trees. Land cover mostly paved. Concrete, steel, stone, and glass construction materials.	6. Open low-rise	Open arrangement of low-rise buildings (1–3 stories). Abundance of pervious land cover (low plants, scattered trees). Wood, brick, stone, tile, and concrete construction materials.
2. Compact midrise	Dense mix of midrise buildings (3–9 stories). Few or no trees. Land cover mostly paved. Stone, brick, tile, and concrete construction materials.	7. Lightweight low-rise	Dense mix of single-story buildings. Few or no trees. Land cover mostly hard-packed. Lightweight construction materials (e.g., wood, thatch, corrugated metal).
3. Compact low-rise	Dense mix of low-rise buildings (1–3 stories). Few or no trees. Land cover mostly paved. Stone, brick, tile, and concrete construction materials.	8. Large low-rise	Open arrangement of large low-rise buildings (1–3 stories). Few or no trees. Land cover mostly paved. Steel, concrete, metal, and stone construction materials.
4. Open high-rise	Open arrangement of tall buildings to tens of stories. Abundance of pervious land cover (low plants, scattered trees). Concrete, steel, stone, and glass construction materials.	9. Sparsely built	Sparse arrangement of small or medium-sized buildings in a natural setting. Abundance of pervious land cover (low plants, scattered trees).
5. Open midrise	Open arrangement of midrise buildings (3–9 stories). Abundance of pervious land cover (low plants, scattered trees). Concrete, steel, stone, and glass construction materials.	10. Heavy industry	Low-rise and midrise industrial structures (towers, tanks, stacks). Few or no trees. Land cover mostly paved or hard-packed. Metal, steel, and concrete construction materials.

Figure 1. Abridged definitions for LCZ built types. Source: Stewart and Oke [14]—©American Meteorological Society, used with permission.

The Paris Region Urban and Environmental Agency IAUIdF (https://www.iau-idf.fr/en/about-us/who-we-are/a-foundation.html—accessed in April 2018) has manually attributed a LCZ type to each Urban Morphological Blocks (UMB) of the Paris region More information about the methodology used at http://www.iau-idf.fr/en/know-how/environment/vulnerability-of-towns-and-cities-to-rising-temperatures-assessed-using-the-local-climate-zones.html–accessed in April 2018. The boundaries of each UMB are built using a complex combination of data (road and rail networks, water surfaces and a land-use database [26], but they are most of the time very close to the urban block defined by the road network. In order to test the algorithms on each LCZ built type, ten urban blocks are randomly selected within the Paris commune for each LCZ type (except for LCZ 1, 7, 10 having, respectively, just 1, 1 and 0 individuals in the entire Paris territory). For the needs of this study, the buildings included within a 300 m radius around each urban block are conserved. The building data consist of the building footprint and height and they are provided by the French National Geographical Institute (more information about the data at http://professionnels.ign.fr/bdtopo—accessed in April 2018). The choice of the 300 m radius is based on preliminary results obtained by Chen et al. [23]: they showed that, in most of the cases, considering the buildings located further than 300 m from their calculation point does not bring much accuracy to the SVF calculation. Figure 2 shows the urban blocks and the buildings selected for the study.

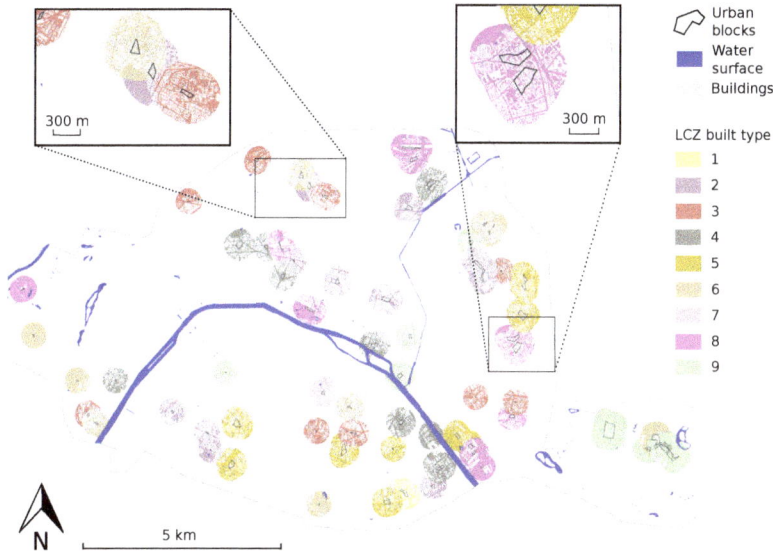

Figure 2. Selected 10 Paris urban blocks by LCZ type (defined Figure 1) and the corresponding buildings located within a 300 m radius. Based on the LCZ database produced by the IAUIdF. Note that the LCZ corresponding to each urban block has been colored at the 300 m buffer scale in order to make them identifiable.

3. Methodology

3.1. Presentation of the SVF Algorithm Used in OrbisGIS

The SVF is calculated from a given point considering all surrounding obstacles to the sky hemisphere. The points and obstacles are vector based data (geometries–polygons or lines) and should have vertical level information (three-dimensional geometry—if none, the level is considered to be 0 m from sea level).

For calculation purposes, the sky hemisphere is sliced into a given number of sectors (number of directions of analysis). A horizontal ray [R] of a given length is associated with each sector, starting from the center of the sphere and heading toward one of the sphere meridian confining the sector (Figure 3).

For each of the building obstacles B_i encountered by the ray, the angle Θ_i is calculated thanks to Equation (3):

$$\Theta_i = \mathrm{atan}\left(\frac{H_i}{L_i}\right), \tag{3}$$

where H_i the height of the obstacle B_i and L_i the horizontal distance from the center of the sphere and the obstacle B_i.

The highest angle Θ_m made along the ray [R] is used to calculate the corresponding surface hemisphere S_H hidden in this given direction (the sector is considered hidden all along its azimuthal angle—Figure 3). For a one meter radius sphere, S_H is given by Equation (4):

$$S_H = \sin^2(\Theta_m) \cdot \frac{2\pi}{ND}, \tag{4}$$

where ND is the number of sky hemisphere sectors (number of ray directions)

The sky view factor is then calculated using Equation (5):

$$SVF = 2\pi - \sum_{d}^{ND} S_{H_d},$$ (5)

where d is a given azimuthal ray direction and S_{H_d} is the surface of sky hemisphere hidden by obstacles in the azimuthal direction d.

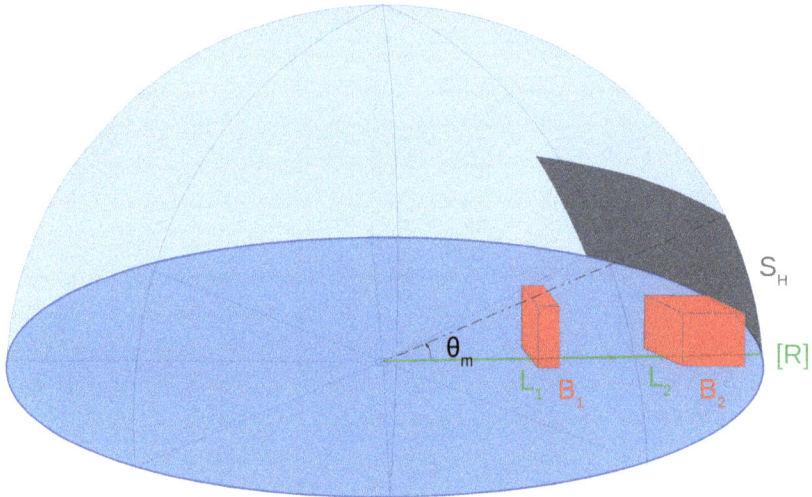

Figure 3. Calculation of the approximated surface hemisphere (blue) hidden by the buildings' obstacles (red) in a given direction (green line). In this example, the hemisphere is split into height sectors (number of directions) and the surface of the hemisphere considered as hidden by the buildings in the given direction is shown in grey.

Three input parameters are needed for the SVF calculation: the number of directions ND (number of sectors), the ray length RL and the ray step length RSL (size of an elementary ray subdivision). ND and RL influence the processing time duration but also the accuracy of the calculation, whereas RSL is only used to optimize the processing time duration. Indeed, the algorithm is based on the following procedure (simplified further details can be found on the function documentation page http://www.h2gis.org/docs/dev/ST_SVF/ and the source code is available at https://github.com/orbisgis/h2gis/blob/master/h2gis-functions/src/main/java/org/h2gis/functions/spatial/earth/ST_Svf.java—accessed in April 2018):

1. for each ray (each direction);

    - the ray is split into segments of $RSL = 5$ m length in order to optimize the calculation speed (according to a preliminary study, this parameter has very little effect on the calculation speed when its value is included within a (3–7) m range),
    - the obstacles intersecting each of these segments are identified,
    - for each intersection, the angle Θ_i is calculated by means of Equation (3),
    - the highest Θ_m value encountered along the ray is conserved and used to calculate the corresponding sky hemisphere hidden in this direction (Equation (4)),

2. the SVF is finally calculated subtracting the sum of the hidden surfaces S_{H_d} in the ND directions to the total surface hemisphere (Equation (5)).

The SVF is implemented as a SQL function in H2GIS (the spatial database used by OrbisGIS to store and process GIS data [27]).

3.2. Global Methodology for Algorithm Evaluation

Figure 4 summarizes the global methodology used in the algorithm evaluation process.

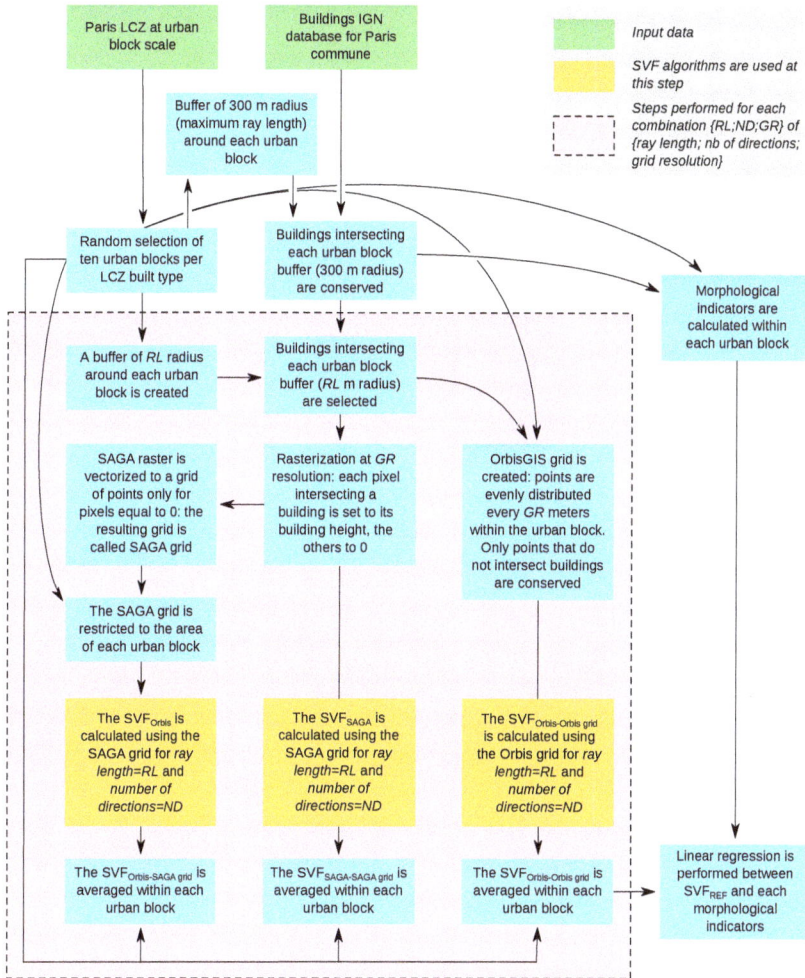

Figure 4. General data flow of the study.

The algorithms are compared based on SVF values, both calculated at point and block scales. To compare points calculated by two algorithms one by one, there is a need to have the same set of points for SAGA-GIS and OrbisGIS. SAGA-GIS is more constrained by the buildings since they need a rasterization pre-process step: the value of a pixel partly covered by a building is set to its building height (cf. Figure 5—this is an arbitrary choice, the value could have been set to ground level whenever the building do not cover entirely the pixel). Then, since the SVF defined in the LCZ classification should be calculated at ground level, only the SAGA pixels that do not intersect buildings will be used.

These pixels are vectorized to points, their value being attributed to their central point. The resulting grid of points is called SAGA-grid and is used by the OrbisGIS algorithm (Figure 4).

The SAGA-GIS grid is very constrained by the building location and may be biased by the rasterization process (no point can be closer to a building than half the grid resolution—Figure 5). For this reason, the OrbisGIS algorithm will also be used on a second grid called Orbis grid, which is not affected by the rasterization process. Figure 5 shows the differences between the SAGA-GIS grid and the OrbisGIS grid.

Figure 5. The SAGA-GIS and OrbisGIS grids and the corresponding buildings used for the SVF calculation.

One of the objectives of this study is to identify what the input parameters values are in order to both minimize the algorithm processing time and to maximize its accuracy. The parameter values given in Table 1 will be tested.

Table 1. Value range of each parameter tested herein.

Parameter to Evaluate	Values to Test
Ray length (m)	100, 200, 300
Number of directions	60, 120, 180
Grid resolution (m)	1, 2, 5, 10

The maximum values have been chosen based on Chen et al. [23] study. They calculated a SVF value for about a million of points in the city of Hong Kong testing different values for the ray length and the number of directions. Results showed that:

- when they compared values based on a 300 m ray length to the ones obtained with a 500 m ray length, the largest difference reported in their domain was just 0.031,
- no difference was reported in their domain when they used either 180 or 360 numbers of directions.

Thus, the maximum ray length has been set to 300 m and the number of directions to 180 (note that Gál et al. [28] chose for their calculations, respectively, 200 m and 360). Regarding the grid resolution, Chen et al. [23] set arbitrarily the value to 2 m since it is not a sensible parameter in their algorithm. We tested a finer resolution (1 m) to observe if it brings any improvement to the resulting SVF for the raster based algorithm.

3.3. Identification of the Reference Algorithm

The SVF value that will be taken as the reference at point or block scale should be the closest to the real SVF value.

The accuracy of the two algorithms is tested at point scale setting their parameters at the finest resolution (the ray length is 300 m, the number of directions is 180 and grid resolution is 1 m). Two academic building configurations having an analytical solution for the SVF calculation are then used to test the accuracy of each algorithm. The first configuration is a square building extruded in the middle with a circular hole (Figure 6a). The SVF is then calculated from the centre of the courtyard and the analytical solution is given by Equation (6) [1]:

$$SVF_{ext-bu} = \cos^2(\text{atan}(\frac{2 \cdot H_{ext-bu}}{D_{ctyd}})),\tag{6}$$

where SVF_{ext-bu} is the SVF calculated at the center of the courtyard; H_{ext-bu} is the height of the extruded building (m); and D_{ctyd} is the diameter of the circular courtyard (m).

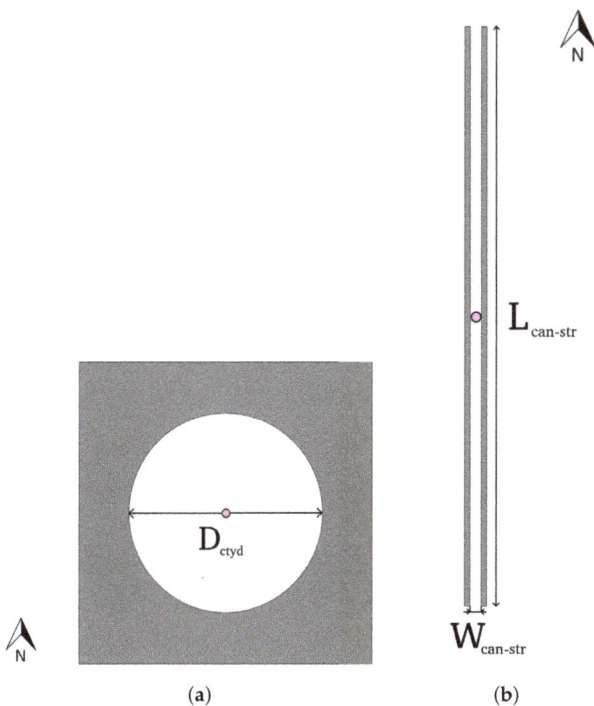

Figure 6. Building configuration for the analytical solution calculation: (**a**) configuration 1: circularly extruded building; (**b**) configuration 2: street canyon.

The second configuration is an infinite street canyon (Figure 6b), the SVF being calculated at the middle of the street by means of the Equation (7) [15]:

$$SVF_{canyon} = \cos(\text{atan}(\frac{2 \cdot H_{can-bu}}{W_{can-str}})),\tag{7}$$

where SVF_{canyon} the SVF calculated at the middle of the street canyon, H_{can-bu} the building height of the street canyon (m), and $W_{can-str}$ the width of the street canyon (m).

Each configuration is stored in a geographical database and can be displayed in a GIS (Figure 6).

Two scenarios are proposed in order to test the effect of specific dimensions on the raster algorithm (Table 2).

Table 2. Geometric dimensions for each combination of configuration/scenarios.

Scenario	Configuration 1		Configuration 2		
	H_{ext-bu}	D_{ctyd}	H_{can-bu}	$W_{can-str}$	$L_{can-str}$
1	20 m	39 m	20 m	39 m	2000 m
2	20 m	40 m	20 m	40 m	2000 m

The only difference between these scenarios concerns the horizontal distances. For the scenario 1, D_{ctyd} and $W_{can-str}$ are chosen in order to match the pixel boundaries and the building walls exactly, limiting the rasterization effect. In the second scenario, the buildings are shifted from half the grid resolution (0.5 m) in order to cover only partially some of the cells. In this way, the effect of the rasterization can be evaluated (each cell covered partially by buildings is transformed to a building cell). Note that, for both scenarios, the configurations do not exactly respect the ideal cases defined by [1,15]. However, we tried to limit the adverse aspects of the real cases. The courtyard of the extruded building is not a perfect circle, but it is defined by 720 points (four times the number of directions of the algorithms). The canyon street does not have an infinite length, but it is 2000 m long (longer than the ray length of the algorithm set to 300 m).

The accuracy of the SVF averaged at block scale depends on the accuracy of the SVF calculation at point scale, on the number and on the position of the points within the block. The accuracy at point level is evaluated using the scenarios described in Table 2, whereas the number and position of the points give a clear advantage to the vector approach due to the lack of flexibility induced by the raster based approach: the point position is constrained by the raster (points are equally distributed within the block). Moreover, the number of points will be limited since, in very dense built-up areas, most of the grid cells will be covered (even partially) by a building (see, for example, Figure 5). Consequently, except if the SAGA-GIS shows a higher accuracy at point level, the OrbisGIS grid and algorithm are selected to calculate the SVF value that will be used as a reference at block scale.

3.4. Empirical Relationship between Block Mean SVF and Morphological Indicators

Four morphological indicators are proposed to estimate the block mean SVF. They are calculated based on the building and the urban block (defined by the road network) geometries. Their definition and calculation method are given in Table 3.

The empirical model is built considering a simple linear relationship between the reference SVF calculated at block scale (SVF_{REF}) and each of the morphological indicator I_M (Equation (8)):

$$SVF_{REF} = a_0 + a_1 \cdot I_M,$$ (8)

where a_0 is the regression intercept, and a_1 is the regression coefficient.

Table 3. Morphological indicator definition.

Notation	Name	Description	Formulae	Reference
D_F	Facade density	The building external free facades area (S_F) is divided by the block surface (S_{REF})	$\frac{S_F}{S_{REF}}$	Bocher et al. [29]
F_F	Facade density fraction	The building external free facades area is divided by the block surface plus the building free facades area	$\frac{S_F}{S_{REF}+S_F}$	Bernabé et al. [16]
H/W	Aspect ratio	The street canyon definition is extended to any urban tissue using the ratio of the building external free facades and the land free surface (urban block area minus building footprint area–S_B)	$\frac{S_F}{S_{REF}-S_B}$	-
SVF_{CAN}	Canyon street SVF	The SVF is calculated using the aspect ratio considering all the streets of the urban block as street canyons	$cos(atan(2 \cdot H/W))$	Oke [15]

4. Results

4.1. Identification of the Reference Algorithm

The results of the analysis of SVF calculation at point scale shows a slight advantage of the OrbisGIS algorithm (Table 4).

Table 4. SVF values calculated for each combination of method (algorithm or analytical solution), scenario and building configuration.

Scenario	Building Configuration	OrbisGIS	SAGA-GIS	Analytical Solution
1	1 (circular courtyard)	0.50000 ($\simeq$0%)	0.50000 (=0%)	0.5
	2 (street canyon)	0.70713 (+0.004%)	0.70608 (−0.15%)	0.70711
2	1 (circular courtyard)	0.52199 ($\simeq$0%)	0.52438 (+0.45%)	0.52199
	2 (street canyon)	0.72252 (+0.003%)	0.71993 (−0.35%)	0.72249

(%) the number given inside the parenthesis is the relative error percentage.

The relative error of the OrbisGIS SVF is considered as negligible (always <0.01%) for any scenario and any configurations. Concerning the SAGA-GIS one, it is under 0.1% only for the circular courtyard case (configuration 1), when a given number of pixels constitutes the vertical or the horizontal building courtyard diameter (scenario 1—cf. Figure 7).

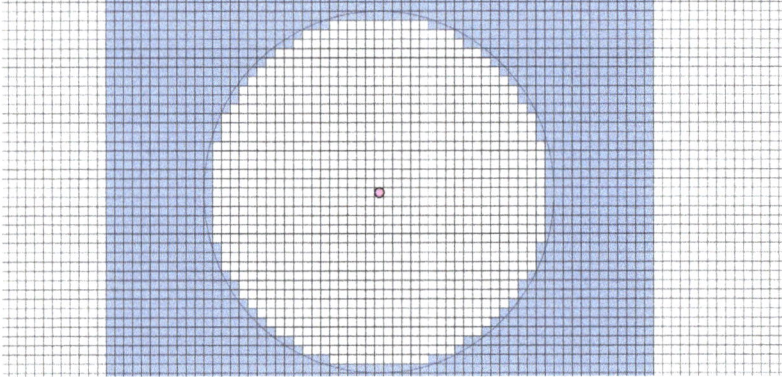

Figure 7. Scheme of the raster-based approach (SAGA software) for the configuration 1–scenario 1. The squares represent the pixels of the raster: the blue ones are considered as building surfaces, the white ones as ground surfaces. The SVF is calculated from the pixel covered by the pink circle.

For this pixel resolution (1 m) and whenever the buildings do not exactly cover the raster pixels, the OrbisGIS algorithm seems to be more accurate than the SAGA-GIS one. Thus, for a given combination of ray length/number of direction, the SVF values obtained with the OrbisGIS algorithm will be considered as reference values.

As stated in Section 3.3, the SVF averaged at block scale will be more accurate when calculated using a vector-based approach rather than a raster-based approach. Therefore, the SVF calculated with OrbisGIS is considered as the reference at block scale for two reasons: the algorithm is based on a vector approach and the SVF calculated at point scale is more accurate than the SAGA-GIS one. The reference case used for block scale comparisons will then be the SVF calculated using the OrbisGIS algorithm, the OrbisGIS grid and with the following combination of parameter values: ray length equals to 300 m, number of direction equals to 180 and grid resolution equals to 1 m.

4.2. Effect of the Algorithm Input Parameters

For each point of the SAGA-GIS grid, the SVF has been calculated using the OrbisGIS algorithm and the SAGA-GIS algorithm. The effect of the input parameters (ray length, number of directions and grid resolution) is then investigated subtracting for each point the SVF values calculated using SAGA-GIS to the one calculated using OrbisGIS. The analysis of the results shows that SVF values calculated using OrbisGIS are most of the time higher than the one calculated using SAGA-GIS (Figure 8). The effect of the grid resolution is of major importance: the larger the grid resolution, the higher the OrbisGIS to SAGA-GIS SVF difference (Figure 8a). The number of direction and the ray length seem to have a similar influence on the two algorithms since no change are highlighted by the modification of their values (Figure 8b,c).

The reason of the low SAGA SVF values results from the raster-based method, in particular from the method chosen for pixel value attribution. Figure 5 illustrates well the problem: as long as a building covers a part of a raster pixel, the entire pixel is considered as a building. Then, the buildings are closer to the calculated point in the SAGA-GIS case than in the OrbisGIS case, thus resulting in a lower SVF value. The intensity of this phenomenon is increased as the grid resolution increases.

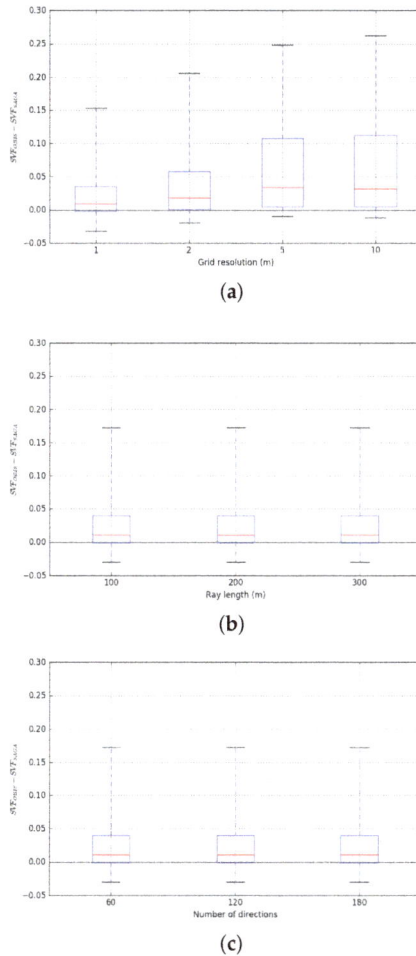

Figure 8. Effect of the input parameters value on the SVF difference recorded at point scale using OrbisGIS or SAGA-GIS algorithm: (**a**) grid resolution; (**b**) ray length; and (**c**) number of directions. The red line is the median point, the upper and lower box limits are, respectively, the first and third quartiles and the lower and upper whisker are the 5 and 95 percentiles.

The SVF calculated using the two algorithms converges towards a same value when the grid resolution decreases. However, even for a 1 m grid resolution, it remains locally high for some points. The maximum observed is 0.65: the SAGA SVF value at this point is negative (-0.104), whereas the one calculated using OrbisGIS is 0.549. This value and most of the largest observed differences are located near the building walls (Figure 9). For such points, in the rasterization case, the relative error of distance between the point and the wall could be very high. This phenomenon is especially exacerbated when a point is located near a building corner facing one of the four cardinal points. Therefore, more investigation is needed to understand the reason of the negative values obtained with the SAGA-GIS software.

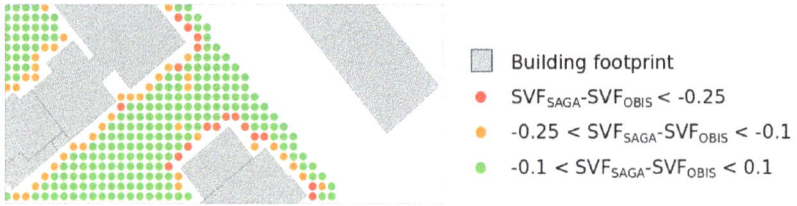

Figure 9. SVF difference between SAGA and OrbisGIS values.

4.3. Ideal Input Parameters for Block Mean SVF Calculation

The block mean SVF has been calculated for each combination of *RL*, *ND*, *GR*, algorithm and grid. The computation duration of each combination has been measured. The error is defined as the difference between the block SVF calculated using a given combination and the one calculated using the reference combination (300 m, 180, 1 m, OrbisGIS, OBIS grid). The combination both minimizing the error and computational duration is identified in Figure 10.

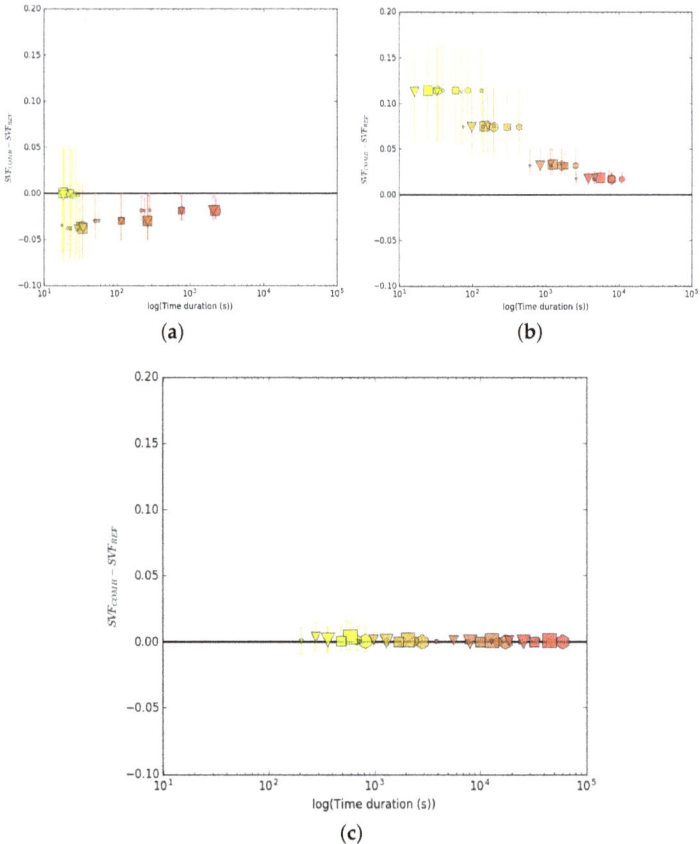

Figure 10. Block mean SVF error versus processing time duration using (**a**) SAGA-GIS algorithm and SAGA grid; (**b**) OrbisGIS algorithm and SAGA grid; and (**c**) OrbisGIS algorithm and Orbis grid. The marker is the median value of the 72 blocks and the whiskers are the lower and upper quartile.

Concerning time duration, the SAGA algorithm seems to be more impacted by the grid resolution and the ray length change rather than by the number of directions (Figure 10a). On the contrary, all parameters almost equally affect the OrbisGIS performance (Figure 10b,c). The time duration comparison will be further investigated in the discussion section.

Concerning the SVF error, three distincts behaviours are identified. The SAGA error is almost null for the theoretically worst combination of input parameters, it becomes negative when the combination is getting better and finally converges towards zero for the theoritically best combination. The Orbis/SAGA (OrbisGIS algorithm using SAGA grid) values are always overestimated but the error converges toward zero as the resolution decreases. The Orbis/Orbis (OrbisGIS algorithm using OrbisGIS grid) error very quickly decreases to zero as the combination of input parameter is getting better. An interesting fact is that it seems impossible to lower the error under a certain threshold (about 0.002) if the ray length is kept under 200 m. The difference between the Orbis/SAGA and the Orbis/Orbis cases is solely attributable to the grid fault. The minimum distance between any point of the SAGA grid and a building can not be lower than half the grid resolution, whereas it can be any distance for the points of the Orbis grid (cf. Figure 5). Thus, the use of the SAGA grid tends to underestimate the SVF value at block scale. We previously showed in Section 4.2 that the SAGA algorithm tends to slightly overestimate the SVF calculated at point level. For low grid resolution, the negative bias induced by the SAGA grid at block scale almost annihilates its positive bias observed at point scale (Figure 10a).

Based on the analysis of Figure 10, four combinations are selected to further investigate the block scale results (note that the combinations using the OrbisGIS algorithm combined to the SAGA grid are not considered since they are interesting only at point scale):

- C1: (RL = 300 m, ND = 180, GR = 10 m, SAGA algorithm, SAGA grid): it is the fastest of the (almost) no biased SAGA combination,
- C2: (RL = 100 m, ND = 60, GR = 2 m, SAGA algorithm, SAGA grid): the combination C1 has no bias but a high interquartile, whereas this combination has a limited bias and is one of the fastest having one of the lowest interquartiles,
- C3: (RL = 100 m, ND = 60, GR = 5 m, OrbisGIS algorithm, Orbis grid): it is the fastest of the almost no biased OrbisGIS combinations (the median error is less than 0.001),
- C4: (RL = 100 m, ND = 60, GR = 10 m, OrbisGIS algorithm, Orbis grid): the fastest combination having about the same bias and interquartile error as combination C2.

4.4. Sensitivity of the Block Mean SVF Calculation to the LCZ Built Type

The block mean SVF error of each combination is analyzed sorting the results by LCZ built type (Figure 11), illustrated in Figure 1.

The combinations using the SAGA-GIS algorithm (C1 and C2) seems to be sensitive to the LCZ built type: the compact built types (LCZ1, 2 and 3) show higher values than the other LCZ. We have seen in Section 4.2 that the method used for the rasterization of building values affects the SAGA results: the closer the point to a building, the higher the overestimation of its SVF value. In the case of compact urban blocks, most of the grid points are located very close to a building. Thus, the SVF error obtained for such block will often be slightly higher than the one obtained for open urban blocks such as LCZ4, 5 or 6. This is especially true for combination 1 when the grid used for the SVF calculation has a high resolution (10 m). For combination 2, this overestimation is compensated by the block scale averaging described in Section 4.3: the SVF values are slightly lower than the reference SVF value for all LCZ built types.

The SVF values, calculated using the OrbisGIS algorithm combinations (C3 and C4), seem to not be impacted by the LCZ built type: there is no clear trend showing a larger or lower SVF bias for a given set of LCZ type. However, there is a higher error variability for LCZ3 and LCZ6, which are low-rise tissues, than for the others. The probability to meet a building located further than 100 m from the calculation point that may hide a part of the sky hemisphere is higher when an urban block is composed of low-rise buildings than when it is composed of high-rise buildings.

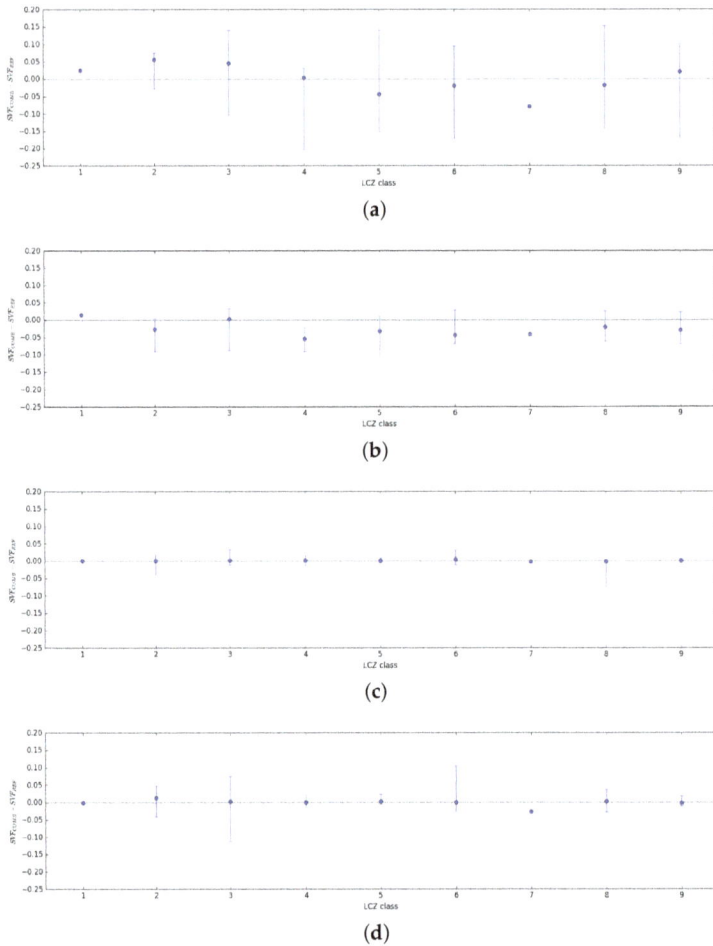

Figure 11. Block mean SVF error regarding the LCZ built type for (**a**) combination C1; (**b**) combination C2; (**c**) combination C3; and (**d**) combination C4. The blue dot represents the median SVF error, the blue segment is the range of the SVF error. Based on the LCZ database produced by the IAUIdF.

4.5. Uncertainty on the SVF Value Predicted by the Algorithms or by the Morphological Indicators

Linear relationships between SVF_{REF} and each morphological indicator are established. To improve the performances of the regression, the nine urban blocks without buildings are removed from the analysis. Indeed, a LCZ type has been attributed to each of them although they are very small. Thus, the urban blocks that surround them could have a high building density and consequently decrease the percentage of sky hemisphere visible from the empty urban block. The final formulas related SVF_{REF} to each morphological indicator are given in Equations (9) ($F_F - R^2 = 0.76$), (9) ($D_F - R^2 = 0.64$), (11) ($H/W - R^2 = 0.32$) and (12) ($SVF_{CAN} - R^2 = 0.83$). Note that all *p*-values are much lower than 0.001:

$$SVF = 1.021 - 0.969 \cdot F_F, \tag{9}$$

$$SVF = 0.760 - 0.185 \cdot D_F, \tag{10}$$

$$SVF = 0.603 - 0.029 \cdot H/W, \tag{11}$$

$$SVF = 0.318 + 0.742 \cdot SVF_{CAN}. \tag{12}$$

These equations may be used for SVF estimation at block scale. SVF_{CAN} has the highest determination coefficient ($R^2 = 0.83$), but its formulae is not adapted for very low SVF: it is mathematically not possible to reach a SVF value under 0.318 since SVF_{CAN} cannot be negative (Equation (12)). The shape of the scatter points not being linear, a logarithmic function could be used to manage this issue (Figure 12a). The same type of problem is observed for H/W and F_F: the SVF cannot be above 0.603 and 0.760, respectively, and the shape of the scatter is not linear (especially for H/W–Figure 12b). On the contrary, the regression coefficient and the intercept values of Equation (9) are perfectly adapted to cover the entire range of SVF value since both are very close to 1 (or -1 for the regression coefficient). This results complement those of Bernabé et al. [16] and Groleau and Mestayer [30]: they showed that the block mean SVF can be calculated by the simple relation $SVF = 1 - F_F$. In their case, the block mean SVF was calculated using roofs, facades and ground surfaces. Our results highlight that this simple relationship is still valid when the SVF is averaged only within the ground surface (note that the intensity of the relation found by Bernabé et al. [16] is similar to the one obtained herein). The second advantage of this relation is that the scatter points seem almost normally distributed around the regression line (Figure 12d).

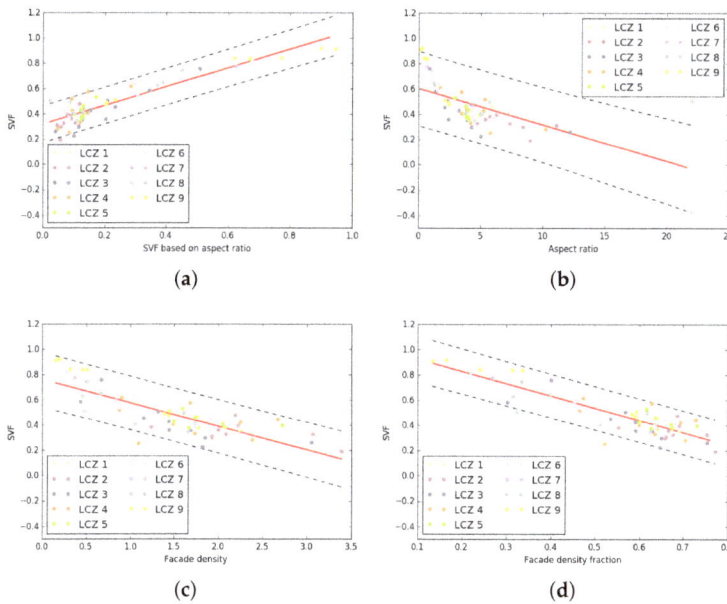

Figure 12. Relationship between block mean SVF value calculated using the reference algorithm combination SVF_{REF} and (**a**) canyon street SVF; (**b**) aspect ratio; (**c**) facade density; and (**d**) facade density fraction. The red line is the regression line and the dotted-lines represent the 95% prediction interval. Based on the LCZ database produced by the IAUIdF.

Linear relationships have also been established between SVF_{REF} and the SVF values calculated by means of each of the combinations selected Section 4.3 (Equations (13)–(16)) having respectively R^2

equal to 0.84, 0.98, 1.00, and 0.98). The objectives were to identify the potential bias related to each combination and calculate the standard error of the estimate. Note that all *p*-values are much lower than 0.001:

$$SVF = 0.066 + 0.892 \cdot SVF_{C1}, \tag{13}$$

$$SVF = 0.041 + 0.977 \cdot SVF_{C2}, \tag{14}$$

$$SVF = -0.002 + 1.00 \cdot SVF_{C3}, \tag{15}$$

$$SVF = 0.004 + 0.987 \cdot SVF_{C4}. \tag{16}$$

Only the combination C1 (Equation (13)) is slightly biased (low SVF values are overestimated whereas high SVF values are underestimated). The others have a negligible bias and a very high determination coefficient ($R^2 > 0.98$). Note that all *p*-values are lower than 0.001.

The standard error of the estimate has been calculated for each formulae (Equations (9)–(16)); the values are gathered in Table 5.

Table 5. Predicted SVF block uncertainty for each of the methods presented herein.

Method	Morphological Indicators				Algorithm Combinations			
	F_F	D_F	H/W	SVF_{CAN}	C1	C2	C3	C4
Standard deviation of the estimate	0.087	0.107	0.147	0.073	0.085	0.033	0.014	0.027

There is a clear advantage to use the SAGA or OrbisGIS algorithms instead of the morphological indicators to accurately predict the SVF (Table 5). Except when using the combination C1, 68% of the predicted values will be within a [−0.033, 0.033] interval around the real SVF value using the software algorithms, whereas the size of the interval would double if the morphological indicators are used ([−0.073, 0.073] in the best case). However, the fastest algorithm combination (C1) has the same level of magnitude than the one calculated using F_F or SVF_{CAN} (through Equations (9) and (12)).

5. Discussion

The IAUIdF did not consider the SVF indicator to attribute a LCZ built type to each urban block. It is then interesting to retrospectively verify whether their results are consistent with the SVF thresholds defined for each LCZ type. The SVF values calculated for each urban block using the OrbisGIS reference combination (Obis grid, 300 m ray length, 180 rays, 1 m grid resolution) are displayed in Figure 13. The results are splitted regarding the LCZ built type of each urban block and they are compared to the lower and upper SVF threshold proper to each LCZ type.

Most of the urban blocks classified in the LCZ 2 and 3 are within the recommended SVF interval of their class. On the contrary, for the other LCZ types, most of the urban blocks seem misclassified regarding the SVF indicator. It would then be interesting to integrate this parameter in the IAUIdF classification algorithm in order to improve their classification accuracy. For this purpose, the SVF of each of the Paris urban blocks has been calculated using the combinations C2 and C4 (cf. Section 4.3) as well as the facade density fraction (using Equation (9)). The urban blocks' boundaries have been produced from the road network and the SVF has been calculated from building footprint and height. The corresponding data have been downloaded through the Open Data website of the city of Paris. The results obtained using the C4 combination are presented in Figure 14a since they are supposed to be the less erroneous (Table 5). In order to evaluate the accuracy of the SVF value attributed to each block, the SVF difference between the reference value and the one obtained with the C2 combination or with the facade density ratio formulae are calculated and displayed (respectively Figure 14b,c). Only the buildings located in the Paris commune territory have been considered for the SVF calculation. Then, the SVF of urban blocks located less than 100 m from the commune boundaries is considered as erroneous (these zones are not considered in the cartography).

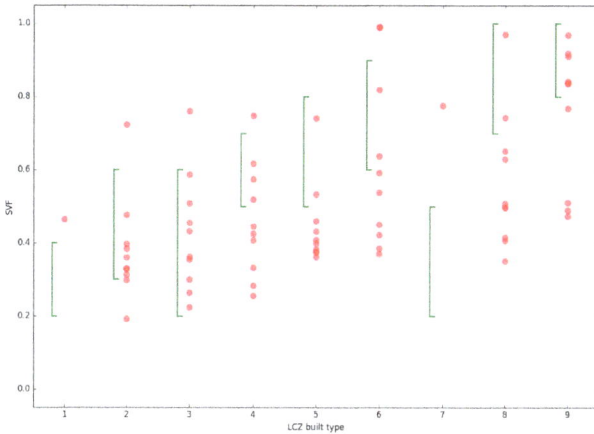

Figure 13. Urban block SVF values (red dots) splitted by LCZ built type and their corresponding SVF lower and upper threshold (green segments), based on the LCZ database produced by the IAUIdF.

The SVF is higher in the northern than in the southern part of Paris (Figure 14a). Much more differences can be observed in Figure 14c (between the C4 combination and F_F method) than in Figure 14b (between the C2 and C4 combinations). Most of the differences observed in Figure 14b correspond to very densely built areas, which is consistent with the observations drawn from Figure 9: the SVF near a facade is often underestimated using SAGA. In a very dense built-up environment, most of the raster pixels are very close to a facade and then the resulting block mean SVF is underestimated. Concerning the SVF obtained using the F_F method, it seems that they are slightly higher than the reference values in the northern part (where the SVF is high) and lower than the reference values in the southern part (where the SVF is low). This phenomenon is exacerbated for very thin blocks having no building and for large urban blocks having a very heterogeneous building distribution. Concerning the first case, the F_F method is biased since the buildings located around the thin block are not considered for the F_F calculation, whereas they are for the SVF calculation of the reference values (such urban blocks were removed from the analysis when establishing the linear relationships between SVF and morphological indicators). To explain the second case issue, two assumptions are proposed. The problem may come from the C4 combination: the 100 m ray length is maybe too short for such urban block configuration (e.g., the value calculated using the C4 combination was 0.1 higher than the reference value for some blocks—Figure 11d). The problem may also come from the F_F method that would not been adapted for urban block having an heterogeneous building distribution. Further investigation is required to verify these assumptions and potentially propose a solution to fix the issue (for example, proposing a grid point distribution proper to a given urban block type).

Eventually, the computational time of each SVF calculation method can be compared: the combination C2, the combination C4 and the facade density fraction methods lasted, respectively, 18,970 s, 2890 s and 250 s. Note that the processed dataset is composed of about 280,000 buildings located within 9000 urban blocks. The calculation has been performed using only one core of one processor (frequency of 2.3 GHz). For the OrbisGIS case, the grid was composed of 500,000 points and the time dedicated to memory access (Solid State Disk—SSD) for each point was not subtracted from the final time. The random access memory dedicated to the OrbisGIS software was 2.4 Mo.

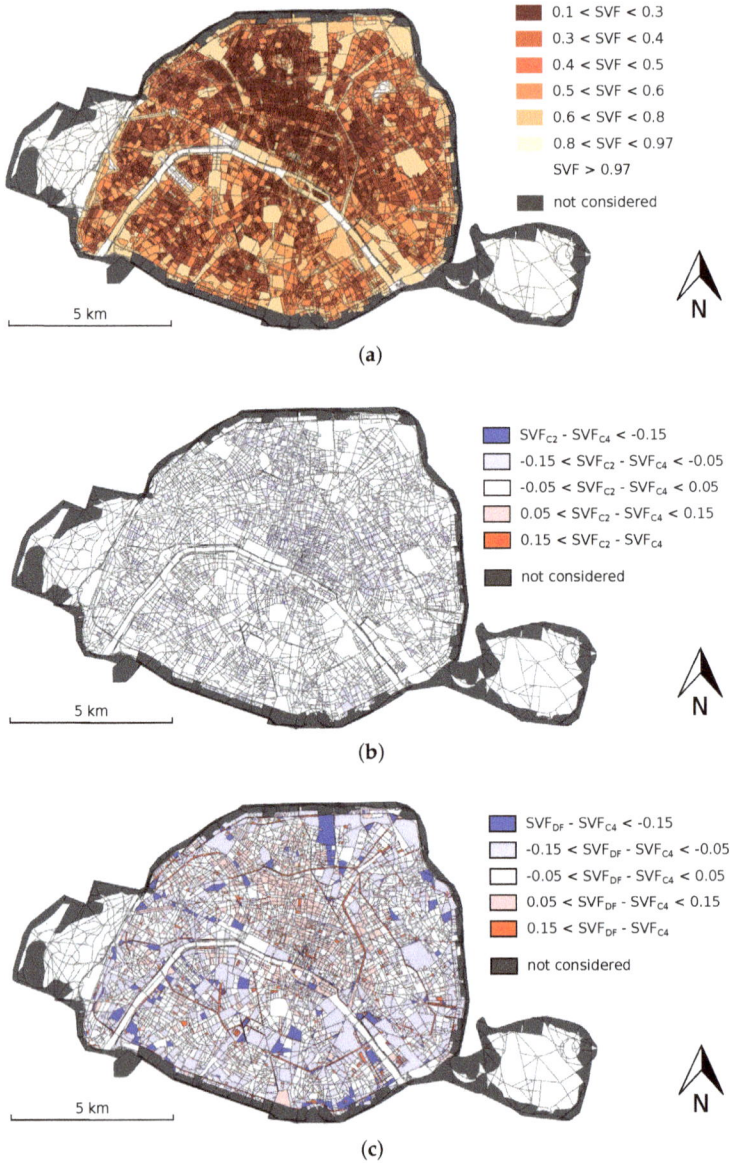

Figure 14. SVF of the Paris urban blocks: (**a**) values obtained using combination C4; (**b**) difference between the values obtained using combination C2 and the ones using combination C4; and (**c**) difference between the values obtained using the facade density ratio formulae and the ones using combination C4.

6. Conclusions

To classify the urban fabric of a city using the LCZ classification system, the ground SVF of each zone should be known. However, the existing algorithms developed to calculate the SVF have not been evaluated at block scale. The main objective of this study was to evaluate the ability to calculate the block SVF of a raster based algorithm (SAGA-GIS) and of a vector based one (OrbisGIS). They both need two input parameters to work: the number of direction of search (or number of rays) and the distance of search (or ray length). Because the SVF is originally a point information, it should be calculated for a greater number of points to be averaged at zone scale. A grid of points linearly spaced is then used.

The vector based algorithm is found to show the best accuracy at point scale and is then considered as the reference in the rest of the study. The grid resolution affects more the SAGA-GIS values: the larger the resolution, the more the SVF is overestimated. Several combinations of number of directions, ray length and grid resolution are tested to evaluate the accuracy and computational time of each algorithm at block scale. Four of them having a relatively low error and computational time are further investigated. Four geographical indicators defining the urban morphology of an urban block are also calculated. They are used in linear regressions as explanatory variables to estimate the reference block SVF (obtained with the OrbisGIS algorithm). The main results (accuracy, computational time, bias due to a specific LCZ) of each method proposed to calculate the block SVF are summarized in Table 6.

For future LCZ classification and according to the current results, the combination C4 is recommended for the calculation of the block SVF. Cities known to have a particularly high number of blocks corresponding to LCZ 3 or 4 should draw attention since the SVF values obtained using combination C4 are then more uncertain than for other LCZ built types.

The results of this study are useful for the urban climate community. On one hand, it demonstrates the advantages of customized parameters for two algorithms available in free and open GIS: the input parameters (number of ray directions, ray length) of each algorithm can be adjusted to find a compromise between the accuracy of the expected result and the processing time. On the other hand, it shows the impact of the point grid: a raster-based algorithm is much more impacted by this parameter than a vector-based algorithm. The advantage of the vector algorithm is that the grid point distribution may be easily adapted to each urban block type (e.g., to the density of the obstacles—denser in the dense built-up blocks).

Overall, the following aspects would need further investigations:

- some of the values obtained using the SAGA software are negative. A further description and analysis of the SAGA algorithm and performance should be conducted and shared with the urban climate community.
- The nonlinearity of the relation between SVF and the SVF_{CAN} indicator may be further investigated, as well as the curious simplicity of the relation between SVF and F_F.
- All results have been produced on a limited dataset. Only the city of Paris was sampled, it would then be interesting to verify that the optimized combinations identified for the city of Paris are identical for other urban configurations and urban tissues (such as American or Asian large cities).
- The influence of the LCZ built type on the SVF calculation error has been investigated. However, the LCZ dataset used seems slightly biased regarding the SVF parameter. Very few LCZ classification algorithms have been elaborated taking into account the SVF values. This issue is addressed to the urban climate community.
- The effect of the RSL parameter (OrbisGIS) on the computation time has been studied on a limited number of points. More investigations could be performed in order to further increase the performance of this algorithm.

Table 6. 1Main results summary.

	Name	C1	C2	C3	C4	F_F	D_F	H/W	SVF_{CAN}
Method	Type	SAGA GIS	SAGA GIS	Orbis GIS	Orbis GIS	Morpho indic	Morpho indic	Morpho indic	Morpho indic
	ND	180	60	60	60	-	-	-	-
	RL (m)	300	100	100	100	-	-	-	-
	GR (m)	10	2	5	10	-	-	-	-
Results	$SVF_{REF} = f(SVF_{CALC})$	OK	OK	OK	OK	Non normal residuals	OK	Non normal residuals	Non normal residuals
	Particular LCZ bias	LCZ1, 2 & 3	LCZ1, 2 & 3	No	No	-	-	-	-
	Particular LCZ uncertainty	No	No	No	LCZ3 & 6	-	-	-	-
	Stand. dev. estimate	0.085	0.033	0.014	0.027	0.087	0.107	0.147	0.073
	Comput. time for Paris (s)	-	18,970	-	2890	-	250	-	-

Author Contributions: Conceptualization, J.B., E.B., G.P. and S.P.; Data curation, J.B. and G.P.; Formal analysis, J.B., E.B. and S.P.; Investigation, J.B., E.B. and G.P.; Methodology, J.B., E.B. and G.P.; Project administration, J.B. and E.B.; Software, J.B., E.B., G.P. and S.P.; Supervision, J.B. and E.B.; Validation, E.B.and G.P.; Visualization, J.B. and G.P.; Writing—Original draft, J.B.; Writing—Review & editing, E.B. and G.P.

Funding: This work has been funded within the following research projects frameworks: MAPuCE (Grant No. ANR-13-VBDU-0004) and URCLIM (Urban CLIMate services) framework (Grant No. 690462).

Acknowledgments: We would like to thank the producers of the data used in this work: the French National Geographical Institute (IGN) for the Paris building dataset, the Mairie de Paris for their buildings and road network open data and the Paris Region Urban and Environmental Agency for the access to their LCZ database. The English proofreading was done by Sarah Botta.

Conflicts of Interest: The authors declare no conflict of interest.

References

1. Oke, T.R. *Boundary Layer Climates*; Routledge: Abington-on-Thames, UK, 2002.
2. Jamei, E.; Rajagopalan, P.; Seyedmahmoudian, M.; Jamei, Y. Review on the impact of urban geometry and pedestrian level greening on outdoor thermal comfort. *Renew. Sustain. Energy Rev.* **2016**, *54*, 1002–1017.
3. Robine, J.M.; Cheung, S.L.K.; Le Roy, S.; Van Oyen, H.; Griffiths, C.; Michel, J.P.; Herrmann, F.R. Death toll exceeded 70,000 in Europe during the summer of 2003. *Comptes Rendus Biol.* **2008**, *331*, 171–178.
4. Jochner, S.; Menzel, A. Urban phenological studies–Past, present, future. *Environ. Pollut.* **2015**, *203*, 250–261.
5. Kaufmann, R.K.; Seto, K.C.; Schneider, A.; Liu, Z.; Zhou, L.; Wang, W. Climate response to rapid urban growth: evidence of a human-induced precipitation deficit. *J. Clim.* **2007**, *20*, 2299–2306.
6. Kolokotroni, M.; Ren, X.; Davies, M.; Mavrogianni, A. London's urban heat island: Impact on current and future energy consumption in office buildings. *Energy Build.* **2012**, *47*, 302–311.
7. Sarrat, C.; Lemonsu, A.; Masson, V.; Guedalia, D. Impact of urban heat island on regional atmospheric pollution. *Atmos. Environ.* **2006**, *40*, 1743–1758.
8. Revel, D.; Füssel, H.M.; Jol, A. *Climate Change, Impacts and Vulnerability in Europe 2012*; European Economic Area (EEA): Oslo, Norway, 2012.
9. Santamouris, M.; Ding, L.; Fiorito, F.; Oldfield, P.; Osmond, P.; Paolini, R.; Prasad, D.; Synnefa, A. Passive and active cooling for the outdoor built environment—Analysis and assessment of the cooling potential of mitigation technologies using performance data from 220 large scale projects. *Sol. Energy* **2017**, *154*, 14–33.
10. Kikegawa, Y.; Genchi, Y.; Kondo, H.; Hanaki, K. Impacts of city-block-scale countermeasures against urban heat-island phenomena upon a building's energy-consumption for air-conditioning. *Appl. Energy* **2006**, *83*, 649–668.
11. Bernard, J.; Rodler, A.; Morille, B.; Zhang, X. How to Design a Park and Its Surrounding Urban Morphology to Optimize the Spreading of Cool Air? *Climate* **2018**, *6*, 10.
12. Ng, E.; Chen, L.; Wang, Y.; Yuan, C. A study on the cooling effects of greening in a high-density city: An experience from Hong Kong. *Build. Environ.* **2012**, *47*, 256–271.
13. Stewart, I.D. Redefining the Urban Heat Island. Ph.D. Thesis, University of British Columbia, Vancouver, BC, Canada, 2011.
14. Stewart, I.D.; Oke, T.R. Local climate zones for urban temperature studies. *Bull. Am. Meteorol. Soc.* **2012**, *93*, 1879–1900.
15. Oke, T.R. Canyon geometry and the nocturnal urban heat island: comparison of scale model and field observations. *Int. J. Climatol.* **1981**, *1*, 237–254.
16. Bernabé, A.; Bernard, J.; Musy, M.; Andrieu, H.; Bocher, E.; Calmet, I.; Kéravec, P.; Rosant, J.M. Radiative and heat storage properties of the urban fabric derived from analysis of surface forms. *Urban Clim.* **2015**, *12*, 205–218.
17. Hämmerle, M.; Gál, T.; Unger, J.; Matzarakis, A. Comparison of models calculating the sky view factor used for urban climate investigations. *Theor. Appl. Climatol.* **2011**, *105*, 521–527.
18. Matzarakis, A.; Matuschek, O. Sky View Factor as a parameter in applied climatology—Rapid estimation by the SkyHelios Model. *Meteorol. Z.* **2011**, *20*, 39–45.
19. Zeng, L.; Lu, J.; Li, W.; Li, Y. A fast approach for large-scale Sky View Factor estimation using street view images. *Build. Environ.* **2018**, *135*, 74–84.

20. Kastendeuch, P.P. A method to estimate sky view factors from digital elevation models. *Int. J. Climatol.* **2013**, *33*, 1574–1578.

21. Chapman, L.; Thornes, J. Real-time sky-view factor calculation and approximation. *J. Atmos. Ocean. Technol.* **2004**, *21*, 730–741.

22. Unger, J. Connection between urban heat island and sky view factor approximated by a software tool on a 3D urban database. *Int. J. Environ. Pollut.* **2008**, *36*, 59–80.

23. Chen, L.; Ng, E.; An, X.; Ren, C.; Lee, M.; Wang, U.; He, Z. Sky view factor analysis of street canyons and its implications for daytime intra-urban air temperature differentials in high-rise, high-density urban areas of Hong Kong: A GIS-based simulation approach. *Int. J. Climatol.* **2012**, *32*, 121–136.

24. Böhner, J.; McCloy, K.R.; Strobl, J. *SAGA: Analysis and Modelling Applications*; Number 115; Goltze: Gottingen, Germany 2006.

25. Bocher, E.; Petit, G. OrbisGIS: Geographical Information System designed by and for research. In *Innovative Software Development in GIS*; Wiley: Hoboken, NJ, USA, 2013; pp. 23–66.

26. Cordeau, E. Les îLots Morphologiques Urbains (IMU). 2016. Available online: https://www.iau-idf.fr/savoir-faire/nos-travaux/edition/les-ilots-morphologiques-urbains-imu.html (accessed on 3 July 2018).

27. Bocher, E.; Petit, G.; Fortin, N.; Palominos, S. H2GIS a spatial database to feed urban climate issues. In Proceedings of the 9th International Conference on Urban Climate (ICUC9), Toulouse, France, 20–24 July 2015.

28. Gál, T.; Lindberg, F.; Unger, J. Computing continuous sky view factors using 3D urban raster and vector databases: Comparison and application to urban climate. *Theor. Appl. Climatol.* **2009**, *95*, 111–123.

29. Bocher, E.; Petit, G.; Bernard, J.; Palominos, S. A geoprocessing framework to compute urban indicators: The MApUCE tools chain. *Urban Clim.* **2018**, *24*, 153–174.

30. Groleau, D.; Mestayer, P.G. Urban Morphology Influence on Urban Albedo: A Revisit with the S olene Model. *Bound.-Layer Meteorol.* **2013**, *147*, 301–327.

climate

MDPI

Article

Leftover Spaces for the Mitigation of Urban Overheating in Municipal Beirut

Noushig Kaloustian [1],*, David Aouad [2],*, Gabriele Battista [3] and Michele Zinzi [4]

[1]   Department of Natural Sciences, Lebanese American University, Beirut 13-5053, Lebanon
[2]   Department of Architecture and Design, Lebanese American University, Beirut 13-5053, Lebanon
[3]   Engineering Department, University of Roma TRE, Via della Vasca Navale 79, 00146 Rome, Italy;
      gabriele.battista@uniroma3.it
[4]   Agenzia Nazionale per le Nuove Tecnologie, L'energia e lo Sviluppo Economico Sostenibile (ENEA),
      Via Anguillarese 301, 00123 Rome, Italy; michele.zinzi@enea.it
*   Correspondence: noushigk@gmail.com (N.K.); david.awad@lau.edu.lb (D.A.);
     Tel.: +961-70-126329 (N.K.); +961-3-721794 (D.A.)

Received: 8 July 2018; Accepted: 16 August 2018; Published: 21 August 2018

Abstract: The Urban Heat Island phenomenon and urban overheating are serious consequences of urbanization resulting in impacts on thermal comfort levels, heat stress and even mortality. This paper builds on previous findings on the topic of non-constructible parcels, small vacant or built spaces in Municipal Beirut, some of which belong to the municipality while others are privately owned and which might be used for different functional purposes. This paper further examines the possibility of implementing cool surface or paving materials and urban vegetation to reduce air urban temperature, especially during the summer period and with the view to project the positive findings of this case study to the entire Municipal Beirut area. A numerical analysis using ENVI-met 4.0 investigates the thermal performance of these non-constructibles further to implementation of high reflective surfaces and urban vegetation on a broad neighborhood scale, taking the Bachoura District as a reference case for a typical summer day. The best air temperature reductions correspond to the use of cool material in areas that are far from buildings where there are no shadow effects. In some cases, the introduction of trees leads to an increase of the air temperature near the ground because they became an obstacle of the natural ventilation. Results show a maximum mitigation effect with the use of cool materials that lead to reductions in air temperatures up to 0.42 °C if used alone and up to 0.77 °C if used in combination with trees. Within the framework of an integrated approach to planning, this form of urban intervention aims for substantial overheating reduction.

Keywords: urban heat island; urban overheating; non-constructible parcels; cool surfaces; urban vegetation; ENVI-met; mitigation measures; Beirut

1. Introduction

Today, 54% of the world's population lives in urban areas, a proportion that is expected to increase to 66% by 2050 [1]. With increasing urbanization trends, where natural surfaces are replaced by impermeable surfaces, stresses on the surrounding urban temperatures are escalating. This urban warming, or urban heat island (UHI), is the increase of urban temperatures with respect to surrounding rural areas and has a strong impact on thermal comfort levels, energy performance of buildings and health [2]. With global warming, associated to climate changes and increasing summer heat waves, the potential means by which the consequences of the UHI and urban overheating can be adapted to and mitigated to build more resilient cities is a topic of much research and discussion worldwide. In fact, counterbalancing the effects of urban heat island is a major priority for the scientific community [3–6].

Starting from as early as 1818 when Luke Howard, a chemist turned meteorologist, published his first documentation on the "Climate of London," many studies have been carried out to detect and assess UHI and today there is an impressive amount of data throughout the world [7–10]. The scientific community produced many studies aiming at assessing the mitigation potentials of technologies and strategies to minimize urban heating. While several technologies and strategies exist for the cooling of urban temperatures, the two major solutions to pursue urban heating are cool materials and urban greenery [11–18] and some of the available literature on these topics is described further below.

Vegetation has been studied in urban climates [19] in order to reduce the urban heat island effect. This type of mitigation technique cools the environment through a higher albedo (typically 0.18–0.22) compared to common pavements such as asphalt (typically 0.05–0.15) [20–24] and by evapotranspiration [25]. An experimental and numerical analysis was developed by Srivanit and Hokao [26] at an institutional campus in the subtropical-humid climate of Saga, Japan. In this work, the average daily maximum temperature decreased by 2.7 °C when the quantity of the trees was increased by 20% in the campus area. In addition, in another study carried out in the city of Lisbon by Oliveira, Andrade and Vaz [27], the thermal performance of a small green space (0.24 ha) and its influence on the surrounding environment of a densely urbanized area, was conducted. Results showed that the garden was cooler than the surrounding areas, either in the sun or in the shade, by up to 6.9 °C of air temperature.

Cool materials are characterized by a high solar reflectance or albedo (α) to the short-wave radiation and a high emissivity (ε) to the long-wave radiation [28]. These materials can be used on building surfaces or on urban pavements in order to reduce the surface temperature. Synnefa et al. in 2008 [29] carried out a numerical study on large areas of Athens, Greece, in order to simulate the impacts of cool materials on the urban heat island effect. It was found that large-scale increases in albedo could lower ambient air temperatures by 2 °C.

There is a strong correlation between albedo increase and drops in average and maximum temperature. Santamouris, in 2014, estimated a decrease of the average and maximum air temperatures close to 0.27 °C and 0.78 °C per 10% increase of the albedo.

In the last few years, the use of ENVI-met tool for the analysis of urban heat island phenomenon has been growing (Section 3). In particular, numerous studies have been conducted to assess the impact of urban vegetation on microclimatic conditions [30–33]. In Hong Kong, Morakinyo and Lam [34] simulated the impact of different configurations of trees in an urban canyon; changing the aspect ratio, leaf area index (LAI), leaf area density (LAD) distribution and trunk height under different wind conditions. Wang and Zacharias [35] studied possible interventions of urban requalification in Beijing (China) estimating a decrease in the air temperature of about 0.5–1 °C due to the substitution of roads with urban greening and permeable soils. In 2015, Amor et.al [36] carried out a numerical analysis by using the ENVI-met tool in order to determine the impact of vegetation, ponds and fountains in a square of the Algerian city of Setif. Other researchers have been more focused on the evaluation of the effect of cool materials using the ENVI-met tool [37–40]. In 2015, Sodoudi [41] conducted a study to assess the effectiveness of three UHI mitigation strategies, such as high albedo materials, greenery on the surface and on the roofs, and a combination of these strategies in the city of Tehran. Wang and Akbari in 2015 [42] developed the "thermal radiative power" (TRP) parameter to investigate the impact of building surface material albedo on urban environment in the central city of Montreal.

Municipal Beirut, with approximately 21,000 inhabitants/km^2 [43,44] is a densely populated city, while covering a relatively small surface area of about 20km^2. Concerning the weather conditions, Beirut is classified as a hot-dry summer Mediterranean Climate, "Csa" class, according to the Köppen-Geiger system. No relevant studies exist on the UHI intensity (UHII) in Beirut, however some quantification can be derived from data acquired by the weather stations in and around the city and the Beirut International Airport. Data from the Fanar weather station, situated a little outside the city (Section 3), includes daily minimum, maximum and average air temperatures. Based on data from this weather station from the year 2011 which consists of the most uninterrupted and therefore continuous data, it is estimated that the maximum UHII has reached 4.56 °C considering the daily maximum air temperature difference and a value of 1.68 °C considering the daily average one.

The major master plans for Beirut were subject to several updates throughout their history, such as the Danger Brothers Scheme in 1932 and the Ecochard schemes in 1943 and 1964, and combined with an outdated laws and regulations system that manages urban operations, have enriched the urban fabric of Municipal Beirut with a non-negligible number of non-constructible parcels (NCs) [45]. These are leftover spaces found in the shape of small vacant or built spaces in between buildings or around corners and belonging to the municipality, which might be used for social purposes. Since 1954, with the introduction of Article 5 of the decree n° 6285/54, Lebanon's Construction's Law has acknowledged and defined non-constructible surfaces as such: (a) residue of an old road after a new alignment and (b) result of land consolidation or left-over spaces after planning. Moreover, Article 5, does not allow construction on narrow parts of parcels for three main reasons: (a) the intent of visual clearance on street corners and intersections; (b) to manage parcel densification; and (c) to avoid transferring its odd forms into the volumetric of the buildings [46].

The surface areas of these NCs do not exceed 250 m², depending on the zone they belong to and according to the zoning scheme introduced in 1973 by decree n° 5550, which divides Beirut into 10 zones. As for their current land uses, they have various current temporary land uses such as parking lots for neighboring buildings, vacant plots, dumpsters, brown land, right-of-ways and so on. Some of them are illegally built and as for the un-built ones, they are rapidly being consolidated with larger adjacent plots by developers who are seeking more space for their projects. Previous research has shown that unfiltered results for non-constructible parcels within Municipal Beirut add up to 6039.

Upon comparing these findings to previous Town Energy Balance (TEB) single layer urban canopy model [47] simulations for Municipal Beirut [48,49], it was found that Zones 3 and 4 of the city had the highest urban warming results, as well as the highest number of NCs, amongst which Bachoura District is situated. As such, Bachoura District was selected as a case study area for the purpose of this paper to be replicated for future similar analyses within Municipal Beirut.

2. Objective and Methods

The objective of this paper is to explore the potential of NCs to reduce urban air temperatures through the case study of the Bachoura District, preparing grounds for future implementation projects on selected NCs throughout Municipal Beirut at a larger scale. This approach is an innovative one within the context of Municipal Beirut, where no thermal mitigation strategies were experienced in the past. The integration of technical and policy aspects to pursue urban mitigation will be also discussed further in this paper.

To achieve the research objective, the study was carried out according to the following procedure:

- Identification and characterization of the case study: this part includes the identification of the main characteristics, including the thermal and solar properties, of the selected district, next transferred on an information model, to carry out successive analyses;
- Identification of mitigation strategies and technologies to be implemented: this part includes the description technologies and strategies, selected to perform mitigation analyses.
- Assessment of mitigation potentials through numerical analyses: this part, implemented in Section 3, includes the calculation, modeling and simulation work carried out to quantify the impact of selected solutions on relevant urban thermal indicators using the software ENVI-met 4.0.
- Derive policy approaches as a consequence of technical results, taking into account potentials and limitations of the latter.

2.1. The Case Study Zone: Bachoura District

The selected district is one of the most affected by urban overheating, thus suitable for the presented study. Previous research (Figure 1) has shown that Bachoura District holds a total 197 NC parcels. Further studies have highlighted the exploitable parcels for the purpose of this paper, which are the vacant lands (VL) and right of ways (ROW). Vacant lands are the un-built lands of various uses such as brown land, junkyards or dumpsites, gardens, entrances for buildings, support for neighborhood generator or parking lots. Right of way are the un-built passages and movement corridors that lead to the inside of blocks. It was surveyed [50] that Bachoura holds 1307 m^2 of VL and 1525 m^2 of ROWs.

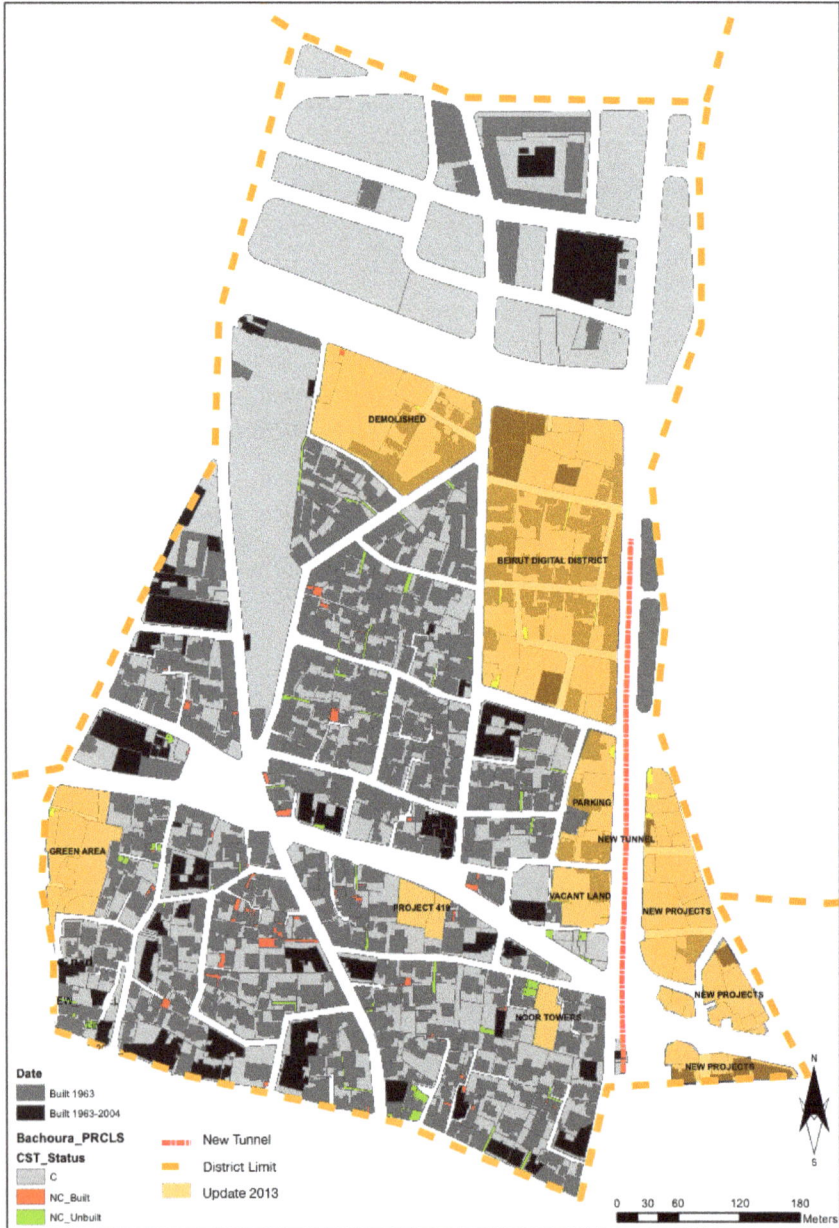

Figure 1. Non-Constructible parcels map within the Bachoura District [45].

2.2. Transferring Geometrical and Construction Properties on an Information Model

Data from previous findings for NCs, as well as the thermal properties of dominating urban surfaces, were combined and updated using the Geographic Information System software (ArcGIS) [51]. This systematic integration of the main features of the urban fabrics has a double value: it helps the simulation and calculation work carried out in this study and, in a broader perspective, it helps to develop an informational database that can be used by Municipal Beirut as an instrument to plan mitigation measures in the future.

All the NCs were characterized in detail in the information system; each of these NCs include the following information:

- Identification code and reference district;
- Surface;
- Nature of the area (e.g., vacant, park, etc.);
- Age of the neighborhood;
- Average albedo of walls of the dominating building group;
- Average emissivity of walls of the dominating building group;
- Average albedo of roofs of the dominating building group;
- Average emissivity of roofs of the dominating building group;
- Average albedo of roads;
- Average emissivity of roads;
- Town cover fraction of buildings;
- Town cover fraction of green areas;
- Town cover fraction of roads

The observation of non-constructible parcels shows that they are primarily impermeable surfaces, roads and buildings account for the majority of the town covers surrounding the identified NCs, while gardens, which include predominantly high vegetation such as scattered urban trees or low vegetation like grass or bushes, account for the lowest town cover fraction thus emphasizing the deficiency of these important urban cooling spaces within the city. For example, in an identified Bachoura neighborhood covering a surface area of 200 × 200 m, it was calculated that the building fraction was 0.48, the garden fraction 0.14 and road fraction 0.38 and this is typical throughout Bachoura and indeed the city which is a predominantly artificial city. Dominating roof surfaces, comprised primarily of grey concrete roof slabs, have a very low albedo (α) of 0.23 as adapted from Oke 1987. Moreover, road surfaces with topmost layer comprised of asphalt, also have a low α of 0.1–0.2. It has to be noted that the albedo values for the purpose of this paper are adapted from Oke, 1987 as they are typically reference points in the available literature. The emissivity (ε) of these non-metallic materials is assumed to be 0.9.

The non-constructible parcels defined as vacant land, brownfields or dumpsters were selected for investigation. Figure 2 therefore shows the selected NCs within a previously defined TEB grid cell within the district of Bachoura.

2.3. Selection and Characterization of the Mitigation Strategies

This paper addresses two potential solutions, which are among the most used for mitigation purposes, due to their performances, reliability and economic feasibility.

Cool materials are the first solution, tailored for the different NCs of the district. Table 1 defines the characteristics of these NCs with proposed types of light-colored materials or reflective coatings and their respective albedos that can be used in this regard within the context of Municipal Beirut [52,53].

Concerning the urban greenery, the type of urban trees proposed for UHI mitigation is the *ficus nitida* species, which is commonly found and planted decoratively throughout municipal Beirut. For the purpose of this paper, *ficus nitida* is considered to have a total height of 10 m, a trunk height of

5 m and a leaf area index (LAI) of 4, the latter quantity defined as the ration of one-sided green leaf area per unit ground surface area, both expressed in square meters.

Figure 2. Non-constructible parcels identified in previous TEB cell(s) in Bachoura District.

Table 1. Characteristics of existing NCs with proposed types of cool materials or reflective coatings and their thermal properties in Bachoura District.

Plot No.	Surf. Area [m²]	Nature of Plot	Existing Material	Initial Albedo [-]	Proposed Cool Material	Final Albedo [-]
706	97	Parking Lot	Asphalt	0.2	White alkyd Chlorine rubber coating	0.5–0.6
707	46.64	ROW	Pavestone	0.2–0.35	White acrylic latex	0.6–0.7
713	33.3	ROW	Pavestone	0.2–0.35	Marble	0.45–0.5
724	30.1	ROW	Pavestone	0.2–0.35	Marble	0.45–0.5
736	68.2	Parking Lot	Asphalt	0.225	White acrylic latex	0.6–0.7
1213	14.7	Vacant Land	Bush/Garden	0.18	Marble	0.45–0.5
1219	36.75	ROW	Pavestone	0.2–0.35	Marble	0.45–0.5
1261	78.6	Parking Lot	Asphalt	0.20	White alkyd chlorine rubber coating	0.5–0.6
1282	57.4	Generator	Concrete Slab	0.10–0.35	White alkyd chlorine rubber coating	0.5–0.6
1287	53.8	Bush	Bush	0.18	White acrylic latex	0.6–0.67
1378	34.6	Parking Lot	Asphalt	0.20	Marble	0.45–0.5
1392	74.2	Parking Lot	Asphalt	0.20	White alkyd chlorine rubber coating	0.5–0.6
1486	9.1	ROW	Pavestone	0.2–0.35	Marble	0.45–0.5

The total surface area of the above-listed NCs in the identified TEB grid is 634 m^2 thus comprising approximately 1.6% of the total TEB grid area of 40,000 m^2. The selected mitigation strategies are applied in this area, individually and in combination.

3. Calculation

The simulation was carried out with ENVI-met 4.0 which is a transient tool able to recreate urban 3D models based on a soil-vegetation-atmosphere transfer scheme realized with deterministic equations that couple thermal and fluid-dynamics processes [54]. This software is widely used in the scientific community to assess thermal conditions in the urban environment and the impact of mitigation strategies and technologies, as addressed in the introduction. The main indicator to assess the mitigation potential is the air temperature.

3.1. Implementation of the Numerical Model in ENVI-Met 4.0

The target area taken into account is the Bachoura District, the implementation of which in ENVI_met is presented in Figure 3, covering a surface area of approximately 130,000 m^2. It is to be noted that this latter surface area represents the domain area as presented in Figure 2, which is used for the purpose of this paper and which is less than the area of the previously mentioned TEB grid cells (40,000 m^2) from which dominating urban morphological characteristics have been used for the purpose of this study. The selected NCs account for approximately 0.5% (650 m^2) of this chosen domain (Figure 2). The three-dimensional model recreates the distribution of structures, pavements and vegetation and it is composed by a mesh of 77 × 105 × 25 square cells. Each cell has a dimension of 4(x) × 4(y) × 2(z) meters.

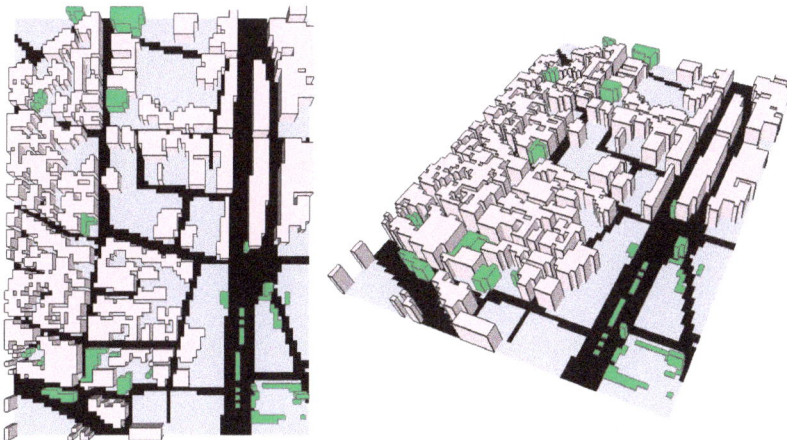

Figure 3. Three-dimensional computational domain of Bachoura District.

3.2. Model Calibration

The thermal conditions inside the Bachoura District are simulated on a typical summer day in Beirut. For this reason, a Beirut weather station situated in Fanar (placed at 1.8 m above the ground) was selected as representative of the mean air temperature condition in Beirut in 2011. The Fanar station is located at approximately 3.5 km to the east of Beirut at Latitude 33°53′02.8″ N and Longitude 35°33′08.3″ E at an elevation of 90m ASL, in a medium density urban area. This station, which is operated by the Lebanese Institute for Agricultural Research (LARI), started its operation in 2009 and continues to present day. Records were collected and accordingly compiled from this station, for the numerical analysis of this research. The minimum, maximum and average air temperature in the summer of 2011 are shown in Figure 4.

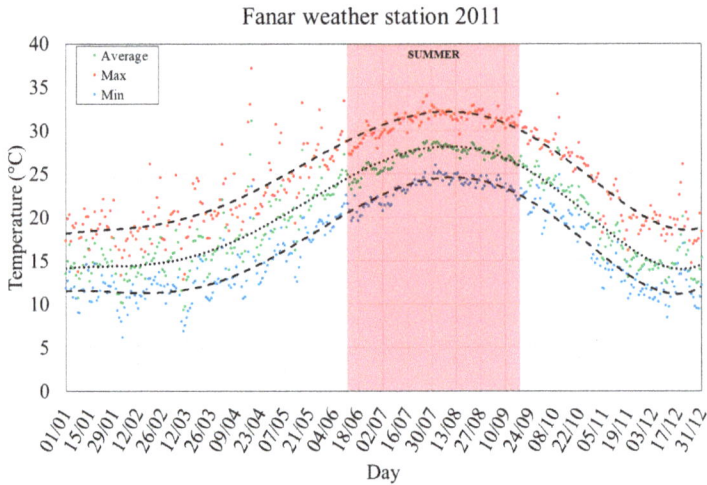

Figure 4. Minimum, maximum and average air temperature in the summer of 2011 of Fanar weather station

The weather boundary conditions were imposed as the average data for a typical summer day as follow: wind speed of 1.24 m/s, wind direction of 142° (south-east wind) and relative humidity of 68%.

In order to calibrate the numerical model, a comparison was carried out between the maximum air temperature recorded at 1.8 m above the ground in the entire numerical domain and the maximum air temperature of 31.2 °C measured by the Fanar weather station in the summer of 2011. In order to reduce the error, the initial air potential temperature was varied at 2500 m. It has to be clearly stated that the calibration work is strictly dependent on the measured data, which in this case were not sufficiently disaggregated to carry out more detailed analyses.

The initial potential air temperature of 20 °C lead to a minimum error of +0.1% between the measured and the model-predicted maximum air temperature.

4. Results

Simulation results are presented in Figures 5–7, where the air temperature variations versus the height from the ground is plotted in order to compare the current situation with the mitigation solutions. The air temperature variations on the X-axes are varied in the graphs so that the temperature differences between the represented configurations can be clearly inferred. The curves are plotted from 0 to 5 m above the ground, since after that value the impact of the proposed solution is not significant; more over the selected height is relevant because of the interaction between the ambient temperature and both, buildings and pedestrians.

Table 2 presents the maximum air temperature reductions for the selected zones, to be noted that this peak is reached always at 2 pm in all cases. To be noted that urban greenery was placed only in 6 out of the 13 NCs listed in Table 1, since it was deemed preferable to place the trees in large spaces like parking lots; as such, right of ways were excluded from the simulations.

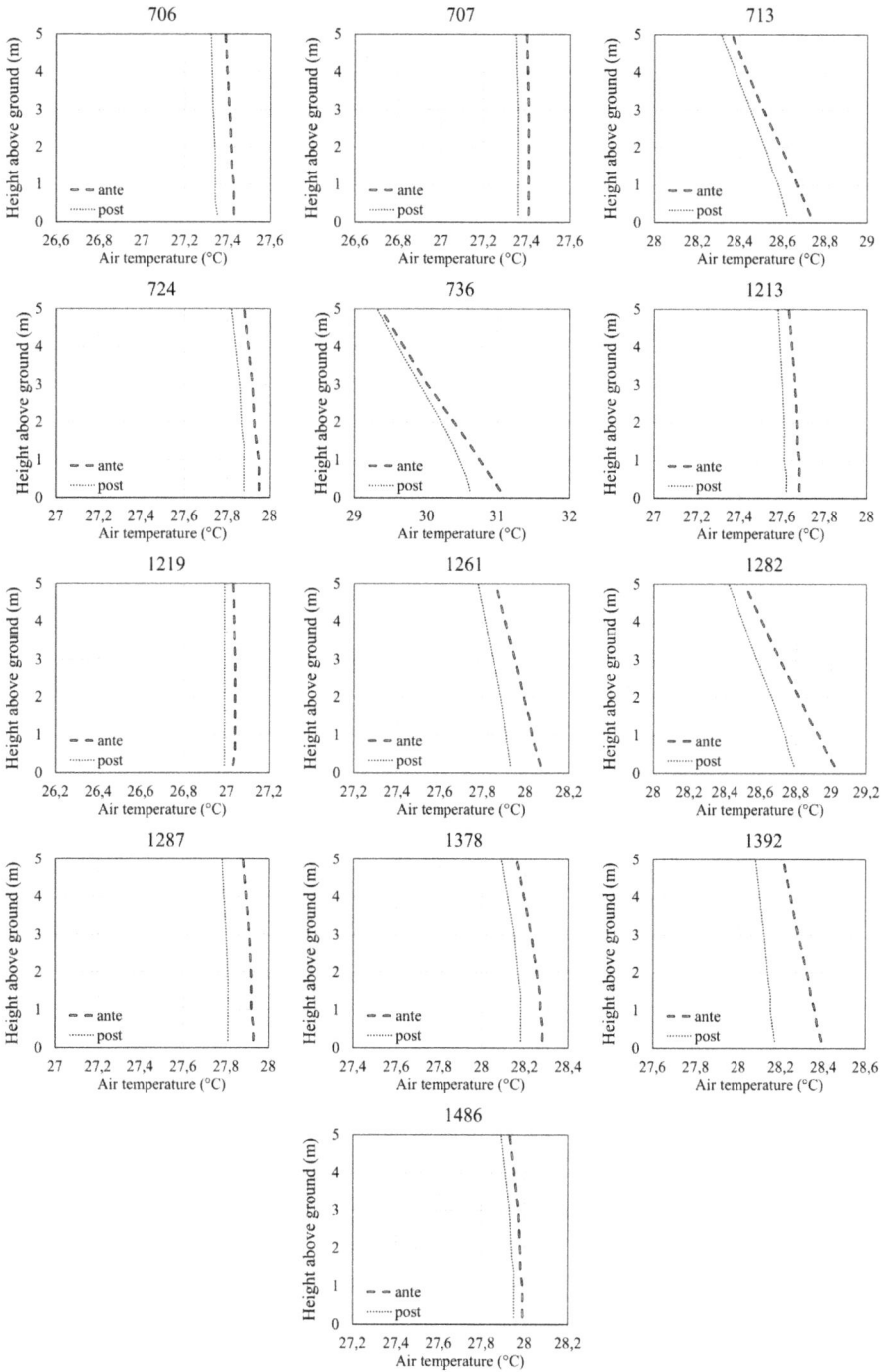

Figure 5. Air temperature variations from ante- to post-operam with cool materials.

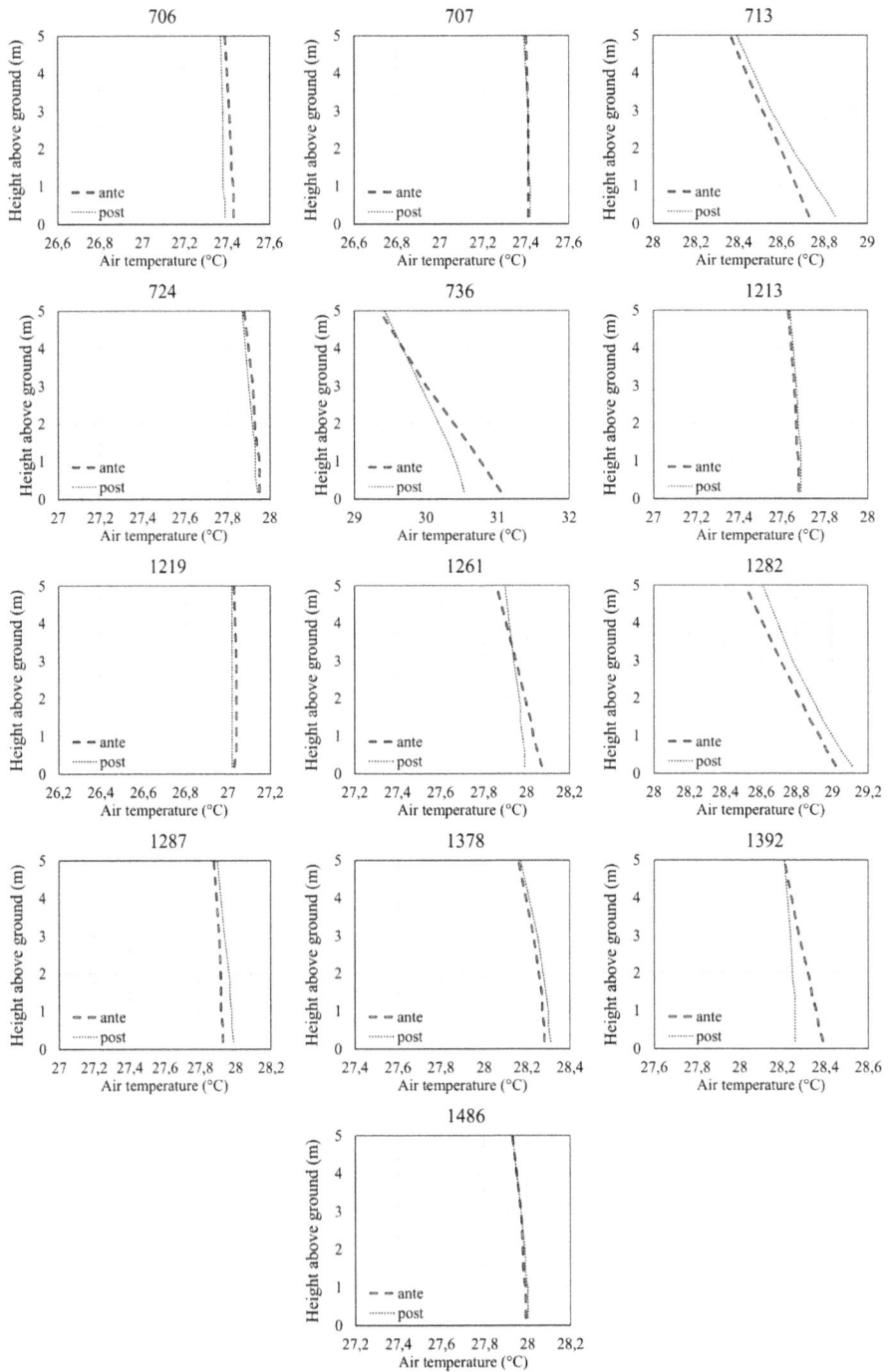

Figure 6. Altitude air temperature variations from ante to post-operam with trees.

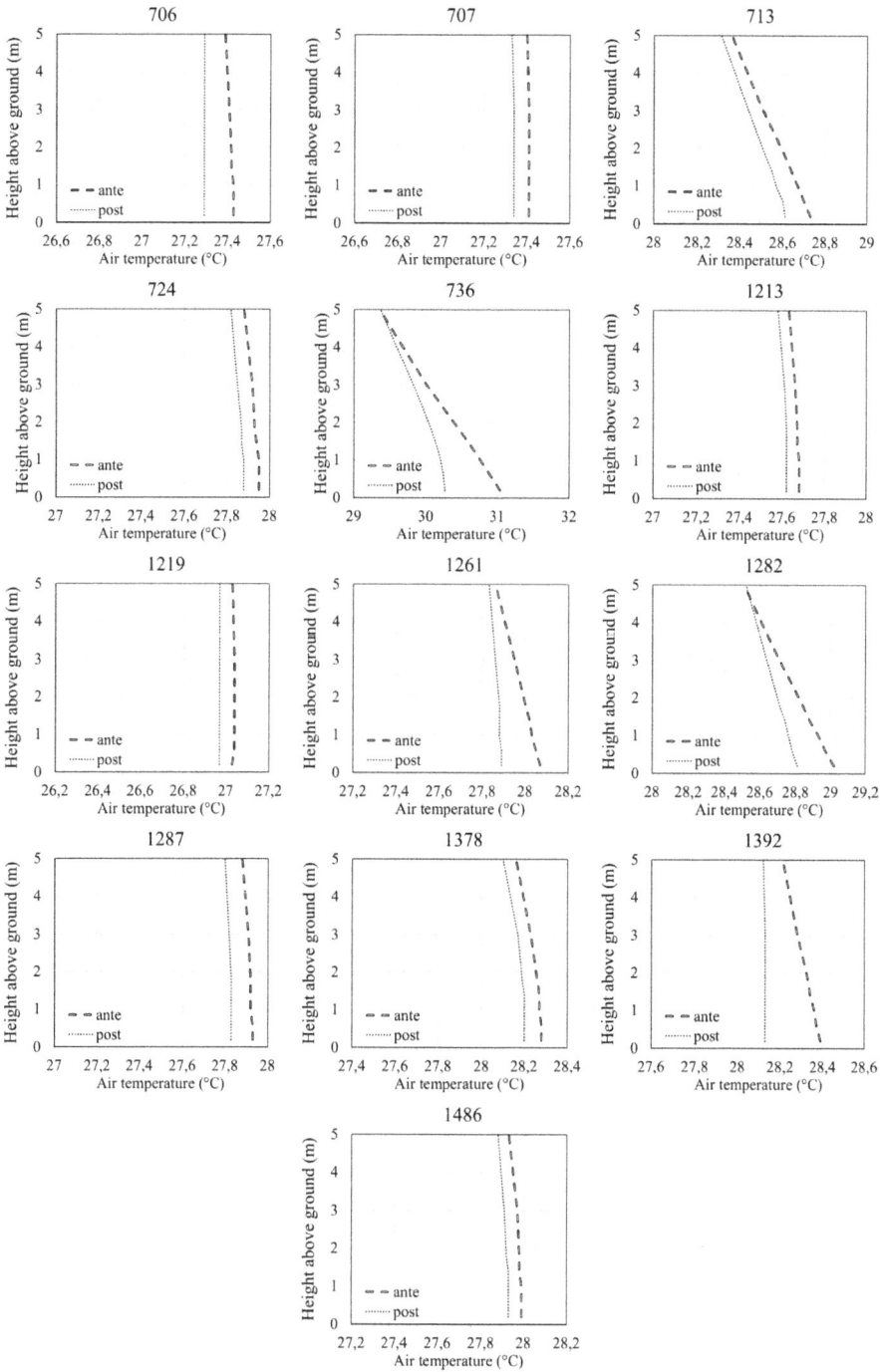

Figure 7. Altitude air temperature variations from ante to post-operam with cool materials and trees.

Table 2. Results of simulations for selected NCs in Bachoura District. Maximum air temperature difference from 0 to 5 m above ground before and after the application of mitigation strategies.

	Maximum Temperature Reduction (°)													
NC Plot Nos.	706	707	713	724	736	1213	1219	1261	1282	1287	1378	1392	1486	*Max*
Cool mat.	−0.09	−0.05	−0.11	−0.07	−0.42	−0.07	−0.05	−0.14	−0.23	−0.12	−0.10	−0.22	−0.04	*−0.42*
Trees	−0.05	+0.01	+0.12	−0.02	−0.51	+0.02	−0.02	−0.08	+0.10	+0.06	+0.03	−0.13	+0.01	*−0.51*
Cool mat. & trees	−0.14	−0.07	−0.12	−0.07	−0.77	−0.06	−0.07	−0.18	−0.21	−0.10	−0.08	−0.26	−0.06	*−0.77*

Figure 5 shows the air temperature variation before and after with the introduction of cool materials listed in Table 1. It is worth noting that in all the NCs the introduction of cool material leads to a decrease of the air temperature along 5 m above the ground. The maximum effects are more relevant near the ground. This effect leads to a decrease of the air temperature from 0.04 for the NC 1486 to 0.42 for the NC 736. The marble material leads to a maximum air temperature decrease from 0.04 for the NC 1486 to 0.11 for the NC 713. Furthermore, the white alkyd chlorine rubber coating lead to a maximum air temperature decrease from 0.09 for the NC 706 to 0.23 for the NC 1282. Finally, the white acrylic latex leads to a maximum air temperature decrease from 0.05 for the NC 707 to 0.42 for the NC 736.

Figure 6 shows the altitude air temperature variations before and after the introduction of trees in the NCs number 706, 736, 1261, 1282, 1378 and 1392. Relevant effects are recorded in the NCs number 736, 1261 and 1392 where it is worth to notice that the decrease of the air temperature is higher near the ground and lower at an altitude ranging from 3 to 4 m.

Figure 7 shows the altitude air temperature variations before and after the introduction with the introduction of cool materials and the trees implemented in the cases of Figure 6. The combining effects are evident in the NCs number 736, 1261, 1282 and 1392 where the decrease of the air temperature is from 0.18 to 0.77 especially in the levels near the ground.

The best air temperature reduction effects are found in the NC number 736 because this is located relatively far from buildings with no shadow effects. On the other hand, NC 1282 with trees showed an increase in simulation results. The reason could be that in this particular area, the trees became an obstacle of the natural ventilation (see Section 5 Discussions). Overall however, the maximum mitigation effects are seen with the use of cool materials.

5. Discussion

Based on the simulation results therefore, the selected mitigation strategies showed that in open areas, where the effect of surrounding buildings is limited, temperature reduction close to 1°C might be reached, however in most cases the mitigation potential appears of limited or null impact in our analysis, when compared with similar studies. This mainly depends on the following factors:

- It was outlined in the previous sections that NC areas in which it was possible to implement the selected solutions were small compared to the district size, by about 1%. In these conditions, the air temperature reduction is limited.
- The urban texture in the selected district is very dense and mainly characterized by high building height to street width ratios; thus, narrow parcels can be easily shaded by surrounding buildings. Under these conditions, direct benefits of cool materials and urban greenery decrease when compared to those detectable in strongly irradiated open urban areas.
- Using trees showed positive as well as negative UHI effects. The reason for this could be that trees could obstruct summer breezes if not strategically placed thus having the reverse desired effect [55]. In fact, trees create perturbation of the thermo fluid-dynamic conditions because they are an obstacle to the natural ventilation, which can accordingly cause a variation of the micro-climate near the trees and in the rest of the domain. It is therefore suggested to have more in-depth knowledge of the plant species and their strategic placement within the NCs proposed as such.

Another relevant aspect of this study is that only air mitigation solutions are addressed, while pedestrian thermal comfort issues are not analyzed as this is not within the scope of this paper. It is, however, well recognized that air temperature is only one of the many physical parameters affecting the comfort conditions; air humidity and velocity, as well as solar radiation all play a crucial role within the context of thermal comfort. The latter, in particular, affects the surface temperature of urban fabrics and, thus, the longwave radiation exchanges. As an example, it was demonstrated in this paper that cool materials while able to mitigate the ambient temperature, could in fact worsen the thermal comfort conditions at the pedestrian level, due to the increase of solar radiation [56,57]. In this sense, this study can be considered a preliminary assessment to be elaborated upon by conducting further, more in-depth, analyses to assess other mitigating strategies and technologies, as well as the impact on the thermal comfort conditions of users.

However, it has to be noted that this study acts on two levels, technical and policy one, where the results achieved in the former acts as a lever for the latter ones. To successfully intervene on such lands of Municipal Beirut and to develop a pilot project based on these findings, different types of strategies could be developed, where technical and policy issues converge to promote solution for urban regeneration. One strategy is a user driven tactical strategy where UH mitigation programs are applied as part of a long-term vision for regeneration and implemented with adequate resources such as the NDSM project in Amsterdam [58] and The High Line Project in New York [59].

Another tactical strategy would be based on the idea of "best practice approach" [60] where local temporary projects are taken as a model for broader policy-making and subsequent implementation. This type of strategy usually uses intermediary agents to find short and medium-term uses for vacant land, disused, or awaiting redevelopment land. In this strategy, applied in the Meanwhile London Project, temporary uses, should they succeed, could become permanent [61].

There are other strategies where temporary uses are applied in an event-like manner and where long-term vision is coupled with limited resources. These project-based strategies, such as the Leipzig plantation project [62], are very important in triggering a more sustained strategy. These event-like projects can attract potential investors and provide resources for future projects, hence move towards a strategy of the first kind. Finally, there are the strategies, where power is kept centralized and no collaboration is envisioned. This strategy does not distribute resources for the implementation of temporary uses and reveals only partial understanding of potential benefits on the authorities' side and leave unclear the will to collaborate further.

The above-mentioned urban strategies for the implementation of urban heating mitigation programs on residual parcels differ in their approaches, they all share the same understanding of the potential of mitigation programs as a catalyst for the regeneration of the city. Whether long or short-term strategies were envisioned; resources were or were not available; leaders were local authorities or intermediary agents; or developments were user driven pushing for collaboration or centralized, they all shared a bottom-up strategy starting with the understanding of user's needs and the acknowledgment of the potential of residual spaces/urban heating mitigation.

In the case of Beirut, no specific urban strategy is modeled to integrate non-constructible parcels and implement mitigation programs. It will be important to define a strategy that embraces a long-term vision for the regeneration of the city, through the integration of these parcels and the implementation of mitigation programs. In a longer perspective, a wider area of the city should be involved in such urban regeneration measures, so to have higher mitigation potentials than those achievable operating only on NCs. By acknowledging user's needs, finding resources and encouraging collaboration with involved actors, the developed user oriented strategy will have better chances of responding to the city's needs rather than becoming a strategy for punctual interventions.

6. Conclusions

Within Municipal Beirut, covering a total surface area of 20 km^2, approximately 300,000 m^2 of non-constructible parcels have been identified. Within the context of resilient cities, there is a potential for re-naturalization of these non-constructible areas by implementation of green and cool urban design. For the purpose of this paper, numerical analyses using ENVI-met 4.0 were carried out to study the implementation of cool materials and urban trees as a sustainable strategy for reductions of the urban air temperature in Bachoura District. However, only 0.5% of the zones in the Bachoura domain were used for the purpose of this paper.

Implementation of cool design approaches showed reductions of urban air temperatures up to 0.42 °C. The introduction of trees may lead, in some cases, to an increase of the air temperature near the ground because they can obstruct natural ventilation. In an investigated area, which is far from the surrounding buildings, the maximum air temperature reduction with the implementation of combined cool materials and trees reached up to 0.77 °C. Hence, the use of mitigation strategies in small non-constructible areas can create small urban oases in which people can benefit from better thermal conditions, especially in the summer. However, this research can be further expanded to examine other mitigating solutions within a similar urban context, especially in the view to assess how such solutions could effectively influence and improve thermal comfort aspects.

This research can be considered to be a milestone for the case of Beirut, since it provides evidence of the importance of making use of not constructible areas for multiple objectives including beautification of the city, providing a 'breathing' space for the local communities, while building a more resilient city within the context of climate change and the urban heat island.

Author Contributions: Conceptualization, N.K. and D.A.; Formal analysis, G.B.; Investigation, G.B. and M.Z.; Methodology, N.K. and D.A., M.Z., G.B.; Project administration, N.K.; Resources, N.K., D.A., G.B. and M.Z.; Software, G.B. and M.Z.; Supervision, N.K. and M.Z.; Visualization, N.K. and D.A.; Writing—original draft, N.K., D.A. and G.B.; Writing—review & editing, N.K., D.A., G.B. and M.Z.

Funding: This research received no external funding. **Conflicts of Interest:** The authors declare no conflict of interest.

References

1. Un-habitat. *State of the World's Cities 2010/2011: Bridging the Urban Divide*; Earthscan: London, UK, 2010.
2. Rizwan, A.M.; Dennis, Y.C.L.; Liu, C. A review on the generation, determination and mitigation of Urban Heat Island. *J. Environ. Sci.* **2009**, *20*, 120–128. [CrossRef]
3. Akbari, H. *Energy Saving Potentials and Air Quality Benefits of Urban Heat Island Mitigation*; Lawrence Berkeley National Lab (LBNL): Berkeley, CA, USA, 2005.
4. Akbari, H. Rapid development of global heat island mitigation studies and countermeasures: A 30 year success story. In Proceedings of the Prepared for: The Third International Conference on Countermeasures to UHI, Venezia, Italy, 13–15 October 2014.
5. Bozonnet, E.; Musy, M.; Calmet, I.; Rodriguez, F. Modeling methods to assess urban fluxes and heat island mitigation measures from street to city scale. *Int. J. Low-Carbon Technol.* **2013**, *10*, 62–77. [CrossRef]
6. Oliveira, S.; Andrade, H.; Vaz, T. The cooling effect of green spaces as a contribution to the mitigation of urban heat: A case study in Lisbon. *Build. Environ.* **2011**, *46*, 2186–2194. [CrossRef]
7. Santamouris, M. Using cool pavements as a mitigation strategy to fight urban heat island—A review of the actual developments. *Renew. Sustain. Energy Rev.* **2013**, *26*, 224–240. [CrossRef]
8. Gartland, L. *Heat Islands*; Routledge: London, UK, 2008.
9. Oke, T.R. *Boundary Layer Climates*, 2nd ed.; Methuen & Co. Ltd.: London, UK, 1987.
10. Stewart, I.D. A systematic review and scientific critique of methodology in modern urban heat island literature. *Int. J. Climatol.* **2011**, *31*, 200–217. [CrossRef]
11. Akbari, H.; Pomerantz, M.; Taha, H. Cool surfaces and shade trees to reduce energy use and improve air quality in urban areas. *Sol. Energy* **2001**, *70*, 295–310. [CrossRef]
12. Rosenfeld, A.H.; Akbari, H.; Romm, J.J.; Pomerantz, M. Cool communities: Strategies for heat island mitigation and smog reduction. *Energy Build.* **1998**, *28*, 51–62. [CrossRef]
13. Synnefa, A.; Saliari, M.; Santamouris, M. Experimental and numerical assessment of the impact of increased roof reflectance on a school building in Athens. *Energy Build.* **2012**, *55*, 7–15. [CrossRef]

14. Taha, H.; Sailor, D.; Akbari, H. *High-Albedo Materials for Reducing Building Cooling Energy Use*; Report Number LBL-31721; Lawrence Berkeley National Laboratory: Berkeley, CA, USA, 1992.

15. Taha, H. Urban climates and heat islands: Albedo, evapotranspiration, and anthropogenic heat. *Energy Build.* **1997**, *25*, 99–103. [CrossRef]

16. Che-Ani, A.I.; Shahmohamadi, P.; Sairi, A.; Mohd-Nor, M.F.I.; Zain, M.F.M.; Surat, M. Mitigating the urban island effect: Some points with altering existing city planning. *Eur. J. Sci. Res.* **2009**, *35*, 204–216.

17. Santamouris, M.; Ding, L.; Fiorito, F.; Oldfield, P.; Osmond, P.; Paolini, R.; Prasad, D.; Synnefa, A. Passive and active cooling for the outdoor built environment–Analysis and assessment of the cooling potential of mitigation technologies using performance data from 220 large scale projects. *Sol. Energy* **2017**, *154*, 14–33. [CrossRef]

18. Santamouris, M. Cooling the cities–a review of reflective and green roof mitigation technologies to fight heat island and improve comfort in urban environments. *Sol. Energy* **2014**, *103*, 682–703. [CrossRef]

19. Oke, T.R.; Crowther, J.M.; McNaughton, K.G.; Monteith, J.L.; Gardiner, B. The micrometeorology of the urban forest [and discussion]. *Philos. Trans. R. Soc. Lond. B Biol. Sci.* **1989**, *324*, 335–349. [CrossRef]

20. Dimoudi, A.; Nikolopoulou, M. Vegetation in the urban environment: Microclimatic analysis and benefits. *Energy Build.* **2003**, *35*, 69–76. [CrossRef]

21. Kleerekoper, L.; van Esch, M.; Salcedo, T.B. How to make a city climate-proof, addressing the urban heat island effect. *Resour. Conserv. Recycl.* **2012**, *64*, 30–38. [CrossRef]

22. McPherson, E.G.; Rowntree, A.R.; Wagar, J.A. Energy-efficient landscapes. In *Urban Forest Landscapes— Integrating Multidisciplinary Perspectives*; Bradley, G., Ed.; University of Washington Press: Seattle, WA, USA; London, UK, 1994.

23. Oke, T.R. The energetic basis of the urban heat island. *Q. J. R. Meteorol. Soc.* **1982**, *108*, 1–24. [CrossRef]

24. Taha, H.; Douglas, S.; Haney, J. Mesoscale meteorological and air quality impacts of increased urban albedo and vegetation. *Energy Build.* **1997**, *25*, 169–177. [CrossRef]

25. Oke, T.R. *Boundary Layer Climates*; Routledge: New York, NY, USA, 1987.

26. Pereira, L.; Perrier, A.; Allen, R.; Alves, I. Evapotranspiration: Concepts and future trends. *J. Irrig. Drain. Eng.* **1999**, *125*, 45–51. [CrossRef]

27. Srivanit, M.; Hokao, K. Evaluating the cooling effects of greening for improving the outdoor thermal environment at an institutional campus in the summer. *Build. Environ.* **2013**, *66*, 158–172. [CrossRef]

28. Doulos, L.; Santamouris, M.; Livada, I. Passive cooling of outdoor urban spaces The role of materials. *Sol. Energy* **2004**, *77*, 231–249. [CrossRef]

29. Synnefa, A.; Dandou, A.; Santamouris, M.; Tombrou, M.; Soulakellis, N. On the use of cool materials as a heat island mitigation strategy. *J. Appl. Meteorol. Climatol.* **2008**, *47*, 2846–2856. [CrossRef]

30. Lee, H.; Mayer, H.; NChen, L. Contribution of trees and grasslands to the mitigation of human heat stress in a residential district of Freiburg Southwest Germany. *Landsc. Urban Plan.* **2016**, *148*, 37–50. [CrossRef]

31. Duarte, D.H.S.; Shinzato, P.; dos Santos Gusson, C.; Abrahão Alves, C. The impact of vegetation on urban microclimate to counterbalance built density in a subtropical changing climate. *Urban Clim.* **2015**, *14*, 224–239. [CrossRef]

32. Tsilini, V.; Papantoniou, S.; Kolokotsa, D.; Maria, E. Urban gardens as a solution to energy poverty and urban heat island. *Sustain. Cities Soc.* **2014**, *14*, 323–333. [CrossRef]

33. Ketterer, C.; Matzarakis, A. Comparison of different methods for the assessment of the urban heat island in Stuttgart, Germany. *Int. J. Biometeorol.* **2015**, *59*, 1299–1309. [CrossRef] [PubMed]

34. Morakinyo, T.E.; Lam, Y.F. Simulation study on the impact of tree-configuration, planting pattern and wind condition on street-canyon's micro-climate and thermal comfort. *Build. Environ.* **2016**, *103*, 262–275. [CrossRef]

35. Wang, Y.; Zacharias, J. Landscape modification for ambient environmental improvement in central business districts—A case from Beijing. *Urban For. Urban Green.* **2015**, *14*, 8–18. [CrossRef]

36. Amor, B.; Lacheheb, D.E.Z.; Bouchahm, Y. Improvement of Thermal Comfort Conditions in an Urban Space (Case Study: The Square of Independence, Sétif, Algeria). *Eur. J. Sustain. Dev.* **2015**, *4*, 407–416.

37. Ambrosini, D.; Galli, G.; Mancini, B.; Nardi, I.; Sfarra, S. Evaluating mitigation effects of urban heat islands in a historical small center with the ENVI-met® climate model. *Sustainability (Switzerland)* **2014**, *6*, 7013–7029. [CrossRef]

38. Razzaghmanesh, M.; Beecham, S.; Salemi, T. The role of green roofs in mitigating Urban Heat Island effects in the metropolitan area of Adelaide, South Australia. *Urban For. Urban Green.* **2016**, *15*, 89–102. [CrossRef]

39. Lobaccaro, G.; Acero, J.A. Comparative analysis of green actions to improve outdoor thermal comfort inside typical urban street canyons. *Urban Clim.* **2015**, *14*, 251–267. [CrossRef]
40. Battista, G.; Carnielo, E.; De Lieto Vollaro, R. Thermal impact of a redeveloped area on localized urban microclimate: A case study in Rome. *Energy Build.* **2016**, *133*, 446–454. [CrossRef]
41. Sodoudi, S.; Shahmohamadi, P.; Vollack, K.; Cubasch, U.; Che-Ani, A.I. Mitigating the urban heat island effect in megacity Tehran. *Adv. Meteorol.* **2014**, *2014*, 547974. [CrossRef]
42. Wang, Y.; Akbari, H. Development and application of 'thermal radiative power' for urban environmental evaluation. *Sustain. Cities Soc.* **2015**, *14*, 316–322. [CrossRef]
43. UNFPA Lebanon. Available online: http://www.unfpa.org.lb/PROGRAMME-AREAS/Population-and-Development/Population-and-Development-Strategies.aspx (accessed on 14 September 2015).
44. Ministry of Environment/Lebanese Environment and Development Observatory (LEDO)/ECODIT. Lebanon State of the Environment Report. Available online: http://www.moe.gov.lb (accessed on 14 January 2018).
45. Aouad, D. Urban Acupuncture as a tool for today's re-naturalization of the city: The non-constructible parcels of Municipal Beirut through the case study of Saifi district. In *Architectural Research Addressing Societal Challenges, Proceedings of the EAAE/ARCC 10th International Conference (EAAE ARCC 2016), Lisbon, Portugal, 15–18 June 2016*; Da Costa, M.J.R.C., Roseta, F., Lages, J.P., Da Costa, S.C., Eds.; Taylor & Francis: Lisbon, Portugal, 2017.
46. El-Achkar, E. *Réglementation et Formes Urbaines: Le cas de Beyrouth*; CERMOC: Beirut, Lebanon, 1998.
47. Masson, V. A physically-based scheme for the urban energy budget in atmospheric models. *Bound.-Lay. Meteorol.* **2000**, *94*, 357–397. [CrossRef]
48. Kaloustian, N.; Diab, Y. Effects of urbanization on the urban heat island in Beirut. *Urban Clim.* **2015**, *14*, 154–165. [CrossRef]
49. Kaloustian, N. On the Urban Heat Island in Beirut. Ph.D. Thesis, Académie Libanaise des Beaux-Arts (ALBA), Balamand University, Beirut, Lebanon, Université Paris-Est, Paris, France, 17 November 2015.
50. Aouad, D. Non-constructible Parcels within the Boundaries of Municipal Beirut: The Case Study of Saifi, Bachoura and Zokak El-Blat. Masters' Thesis, Académie Libanaise des Beaux-Arts (ALBA), Balamand University, Balamand, Lebanon, 2014.
51. ESRI. *ArcGIS Desktop: Release 10*; Environmental Systems Research Institute: Redlands, CA, USA, 2011.
52. Asaeda, T.; Ca, V.T.; Wake, A. Heat storage of pavement and its effect on the lower atmosphere. *Atmos. Environ.* **1996**, *30*, 413–427. [CrossRef]
53. Synnefa, A.; Santamouris, M.; Livada, I. A study of the thermal performance of reflective coatings for the urban environment. *Sol. Energy* **2005**, *80*, 968–981. [CrossRef]
54. ENVI-met 4.0. Available online: www.envi-met.com (accessed on 1 August 2018).
55. Gago, E.J.; Roldan, J.; Pacheco-Torres, R.; Ordóñez, J. The city and urban heat islands: A review of strategies to mitigate adverse effects. *Renew. Sustain. Energy Rev.* **2013**, *25*, 749–758. [CrossRef]
56. Erell, E.; Pearlmutter, D.; Boneh, D.; Bar Kutiel, P. Effect of high-albedo materials on pedestrian heat stress in urban street canyons. *Urban Clim.* **2014**, *10*, 367–386. [CrossRef]
57. Yaghoobian, N.; Kleissl, J.; Scott Krayenhoff, E. Modeling the Thermal Effects of Artificial Turf on the Urban Environment. *J. Appl. Meteorol. Climatol.* **2010**, *49*, 332–345. [CrossRef]
58. Andersson, L. Urban Experiments and Concrete Utopias: Platform4 a 'Bottom Up Approach to the Experience City. In *Architecture and Stages in the Experience City*; Kiib, H., Ed.; Institut for Arkitecktur og Medieteknologi, Department of Architecture and Designs Skriftserie, No. 30; Aalborg University: Aalborg, Denmark, 2008; pp. 84–95.
59. NYC Parks. Available online: http://www.nycgovparks.org/parks/highline (accessed on 24 March 2014).
60. Lehtovuori, P.; Ruoppila, S. Temporary Uses as Means of Experimental Urban Planning. *SAJ—Serbian Archit. J.* **2012**, *4*, 29–53.
61. Killing Architects. Urban Tactics: Temporary Interventions + Long Term Planning. Available online: http://www.killingarchitects.com/urban-tactics-final-repor/ (accessed on 22 May 2013).
62. Heck, A.; Will, H. Interim Use: Opportunity for New Open-Space Quality in the Inner City—The Example of Leipzig. Available online: http://www.difu.de/node/5959 (accessed on 16 January 2014).

MDPI

St. Alban-Anlage 66

4052 Basel

Switzerland

Tel. +41 61 683 77 34

Fax +41 61 302 89 18

www.mdpi.com

Climate Editorial Office

E-mail: climate@mdpi.com

www.mdpi.com/journal/climate

www.ingramcontent.com/pod-product-compliance
Lightning Source LLC
Chambersburg PA
CBHW051711210326
41597CB00032B/5445